T0139932

Lecture Notes in Computer Science 12646

Katharina Toeppe · Hui Yan ·
Samuel Kai Wah Chu (Eds.)

Diversity, Divergence, Dialogue

16th International Conference, iConference 2021
Beijing, China, March 17–31, 2021
Proceedings, Part II

Editors
Katharina Toeppe
Humboldt University of Berlin
Berlin, Germany

Hui Yan
Renmin University of China
Beijing, China

Samuel Kai Wah Chu
The University of Hong Kong
Hong Kong, Hong Kong

ISSN 0302-9743 ISSN 1611-3349 (electronic)
Lecture Notes in Computer Science
ISBN 978-3-030-71304-1 ISBN 978-3-030-71305-8 (eBook)
https://doi.org/10.1007/978-3-030-71305-8

LNCS Sublibrary: SL3 – Information Systems and Applications, incl. Internet/Web, and HCI

This Springer imprint is published by the registered company Springer Nature Switzerland AG
The registered company address is: Gewerbestrasse 11, 6330 Cham, Switzerland

Preface

Since last year's iConference our globalized world has been in turmoil: The Coronavirus, COVID-19 has changed our lives immensely – economically, ecologically, scientifically, and socially. The information sciences have never before been of so much importance. Increasing interest in scientific research by the majority population was just one outcome, leading to misinformation, fake news, and information overload. Our everyday work and social lives shifted as well: Working in home office became a new standard, video chat apps the new way to see friends, family, and coworkers.

Even though the world grew physically apart, new mental connections were formed. The 16th iConference was the second online conference in this series since 2005, but as the first iConference being planned in an online format from an early stage on, a plethora of new possibilities for scientific advancement and networking emerged. This abundance of new opportunities was mirrored by the conference theme of *Diversity, Divergence, Dialogue*.

Under this banner the 2021 iConference took place in Beijing, China, hosted by the Renmin University School of Information Resource Management.

The conference theme attracted a total of 485 submissions with 103 full research papers, 122 short research papers, and 63 posters. In a double-blind review process with an average number of two reviews per submission and two papers per reviewer 242 entries emerged, including 32 full research papers, 59 short research papers, and 48 posters. This selection was the result of a meticulous review process by 296 internationally renowned experts, with only 31% of the full research papers and 48% of the short research papers being chosen for presentation at the conference. 103 additional submissions were selected in case of the Workshops, the Sessions for Interaction and Engagement, the Virtual Interactive Sessions, the Doctoral Colloquium, the Early Career Colloquium, the Student Symposium, the Chinese papers, and the Archival Education.

The full and short research papers are published for the fourth time in Springer's *Lecture Notes in Computer Science* (LNCS). These proceedings are sorted into the following eleven categories, once again depicting the diversity of the iField: AI & Machine Learning, Data Science, Human-Computer Interaction, Social Media, Digital Humanities, Education & Information Literacy, Information Behavior, Information Governance & Ethics, Archives & Records, Research Methods, and Institutional Management. The conference posters are available at the *Illinois Digital Environment for Access to Learning and Scholarship* (IDEALS).

We would like to thank the reviewers immensely for their expertise and valuable review work and the track chairs for their hard work and vast expert knowledge. This conference would not have been possible without them. We wish to extend our gratitude to the full research paper chairs Daqing He from the University of Pittsburgh and

Maria Gäde from Humboldt-Universität zu Berlin, and the short research paper chairs Toine Bogers from Aalborg University and Dan Wu from Wuhan University.

Once again the iConference broadened the field of information science, forming and strengthening connections between scholars and scientists from all around the world in practicing *Diversity, Divergence, Dialogue*.

February 2021

Katharina Toeppe
Hui Yan
Samuel Kai Wah Chu

Organization

Organizer

Renmin University of China, People's Republic of China

Conference Chairs

Yuenan Liu	Renmin University of China, People's Republic of China
Bin Zhang	Renmin University of China, People's Republic of China

Conference Honorary Co-chair

Sam Oh	Sungkyunkwan University, Korea

Program Chairs

Hui Yan	Renmin University of China, People's Republic of China
Samuel Kai Wah Chu	The University of Hong Kong, People's Republic of China

Local Arrangements Chairs

Jian (Jenny) Wang	Renmin University of China, People's Republic of China
Hui Yan	Renmin University of China, People's Republic of China
Minghui Qian	Renmin University of China, People's Republic of China

Proceedings Chair

Katharina Toeppe	Humboldt-Universität zu Berlin, Germany

Full Research Paper Chairs

Daqing He	University of Pittsburgh, USA
Maria Gäde	Humboldt-Universität zu Berlin, Germany

Short Research Paper Chairs

Toine Bogers	Aalborg University, Denmark
Dan Wu	Wuhan University, People's Republic of China

Chinese Paper Chairs

Xiaobin Lu	Renmin University of China, People's Republic of China
Qinghua Zhu	Nanjing University, People's Republic of China
Zhiying Lian	Shanghai University, People's Republic of China
Gang Li	Wuhan University, People's Republic of China

Poster Chairs

Atsuyuki Morishima	University of Tsukuba, Japan
Leif Azzopardi	University of Strathclyde, UK

Workshops Chairs

Yuxiang (Chris) Zhao	Nanjing University of Science and Technology, People's Republic of China
Jiqun Liu	University of Oklahoma, USA

Sessions for Interaction and Engagement Chairs

Ming Ren	Renmin University of China, People's Republic of China
Pengyi Zhang	Peking University, People's Republic of China

Virtual Interactive Session Chairs

António Lucas Soares	University of Porto, Portugal
Chern Li Liew	Victoria University of Wellington, New Zealand

Archival Education Chairs

Jian (Jenny) Wang	Renmin University of China, People's Republic of China
Patricia Whatley	University of Dundee, UK

Doctoral Colloquium Chairs

Xiaomi An	Renmin University of China, People's Republic of China
Anne Gilliland	University of California at Los Angeles, USA

Early Career Colloquium Chairs

Mega Subramaniam	University of Maryland, USA
Sohaimi Zakaria	Universiti Teknologi MARA, Malaysia

Student Symposium Chairs

Ian Ruthven	University of Strathclyde, UK
Yuelin Li	Nankai University, People's Republic of China
Yao Zhang	Nankai University, People's Republic of China

Doctoral Dissertation Award Chairs

Udo Kruschwitz	Universität Regensburg, Germany
George Buchanan	University of Maryland, USA

Conference Coordinators

Michael Seadle	iSchools Organization, USA
Clark Heideger	iSchools Organization, USA
Slava Sterzer	iSchools Organization, USA
Cynthia Ding	iSchools Organization, USA

Reviewers Full and Short Papers iConference 2021

Jacob Abbott
Amelia Acker
Noa Aharony
Shameem Ahmed
Isola Ajiferuke
Dharma Akmon
Bader Albahlal
Michael Albers
Daniel Alemneh
Hamed Alhoori
Wafaa Ahmed Almotawah
Sharon Amir
Lu An
Muhammad Naveed Anwar
Leif Azzopardi
Cristina Robles Bahm
Alex Ball
Syeda Batool
Edith Beckett
Ofer Bergman
Nanyi Bi
Bradley Wade Bishop
Toine Bogers
Erik Borglund
Christine L. Borgman
Ceilyn Boyd
Sarah Bratt
Jenny Bronstein
Jo Ann M. Brooks
Ricardo Brun
Sarah A. Buchanan
Julia Bullard
Christopher Sean Burns
Yu Cao
Daniel Carter
Vittore Casarosa
Biddy Casselden
Niel Chah
Yung-Sheng Chang
Tiffany Chao

Hsin-liang Chen
Hsuanwei Chen
Jiangping Chen
Chola Chhetri
Shih-Yi Chien
Yunseon Choi
Steven Siu Fung Chong
Barry Chow
Rachel Clarke
Johanna Cohoon
Mónica Colón-Aguirre
Anthony Joseph Corso
Kaitlin Costello
Andrew Martin Cox
Hong Cui
Amber L. Cushing
Mats Dahlstrom
Dharma Dailey
Gabriel David
Rebecca Davis
Shengli Deng
Bridget Disney
Brian Dobreski
Philip Doty
Jennifer Douglas
Helmut Hauptmeier
Kedma Duarte
Patrick Dudas
Emory James Edwards
Lesley Farmer
Yuanyuan Feng
Fred Fonseca
Helena Francke
Henry A. Gabb
Maria Gäde
Chunmei Gan
Daniel Gardner
Rich Gazan
Tali Gazit
Yegin Genc
Patrick Golden
Koraljka Golub
Michael Gowanlock
Elke Greifeneder
Melissa Gross
Michael Robert Gryk
Ayse Gursoy
Stephanie W. Haas
Jutta Haider
Lala Hajibayova
Ruohua Han
Susannah Hanlon
Preben Hansen
Jennifer Hartel
Jiangen He
Zhe He
Alison Hicks
Shuyuan Mary Ho
Kelly M. Hoffman
Chris Holstrom
Liang Hong
Jiming Hu
Kun Huang
Yun Huang
Gregory Hunter
Charles Inskip
Joshua Introne
Isa Jahnke
Hamid R. Jamali
David Jank
Wei Jeng
Tingting Jiang
Jenny Johannisson
Michael Jones
Nicolas Jullien
Jaap Kamps
Jeonghyun Kim
Kyung Sun Kim
Vanessa Kitzie
Emily Knox
Kyungwon Koh
Kolina Sun Koltai
Rebecca Koskela
Ravi Kuber
Jin Ha Lee
Kijung Lee
Lo Lee
Myeong Lee
Noah Lenstra
Aihua Li
Daifeng Li
Kai Li

Meng-Hao Li
Louise Limberg
Chi-Shiou Lin
Zack Lischer-Katz
Chang Liu
Elizabeth Lomas
Kun Lu
Quan Lu
Christopher Lueg
Haakon Lund
Jessie Lymn
Haiqun Ma
Lai Ma
Long Ma
Craig MacDonald
Kate Marek
Kathryn Masten
Matthew S. Mayernik
Athanasios Mazarakis
Claire McGuinness
Pamela Ann McKinney
David McMenemy
Shawne D. Miksa
Staša Milojević
Chao Min
Ehsan Mohammadi
Lorri Mon
Camilla Moring
Atsuyuki Morishima
Gustaf Nelhans
Valerie Nesset
David M. Nichols
Hui Nie
Rebecca Noone
Brian O'Connor
Benedict Salazar Olgado
Peter Organisciak
Felipe Ortega
Virginia Ortiz-Repiso
Yohanan Ouaknine
Kathleen Padova
Marco Painho
Britt Paris
Albert Park
Hyoungjoo Park
Min Sook Park
Daniel Adams Pauw
Bo Pei
Diane Pennington
Olivia Pestana
Vivien Petras
Tamara Peyton
Ola Pilerot
Anthony T. Pinter
Alex Poole
Jian Qin
Marie L. Radford
Arcot Rajasekar
Congjing Ran
Susan Rathbun-Grubb
Ming Ren
Colin Rhinesmith
Cristina Ribeiro
Angela U. Rieks
Abebe Rorissa
Vassilis Routsis
Alan Rubel
Melanie Rügenhagen
Ehsan Sabaghian
Athena Salaba
Ashley Sands
Madelyn Rose Sanfilippo
Sally Sanger
Vitor Santos
Maria Janina Sarol
Laura Sbaffi
Kirsten Schlebbe
Kristen Schuster
John S. Seberger
Kalpana Shankar
Patrick Shih
Richard Slaughter
Antonio Soares
Shijie Song
Daniel Southwick
Clay Spinuzzi
Beth St. Jean
Gretchen Renee Stahlman
Hrvoje Stančić
Caroline Stratton
Besiki Stvilia
Tanja Svarre

Sue Yeon Syn
Jian Tang
Rong Tang
Yi Tang
Carol Tenopir
Andrea Karoline Thomer
Kentaro Toyama
Aaron Trammell
Chunhua Tsai
Tien-I Tsai
Yuen-Hsien Tseng
Pertti Vakkari
Frans Van der Sluis
Merce Væzquez
Nitin Verma
Julie Walters
Hui Wang
Jieyu Wang
Lin Wang
Xiangnyu Wang
Xiaoguang Wang
Yanyan Wang
Ian Watson
Jingzhu Wei
Brian Wentz
Michael Majewski Widdersheim
Rachel Williams
R. Jason Winning
Dietmar Wolfram
Maria Klara Wolters
Adam Worrall
Dan Wu
I-Chin Wu
MeiMei Wu
Peng Wu
Qunfang Wu
Lu Xiao
Iris Xie
Juan Xie
Jian Xu
Lifang Xu
Shenmeng Xu
Xiao Xue
Erjia Yan
Hui Yan
Lijun Yang
Qianqian Yang
Seungwon Yang
Siluo Yang
Ying (Fred Ying) Ye
Geoffrey Yeo
Ayoung Yoon
JungWon Yoon
Sarah Young
Liangzhi Yu
Xiaojun Yuan
Marcia L. Zeng
Xianjin Zha
Bin Zhang
Chengzhi Zhang
Chenwei Zhang
Jinchao Zhang
Mei Zhang
Ziming Zhang
Yang Zhao
Yuxiang (Chris) Zhao
Lihong Zhou
Xiaoying Zhou
Qinghua Zhu
Xiaohua Zhu
Zhiya Zuo
林 五

Contents – Part II

Information Behavior

Institutional Management

Contents – Part I

AI and Machine Learning

Data Science

Human-Computer Interaction

Social Media

Digital Humanities

Education and Information Literacy

Information Behavior

"We Can Be Our Best Alliance": Resilient Health Information Practices of LGBTQIA+ Individuals as a Buffering Response to Minority Stress

Valerie Lookingbill(✉), A. Nick Vera, Travis L. Wagner, and Vanessa L. Kitzie

University of South Carolina, Columbia, SC 29208, USA
lookingv@email.sc.edu

Abstract. This article examines the resilient health information practices of lesbian, gay, bisexual, transgender, queer, intersex, and asexual (LGBTQIA+) individuals as agentic forms of buffering against minority stressors. Informed by semi-structured interviews with 30 LGBTQIA+ community leaders from South Carolina, our findings demonstrate how LGBTQIA+ individuals engage in resilient health information practices and community-based resilience. Further, our findings suggest that LGBTQIA+ communities integrate externally produced stressors. These findings have implications for future research on minority stress and resiliency strategies, such as shifting from outreach to engagement and leveraging what communities are doing, rather than assuming they are lacking. Further, as each identity and intersecting identities under the LGBTQIA+ umbrella has unique stressors and resilience strategies, our findings indicate how resilience strategies operate across each level of the socio-ecological model to better inform understanding of health information in context.

Keywords: LGBTQ+ · Health information practices · Minority stress

1 Introduction

Relative to their heterosexual and cisgender peers, LGBTQIA+ individuals experience greater health disparities [1] as a result of internal and external stressors produced by stigma and discrimination. These stressors suggest minority stress theory is informative to how LGBTQIA+ persons relate to health care. Minority stress theory holds that prejudice and stigma directed towards members of marginalized populations bring about unique stressors, and these stressors contribute to adverse health outcomes [2]. Research shows that members of LGBTQIA+ communities often actively manage these adverse health stressors via resilient behaviors [2]. Resiliency buffers the negative effect of stressors and allows LGBTQIA+ communities to avoid adverse health outcomes.

To examine how LGBTQIA+ communities are resilient against adverse health stressors, we used an information practices approach, which understands people's

K. Toeppe et al. (Eds.): iConference 2021, LNCS 12646, pp. 3–17, 2021.
https://doi.org/10.1007/978-3-030-71305-8_1

information-related activities and skills as socially constructed [3]. By this definition, health information practices include creating, seeking, sharing, and using health information. Health information practices may be a key strategy of resilience for LGBTQIA+ populations because health information practices allow individuals to confront and dismantle stressors through collecting, processing, and sharing information to help solve a problem and regain emotional stability [4]. Our research extends minority stress models and examines the resilient health information practices of LGBTQIA+ individuals as agentic forms of buffering against minority stressors. In doing so, our findings challenge views of LGBTQIA+ identities as monoliths and instead recognize that each identity and intersecting identities under the broader LGBTQIA+ umbrella have unique stressors and resilience strategies to buffer against minority stress.

2 Literature Review

2.1 Minority Stress

Minority stress theory describes excess stressors experienced by socially-stigmatized individuals whose position as a social minority results in reduced access to care and increased chronic stress [2]. Stress, defined under this theory, is the result of an "imbalance between the external and internal demands perceived as threatening by an individual and their assessment of the resources available to cope with them" [5] (p. 799). The theory posits that health disparities observed in LGBTQIA+ populations result from persistent stigma directed towards community members. Members of LGBTQIA+ communities experience minority stress through three processes: 1) through the external events that occur in an LGBTQIA+ person's life, such as harassment or discrimination; 2) the anticipation of harassment or discrimination, which leads to increased vigilance or concealment of one's identity, and; 3) the internalization of external negative beliefs and societal prejudice [2]. Through no fault of the LGBTQIA+ individual, such stressors often contribute to poor health.

Minority stress theory accounts for both distal stressors and proximal stressors. Distal stressors encompass objective stressors not dependent on an individual's perceptions, such as prejudice, discrimination, or microaggressions. For example, LGBTQIA+ individuals may experience heteronormativity, the belief that heterosexuality, based on the gender binary, is the default sexual orientation [6]. LGBTQIA+ individuals may experience heteronormative stressors in health care settings. For instance, when a woman who identifies as lesbian confides in her doctor that she is sexually active, the doctor may ask her to take a pregnancy test because they assume she is exclusively engaging in sex with cisgender males.

Proximal stressors are internal processes that follow exposure to these distal stressors, such as the expectation of distal stressors and the vigilance required of this expectation and the internalization of negative societal attitudes [2]. For example, LGBTQIA+ individuals may exhibit proximal stressors by hiding their sexual orientation, as concealment comes through internal psychological processes. The concealment of LGBTQIA+ identities results in significant psychological distress, such as shame, guilt, and isolation from other LGBTQIA+ community members.

Importantly, however, proximal stressors are subjective and related to self-identity with a minority group. Within minority stress theory, distal and proximal stressors exist as chronic and socially-based experiences, resulting from sociocultural rather than individual conditions. Further, these stressors are unique to individuals with a minority status, as these individuals must adapt to stressors at a greater capacity than those who do not have a minority status [1]. Minority stress theory states that both distal and proximal stressors can lead to adverse health outcomes. These outcomes may include poor mental health outcomes, such as depression, anxiety, and substance use disorders, as well as poor physical health outcomes that are responsive to stress (e.g., asthma) [7].

Stigma and minority stress exist at each level across the socio-ecological model, which emphasizes multiple levels of influence on behaviors and holds that behaviors both shape and are shaped by the social environment [8]. The following levels influence behavior:

1. Intrapersonal: characteristics of the individual, including knowledge, attitudes, and skills. Examples on the intrapersonal level may include identifying as a trans woman or as a person of color.
2. Interpersonal: the individual's social network and social support system, including family and friends.
3. Organizational: organizations and social institutions with formal and informal rules and regulations, such as schools, workplaces, and community groups.
4. Community: relationships between organizations, social environments, and cultural norms.
5. Public policy: local, state, national laws and policies, and the media [9]

[1] examined the health consequences of minority stress on LGBTQIA+ youth across these levels and determined that experiences with distal stressors cause LGBTQIA+ individuals to be vigilant of their social environment to anticipate and avoid stigmatizing encounters. Repeated encounters with distal stressors led LGBTQIA+ individuals to ruminate, a maladaptive emotion regulation strategy characterized by a repeated focus on the causes and symptoms of distress. Individuals with a high degree of life stress develop increasingly ruminative tendencies, and LGBTQIA+ individuals tend to ruminate more than their cis, heterosexual peers. Further, [1] found that proximal stressors on the intrapersonal level included LGBTQIA+ individuals engaging in concealment behaviors, wherein the individual hides their identity to avoid future victimization. In addition to encountering minority stress at the intrapersonal level, LGBTQIA+ individuals face distal stressors on the interpersonal level. Stressors on this level may include intentional prejudice and discrimination but may also include unintentional actions such as microaggressions. For instance, when an individual asks a man who identifies as gay if he has a girlfriend, that individual engages in a microaggression as they are endorsing heteronormative culture and thus reinforcing heterosexuality as a cultural default.

LGBTQIA+ individuals also experience distal stressors on the organizational, community, and public policy levels. For instance, on the organizational level, LGBTQIA+ youth may experience discrimination at school when LGBTQIA+ student organizations are not permitted to form and operate on the same terms as all other student organizations. Further, LGBTQIA+ individuals may experience stressors resulting from cultural norms, such as heteronormativity, on the community level. Lastly, stressors on the public policy level, such as laws that prohibit public schools from including same-sex health topics into the curriculum, constrain the opportunities, resources, and well-being of LGBTQIA+ individuals.

2.2 Resiliency of LGBTQIA+ Individuals

[2] suggested that LGBTQIA+ individuals respond to minority stress with resilience. Resilience is the "process of positive adaptation to significant threats to well-being" [10] (p. 1436). Thus, resilience relies on the availability, accessibility, and strategic use of resources. These resources may exist at any level of the socioecological model, and factors promoting resilience may include the interaction of resources among the five levels. LGBTQIA+ individuals employ resilient strategies that are sustainable, developmentally appropriate, and reinforced by the environment [10].

It is further imperative to distinguish between individual and community-based resilience within minority stress frameworks. Individual-based resilience emphasizes personal agency concerning the qualities an individual possesses that help them cope with stress (e.g., sense of coherence and hardiness). Focusing exclusively on individual-based resilience limits efforts to develop effective interventions and policies, as not every LGBTQIA+ individual has the same resources or opportunities to enact resilience. Underlying social structures are often unequal, as the social, economic, and political structures that create opportunities for success in society are not equally distributed [7]. Thus, social disadvantages such as racism, socioeconomic status, and sexism limit individual resilience.

On the other hand, community-resilience refers to how communities provide resources that help individuals develop and sustain well-being [7]. Community-resilience focuses on resilience in ecological contexts, emphasizing social resources, such as friends who also identify as LGBTQIA+ or information sources developed by grass-root LGBTQIA+ communities, rather than individual traits. However, members of specific segments within LGBTQIA+ communities may not benefit equally from community-resilience due to structural inequalities within the community itself. For instance, racism, biphobia, and transphobia deprive individuals with select identities of community resilience. As our research will show, it is crucial to understand that LGBTQIA+ experiences intersect alongside other lived experiences, many of which face their own minority-based stressors [11]. As with all discussed forms of resiliency, several intervening factors on the intrapersonal, interpersonal, organizational, community, and public policy levels shape resilient health information practices. As such, the social and cultural environment affects the information channels and sources an individual uses [12]. Thus, individual knowledge and attitudes, relationships with friends and family, socioeconomic circumstances, and physical and social environments affect individuals' health information practices.

2.3 Health Information Practices of LGBTQIA+ Individuals

LGBTQIA+ people face significant social and discursive barriers due to heteronormativity (i.e., the presumption that all people identify as heterosexual) and cisnormativity (i.e., the presumption that all people identify as a gender that matches their sex-assigned-at-birth) [13]. Furthermore, it is essential to attend to intersectionality and its relationship to the presumed monolith of the LGBTQIA+ identity. According to [11], intersectionality acknowledges that individuals do not experience a given identity singularly. Instead, they live with various experiences of social differences grounded within identities such as class, race, ability, and age. As a result of these barriers and intersectionalities, LGBTQIA+ populations often postpone seeking treatment or healthcare from health resources and health professionals at a 30% greater rate than cisgender individuals, thus increasing LGBTQIA+ individuals' risks to health and well-being [14]. These delays in seeking help often result in LGBTQIA+ individuals engaging in resilient health information seeking practices. LGBTQIA+ individuals seek, share, and use health information in various ways and for many different reasons; however, due to their marginalization and lack of acceptance from the larger society, there are barriers to information. These barriers are perpetuated by systems that discriminate against LGBTQIA+ individuals, driving them to consider new ways of seeking information that circumnavigates oppressive systems [14]. For instance, previous research focusing on health information practices and behaviors found that LGBTQIA+ communities engage with health information and health resources in ways that are socially and medically understood as harmful to one's health and well-being [15]. Additionally, while these engagements are distinct and vary by community, nearly all are produced by the ongoing marginalization and discrimination of LGBTQIA+ individuals [15].

There is limited research on the health information practices of LGBTQIA+ individuals. Thus, it is imperative to further research in the area using minority stress theory and models to examine how LGBTQIA+ individuals engage in resilient health information practices to adapt to and overcome minority stressors in their environments. The research questions below informed our approach to examining resilient health information practices of LGBTQIA+ individuals,

2.4 Research Questions

R 1. How do members of LGBTQIA+ communities experience minority stress on each level of the socio-ecological model?
R 2. How do members of LGBTQIA+ communities engage in resilient health information practices on each level of the socio-ecological model as a response to minority stress?

Semi-structured interviews with 30 community leaders in South Carolina informed our research.

3 Methods

This research is part of a more extensive investigation (University of South Carolina IRB approval number Pro0008587) funded by an Institute for Museum and Library Services Early Career Development Grant that examines the health information practices of LGBTQIA+ communities. As such, the methodology and findings discussed in this paper specifically focus on applying deductive codes to the data after we conducted semi-structured interviews with 30 LGBTQIA+ community leaders from South Carolina.

Speaking with individuals in South Carolina elicits the social and structural barriers that are distinct to the area.

We selected community as our unit of analysis as LGBTQIA+ individuals are more effective when exhibiting community resilience. For the study, we defined community as possessing three criteria: 1) community members conduct the majority of their work in South Carolina; 2) their work is social and involves group-oriented engagements; and 3) members collectively possess LGBTQIA+ identities [16, 17].

Informed by these criteria, we engaged in purposive sampling by identifying over 100 LGBTQIA+ groups and organizations in South Carolina and then asking them to self-nominate leaders for participation in the study. During the interview process, we engaged in snowball sampling by asking participants to recommend additional participants. Finally, we used theoretical sampling to identify participants from informal communities, such as social media-based LGBTQIA+ groups, which we may not have identified in our initial purposive sampling.

Before interviews, participants filled out a pre-interview questionnaire, providing demographic information. During interviews with participants, we asked about their involvement with their communities, their personal and community identities, and how they and their communities addressed health questions and concerns. We then asked participants to partake in an information world mapping exercise where participants drew people, places, and things that helped or did not help them address their health questions and concerns [18].

We used interview transcripts as the data source for this article. We analyzed the data using a deductive coding process to develop a provisional list of primary codes established in minority stress theory literature and informed by the above literature review. These primary codes include stressors, distal stressors, proximal stressors, and resilient health information practices, using definitions from [7]. We further coded these four primary codes according to the appropriate level of the socioecological model in which they occur (e.g., intrapersonal, interpersonal, organizational, community, and public policy). We applied these codes to the interview transcripts using sentences as our unit of observation.

4 Findings

4.1 Participant Demographics

The majority of our study's participants were young adults between the ages of (18–25: n = 11; 36.7%) and middle-aged adults between the ages of (35–54: n = 7; 23.3%). The remainder of our study's participants were adults aged 55 and older (n = 5; 16.7%) and teenagers between the ages of (13–17, n = 4; 13.3%). For more information regarding working with LGBTQIA+ teens for this project, see [15]. Participants selected from a series of racial and ethnic identities with the ability to add identities not listed. The majority of participants identified as white (n = 18; 60%), while (n = 7; 24%) identified as Black, (n = 2; 7%) identified as Black and white, (n = 1; 4%) identified as Black and Afro-Caribbean, (n = 1; 4%) identified as Aboriginal, Arab/West Asian, Black and white, and (n = 1; 4%) as Black, white, and Egyptian. The majority of participants lived in the Upstate and Midlands regions of South Carolina.

Participants self-labeled their LGBTQIA+ identity. Among the identities participants labeled themselves as lesbian, gay, queer, transgender, genderqueer, and bisexual were most prevalent. For more information on demographics, refer to [15].

Participant narratives illustrated three significant findings:

1. LGBTQIA+ individuals engage in resilient health information practices on all socio-ecological levels
2. Community-based resilience characterizes collective health information practices
3. LGBTQIA+ sub-communities and LGBTQIA+ individuals with intersecting minority identities integrate externally produced stressors.

We will illustrate these findings using participant narratives that exemplify our four deductive codes. We refer to participants in this section using their provided pronouns and self-selected pseudonyms to protect individual privacy.

4.2 Finding 1: LGBTQIA+ Individuals Engage in Resilient Health Information Practices on All Socio-Ecological Levels

In the context of their health information practices, participants engaged in resiliency against stressors on every level of the socio-ecological model. On the intrapersonal level, participants engaged in resilient health information practices by successfully adapting to knowledge gaps. Annalisa, a white, young adult who identifies as a cisgender female lesbian, explains how she successfully adapts to her knowledge gap by:

> *just doing research on [my] own to try to make sure you're getting the correct standard of care. See what you actually need and try to talk to different people who have maybe gone through it and had a better experience than you to see what your baseline should be.*

In this instance, Annalisa is showing resilience against an intrapersonal minority stressor, lack of knowledge, "because maybe you don't know and you think the doctor's doing things correctly, but maybe they're not addressing something." By researching to determine the general standard of health care "that a straight person could go in and get," Annalisa is displaying resilient health information practices on the intrapersonal level through the reliance on the availability, accessibility, and strategic use of resources, such as the internet and other community members to find answers to her health care questions.

On the interpersonal level, participants engaged in resilient health information practices by successfully adapting to interpersonal minority stressors. Whitney, a white young adult who identifies as a cis-woman lesbian, described the interpersonal minority stressor of LGBTQIA+ individuals having "families that aren't as accepting." However, Whitney engages in resilient health information practices on the interpersonal level by:

> *Being someone for someone. People have come out to me throughout the years. They've talked to me about their home life. I've met people's parents. They've all met my mom [...] being able to bring my mom down and have them [...] have*

someone that is a motherly figure, that does accept them, that was really amazing. Just seeing people feel the safety and acceptance that I feel.

While Whitney's community members may experience minority stress related to lack of acceptance by their families, Whitney helps them engage in resilient health information practices by encouraging interaction between members of their social network.

Other participants engaged in resilient health information practices by successfully adapting to stressors on the organizational level through their community organizations. Justin explained how his organization engages in resilient health information practices on this level. Salient identities of Justin include white, middle-aged, cis-male, and gay. During his interview, Justin said:

That's one thing that we have done is try to start providing resources and lists of things like friendly churches but also friendly health care providers, and especially addressing the needs of the trans community…so we are not trying to be the resource for everything. But we're trying to be the conduit to get people to the resource.

Through this example, Justin shows how his organization engages in resilient health information practices on the organizational level. His community group successfully adapts to stressors and provides community members with resources that sustain their well-being, such as lists of LGBTQ-friendly doctors, or refers them to resources that can help them maintain it.

On the community level, participants engaged in resilient health information practices by successfully adapting to community-level minority stressors. During her interview, Pat, a young adult who identifies as a trans woman of color, discussed cultural stigma surrounding queer girls, saying, "There is already shame around being a sexually promiscuous young person, but then to also add queering that just compounds the struggle." As a result of these cultural norms surrounding queer youth, LGBTQIA+ people must engage in resilient health information practices. Pat explains, "I think that a lot of the structural prejudices that we face contributes most to why, or at least the specific kinds of ways, that we engage in unhealthy practices or health practices that are detrimental to us."

Finally, participants engaged in resilient health information practices by successfully adapting to public policy-level minority stressors. One public policy stressor that arose in a participant's interview was same-sex education in schools. Vada, a white young adult who identifies as lesbian, explained that same-sex education is illegal to discuss in South Carolina. She detailed an example in which she was trying to "explain how sex education is better and it needs to be done in schools, and older women, in particular, were like 'Well, we can't legally do that.'" As a response to this stressor, Vada conducted panels for same-sex education in schools, which shows resiliency as she adapted to these constricting laws and took it upon herself to educate her LGBTQIA+ peers on sexual health.

4.3 Finding 2: Community-Based Resilience Characterizes Collective Health Information Practices

Participants engaged in community-based resilience through collective health information practices by utilizing social resources within their LGBTQIA+ communities rather than relying on individual traits. The health information practices participants engaged in are collective in that participants and their community members work together to adapt to stressors in their environments positively. For instance, Shateria Cox, a Black, white, and Egyptian gender non-conforming youth, recounted how she and her community members engaged in community-based resilience by providing social and informational resources that help other community members cope with stress and maintain their well-being. She stated, "We can be our worst enemy, and we can be our best alliance." Shateria Cox expands on this by saying, "There's a lot of websites run by our community that have information or values that may teach or help others learn about our healthcare." She then goes on to include "friends, acquaintances from people that we're aligned with." Shateria Cox and her community members "use our own resources" and "ask somebody that's already part of the community" as resilient health information practices. Shateria Cox continues to show community-based resilience as she says.

I'm not afraid to ask questions, put myself in the conversation, or ask questions to somebody and teach somebody about - because I have something that you may not know, and you have something that you could teach me. I could teach you something. You can teach me. We can all teach each other something. And I'm not afraid to ask questions and teach somebody. If they need to know something and they ask me, I'm going to tell them about something.

Engaging in dialogue and teaching others is an example of community-based resilience as Shateria Cox's community works together to provide and share resources (i.e., shared knowledge) regarding mental health. In this example, community members act as social resources. This community's health information practices are collective as community members are motivated by one common issue, an information need, and work together to bridge that gap by sharing the community's collective knowledge among one another. The use and sharing of social resources allow Shateria Cox and her community to cope with stress and develop and sustain well-being within their environment.

Alternatively, Tony Solano, a white young adult who identifies as a gay man, explains how his organization uses community-resilience to address minority stressors, such as stigma and access, to PrEP. His organization engages in this style of resilience by.

Doing events where we would bring everybody together and would have a quick campfire about HIV and other STDs. And how to prevent them, stay healthy, how to recognize them, and most importantly, to be open with your sexual partner that you do have something you guys can discuss that. We also work with a couple of different vendors that give us condoms and other products that we're able to give out for free. So, if cost is an issue, we just remove that all together.

Tony Solano and his organization engage in community resilience as the organization collects social resources to address an identified stressor. They use social resources on the interpersonal level, in which members of the organization utilize collective knowledge to discuss prevention and recognition of HIV and STDs. This, in turn, allows Tony Solano and his community to mitigate both their community and individual risk of transmitting or contracting STDs. Lastly, they also use resources on the community level, where the organization works in partnership with other vendors to provide community members with condoms and other preventative measures against STDs.

4.4 Finding 3: LGBTQIA+ Sub-Communities and LGBTQIA+ Individuals with Intersecting Minority Identities Integrate Externally Produced Stressors

Our findings indicate that LGBTQIA+ individuals experience minority stressors that begin externally and become integrated at the community level. Stressors emerge iteratively so that a stressor's production on one level influences a stressor's production on another level, creating a blend of intrapersonal, interpersonal, organizational, community, and public policy level minority stressors.

Following Vada's discussion of no promo homo laws, Vada described a distal stressor on the public policy level regarding sex education in public schools. Vada states, "I also know that people who were raised here don't get a very good sex education in school…it's illegal to discuss same-sex relationships" in South Carolina. This stressor on the public policy level then trickles down to an organizational level, as public schools cannot incorporate same-sex education into their curriculum. This stressor then integrates into the LGBTQIA+ community on an intrapersonal level as LGBTQIA+ youth may lack knowledge about safe sex practices, "meaning that they have no idea how to be safe about it."

Tessie, a white, middle-aged female who identifies as a lesbian, also identified barriers as those impacting her and "barriers for the community," explaining why her community no longer hosts LGBTQIA+ resiliency groups. Tessie states, "If you're dealing with parents that tell you that you are an abomination, why would kids want to gather around that? That's internalized." In this narrative, Tessie indicates that prejudice - a distal stressor on the interpersonal level - integrates into the LGBTQIA+ community in the form of internalized homophobia - a proximal stressor on the intrapersonal level.

Other examples of stressors externally produced community-level health information practices include division between different identities within LGBTQIA+ communities. Shannon, a middle-aged woman who identifies as a Black and lesbian, noted this integration in her interview. She recounted an example of a woman at a PFLAG meeting who experienced discrimination due to her bisexual identity. Shannon recounted how this woman.

> *went on this crazy tirade about how there's so much bi-visibility and bi-erasure, and she wasn't going to stand for it, and the gay and lesbian community always got all the resources and always got this, and bisexual people and pansexual people and other more sexually fluid people were always left out.*

Distal stressors work at the interpersonal level in this narrative. The woman Shannon described experienced a distal stressor (i.e., an external event) in the form of erasure, as the social network excluded her for not identifying as either gay or lesbian. The women implied in this account that she was denied the resources afforded to other members in the social network who identified as either gay or lesbian.

Further, Whitney explains how externally produced stressors manifest within her community.

> *We get backlash or discrimination based on who we're dating, who we're with, our preference. A lot of ways where LGBTQ people can't feel safe comes from people outside of that community. It happens in the community as well.*

Whitney notes that distal stressors occur on multiple levels of the socio-ecological model, such as interpersonal, and are then integrated into the community itself. Whitney expands on this, saying.

> *There's so much transphobia within the LGBTQ community, biphobia within gays and lesbians, homophobia from other places. It can be pretty polarizing at times when people grab onto their identity and don't support the other ones within that community.*

In this example, Whitney indicates how externally produced distal stressors of prejudice become integrated into the LGBTQIA+ community, resulting in community members displaying prejudice against members of different identities even if those identities are within the LGBTQIA+ spectrum. Thus, while community members fall under the larger umbrella of LGBTQIA +, Whitney addresses the issue of intersectionality, in that falling under this broader umbrella does not guarantee that all needs and experiences are the same for each unique identity.

5 Discussion and Implications

5.1 Stressors Are Integrated into LGBTQIA+ Communities

An emergent theme from our research indicates that while communities are central for collective, community-based resilience, stressors still exist within them. While LGBTQIA+ individuals experience stressors within their communities, these communities are not the producer of stressors. Intersectionality is critical to note as different identities under the broader LGBTQIA+ umbrella will experience different stressors and engage in different health information practices to adapt to these stressors. For example, racial identities produce different stressors, though individuals may share the same LGBTQIA+ identity. Shateria Cox noted this in her interview, saying.

> *Since racism is a system, systematically Black people don't have any power in the United States…the economy was built to support the well-being of white folks and white supremacists. Very rarely see a black folk that has money or power or in office.*

These unique stressors do not disappear in the LGBTQIA+ community simply because members share a common gender identity or sexual orientation. Instead, they remain rather prominent either through ignoring the needs of systemic racism or ignoring the racist histories latent within LGBTQIA+ activism. While members may share a larger identity of LGBTQIA+, their intersecting identities, such as race, differ, and so do their stressors. Shateria Cox highlighted the importance of addressing intersectionality within the larger LGBTQIA+ community, specifically naming Kimberlé Williams Crenshaw, the developer of the theory of intersectionality, to explain how "Black, queer, and trans folk are the ones that set everything out for everybody…regular gays ain't do shit for nobody. That was Black, queer, and trans folks."

As demonstrated by this quote, it is critical to understand that LGBTQIA+ people are not monoliths. Different communities, with varying intersecting experiences of social difference, will create, seek, use, and share health information in various ways. Further, the types of communities LGBTQIA+ individuals are involved in can factor into their information practices. For example, leaders of informal LGBTQIA+ communities, such as ones that do not receive grant funding, might not have to follow specific rules within how they do information practices as they can do so without a potential threat to funding. In contrast, receiving grant funding might prohibit certain information practices for other communities. It is important to note that neither is right nor wrong per se, but these contexts inform their information practices.

For these reasons, practitioners must get a sense of the important identities other than LGBTQIA+, as these inform people's engagements with health information, and tailor information-based solutions to address stressors based on our knowledge that they are intersectional informed. For instance, when a trans individual seeks information from a physician, the physician should also consider other important identities, such as race and socioeconomic status. The physician should identify all important identities to ensure that the information they share is racially and financially appropriate, as well as appropriate to their trans identity.

5.2 Relationship Between Health Information Practices and Resilience

Another emergent theme from our research demonstrates the relationship between health information practices and resilience. This finding shows that resilience operates agentically at every level of the socio-ecological model. This study furthers [19]'s work pushing back against deficit orientations within information practices and behavior research [20]. In their work, [19] emphasized the need to resituate the concept of information poverty to refocus the blame away from individuals experiencing marginalization. Instead, we should focus on the contextual conditions that create information poverty, thus acknowledging individual and community health information practices as responses to marginalization. Our findings revealed that LGBTQIA+ individuals and communities are already employing resilient health information practices. Thus, our findings show that there is nothing inherently wrong with LGBTQIA+ individuals, both medically and informationally. Therefore, the locus of blame should not be placed on the communities but on the social and structural factors elicited by the socio-ecological model that produces stressors and information barriers.

Understanding how these stressors and barriers operate will allow us to identify potential information-based interventions. One such intervention could involve shifting from outreach to engagement, as providing access to information on its own will not rectify information inequities experienced by LGBTQIA+ communities. Instead, practitioners should employ engagement models to partner with LGBTQIA+ communities, allowing the communities to dictate how practitioners can leverage expertise to promote community health. Additionally, interventions should leverage what LGBTQIA+ communities are already doing, rather than assuming that the communities lack information or resources. Thus, practitioners should serve as a connector - connecting LGBTQIA+ individuals to not only health resources and information but to one another. Finally, when developing interventions, practitioners should understand health information in context. For instance, when a person shares their LGBTQIA+ identity, people should consider how these identities may inform what types of health information are relevant to them. Further, practitioners should recognize that when an LGBTQIA+ individual receives health information, the practitioners' information and how affirming they are of LGBTQIA+ individuals and health concerns are likely to be shared within and among communities. Moreso, when understanding health information in context, it is imperative to understand that LGBTQIA+ people are not monoliths. Different communities, with varying intersecting identities, under the broader LGBTQIA+ umbrella, will have different health information practices. Thus, practitioners need to recognize the unique, intersecting identities of LGBTQIA+ individuals to inform people's engagements with health information. These findings support previous research on community resilience and efficacy wherein efficacy frameworks traditionally influenced how health information professionals implement changes in behavior for LGBTQIA+ communities while failing to understand the complex experiences informing LGBTQIA+ health practices [21].

6 Limitations

As mentioned in our methods section, we interviewed 30 LGBTQIA+ community leaders from various age groups, racial and ethnic backgrounds, educational levels, regions, and identities within the broader LGBTQIA+ community. However, we do not have accurate representation from community leaders from "hidden" communities, such as Latinx LGBTQIA+ groups, working-class LGBTQIA+ individuals, and queer sex workers. Another limitation of our study is that we did not ask participants about minority stress in general during interviews; instead, we applied the codes deductively to the larger study. In future work, we could more deliberately ask these questions instead of framing them.

7 Conclusion

Health information practices are a vital strategy of resilience for LGBTQIA+ communities. As evidenced in minority stress literature and our findings, LGBTQIA+ individuals experience minority stressors on every level of the socio-ecological model. In turn, these stressors, which are externally produced, are integrated into LGBTQIA+ communities. These stressors emerge iteratively so that when a stressor is produced on one level, it influences a stressor's production on another level, thus creating a blend of intrapersonal, interpersonal, organizational, community, and public policy level minority stressors. However, members of these communities survive and even thrive despite this stress. Such survival in the face of layered and routinized stressors suggests a necessary reconstitution of how we understand the failure of health information practices to exist within LGBTQIA+ communities. We must note that the failure is not of their doing, but a failure to be seen and represented adequately by society at all levels of the socio-ecological model [22]. In turn, we must ask how health information systems can better facilitate the needs of LGBTQIA+ persons instead of assuming their failure to fit within such frameworks necessitates correction on the part of the communities themselves.

References

1. Hatzenbuehler, M.L., Panchankis, J.E.: Stigma and minority stress as social determinants of health among lesbian, gay, bisexual, and transgender youth: Research evidence and clinical implications. Pediatric Clinics **63**(6), 985–997 (2016)
2. Meyer, I.H.: Prejudice, social stress, and mental health in lesbian, gay, and bisexual populations: conceptual issues and research evidence. Psychol. Bull. **129**(5), 674–697 (2003)
3. Savolainen, R.: Everyday information practices: a social phenomenological perspective. Scarecrow Press, Lanham (2008)
4. Toomey, R.B., Ryan, C., Diaz, R.M., Russell, S.T.: Coping with sexual orientation-related minority stress. J. Homosex. **65**(4), 484–500 (2018)
5. Martinez-Donate, A.P., et al.: Does acculturative stress influence immigrant sexual HIV risk and HIV testing behavior? Evidence from a survey of male Mexican migrants. J. Racial Ethnic Health Disparities **5**(4), 798–807 (2018)
6. Utamsingh, P.D., Richman, L.S., Martin, J.L., Lattanner, M.R., Chaikind, J.R.: Heteronormativity and practitioner–patient interaction. Health Commun. **31**(5), 566–574 (2016)
7. Meyer, I.H.: Resilience in the study of minority stress and health of sexual and gender minorities. Psychol. Sex. Orientation Gender Diversity **2**(3), 209–213 (2015)
8. Glanz, K., Bishop, D.B.: The role of behavioral science theory in development and implementation of public health interventions. Annu. Rev. Public Health **31**, 399–418 (2010)
9. McLeroy, K.R., Bibeau, D., Steckler, A., Glanz, K.: An ecological perspective on health promotion programs. Health Educ. Q. **15**(4), 351–377 (1988)
10. Bry, L.J., Mustanski, B., Garofalo, R., Burns, M.N.: Resilience to discrimination and rejection among young sexual minority males and transgender females: a qualitative study on coping with minority stress. J. Homosex. **65**(11), 1435–1456 (2018)
11. Crenshaw, K.: Mapping the margins: Intersectionality, identity politics, and violence against women of color. Stanford Law Rev. **43**, 1241–1300 (1990)
12. Pohjanen, A.M., Kortelainen, T.A.M.: Transgender information behaviour. J. Doc. **72**(1), 172–190 (2016)
13. Colpitts, E., Gahagan, J.: The utility of resilience as a conceptual framework for understanding and measuring LGBTQ health. Int. J. Equity Health **15**(1), 1–8 (2016)
14. Lerner, J.E., Robles, G.: Perceived barriers and facilitators to health care utilization in the United States for transgender people: a review of recent literature. J. Health Care Poor Underserved **28**(1), 127–152 (2017)
15. Kitzie, V.L., Wagner, T.L., Vera, A.N.: In the beginning, it was little whispers…now, we're almost a road": conceptualizing a model for community and self in LGBTQ+ health information practices. In: Sundqvist, A., Berget, G., Nolin, J. (eds.) iConference 2020, LNCS, vol. 12051, pp. 15–31. Springer, Cham (2020). https://doi.org/10.1007/978-3-030-43687-2_2

16. Hillary, G.A.: Definitions of community: areas of agreement. Rural Sociol. **20**, 111–123 (1995)
17. McMillan, D.W., Chavis, D.M.: Sense of community: a definition and theory. J. Community Psychol. **14**(1), 6–23 (1986)
18. Greyson, D., O'Brien, H., Shoveller, J.: Information world mapping: a participatory arts-based elicitation method for information behavior interviews. Libr. Inf. Sci. Res. **39**(2), 149–157 (2017)
19. Gibson, A.N., Martin III., J.D., : Re-situating information poverty: Information marginalization and parents of individuals with disabilities. J. Am. Soc. Inf. Sci. **70**(5), 476–487 (2019)
20. Floegel, D., Costello, K.L.: Entertainment media and the information practices of queer individuals. Libr. Inf. Sci. Res. **41**(1), 31–38 (2019)
21. Vera, A.N., Wagner, T.L., Kitzie, V.L.: When it's time to come together, we come together": reconceptualizing theories of self-efficacy for health information practices within LGBTQIA+ communities. In: Jean, B.S., Jindal, G., Liao, Y., Jaeger, P.T. (eds.) Roles and Responsibilities of Libraries in Increasing Consumer Health Literacy and Reducing Health Disparities, vol. 47, pp. 263–282. Emerald Publishing Limited (2020)
22. Halberstam, J.: The queer art of failure. Duke University Press, Durham (2011)

Pregnancy-Related Information Seeking in Online Health Communities: A Qualitative Study

Yu Lu[1,2](✉), Zhan Zhang[2], Katherine Min[1], Xiao Luo[3], and Zhe He[1]

[1] Florida State University, Tallahassee, FL 32306, USA
yl20@my.fsu.edu, kmin@outlook.edu, zhe@fsu.edu
[2] Pace University, New York, NY 10038, USA
zzhang@pace.edu
[3] Indiana University - Purdue University Indianapolis, Indianapolis, IN 46202, USA
luo25@iupui.edu

Abstract. Pregnancy often imposes risks on women's health. Consumers are increasingly turning to online resources (e.g., online health communities) to look for pregnancy-related information for better care management. To inform design opportunities for online support interventions, it is critical to thoroughly understand consumers' information needs throughout the entire course of pregnancy including three main stages: pre-pregnancy, during-pregnancy, and postpartum. In this study, we present a content analysis of pregnancy-related question posts on Yahoo! Answers to examine how they formulated their inquiries, and the types of replies that information seekers received. This analysis revealed 14 main types of information needs, most of which were "stage-based". We also found that peers from online health communities provided a variety of support, including affirmation of pregnancy, opinions or suggestions, health information, personal experience, and reference to health providers' service. Insights derived from the findings are drawn to discuss design opportunities for tailoring informatics interventions to support consumers' information needs at different pregnancy stages.

Keywords: Information seeking · Needs assessment · Consumer health information · Pregnant women · Online community

1 Introduction

Pregnancy is an important phase of women's life in which women often experience a variety of physiological, emotional, and medical issues [42]. For many women, pregnancy makes them more intensively interact with the healthcare system, which is fragmented and challenging to navigate in many countries [25]. According to the Healthcare Cost and Utilization Project of the Agency for Healthcare Research and Quality [17], 94.1% of the deliveries in the U.S. in 2008 had experienced complications, such as preterm labor, hemorrhage, and low birth

K. Toeppe et al. (Eds.): iConference 2021, LNCS 12646, pp. 18–36, 2021.
https://doi.org/10.1007/978-3-030-71305-8_2

weight, imposing great risks to women's health as well as financial burden to the society and the families. Due to insufficient information obtained from health providers, women and their caregivers often proactively search and consume health information by themselves. Addressing their pregnancy-related information needs is of utmost importance as appropriate prenatal education improves maternal and fetal outcomes [32,37]. The literature points out that the internet has been widely used as an effective tool by women to fulfill their information needs related to pregnancy [47,49]. In particular, social media, including online health forums and social Q&A sites, has become hugely popular as an alternative source for searching and receiving pregnancy-related information, sharing concerns and experiences felt during pregnancy, and communicating with peers who share similar conditions [15,31,39,48]. Compared with the traditional, face-to-face consultation with obstetricians, inquiring about pregnancy-related information on online health forums is *"convenient, detailed, practical, customized, and unbiased"* [55]. However, the information online often lacks credibility and is not well organized [4]. In both means of consultation, consumers may experience health information overload [34], which may have negative effects, such as stress, confusion, and cognitive strain [19].

Similar to other chronic conditions such as diabetes, pregnancy is a "stage-based" health condition, including pre-pregnancy, during-pregnancy, and post-partum [2]. In particular, during-pregnancy can be further categorized into three gestational trimesters based on the week of pregnancy. Women during the entire pregnancy trajectory may face different challenges and uncertainties, thereby having evolving information needs. For example, women who are considering pregnancy have different questions and needs from those who are pregnant. However, previous studies about pregnancy-related information-seeking mostly treat pregnancy as an uniform condition and have not adequately examined women's information needs in different stages [24]. To design health information technologies for supporting information needs in different pregnancy stages for consumers, it is crucial to obtain an in-depth understanding of their information needs throughout the pregnancy trajectory and how these information needs vary across different stages. However, existing studies were primarily focused on information needs of pregnant women during the trimesters of pregnancy, with little attention paid to before or after pregnancy stages [10,20,24,28,33,48,50,55]. Thus, a study examining the comprehensive typology of consumers' information needs throughout their pregnancy trajectory is necessary.

To fill this research gap, we investigated consumers' pregnancy-related information seeking in different stages in online health communities. The reason for choosing online health communities is that they are common channels for people to seek and share health-related information. Also, online health communities are advantageous in facilitating peer support as they enable consumers to post and answer questions in natural language and include contextual information, such as specific complications they experience during or after pregnancy, period cycles, and medical histories. In this paper, we conducted an in-depth content analysis on 6,713 question posts collected from the Pregnancy and Parenting category of Yahoo! Answers, one of the most well-known social Q&A websites. The analyzed posts covered the pre-pregnancy, during pregnancy, and postpartum stages. Our analysis uncovers the types of information or support consumers in different stages of pregnancy seek, how different they are, and what types of responses they receive from peers. We then draw our findings to discuss practical and theoretical implications for online support interventions. This study is a preliminary study of a large research effort to design and develop a computational approach for delivering tailored information support to healthcare consumers, who are either trying to get pregnant, experiencing pregnancy, or parenting. The contributions of the study are two-fold:

1. An in-depth, comprehensive understanding of consumers' information needs throughout the entire course of pre-pregnancy, during-pregnancy, and postpartum;
2. Practical and theoretical implications for technology support to aid pregnancy-related information seeking and retrieval.

2 Background and Related Work

Online health communities are digital platforms where patients connect with others in similar scenarios to discuss health concerns [43]. Seminal studies have investigated the information seeking behavior in online health communities and how to support such online information-seeking activities (e.g., [7,14,22,29,30,46,53,54]. A well-recognized advantage of online health communities is that patients can obtain valuable experiential knowledge and easy-to-understand medical information from their peers, which the clinicians may not sufficiently provide [26]. Prior work revealed that many patients shared their personal health information and even illness trajectory to contextualize their information seeking in order to elicit personalized advice and suggestions [29]. In the case of pregnancy, existing literature shows that consumers have a variety of questions and confusion related to pregnancy [18,36,44]. As such, they often turn to online health forums to obtain experiential knowledge from their peers pertaining to, for example, how to conceive, how to address some experienced symptoms during pregnancy, and how to deal with complications after giving birth. More importantly, existing research indicates that online health communities provide an anonymous space for consumers to share intimate details and disclose sensitive information, without the concern of being judged by others [40].

A large body of research has examined how consumers seek information and peer support during pregnancy (e.g., [6,14,28,35,41]. For instance, Zhu et al. [55] analyzed pregnancy-related information-seeking on social media among expectant mothers in China; the results revealed that social media increasingly empowers expectant mothers to make more informed decisions during pregnancy. Robinson et al. [45] assessed pregnant women and caregivers' information needs to discover the types of needs that were not met. Gui et al. [24] investigated support seeking for pregnancy in Babycenter along the three gestational stages during pregnancy. Some work also investigated peer support seeking during postpartum and reported that new parents benefited from the experience of their peers in a wide range of topics related to postpartum such as parenting newborns, postpartum depression, and breastfeeding [14,22,29]. We also found a few research projects that examined information-seeking about maternal risk factors, such as infertility [10].

While these studies examined the support sought by consumers about pregnancy or postpartum, they only focused on one stage of pregnancy or mixed multiple groups without studying distinct groups in depth [16,28]. To the best of our knowledge, no prior work has attempted to thoroughly examine consumers' information needs throughout the trajectory of pregnancy, including before, during, and after gestation. Existing literature points out that pregnancy is a unique health condition that is "staged", with each stage presenting different physiological, emotional, and informational needs [24]. Therefore, it is imperative to examine the information needs of women over the course of pregnancy and how online communities respond to these needs. An in-depth understanding of these aspects can shed light on designing personalized online support interventions to provide stage-based, tailored, and more suitable pregnancy care support. Our study aims to fill this knowledge gap.

3 Methods

3.1 Data Collection

Approved by the Institutional Review Board at Pace University, we used a secondary dataset collected from the "Pregnancy" and "Parenting" category of Yahoo! Answers between 2009 and 2014. The dataset consists of a total of 14,646 question posts. As many communities under this category are not health-related, we filtered out these irrelevant ones (e.g., "Adoption", "Baby Names", "Toddler & Preschooler", and "Parenting") and extracted four communities that are highly relevant to pregnancy experience, including "Pregnancy & Parenting", "Trying to Conceive", "Pregnancy", and "Newborn & Baby". After this step, 6,713 posts were retained for further analysis. The name of these four forum communities and the number of posts retrieved from each community are summarized in Table 1.

3.2 Data Analysis

Three researchers performed a content analysis on the entire 6,713 posts. The analysis consists of the following steps (shown in Fig. 1): First, we developed a codebook using 200 posts randomly sampled from the corpus. More specifically, the sampled posts were independently reviewed by two coders, C1 and C2, for relevance judgment. For posts that were considered relevant, we used an open coding technique to categorize posts and their chosen replies (also known as the

Table 1. Relevant online communities and the number of retrieved posts from each community

Community	Number of retrieved posts
"Pregnancy & Parenting"	1,822
"Trying to Conceive"	1,449
"Pregnancy"	1,552
"Newborn & Baby"	1,890

"Best Answer"). As the quality of the post answers varies, the poster often chose a reply which fulfilled his/her information needs as the "Best Answer". Therefore, examining the best answer responses could help us not only understand what types of support were provided but also confirm the poster's information needs.

Once an initial version of the codebook was established, the researchers met to discuss whether to keep, remove, merge, or replace any codes. After the researchers reached an agreement about every code in the codebook, the codebook was then utilized to standardize the coding procedure. The finalized codebook contained 14 themes, which were organized by three general stages of pregnancy: pre-pregnancy, during-pregnancy, and postpartum (shown in Table 2).

Next, the third coder (C3) coded another 100 randomly sampled posts to examine the exhaustiveness of the codebook. Through this process, no new codes merged, confirming that the codebook was comprehensive enough. Following that, C1 and C2 coded another random sample of 200 posts and applied Cohen's Kappa Coefficient to test their inter-rater reliability (Fig. 1, Step 3). Two coders achieved a substantial agreement (kappa's value is 0.80). Once we resolved the disagreements, C1 and C2 coded the remaining posts to conclude the analysis (Fig. 1, Step 4).

4 Results

4.1 General Observations

Overall, we found that posters sought various informational supports on the forum, which can be roughly categorized into three categories, including *advice, pregnancy-related medical knowledge* and *experiential knowledge*. For example, posters explicitly asked for peers' advice and actionable suggestions about what they should do next or how they should handle a situation. A typical example of this type of question is *"I am now 4 days late on my period, however, the pregnancy test still shows negative. I have been trying (to conceive) for a while now, does anyone have any suggestions?"* Furthermore, many posters inquired about pregnancy-related medical knowledge, such as commonly seen pregnancy symptoms, effects of a specific medication, and diagnostic abilities of lab tests. Lastly, they also sought experiential knowledge from others who had similar

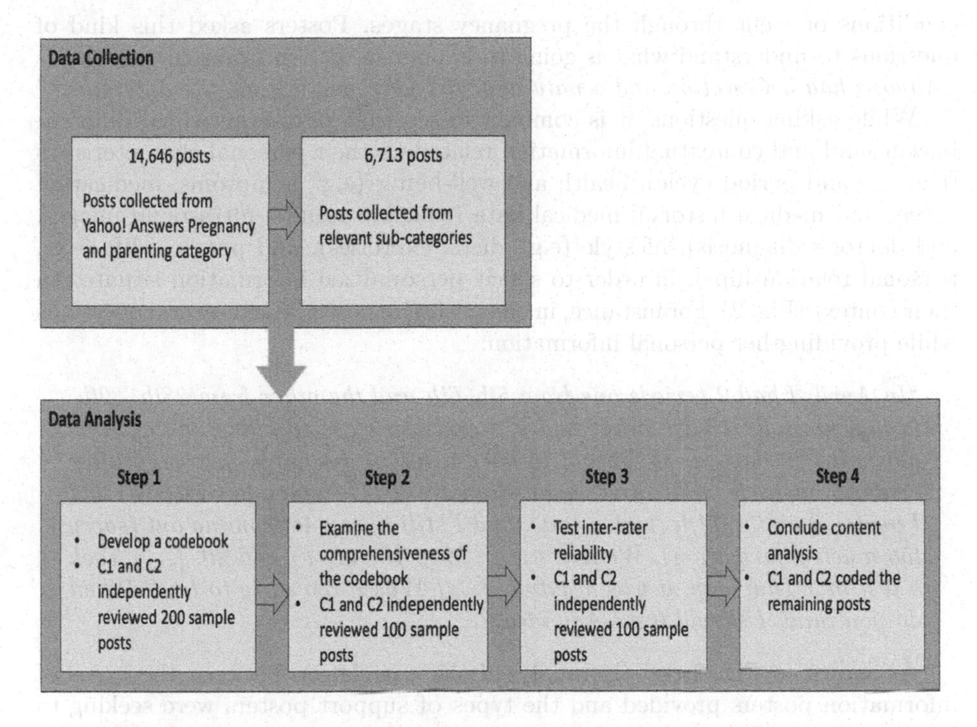

Fig. 1. Data collection and analysis steps

Table 2. Summary of themes for information needs. Some posts fell into multiple themes, so percentages add up to more than 100%. Percentage equal to the number of occurrence divided by the totoal posts analyzed.

Stage	Theme	N (%)
Pre-Pregnancy	Chances of pregnancy	571 (20.4%)
	Conceive or implantation	231 (8.3%)
	Medication or treatment	132 (4.8%)
	Period or ovulation	154 (5.6%)
	Symptoms of pregnancy	141 (5.1%)
	Miscarriage	40 (1.4%)
During-Pregnancy	Health concerns & Disease of pregnancy	434 (15.7%)
	Knowledge and Prognosis of Pregnancy	135 (4.9%)
	Labor	74 (2.7%)
Postpartum	Breastfeeding	68 (2.5%)
	Health Concerns of Newborn	196 (7.1%)
	Nurturing Newborn	124 (4.5%)
	Medication or Treatment	132 (5.8%)
Across Pregnancy Trajectory	Lab Test	335 (12.1%)

conditions or went through the pregnancy stages. Posters asked this kind of questions to understand what is going to happen next. An example question is *"Anyone had a C-section and a natural birth? Can you tell me the difference?"*

While asking questions, it is common to see that posters provided different background and contextual information related to their personal characteristics (e.g. age and period cycle), health and well-being (e.g., symptoms, medication taken, and medical history), medical data (e.g., lab results, ultrasound images, and doctor's diagnosis), lifestyle (e.g., diets, exercises), and personal life (e.g., personal relationships), in order to solicit personalized information situated in their context (Fig. 2). For instance, in one post, the poster asked several questions while providing her personal information:

> *"In April, I had 2 periods one from 5th–6th and the other from 28th–30th. Is this normal? I have never had 2 periods so close together, my cycle is normally 29 days. Is it normal to have a lot of discharge one week after ovulation day? [...] Me and my husband are TTC (trying to conceive) and I ovulated on the 11th (took a test), and I still have a lot coming out (sorry too much information). We had sex on May 10, 11, 13 and 20. [...] Took a test this Thursday, it was negative. [...] Was it too early to test? When do you think I should test? Any ideas?"*

As shown in Fig. 2, we also analyzed the correlation between the types of information posters provided and the types of support posters were seeking to better understand how consumers formulated their questions to fulfill their information needs. For example, in the pre-pregnancy stage, posts inquired about the chances of getting pregnant frequently provided information about demographics, menstrual cycle, symptoms, conception details, or lab results. People during pregnancy often asked questions by providing information about demographics, due dates, or symptoms, while postpartum posts often disclosed newborn's symptoms and demographics, as well as lab results.

4.2 Information Needs at Each Stage of Pregnancy

Although the types of information support (e.g., advice, medical knowledge, and experiential knowledge) consumers sought on the forum were relatively consistent across the entire pregnancy trajectory, the specific information needs expressed by posters varied across different stages of pregnancy. In this section, we present posters" information needs for different stages of pregnancy. The number and percentage of each type of information needs, categorized by the stage of pregnancy, are shown in Table 2. The themes were categorized into pregnancy stage if the type of information need was pertinent to that particular stage.

Pre-pregnancy Stage. In this stage, consumers often raised questions related to the chances and symptoms of pregnancy, conception and implantation, issues related to period or ovulation, and medication or treatment for certain health concerns. In particular, requests for help determining the chances of pregnancy

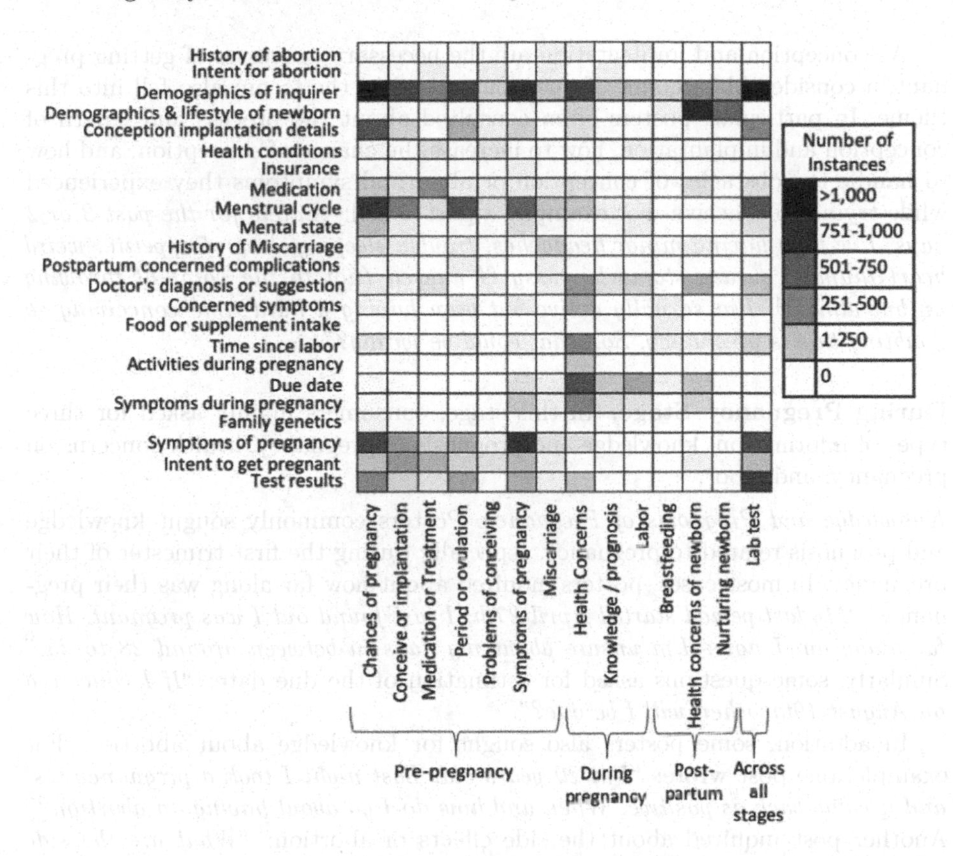

Fig. 2. Co-occurrence of information needs at different stages. x-axis represents the types of information needs in different pregnancy stages and y-axis lists the types of information provided by the posters in a question post. The legend dictates the number of co-occurred instances among the total coded posts.

were by far the most common theme in the corpus. Posters often disclosed their recent history of sexual behavior and symptoms that were signs of pregnancy to obtain more accurate estimation. For example, a post provided relevant information and asked about the chance of getting pregnant:

> *"Ok so I had sex 5 days after my period stopped with my bf and according to my cycle I wasn't supposed to ovulate for another week but 3 days after having sex I felt ovulation coming on. It's now been 12 days since I had sex and my breasts are really tender, cramping and have a lotion-like white discharge when I use the bathroom, I felt sick to my stomach almost to the point of throwing up and my period isn't due for another week and a half [...]. Is it too early to be feeling symptoms? I stress over it every day [...]. Can anyone tell me what you think my chances of being pregnant is? If I'm not pregnant what could it be?"*

As conception and implantation are the necessary processes of getting pregnant, a considerable amount of questions raised in the forum also fell into this theme. In particular, posters often consulted about the process and length of conception and implantation, how to increase the chance of conception, and how to handle the obstacles of conception or abnormal symptoms they experienced while trying to conceive. For example, a post asked: *"Lately for the past 3 or 4 days, I've been having minor headaches, trouble sleeping, loss of appetite, weird heart thumps, queasy stomach, gassy & nausea (not to the point of throwing up but almost). I'm sexually active but been having a hard time conceiving so doubting this is pregnancy. So, what could be wrong?"*

During-Pregnancy Stage. In this stage, consumers mainly asked for three types of information: knowledge and prognosis of pregnancy, health concerns on pregnancy, and labor.

Knowledge and Prognosis of Pregnancy. Posters commonly sought knowledge and prognosis related to pregnancy, especially during the first trimester of their pregnancy. In most cases, posters inquired about how far along was their pregnancy: *"My last period started April 27th. I just found out I was pregnant. How far along am I now? I'm unsure about my days in between around 28 to 33."* Similarly, some questions asked for estimation of the due date: *"If I conceived on August 19th, when will I be due?"*

In addition, some posters also sought for knowledge about abortion. For example, one post wrote: *"I'm 20 years old. Last night I took a pregnancy test and it came back as positive. When and how do I go about having an abortion?"* Another post inquired about the side effects of abortion: *"What are the side effects after having an abortion at age 32?"*

In other cases, posters would like to know the proper diet during pregnancy. For example, a post asked: *"Is it safe to eat black berries during pregnancy? 35 weeks pregnant."* Lastly, questions related to gender prediction were raised by many posters. As one question wrote: *"I had a girl born first and then a boy. Thinking about another baby, what are my chances that it will be another girl???"*

Health Concerns on Pregnancy. Once a woman gets pregnant, her body undergoes hormonal changes which could trigger symptoms ranging from morning sickness, headache, ligament pain, to shortness of breath, contractions, and general discomfort [13,27]. Some symptoms are more common in the first trimester, whereas others are often seen in the second and third trimesters. Nevertheless, we observed that posters raised a variety of health concerns during pregnancy, across the three trimesters. For example, many posts asked about how to handle concerned symptoms they had during pregnancy (31.8%), whether the effects of symptoms during pregnancy can still sustain after giving birth (48.6%), and what consequence of prenatal exposure to certain substances (e.g., alcohol and smoking) could be (2.8%). For example, a post wrote: *"I have had a headache for over 24 h now [...] I'm pregnant, so I can't take Aleve or Ibuprofen. Any suggestions on how to get rid of it?"*

Labor. We found that posters often raised questions about giving birth, especially during their third trimester (which usually lasts from week 29 of pregnancy through birth). The most frequently asked questions under this theme included identifying signs of labor and deciding when to go to the hospital to give birth, knowing the process of delivery and what to prepare, and understanding the pros and cons of giving birth naturally compared to assisted delivery. It is also worth noting that a number of posts are related to emotional state. For example, one post asked for other people's experience of giving birth in order to prepare themselves for the upcoming delivery, both physically and mentally: *"How many weeks were you when you gave birth and how badly did the contractions hurt?"*

Postpartum Stage. For posters at the postpartum stage, they commonly asked questions concerning the health issues of their newborns (7.1%). In particular, new mothers inquired a lot about how to handle the symptoms of their newborns, what might cause the health issues, and what they should do and/or what medications and treatment options are available to deal with certain health issues. As new mothers often experience anxiety or frustration when their children present health issues, it is common to see their posts containing emotional stress, signaling the importance of providing emotional support:

> *"My daughter was 7 weeks old yesterday and she has been crying ALL day for the pat 3 days. She doesn't have a fever and she is exclusively breast fed so it's not constipation. I have no idea if something is wrong or if it is just colic. The only time she is okay is if I wrap her up real tight and turn the vent on in the bathroom with the light off and bouncing her. I'm exhausted! Any suggestions? Encouragement?"*

In addition, some posts are concerned about how to nurture and more broadly, take care of their newborn, including food, sleeping, and breastfeeding. For example, a post inquired about the appropriate amount of formula intake: *"How much formula does your 3-month old baby eat per day?"* In another post, the mother inquired the effect of her medication taken on breastfeeding: *"Can you take some kind of birth control pill while breastfeeding?"*

Across Pregnancy Trajectory. We found that questions pertaining to lab tests were frequently raised by posters across the entire pregnancy trajectory. For example, regardless of the pregnancy stage, women usually take a set of laboratory tests, such as blood tests and Pap smear, to make sure that there is no medical condition that could affect pregnancy. However, as highlighted in previous work [10,52,53], people often have difficulties comprehending and acting upon their lab results. This is more evident for people who have lower health and numeracy literacy [56]. Therefore, they often turn to online forums to seek information or second opinion. The most commonly asked questions related to lab tests include when to take a test, the meaning of lab results, and the accuracy and reliability of lab results. For example, one post sought help clarifying whether a specific lab result over a particular pregnancy period was considered normal: *"I'm about 4–5 weeks (pregnant), I took a HCG test and its level is 1000 something, 2 days later I took a second and it showed a little over 2000. Is that normal?"* Or *"My wife delivered a baby 21 days before when we tested to thyroid past 5 days before her tsh level is 5.2 is this normal level or she want to take some treatment?"*. In another case, a poster suspected that she might have gotten a false negative result since the lab results she received did not align with her symptoms, motivating her to come to the forum to seek help: *"Could I have gotten a false negative result? I'm 10 days late in my period. That NEVER happens to me. I've taken a blood pregnancy test, but it was negative. Could it be wrong?"*

Since there are many known and unknown factors that could affect the implication of lab tests for a person [51], posters often provided contextual information to solicit personalized suggestions pertaining to the indications of lab results. For example, they usually provided demographics, the results of current and previous tests, medical history, and doctors' diagnosis or suggestions to contextualize their inquiries: *"Me ([age] 24) and my husband ([age] 26) are TTC 3 and we had sex on May 10,11,13 20. I ovulated May 12 (took test) for the last couple of days. I have been having really watery/sticky discharge [...] When would be the best time to take a test? [...] My previous pregnancies were surprises so I'm not quite sure what I should be looking for."*

4.3 Types of Replies Posters Received

Among the replies directly addressed posters' information needs, we categorized them into five types: *affirmation of pregnancy*, *opinions or suggestions*, *health information*, *personal experience*, and *reference to healthcare providers' advice or service*. The number of each response type is shown in Table 3. Below we describe the replies chosen by consumers in details.

Affirmation of Pregnancy. Very often, responders replied to affirm whether or not a poster was pregnant based on the information that the poster had provided, such as period cycle, recent sexual activity, and symptoms: *"That does indeed sound like pregnancy symptoms. Since you've already had some period signs, you're mostly like far enough to take a test. [....] But yeah, sore boobs, exhaustion, frequent urination, and the sudden stop of an otherwise normal menstrual cycle...those definitely sound like pregnant symptoms."*

Opinions or Suggestions. Peers on the forum also replied posters with their opinions or suggestions to help posters make better decisions. For example, a post expressed concerns about whether to have a baby if parents have a 25% chance of passing on a severe genetic condition; the chosen answer replied: *"Personally, no. 1 in 4 odds of not living a normal life is not something I would chance. I would rather be childless than to have to bury my children."*

Some responders also suggested what the posters can do given their situations at a specific pregnancy stage. For example, for a poster who was confused about how soon to test for pregnancy, a person replied: : *"Symptoms will show up only after implantation, that takes 12 days from the day you have had sex, if you have conceived. Sometimes it takes longer for symptoms to surface. Wait until you miss your periods and then take a pregnancy test."*

Personal Experience. A large proportion of the responders answered questions by reflecting on their own experience to help the posters better understand their situation and determine what to do. For example, a poster was concerned about the safety of epidural and asked for opinions about it. The responder of the chosen answer offered personal experience about epidural: *"I had four kids. With three of them I had an epidural and with one I had natural child birth. The natural childbirth wasn't exactly by choice. Lol. I got to the hospital too late to get the epidural and had the baby in less than hour after I got there".*

Reference to Providers' Advice or Service. In some cases, responders realized that answering certain questions may require professional expertise, subsequently they would suggest that the posters should refer to doctors' advice. For example, a poster experienced some unusual symptoms and suspected if those symptoms indicate pregnancy, a responder replied: *"To me it sounds like a two-week miscarriage but I'd check with the doctor just to make sure".*

Table 3. Summary of themes for chosen reply

Theme	N
Affirmation of Pregnancy	138
Opinions or Suggestions	941
Health Information	802
Personal Experience	664
Reference to Providers' Advice or Service	138

5 Discussion

In the study, we investigated the information needs throughout the pregnancy trajectory and the support they received from the peers online. We found clear distinctions of information needs in different stages. Below we draw on our findings to discuss the design implications for helping consumers better manage personal health, search information, and make decisions in different stages of their pregnancy trajectory.

As consumers filling their knowledge gap about pregnancy, they engage in self-observation, self-judgement, and self-reaction of their health conditions, which are three sub-processes in the social-cognitive view of self-regulation [8]. Such self-regulation is initiated with the exertion of a common goal to achieve better health outcomes for both the mother and the newborn. As a result, consumers feel more empowered. According to existing literature [5], psychological empowerment has two dimensions: intrapersonal empowerment and interactional empowerment. On one hand, consumers are intrapersonally empowered as they obtain more control and competence by learning knowledge about and experience of pregnancy in online communities. On the other hand, consumers are interpersonally empowered as they actively involve in communicating and exchanging experience of and advice for pregnancy with peers who are in similar situations in the *small world* of social groups [11].

Hence, they come to online forums to seek advice. Online forums offer a wide range of topics to consumers. However, these forums may not use a well-designed typology, classification, or taxonomy to organize information. As a result, it is difficult for consumers to sort out which topic group to post questions. For example, Baby Center has 2543 groups for the topic "pregnancy", consumers may not know where to post questions, in a trimester-specific group or a problem-specific group. In the pre-pregnancy stage, infertile consumers may encounter different problems than those initiating normal preparation for pregnancy. They may not find enough support from communities for regular birth preparation and may need to visit infertility-specific communities. As it is still challenging to address the problem of how to organize pregnancy-related topics in online forums, a comprehensive typology is needed for finding consumers' target information.

We found that consumers made different inquiries as they transit from one stage in pregnancy to another. Therefore, online technology interventions should provide stage-based and customized support instead of providing uniform support for different foci during different stages. In other words, healthcare systems designed to support consumers' information needs should take into consideration the evolving circumstances of consumers' physiological, informational, and emotional needs throughout their pregnancy journey. This study assessed the stages of and types of questions related to pregnancy, which can be used to better categorize pregnancy related information so that consumers can find relevant questions and solutions more easily.

Also, we found posters sought experiential knowledge from their peers with similar conditions. Therefore, technology interventions could incorporate experiential knowledge of peer patients. For instance, similar or relevant experience of peers could become an effective alternative for consumers to fulfill their information needs. This population often expressed stress (e.g., worry, anxiety, and disappointment) while making inquiries, thus, receiving information from peers who have gone through similar experiences and emotions could be a great help.

Supplying health information that is reliable, operationalizable, and suitable for the personal circumstances of the specific health consumer is crucial. In our analysis, posters' inquiries indicated that they often encountered individualized health issues concerning pregnancy. However, as health providers may offer one-size-fits-all information, sometimes these individualized problems may not receive information that can be operationalized within their personal contexts. Therefore, it is important that providers develop systematic approaches to better operationalize and personalize pregnancy-related information that suits the individualized needs and provide such information at the right time. Technology designers could consider the impact of patients' contextual information (e.g., demographic information and medical / medication history) on the appropriate medical suggestions these patients receive. The contextual information could be modeled as groups of useful features to retrieve medical information and health suggestions relevant to consumers' health situations. For example, the information a 25-years old black woman with family inherited complications, no health insurance, no history of miscarriage, previously delivered one child, search for during her pregnancy is very likely different from a 32-years old white woman at the postpartum stage, with insurance coverage and history of miscarriage, no child previously delivered.

It is also worth noting that mobile health (mHealth) applications have been gaining momentum in supporting pregnancy-related information seeking [3,12,38]. For example, Ainscough et al. [3] investigated the impact of mHealth applications on behavioral change in overweight and obese pregnant women. Their results revealed that mHealth interventions could help overweight and obese women shift from contemplation/preparation to maintenance of healthy lifestyle in pregnancy, and could potentially have positive impact on pregnancy outcome, maternal and newborn's health [3]. Choi et al. [12] assessed the feasibility and efficacy of mHealth intervention in promoting physical activity to inactive pregnant women. The results demonstrated that mHealth is feasible in promoting physical activity of pregnant women [12]. Ledford et al. [38] compared the effectiveness of mHealth applications in prenatal education and engagement versus a spiral-note guide. The results showed that patients using mHealth to record information about their health developed greater patient activation than patients who used notebooks, indicating that mHealth can result in more activated self-management of prenatal health [38]. Despite those efforts, as Robinson et al. [45] pointed out, information needs of consumers are still largely unfilled by mHealth applications, highlighting the importance of examining how to better address the unmet needs.

Supportive applications (e.g. Pregnancy Tracker by *BabyCenter* [1]) now can provide immense information to consumers. Despite so, certain challenges still persist. For instance, throughout childbearing consumers' information encountering online [21], a large quantity of diverse information has been provided. Consequently, it could result in information overload and information anxiety [9], making it hard for consumers to identify the information they need. Furthermore, a lot of static information was provided with no or little customization based on the personal situation. Although such information is credible and well-organized, it is hard for consumers to quickly identify the information they need [23]. Hence, it is important that supportive tools are providing better ways of organizing and categorizing information to ensure that consumers find the information they need.

Taken together, we propose several recommendations for patient portals, which are summarized in Table 4. For pre-pregnancy consumers, portals should list symptoms of pregnancy, and provide tips for preparing for pregnancy and coping with miscarriage. During pregnancy, portals should offer antenatal knowledge, information about nutrition, healthy lifestyle, and labor. For postpartum care, portals could list common issues with the newborns and breastfeeding.

Table 4. Design recommendations

Stage	Design recommendation
Pre-Pregnancy	List symptoms of pregnancy
	Provide tips for preconception preparation
	Offer information related to period and ovulation
	Supply tips for preventing miscarriage
During-Pregnancy	Provide antenatal knowledge to patients
	Introduce information about health lifestyle for pregnancy
	Supply strategies for labor preparation and due date calculation
Postpartum	Provide guidance for newborn care to first-time parents
	List common symptoms of newborns along with treatment options
	Supply treatment options for concerned symptoms during breastfeeding along with treatment options
Across Pregnancy Trajectory	Provide adequate information by the stage of pregnancy and at the right time
	Incorporate information regarding patient's unmet needs (e.g. dietary information)
	Incorporate peer experience and coping strategies if needed
	Provide personalized information and suggestions based patient's demographic information

Across the pregnancy trajectory, portals should provide adequate information at the right time, incorporate peer experience and coping strategies where appropriate, and personalize information and suggestions to new parents.

6 Limitations and Future Work

This study has certain limitations. First, we only focused on question posts from Yahoo! Answers, so the results may not be generalizable to other online health communities or the broader population. Also, due to the limited amount of data available, tracing how the type of inquiries a specific user made evolves across different stages became difficult, which prevents our investigation of the evolution of consumers' information needs in a case-by-case setting. In our future work, we will investigate the evolution of consumers' information needs about pregnancy. We will also compare consumers' pregnancy-related information seeking on our study site with other professional health forums, such as MedHelp. Second, this dataset was collected between 2009 and 2014. With the emergence of new tools and applications, some of the information needs or concerns might have been addressed. However, new challenges can be introduced by these tools. For example, when using mobile health applications that are not designed to support "stage-based" information seeking, there is a potential risk that consumers may be confused by the inconsistent information or misinformed with inaccurate or unreliable information. Hence, our future work includes conducting user studies to investigate their needs and faced challenges in the current context.

7 Conclusions

In this study, we investigated the comprehensive typology of posters' information needs throughout the pregnancy trajectory by analyzing forum posts collected from a social Q&A site. Findings of this study implied that doctors need to know what responses consumers may receive regarding their questions online. The findings can also shed light on the design of personalized pregnancy-related technologies to accommodate different physical and informational needs that accompany different stages of pregnancy to provide personalized, stage-based, and situated care support. For future work, we will investigate the credibility of or perform fact-checking on the responses that health consumers receive regarding their questions online.

Acknowledgement. We thank Dr. Sanghee Oh for sharing the data collected from Yahoo!Answers. This work was supported in part by University of Florida-Florida State University Clinical and Translational Science Award funded by National Center for Advancing Translational Sciences under Award Number UL1TR001427. The content is solely the responsibility of the authors and does not necessarily represent the official views of the NIH.

References

1. babycenter. https://www.babycenter.com
2. Tips for first-time moms on pre-pregnancy, pregnancy and postpartum. https://www.hopkinsallchildrens.org/ACH-News/General-News/Tips-for-First-time-Moms-on-Pre-pregnancy,-Pregnan
3. Ainscough, K., Kennelly, M., Lindsay, K., O'Sullivan, E., McAuliffe, F.: Impact of an mhealth supported healthy lifestyle intervention on behavioural stage of change in overweight and obese pregnancy. Proc. Nutrition Soc. **75**(OCE3) (2016)
4. Al Wattar, B.H., Pidgeon, C., Learner, H., Zamora, J., Thangaratinam, S.: Online health information on obesity in pregnancy: a systematic review. Eur. J. Obstet. Gynecol. Reprod. Biol. **206**, 147–152 (2016)
5. Atanasova, S., Petric, G.: Collective empowerment in online health communities: scale development and empirical validation. J. Med. Internet Res. **21**(1), e14392 (2019)
6. Baker, B., Yang, I.: Social media as social support in pregnancy and the postpartum. Sex. Reprod. Healthc. **17**, 31–34 (2018)

7. Balka, E., Krueger, G., Holmes, B.J., Stephen, J.E.: Situating internet use: Information-seeking among young women with breast cancer. J. Comput. Mediated Commun. **15**(3), 389–411 (2010)
8. Bandura, A.: Social foundations of thought and action: a social cognitive theory. Social foundations of thought and action: A social cognitive theory., Prentice-Hall Inc, Englewood Cliffs, NJ, US (1986). pages: xiii, 617
9. Bawden, D., Robinson, L.: The dark side of information: overload, anxiety and other paradoxes and pathologies. J. Inf. Sci. **35**(2), 180–191 (2009)
10. Brochu, F., et al.: Searching the internet for infertility information: a survey of patient needs and preferences. J. Med. Internet Res. **21**(12), e15132 (2019)
11. Burnett, G., Jaeger, P.T.: Small worlds, lifeworlds, and information: the ramifications of the information behaviour of social groups in public policy and the public sphere. Inf. Res. Int. Electron. J. **13**(2), 1–18 (2008)
12. Choi, J., Hyeon Lee, J., Vittinghoff, E., Fukuoka, Y.: mHealth physical activity intervention: a randomized pilot study in physically inactive pregnant women. Mat. Child Health J. **20**(5), 1091–1101 (2016)
13. Clinic, M.: Morning sickness (2020). https://www.mayoclinic.org/diseases-conditions/morning-sickness/symptoms-causes/syc-20375254
14. Cowie, G.A., Hill, S., Robinson, P.: Using an online service for breastfeeding support: what mothers want to discuss. Health Promot. J. Austr. **22**(2), 113–118 (2011)
15. Declercq, E.R., Sakala, C., Corry, M.P., Applebaum, S.: Listening to mothers II: report of the second national us survey of women's childbearing experiences. J. Perinat. Educ **16**(4), 9–14 (2007)
16. Dufur, M.J., Parcel, T.L., McKune, B.A.: Capital and context: Using social capital at home and at school to predict child social adjustment. J. Health Soc. Behav. **49**(2), 146–161 (2008)
17. Elixhauser, A., Wier, L.M.: Complicating conditions of pregnancy and childbirth, 2008: Statistical brief# 113 (2011)
18. Enquist, H., Tollmar, K.: The memory stone: a personal ICT device in health care. In: Proceedings of the 5th Nordic Conference on Human-Computer Interaction: Building Bridges, pp. 103–112 (2008)
19. Eppler, M.J., Mengis, J.: The concept of information overload-a review of literature from organization science, accounting, marketing, MIS, and related disciplines (2004). In: Meckel, M., Schmid, B.F. (eds.) Kommunikationsmanagement im Wandel, pp. 271–305. Springer, Heidelberg (2008). https://doi.org/10.1007/978-3-8349-9772-2_15
20. Epstein, D.A., et al.: Examining menstrual tracking to inform the design of personal informatics tools. In: Proceedings of the 2017 CHI Conference on Human Factors in Computing Systems, pp. 6876–6888 (2017)
21. Erdelez, S.: Information encountering: it's more than just bumping into information. Bull. Am. Soc. Inf. Sci. Technol. **25**(3), 26–29 (1999)
22. Evans, J.R., Selstad, G., Welcher, W.H.: Teenagers: fertility control behavior and attitudes before and after abortion, childbearing or negative pregnancy test. Family Planning Perspect. **8**(4), 192–200 (1976). http://www.jstor.org/stable/2134210
23. Fox, S., et al.: The Social Life of Health Information, 2011. Pew Internet & American Life Project, Washington, DC (2011)
24. Gui, X., Chen, Y., Kou, Y., Pine, K., Chen, Y.: Investigating support seeking from peers for pregnancy in online health communities. Proc. ACM Hum. Comput. Interact. **1**(CSCW), 1–19 (2017)

25. Gui, X., Chen, Y., Pine, K.H.: Navigating the healthcare service "black box" individual competence and fragmented system. Proc. ACM Hum. Comput. Interact. **2**(CSCW), 1–26 (2018)
26. Hartzler, A., Pratt, W.: Managing the personal side of health: how patient expertise differs from the expertise of clinicians. J. Med. Internet Res. **13**(3), e62 (2011)
27. of Health & Human Services, D.: Pregnancy stages and changes (2020). https://www.betterhealth.vic.gov.au/health/healthyliving/pregnancy-stages-and-changes
28. Holtz, B., Smock, A., Reyes-Gastelum, D.: Connected motherhood: social support for moms and moms-to-be on facebook. Telemed. e-Health **21**(5), 415–421 (2015)
29. Huh, J., Ackerman, M.S.: Collaborative help in chronic disease management: supporting individualized problems. In: Proceedings of the ACM 2012 Conference on Computer Supported Cooperative Work, pp. 853–862 (2012)
30. Huh, J., McDonald, D.W., Hartzler, A., Pratt, W.: Patient moderator interaction in online health communities. In: AMIA Annual Symposium Proceedings. vol. 2013, p. 627. American Medical Informatics Association (2013)
31. Johnson, S.A.: "maternal devices", social media and the self-management of pregnancy, mothering and child health. Societies **4**(2), 330–350 (2014)
32. Kaempf, J.W., Tomlinson, M.W., Campbell, B., Ferguson, L., Stewart, V.T.: Counseling pregnant women who may deliver extremely premature infants: medical care guidelines, family choices, and neonatal outcomes. Pediatrics **123**(6), 1509–1515 (2009)
33. Kahn, R.S., Zuckerman, B., Bauchner, H., Homer, C.J., Wise, P.H.: Women's health after pregnancy and child outcomes at age 3 years: a prospective cohort study. Am. J. Public Health **92**(8), 1312–1318 (2002)
34. Khaleel, I., et al.: Health information overload among health consumers: a scoping review. Patient Educ. Couns. **103**(1), 15–32 (2020)
35. Kouri, P., Turunen, H., Tossavainen, K., Saarikoski, S.: Pregnant families' discussions on the net-from virtual connections toward real-life community. J. Midwifery Women's Health **51**(4), 279–283 (2006)
36. Kumar, N., Anderson, R.J.: Mobile phones for maternal health in rural India. In: Proceedings of the 33rd Annual ACM Conference on Human Factors in Computing Systems, pp. 427–436 (2015)
37. Lazarus, E.S.: What do women want?: Issues of choice, control, and class in pregnancy and childbirth. Med. Anthropol. Quart. **8**, 25–46 (1994)
38. Ledford, C.J., Canzona, M.R., Cafferty, L.A., Hodge, J.A.: Mobile application as a prenatal education and engagement tool: a randomized controlled pilot. Patient Educ. Couns. **99**(4), 578–582 (2016)
39. Lupton, D.: The use and value of digital media for information about pregnancy and early motherhood: a focus group study. BMC Pregnancy Childbirth **16**(1), 1–10 (2016)
40. Madge, C., O'Connor, H.: Life in happy land: using virtual space and doing motherhood in Hong Kong. Soc. Cult. Geogr. **7**(2), 199–220 (2006)
41. McLeish, J., Redshaw, M.: Peer support during pregnancy and early parenthood: a qualitative study of models and perceptions. BMC Pregnancy Childbirth **15**(1), 257 (2015)
42. Mitchell, C.L., Dorian, E.H.: Police Psychology and its Growing Impact on Modern Law Enforcement. IGI Global, Hershey (2016)
43. Neal, L., Oakley, K., Lindgaard, G., Kaufman, D., Leimeister, J.M., Selker, T.: Online health communities. In: CHI 2007 Extended Abstracts on Human Factors in Computing Systems, pp. 2129–2132 (2007)

44. Perrier, T., et al.: Engaging pregnant women in Kenya with a hybrid computer-human SMS communication system. In: Proceedings of the 33rd Annual ACM Conference on Human Factors in Computing Systems, pp. 1429–1438 (2015)
45. Robinson, J.R., et al.: Consumer health-related needs of pregnant women and their caregivers. JAMIA Open **1**(1), 57–66 (2018)
46. Salonen, A.H., Kaunonen, M., Åstedt-Kurki, P., Järvenpää, A.L., Isoaho, H., Tarkka, M.T.: Effectiveness of an internet-based intervention enhancing finnish parents' parenting satisfaction and parenting self-efficacy during the postpartum period. Midwifery **27**(6), 832–841 (2011)
47. Sayakhot, P., Carolan-Olah, M.: Internet use by pregnant women seeking pregnancy-related information: a systematic review. BMC Pregnancy Childbirth **16**(1), 65 (2016)
48. Shieh, C., Mays, R., McDaniel, A., Yu, J.: Health literacy and its association with the use of information sources and with barriers to information seeking in clinic-based pregnant women. Health Care Women Int. **30**(11), 971–988 (2009)
49. Sinclair, M., Lagan, B.M., Dolk, H., McCullough, J.E.: An assessment of pregnant women's knowledge and use of the internet for medication safety information and purchase. J. Adv. Nurs. **74**(1), 137–147 (2018)
50. Wallace, L.S., Zite, N.B., Homewood, V.J.: Making sense of home pregnancy test instructions. J. Womens Health **18**(3), 363–368 (2009)
51. Witteman, H.O., Zikmund-Fisher, B.J.: Communicating laboratory results to patients and families. Clin. Chem. Lab. Med. (CCLM) **57**(3), 359–364 (2019)
52. Zhang, Z., Citardi, D., Xing, A., Luo, X., Lu, Y., He, Z.: Patient challenges and needs in comprehending laboratory test results: mixed methods study. J. Med. Internet Res. **22**(12), e18725 (2020)
53. Zhang, Z., Lu, Y., Kou, Y., Wu, D.T., Huh-Yoo, J., He, Z.: Understanding patient information needs about their clinical laboratory results: a study of social Q&A site. Stud. Health Technol. Inf. **264**, 1403 (2019)
54. Zhou, X., Sun, S., Yang, J.: Sweet home: understanding diabetes management via a Chinese online community. In: Proceedings of the SIGCHI Conference on Human Factors in Computing Systems, pp. 3997–4006 (2014)
55. Zhu, C., Zeng, R., Zhang, W., Evans, R., He, R.: Pregnancy-related information seeking and sharing in the social media era among expectant mothers: qualitative study. J. Med. Internet Res. **21**(12), e13694 (2019)
56. Zikmund-Fisher, B.J., Exe, N.L., Witteman, H.O.: Numeracy and literacy independently predict patients' ability to identify out-of-range test results. J. Med. Internet Res. **16**(8), e187 (2014)

COVID-19 Epidemic Information Needs and Information Seeking Behavior of Overseas Chinese Students

Lin Wang[1], Ziqiao Ma[2], and Yuwei Jiang[2](✉)

[1] Academy of Chinese Science and Education Research, Hangzhou Dianzi University, Hangzhou 310018, China
[2] School of Management, Tianjin Normal University, Tianjin 300387, China
jiangyw_xpp@163.com

Abstract. This paper aims to explore the COVID-19 epidemic information needs and information seeking behavior of overseas Chinese students. The questionnaire survey method was adopted to collect data. The results show that personal health protection knowledge, Chinese and foreign governments' countermeasures, control effects and future plans were the most important information needs of overseas Chinese students. The information need, quality of information sources, information source accessibility and satisfaction of information source update rate have significantly positive effects on information seeking frequency. The relationship between information needs and information seeking frequency is positively moderated by both quality of information source and satisfaction of information source update rate. On the basis of these results, this paper makes suggestions for health information services to overseas Chinese students.

Keywords: The COVID-19 epidemic · Overseas Chinese students · Information needs · Information seeking behavior

1 Introduction

The COVID-19 epidemic broke out at the beginning of 2020 and spreads rapidly all over the world. This disease was declared as a global public health emergency by the World Health Organization (WHO) and aroused the high attention globally [1]. China is one of the largest source of international students in the world. When the epidemic in China is under control, the epidemic abroad is becoming increasingly serious. This large group of overseas Chinese students is under more close attention. Overseas Chinese students attach great importance to the epidemic for the sake of safety, situation change, plans of returning to China etc. Therefore, it is of significance to investigate the epidemic information needs and information seeking behavior of overseas Chinese students. The results can enrich the existing theories of health information behavior. It can also help the students be well-informed about COVID-19 epidemic, thus reducing their possibility of having mental health problems.

K. Toeppe et al. (Eds.): iConference 2021, LNCS 12646, pp. 37–47, 2021.
https://doi.org/10.1007/978-3-030-71305-8_3

2 Literature Review and Research Hypotheses

Information can reduce uncertainty, thus reducing the risk of information users and make them perceive a sense of security [2, 3]. Wilson conducts pioneering research on information behavior [4]. He claims that information seeking behavior originates from information needs. It is the process of finding information purposefully to reduce uncertainty. Therefore, it is critical to fully comprehend users' information needs in information behavior research.

There are increasing researches on the health information needs during public health emergencies. Research on the health information needs of the Vietnamese people in the COVID-19 epidemic shows that the most important health information needs during the period of national lockdown were "the latest information of disease and treatment", "disease transmission mechanism and specific precautions" and "epidemiological symptoms, treatment and prevention" [5]. Majid et al. found that during the H1N1 pandemic in Singapore, the public's information needs mainly include H1N1 symptoms, the causes of infection, preventive measures and possible treatment. They also demonstrated that mass media such as television, newspapers and radio were the most commonly used channels to address the information need [6]. Randle et al. report that when the ZIKV epidemic is reported by the media, the public will actively seek relevant information and health services, such as calling the epidemic hotline and requesting information about virus infection and its testing [7]. Qiao claims that information needs are the basis and motives of information behavior, and the intensity of information needs determines the possibility of information behavior [8]. Thus, we make the following hypothesis:

H1: The epidemic information needs of overseas Chinese students have a significantly positive impact on epidemic information seeking frequency.

Auster and Choo measure the quality of information sources from two dimensions: relevance and reliability [9]. The information quality tends to deteriorate in the information explosion – fake news, misinformation and disinformation are common. However, the information processing ability of users has not been improved accordingly [10]. Zhang et al. believe that there are differences in the quality of health information sources on different medical websites [11]. It is because people face such a wide range of information sources that makes it difficult for them to fully apprehend the accessed information or to identify the most reliable information source [12, 13]. The emergence of various social media has generated new challenges to the health information quality during the outbreak of the epidemic [14]. Information sources being reliable and high-quality is considered to be an important prerequisite for information users' satisfaction [15]. Li concluded that when facing the choice of information sources, users mostly value the timeliness, authenticity, reliability [16]. Zha et al. found that the quality and credibility of information sources significantly and positively affect academic users' search frequency [17]. The "Moore's law" showed that accessibility influenced users' choice of information sources [18]. Users always choose information on the psychological basis of obtaining maximum benefits with the least effort. Jin et al. [19] reveal that during the COVID-19 epidemic, amplifying the amounts of information access channels for community residents can partly alleviate the psychological problems of community residents such as blind disinfection and sleep disorders. Therefore, we hypothesize:

H2a: Information source quality has a significantly positive impact on epidemic information seeking frequency of overseas Chinese students.

H2b: Information source quality positively moderates the relationship between information needs and information seeking frequency of overseas Chinese students.

H3a: Information source accessibility has a significantly positive influence on the information seeking frequency of overseas Chinese students.

H3b: Information source accessibility positively moderates the relationship between information needs and information seeking frequency of overseas Chinese student.

Li and Yan found that the timeliness of information update was the primary impact factor in the information seeking behavior of university graduates [20]. Based on online interview with some overseas Chinese students, we assume that the satisfaction of information source update rate is an impact factor of their information seeking behavior. We thus hypothesize:

H4a: Satisfaction of information source update rate has a significantly positive impact on the epidemic information seeking frequency of overseas Chinese students.

H4b: The relationship between epidemic information needs and information seeking frequency of overseas Chinese students is positively moderated by satisfaction of information source update rate (Fig. 1).

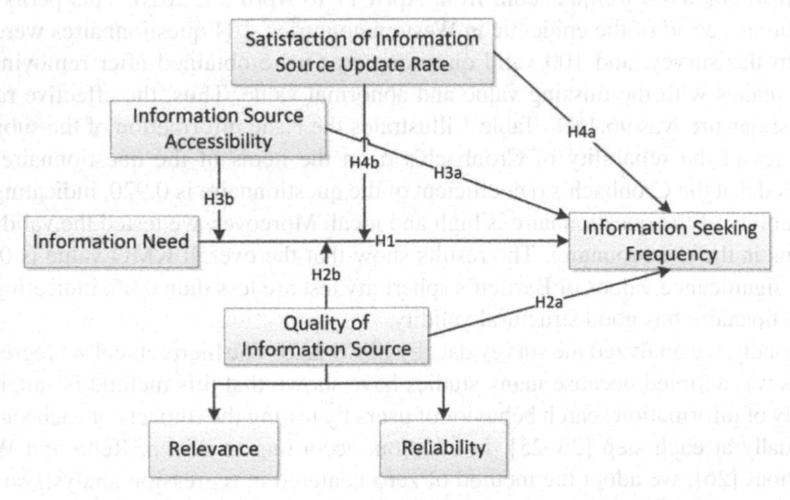

Fig. 1. Research model of epidemic information seeking behavior

3 Research Methods

3.1 Questionnaire Design

The questionnaire of this study has two parts: the first part is to investigate the demographic characteristics of the subjects and other specific personal information, including the participants' gender, age, the length of overseas residence, current status; the second part is the investigation of epidemic information needs and information seeking behavior. It consisted of 12 items.

The variables measured in the second part were composed of information needs, satisfaction of information source update rate, information source accessibility, quality of information source and information seeking frequency. The items and calculation method of the information needs part are based on the Health Information Wants Questionnaire scale [21]; the satisfaction of information source update rate is a measurement based on interview; the items of information source accessibility were revised based on the research of Yitzhaki and Hammershlag [22]; As for the quality of information sources, it equals to relevance plus reliability [9]. The 5-point Likert scale was used for the items of each variable. For instance, for the items of information need, 1 means "completely unnecessary", and 5 means "very necessary".

3.2 Survey and Data Collection Procedure

Given the relatively similarity between different national healthcare policies, the participants of this study are overseas Chinese students and visiting scholars, who mainly study in Western countries like US and UK, etc. The questionnaire was collected on the platform of www.wenjuan.com from April 17 to April 25, 2020. This period was the outbreak period of the epidemic in Western countries. 104 questionnaires were collected in this survey, and 100 valid questionnaires were obtained after removing the questionnaires with the missing value and abnormal value. Thus, the effective rate of the questionnaire was 96.15%. Table 1 illustrates the basic information of the subjects.

We tested the reliability of Cronbach's α on the items of the questionnaire, and concluded that the Cronbach's α coefficient of the questionnaire is 0.920, indicating that the reliability of the questionnaire is high and ideal. Moreover, we tested the validity of the items in the questionnaire. The results show that the overall KMO value is 0.822, and the significance values of Bartlett's sphericity test are less than 0.05, indicating that the questionnaire has good structural validity.

Secondly, we analyzed the survey data by SPSS 22.0. The hierarchical set regression analysis was adopted because many studies have shown that this method is suitable to the study of information search behavior of users by testing the impacts of each variable individually at each step [23–25]. In addition, according to Aiken, Reno and West's suggestions [26], we adopt the method of zero-centered in regression analysis so as to avoid multi-collinearity among independent variables.

Table 1. Demographic data

Variable	Subdivision variable	Number	Percentage (%)
Gender	Male	34	34.00
	Female	66	66.00
Age	18 and under	2	2.00
	19–25	60	60.00
	26–30	31	31.00
	31–35	7	7.00
Length of overseas residence	Less than a year	24	24.00
	1 to 2 years	22	22.00
	3 to 4 years	20	20.00
	More than 4 years	34	34.00
Current status	No intention of returning to China	47	47.00
	Unable to return for special reasons	24	24.00
	Has returned to China	29	29.00

4 Analysis Results

4.1 Epidemic Information Needs of Overseas Chinese Students

The participants' epidemic information needs are shown in Table 2. The top two information needs are personal health protection knowledge, Chinese and foreign governments' countermeasures, control effects and future plans. They are the students' most important information needs. Information needs with relatively lower value are punishment for illegal behaviors related to the epidemic and information about hospitals capable of treating COVID-19 disease. Overall, overseas Chinese students have great epidemic information needs.

4.2 Regression Analysis

Since there are positive correlations among variables ($P < 0.01$) as shown in Table 3, we conducted a hierarchical regression analysis. Because causality may be explained well by demographic characteristics and other specific individual variables in behavior studies, we put this kind of variables as control variables into the model at the first step [27]. After that, in each hierarchic level, we put the centered variables into the model in turn according to the conceptual model. With every step for regression, we recorded regression coefficient and standard deviation which are newly included in variables. Meanwhile, T test was carried out to observe the change of R^2. We also conducted the F test. The results are summarized in Table 4.

Table 2. Information Needs Statistics (N = 100)

Ranking	Information needs	Importance level	
		Mean score (1–5)	SD
1	Personal health protection knowledge	4.30	0.88
2	Chinese and foreign governments' countermeasures, control effects and future plans	4.23	0.95
3	Timely clarification to rumors and misunderstandings of the epidemic	4.16	1.03
4	Epidemiological features for COVID-19	4.08	1.01
5	Logistics information for epidemic controlling	3.90	1.00
6	Daily information for epidemic situation	3.87	0.99
7	Progress of COVID-19 scientific research	3.82	1.06
8	Traveling routes for affirmed cases	3.80	1.14
9	Hospitals capable of treating COVID-19 disease	3.78	1.09
10	Punishment for illegal behaviors related to epidemic	3.67	1.09

The results show that the current status of overseas Chinese students has significantly positive influence on epidemic information seeking frequency ($P < 0.05$), whereas other control variables do not have such effect. Epidemic information needs have a positive influence on epidemic information seeking frequency ($P < 0.01$) and R2 value has risen from 0.193 to 0.341. Hypothesis 1 is supported. The quality of information sources has a significantly positive impact on information seeking frequency ($P < 0.01$). In addition, the quality of information sources significantly positively moderates the relationship between epidemic information needs and information seeking frequency ($P < 0.05$). Therefore, both H2a and H2b are supported. Information source accessibility has a significantly positive impact on epidemic information seeking frequency ($P < 0.01$). However, the information source accessibility does not significantly moderate the relationship between epidemic information needs and information seeking frequency. Thus, H3a is supported and H3b not. H4a and H4b are supported by evidences that satisfaction of information source update rate has significantly positive impact on epidemic information seeking frequency ($P < 0.05$), and it also significantly positively moderates the relationship between epidemic information needs and information seeking frequency ($P < 0.01$). The revised model is shown in Fig. 2.

Table 4. Results of hierarchical regression analysis

	Step 1			Step 2			Step 3			Step 4			Step 5		
	B	S.E.	*t*	B	S.E.	*t*	B	S.E.	*t*	B	S.E.	*t*	B	S.E.	*t*
Constant	0.462	0.365	1.267	0.419	0.332	1.264	0.423	0.302	1.400	0.286	0.287	0.996	0.265	0.265	1.003
Sex	-0.211	0.128	-1.641	-0.136	0.118	-1.152	-0.157	0.108	-1.464	-0.106	0.102	-1.046	-0.091	0.093	-0.970
Age	-0.042	0.107	-0.396	-0.090	0.098	-0.919	-0.083	0.089	-0.933	-0.061	0.084	-0.729	-0.103	0.078	-1.325
Living style	-0.113	0.067	-1.687	-0.095	0.061	-1.554	-0.083	0.056	-1.492	-0.076	0.052	-1.474	-0.031	0.049	-0.643
LOR	-0.027	0.102	-0.270	-0.007	0.093	-0.073	-0.012	0.085	-0.144	0.011	0.079	0.139	0.005	0.073	0.065
Current status	0.185	0.073	2.531*	0.157	0.067	2.363*	0.150	0.061	2.455*	0.104	0.058	1.787	0.108	0.054	1.997*
IN				0.336	0.074	4.571**	0.324	0.067	4.818**	0.283	0.066	4.287**	0.232	0.062	3.759**
QIS							0.491	0.113	4.347**	0.303	0.118	2.564*	0.402	0.113	3.554**
QIS* IN							0.306	0.128	2.394*	0.102	0.171	0.598	0.160	0.168	0.955
ISA										0.228	0.066	3.435**	0.072	0.084	0.865
ISA * IN										0.159	0.098	1.626	-0.328	0.148	-2.222*
SISUR													0.215	0.099	2.176*
SISUR * IN													0.519	0.142	3.656**
R^2		0.193			0.341			0.464			0.548			0.628	
ΔR^2		0.193			0.148			0.123			0.085			0.080	
F		4.487			8.014			9.839			10.771			12.230	

Table 3. Pearson correlation results

	ISF	IN	SISUR	ISA	QIS
ISF	1				
IN	.449**	1			
SISUR	.586**	.227*	1		
ISA	.584**	.287**	.772**	1	
QIS	.382**	.070 ns	.413**	.480**	1

Note: **p < 0.01, *p < 0.05, ns: not significant, ISF: Information seeking frequency, IN: Information need, QIS: Quality of information source, ISA: Information source accessibility, SISUR: Satisfaction of information source update rate

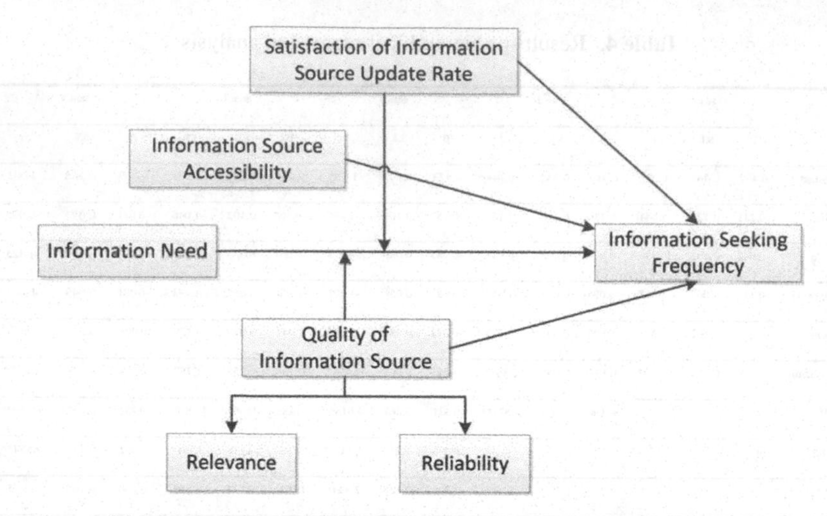

Fig. 2. Revised model of epidemic information seeking behavior

5 Discussion

The results show that hypotheses for H1, H2a, H3a, H4a are supported. Under the COVID-19 epidemic situation, information needs, quality of information sources, information source accessibility and satisfaction of information source update rate have significantly positive influences on information seeking frequency of overseas Chinese students. Qiao [8] believed that information needs are the basic motives for information seeking behavior. Our study has affirmed this statement in the pandemic context. Information is mixed with misinformation and disinformation in the infodemic. Therefore, overseas Chinese students are inclined to actively access information sources with high quality, reliability and timeliness. According to Mooers' Law [18], the higher the information source accessibility, the more information users tend to obtain to thoroughly apprehend and grasp the epidemic situation, thus the information seeking frequency is high. The higher satisfaction of information source update rate indicates the epidemic information is timely to overseas Chinese students so that information seeking frequency becomes higher [20].

Two hypotheses of moderating effects are supported in this research. The results mean that when the epidemic information needs are given, the higher level of information source quality and satisfaction of information source update rate, information needs have more significant positive impact on information seeking frequency. During the epidemic, there were negative public opinions on overseas Chinese students both in China and abroad: Racist reports on Asian groups abroad were frequently reported in the media; In China, some people bitterly satirized overseas Chinese students who had returned through social media as well. In such situation, objective and unbiased information sources are highly appreciated by these students, which is the indicator of information quality. High quality information sources help them to maintain sense of self-identity and thus inspire them to have motives to seek more information. Information source accessibility does not positively moderate the relationship between information needs and information seeking frequency. The possible explanation may be that for overseas Chinese students, epidemic information needs are vital for their life and health. No matter how difficult for accessing information, they will do it actively.

6 Conclusions and Limitations

This paper investigated the epidemic information needs and information seeking behavior of overseas Chinese students during COVID-19 epidemic. It found that their most important epidemic information needs are personal health protection knowledge, Chinese and foreign governments' countermeasures, control effects and future plans. Information needs, quality of information source, information source accessibility and satisfaction of information source update rate have significantly positive influences on epidemic information seeking frequency. Meanwhile, both the quality of information sources and satisfaction of information source update rate significantly moderate the relationship between information needs and information seeking frequency.

Our findings have several implications for overseas Chinese students and epidemic information service provision. First, it is suggested that overseas Chinese students should improve their personal information literacy by self-learning and training. They should be aware of their information needs and express them clearly and accurately. They should also have knowledge of authoritative information sources and rely on these channels to access epidemic information. It is important to enhance their skills of distinguishing truth from fake news. Secondly, Chinese embassies and consulates should update the consular reminder and epidemic prevention and control information on time to ensure the students to understand policies and plans of mother-country. Thirdly, information providers should pay close attention to the quality of information sources in the infodemic and eliminate the misinformation and disinformation dissemination. They should also keep high update rate of epidemic data to help people avoid psychological problems such as anxiety and stress caused by information asymmetry or delay. Due to the limitations of sampling and other factors, there is a lack of comparison between overseas Chinese students and other countries students as far as information needs and information seeking behavior are concerned, which should be further investigated in the future.

References

1. World Health Organization. https://www.who.int/news-room/detail/30-01-2020-statement-on-the-second-meeting-of-the-international-health-regulations-(2005)-emergency-committee-regarding-the-outbreak-of-novel-coronavirus-(2019-ncov). Accessed 20 Apr 2020
2. Fodness, D., Murray, B.: A model of tourist information search behavior. J. Travel Res. **37**(3), 220–230 (1999). https://doi.org/10.1177/004728759903700302
3. Li, D.J.: How Chinese consumers look for commercial information: the case of Tianjin. Nankai J. **1**(2), 30–35 (2001). https://doi.org/10.3969/j.issn.1001-4667.2001.02.005
4. Wilson, T.D.: On user studies and information needs. J. Document. **37**(1), 3–15 (1981)
5. Le, H.T., Nguyen, D.N., Beydoun, A.S., Le, X.T.T., Nguyen, T.T., Pham, Q.T., et al.: Demand for health information on COVID-19 among Vietnamese. Int. J. Environ. Res. Public Health **17**(12), 4377 (2020). https://doi.org/10.3390/ijerph17124377

6. Majid, S., Rahmat, N.A.: Information needs and seeking behavior during the H1N1 virus outbreak. J. Inf. Sci. Theory Pract. **1**(1), 42–53 (2013). https://doi.org/10.1633/JISTaP.2013.1.1.3
7. Randle, J., Nelder, M., Sider, D., Hohenadel, K.: Characterizing the health and information-seeking behaviours of Ontarians in response to the Zika virus outbreak. Can. J. Public Health **109**(1), 99–107 (2018). https://doi.org/10.17269/s41997-018-0026-9
8. Qiao, H.: Information Behavior, 2nd edn. Beijing Normal University Publishing House, Beijing (2010)
9. Auster, E., Choo, C.W.: Environmental scanning by CEOs in two Canadian industries. J. Am. Soc. Inf. Sci. **44**(4), 194–203 (1993). https://doi.org/10.1002/(SICI)1097-4571(199305)44:43.0.CO;2-1
10. Klapp, O.E.: Overload and Boredom: Essays on the Quality of Life in the Information Society. Greenwood Publishing Group Inc, Westport (1986). https://doi.org/10.2307/4308131
11. Zhang, Y., Sun, Y., Xie, B.: Quality of health information for consumers on the web: a systematic review of indicators, criteria, tools, and evaluation results. J. Am. Soc. Inf. Sci. **66**(10), 2071–2084 (2015). https://doi.org/10.1002/asi.23311
12. Beltrán, C., Sánchez, S., Gómez, S., Duque, C., Sukkarie, S., Sukkarieh, L.: Update on sources of health education in pregnant women. Rev Paraninfo Digit **12**, 1–7 (2011)
13. Carolan, M.: Health literacy and the information needs and dilemmas of first-time mothers over 35 years. J. Clin. Nurs. **16**(6), 1162–1172 (2007). https://doi.org/10.1111/j.1365-2702.2007.01600.x
14. Macario, E., Ednacot, E.M., Ullberg, L., Reichel, J.: The changing face and rapid pace of public health communication. J. Commun. Healthcare **4**(2), 145–150 (2011). https://doi.org/10.1179/175380611X13022552566254
15. Devaraj, S., Fan, M., Kohli, R.: Antecedents of B2C channel satisfaction and preference: validating e-commerce metrics. Inf. Syst. Res. **13**(3), 316–333 (2002). https://doi.org/10.1287/isre.13.3.316.77
16. Li, J.: Study on the discrepancies between the perception of information quality and the use of information sources. J. Inf. Resources Manage. **4**(4), 17–23,68 (2014). https://doi.org/10.13365/j.jirm.2014.04.017
17. Zha, X.J., Zhang, J.C., Yan, Y.L.: Impacting factors of users' academic information seeking behavior in the context of microblogs: a dual-route perspective of information quality and information source credibility. J. Library Sci. China **41**(217), 71–85 (2015). https://doi.org/10.13530/j.cnki.jlis.150015
18. Moore, C.N.: Mooers' Law or why some retrieval systems are used and others are not. Bull. Am. Soc. Inf. Sci. Technol. **23**(1), 22–23 (1996). https://doi.org/10.1002/bult.37
19. Jin, Y.L., Jiang, M.M., Chen, Y., Zhu, L.J., Fang, Z.M., Wu, N., et al.: Association of acquisition path of epidemic information with psychological problems during period of novel coronavirus disease 2019 epidemic among community residents in Anhui province. Chinese J. Public Health **36**(5), 665–667 (2020). https://doi.org/10.11847/zgggws1128505
20. Li, Y.L., Yan, X.M.: College students' information seeking behavior during job hunting: selection and use of information sources. Document. Inf. Knowl. **5**, 57–65 (2015). https://doi.org/CNKI:SUN:TSQC.0.2015-05-008
21. Xie, B., Wang, M., Feldman, R., Zhou, L.: Internet use frequency and patient-centered care: measuring patient preferences for participation using the health information wants questionnaire. J. Med. Internet Res. **15**(7), e132 (2013). https://doi.org/10.2196/jmir.2615
22. Yitzhaki, M., Hammershlag, G.: Accessibility and use of information sources among computer scientists and software engineers in Israel: academy versus industry. J. Am. Soc. Inform. Sci. Technol. **55**(9), 832–842 (2004). https://doi.org/10.1002/asi.20026
23. Boyd, B.K., Fulk, J.: Executive scanning and perceived uncertainty: a multidimensional model. J. Manage. **22**(1), 1–21 (1996). https://doi.org/10.1016/S0149-2063(96)90010-0

24. Dong, X.Y., Yan, F., Liu, Q.Q., Zhang, J.N.: Perceived environmental uncertainty and environmental scanning of executives in China: an empirical study. Manage. World **6**, 127–135 (2008). https://doi.org/10.1201/9781420009521.ch3
25. May, R.C., Stewart Jr., W.H., Sweo, R.: Environmental scanning behavior in a transitional economy: evidence from Russia. Acad. Manag. J. **43**(3), 403–427 (2000)
26. Aiken, L.S., West, S.G., Reno, R.R.:Multiple Regression: Testing and Interpreting Interactions. SAGE Publications, New Delhi (1991). https://doi.org/10.2307/2583960
27. Cohen, J., Cohen, P., West, S.G., Aiken, L.S.: Applied Multiple Regression/Correlation Analysis for the Behavioral Sciences, 2nd edn. Earlbaum, Hillsdale (1983)

Demographic Factors in the Disaster-Related Information Seeking Behaviour

Rahmi Rahmi[1](✉) and Hideo Joho[2]

[1] Department of Library and Information Science, Faculty of Humanities, Universitas Indonesia, Depok, Indonesia
rahmi.ami@ui.ac.id

[2] Faculty of Library, Information and Media Science, University of Tsukuba, Tsukuba, Japan
hideo@slis.tsukuba.ac.jp

Abstract. Although demographics are important factors in the investigation of disaster studies, existing work on disaster-related information seeking behaviour (ISB) does not offer quantitative insight into the relationship between them. This paper investigates the demographic factors such as age, gender, location, and occupation, and their relationship with information needs, information sources, and information channels. Content analysis was performed on the testimony of 262 people who experienced the Great East Japan Earthquake and Tsunami in 2011. The results suggest that a large effect was observed between (1) age and active information needs and (2) gender and information sources. Our findings can be useful for designing policies and programmes at risk of major disaster events as they offer multiple ideas about how to optimise disaster-coping plans for diverse communities.

Keywords: Disaster-related information seeking behaviour · Demographic factors

1 Introduction

Natural disasters do not stop for an epidemic. Several studies identify infectious-disease outbreaks that happened following natural disasters [14, 36, 37]. For example, a few weeks after the Great East Japan Earthquake and Tsunami in 2011, announcements of epidemics of influenza-infection cases reached evacuation shelters [14]. Accordingly, confronting information science is researchers need to pay more attention to information behaviours and their environment to understand the abilities of individuals to obtain information during pandemics [19, 38]. Individuals, social networks, situations, and context shape information-behaviour during natural disasters [24, 32]. These have also created an urge to understand demographic factors in disaster-related context of research.

Previous research have identified instances of disaster-related information seeking behaviour, such as passive and active information needs, information sources and information channels. Their studies also identify two factors that influence the behaviour, namely, temporal stages of a disaster and human senses [23, 24]. The findings suggest

K. Toeppe et al. (Eds.): iConference 2021, LNCS 12646, pp. 48–65, 2021.
https://doi.org/10.1007/978-3-030-71305-8_4

that demographic characteristics might influence disaster-related information seeking behaviour and thus, represent an important factor to investigate [23, 24]. For instance, the risk of death may vary with such demographic factors as age and gender, which reflect differences in vulnerability resulting from physical differences or the likelihood of exposure [8]. Other demographic factors, such as location and occupation, may also carry risks of exposure—for example, when the poorest people live in particularly vulnerable areas (e.g. during a flood) or when damage that the disaster causes depends on the quality of housing (as in earthquakes). Studies also have examined disasters that cause substantial numbers of deaths, on a scale with the potential to affect regional or national populations [8, 20, 31]. Key parameters of interest include age, gender, location and occupation, which, in turn, have implications for population size and composition. Previous studies did not obtain the quantitative evidence from measuring and evaluating the effects of demographics on disaster-related information seeking behaviour [24, 28, 29, 31]. Quantitative indexes to measure vulnerabilities across scales using demographic data are useful planning tools to characterize regions at risk of major disaster events [40].

We aim to use quantitative methods to analyse the effect of each demographic characteristics in the disaster-related information seeking behaviour instances. Hence, we collected datasets from a leading broadcasting company in Japan that would allow us to examine effects of demographic factors, such as age, gender, location and occupation, on disaster-related information seeking behaviour, including active and passive information needs, information sources and information channels [23]. This study used datasets by analysing 1,936 sentences from the testimony of 262 local people in areas affected by the Great East Japan Earthquake and Tsunami in 2011. Our investigation in data collection is expected to provide evidence broadens the depth and scope of disaster research, advances understanding of demographic factors and informs policy interventions [8]. As a result, disaster specialists can tailor their design of interventions towards specific demographic groups.

This paper is organised as follows. Section 2 describes the related work used to identify demographic characteristics of disaster research. Section 3 introduces the methodology used throughout this paper. Sections 4 describe the effect sizes of demo-graphic factors and instances of disaster-related information-seeking behaviour. Section 5 then discusses the findings, implications, and limitations. Finally, Sect. 6 presents our conclusions.

2 Literature Review

Studies on demographic consequences of natural disasters have a 40-year history [31]. For example, studies have been shed light on differential impacts of disaster associated with demographic characteristics that include age, gender, location and occupation that have changed over the years [9, 29, 31]. This knowledge helps to inform disaster preparedness, response, recovery and mitigation activities. However, these studies have also created a blind spot in our understanding of demographic change and disaster-related information seeking behaviour.

Among other demographic factors, age and gender are effective at classifying and differentiating factors [28]. Gender norms influence disaster studies on morbidity and mortality [1]. For instance, in the 2004 tsunami, females died three times more than males, disaster-related suicide rates were higher among men than females [3]. Due to gendered skill sets and division of labor roles in local economies, physical location at the time of the tsunami, females suffered higher mortality rates that limit by their mobility [11].

Another example is the investigation of media use during the I-35W Mississippi River bridge collapse in 2007, which determined that young people were more likely to use social media than older people, while people in urban areas, particularly females, were more likely to use television [15]. The effects of gender and age on disaster-related information seeking behaviour have been suggested that females regard television and radio as more useful than males do, while males were more likely to use the Internet in the aftermath of the 2001 World Trade Center (WTC) disaster [33]. Another study has also investigated demographic characteristics, such as age, gender and location, as well as the use of sight, hearing and touch in the context of disaster-related information seeking behaviour [23]. Her findings show a significant effect of the senses of sight and hearing on gender, yet no significant association between senses and age. According to her results, the proportion of sight seems to decrease as age increases and, on the contrary, the proportion of hearing increases. Researchers also have examined age relative to a variety of cognitive (memory impairment), psychomotor (vision loss), physical (mobility decline), economic (increased poverty) and social (increased social isolation) resources [21, 34].

Other demographic factors that the present study analyses are location and occupation. Location in disaster prone areas centre more on studying one or a small number of communities and has varied over the years, including displacement [6]. Relocation to distant sites can lead to the loss of social networks, access to employment, healthcare and other services. For example, a visitor to the Hiroshima Peace Memorial Museum today can see in the windows behind the circular photographs many signs of a bustling city and its population, based on the 1945 atomic bombings of Hiroshima [22]. Furthermore, it is common in the social sciences for people to be assigned a class position based on a variety of quantified indicators, including occupation [40], the activity in which people engage for pay. Those people who generate their income directly from their own business, trade or profession are led to higher levels of risk-taking, compared to the people who receive a straight salary working for others and who have low risk-taking ability, who choose low-ranked professions [17, 26, 41]. Dynamic pressures such as occupation related to income inequality have aggravated disaster vulnerability among the population in unprecedented and profound ways.

The limitation was identified from those existing works mentioned above. For example, previous studies on demographic aspects have been identified from the fields of disaster-related information seeking behaviour research [16, 23, 33], yet they have not been examined quantitative assessment of the 2011 Great East Japan earthquake and tsunami. Although demographic data are of a quantitative nature, we use crowdsourcing platform to assign tasks-annotation fit of the information seeking behaviour in disaster-related context [23, 27]. Thus, empirical studies must establish behavioural differences regarding such demographic characteristics as gender, age, location and occupation, in the context of disaster-related information seeking behaviour.

3 Methods

Our previous work has established the following instances of information seeking behaviour (ISB) in the context of natural disaster: active and passive information needs, information sources, and information channels [24]. As for information needs, two prominent modes of fulfilling them emerged from the analysis of instances of disaster-related information seeking behaviour. One was through purposeful, active seeking of information [42]; the other was a result of passively receiving or encountering the information [43]. 'Active information needs' refers to a set of needs inferred from the description of purposeful ISB, to answer a specific query [42]. On the other hand, 'passive information needs' refers to a set of needs inferred from the description of the information that people passively receive or encounter when they are not actively seeking [24]. Readers should refer to Rahmi, Joho and Shirai (2019) for detailed description of these ISB concepts.

Table 1. A summary of the testimony collections ($N = 262$)

ID	Document title	Publication date	Number of testimonies
1	Record of Testimony on The Great East Japan Earthquake: 1	February, 2013	144
2	Record of Testimony on The Great East Japan Earthquake: 2	February, 2014	68
3	Record of Testimony on The Great East Japan Earthquake: 3	February, 2015	50

Although the previous study, the testimony of several people was sufficient to establish a set of concepts related to disaster-related information seeking behaviour (e.g. passive and active information needs, information sources and information channels) in the distribution of such demographic characteristics as age and location was skewed, and thus, excluded from research questions a larger sample. In this work, however, we built and analysed a much bigger sample to investigate the demographic factors. This study provides evidence from the content analysis of 1,936 sentences, retrievable from the NHK digital archive (nhk.or.jp/archives/311shogen/) in testimony collected from 262 people who were affected by the 2011 Great East Japan Earthquake and Tsunami (see Table 1).

In our investigation the next step involved coding instances of disaster-related information seeking behaviour in a set of the testimony texts, using a crowdsourcing service (lancers.jp) from July 2017 until August 2018. We followed the current best practice of crowdsourcing for the annotation [30], as follows. First, we ran screening tasks on instances of disaster-related information seeking behaviour separately, before labelling 262 people's testimony. The goal of this screening was to find reliable crowd workers, i.e. individuals who achieved test accuracy with a score of 80% or above, before annotating our dataset. Screening and project tasks had similar instructions.

Table 2. Age, gender, location and occupation group testimony ($N = 262, n = 1,936$)

Variables		N	%	n	%
Gender	Male	191	73	1,499	77
	Female	71	27	437	23
Age	Adolescence (13–19)	4	2	26	1
	Early adulthood (20–39)	32	12	216	11
	Adulthood (40–64)	144	54	1,165	60
	Maturity (>65)	78	30	513	26
	Not available	4	2	16	1
Location	Iwate Prefecture	90	34	620	32
	Miyagi Prefecture	93	36	643	33
	Fukushima Prefecture	79	30	673	35
Occupation	Managers	35	13	261	14
	Professionals	45	18	366	19
	Technicians and Associate Professionals	8	3	74	4
	Clerical Support Workers	10	4	66	3
	Services and Sales Workers	17	6	138	7
	Skilled Agricultural, Forestry and Fishery Workers	5	2	43	2
	Craft and Related Trades Workers	3	1	16	1
	Plant and Machine Operators and Assemblers	6	2	57	3
	Elementary Occupations	20	8	117	6
	Not available	113	43	798	41

In the screening task, we invited 200 crowd workers for each round of information seeking behaviour concepts. Information channels were easy to categorise; passive and active information needs and information sources were difficult category screening tasks to pass. Other screening tasks were needed for passive and active information needs and information source, until we fulfilled the adequate number of workers to assign to project tasks. We paid incentives of 2.8 USD per person, per screening task.

Those workers who passed the screening test were then invited to code people's testimony texts, the project task. A total of 6,566 sentences from 269 people's testimony were divided into 300 lines on 20 pages in a.pdf file. The labelling results were written in the given.xls file, and diverse workers were assigned to separately label a set of disaster-related ISB instances. Those workers had to complete the task within five days, to receive payment of 17.3 USD. To annotate sentences with a reliable label, we recruited three workers per sentence. Researchers commonly hire two or more coders to ensure consistency and reproducibility of labelling in content analysis [10]. Voting was used to determine the final label suggested by the crowd [39]. If agreement among the three workers was not achieved, we recruited another worker to annotate until a majority consensus was reached. We repeated this process for all ISB labels across all sentences. In our case, one sentence was judged by three crowd workers' votes for four information seeking behaviour instances; thus, a total of approximately 96,660 votes were collected. The use of crowd intelligence makes it possible to complete tasks that cannot be automated, such as providing input labels for initial training [39, 44].

We then removed those sentences without disaster-related ISB instances from the analysis. Of those, 262 people's items of testimony, consisting of 1,936 sentences, were selected for analysis. Table 2 shows the age, gender, location and occupation group of 262 people, individually checked as independent factors, while Table 3 shows disaster-related information seeking behaviour instances, such as active and passive information needs, information sources, and information channels, considered dependent variables.

Table 3. Disaster-related information seeking behaviour instances

Variables		*n*	%
Active Information Needs ($n = 505$)	Current Status	351	70
	Disaster Information	50	10
	Nuclear Explosion	35	7
	Evacuation Instruction	32	6
	Post-Disaster Supplies	19	4
	Warning	15	3
	Transportation	3	1
Passive Information Needs ($n = 977$)	Current Status	494	51
	Evacuation Instruction	136	14
	Story	115	12
	Warning	63	6
	Disaster Information	62	6
	Nuclear Explosion	57	6
	Post-disaster Supplies	50	5
Information Sources ($n = 832$)	Family/Neighbourhood	390	47
	Work community/Colleagues	197	24
	Health and Safety	88	11
	Local Government	70	8
	Broadcast Media	64	8
	JMA (Japan Meteorological Agency)	23	3
Information Channels ($n = 760$)	Face to Face	583	77
	Phone	59	8
	Speakers and Signage	43	6
	Radio	35	5
	Television	24	3
	E-mail	15	2
	Internet	1	0

For age, we used Erikson's lifespan-stages psychosocial-development theory to divide the age categories into adolescence (13–19), early adulthood (20–39), adulthood (40–64) and maturity (>65) [4, 5], since this is one of the most well-known categorisations of ages in social studies. For location, the datasets were sampled from three affected areas: Iwate, Miyagi and Fukushima Prefectures, respectively. For occupation, the International Labour Organisation (ILO) classification structure for organising information on labour and jobs, called The International Standard Classification of Occupations (ISCO-08), was used for the analysis [12]. The ISCO-08 structure divides jobs into ten major groups: (1) Managers, (2) Professional, (3) Technicians and associate professionals, (4) Clerical support workers, (5) Service and sales workers, (6) Skilled agricultural, forestry and fishery workers, (7) Craft and related trades workers, (8) Plant and machine operators and assemblers, (9) Elementary occupations and (10) Armed forces occupations.

We employed descriptive analysis between the demographic characteristics and instances of disaster-related information seeking behaviour, and a *t*-test statistical analysis to uncover any associations between demographic characteristics and those instances using the IBM SPSS Statistics software [7]. For significant results ($p < 0.05$), we inveastigate the effect sizes through the paired sample *t*-test (Cohen's *d*), also using the IBM SPSS Statistics, following the steps described in Field (2009).

4 Findings

4.1 The Effect Sizes of Demographic Factors and Instances of Disaster-Related Information Seeking Behaviour

A paired-samples *t*-test was conducted with all significant results ($p < 0.05$) between demographic characteristics and instances of disaster-related information seeking behaviour. Furthermore, the effect size of a paired sample *t*-test, known as Cohen's *d*, was employed [7, 13, 25].

Table 4 shows the effect sizes of demographic factors and disaster-related information seeking behaviour. If the value of *d* equals 0, then the difference scores equal 0. However, the greater than 0 the *d* value is, the greater is the effect size. The value of *d* is usually categorised as $d = 0.8$ and higher, large effect (green); $d = 0.5$ to 0.8, medium effect (yellow); and $d = 0.2$ to 0.5, small effect (red) [7, 13, 25].

Table 4 shows that large effects were identified between (1) age and active information needs and (2) gender and information sources. Medium effects were identified between (1) age and information channels, (2) gender and passive information needs, (3) location and information sources and (4) occupation and passive information needs and information channels. Small effects were identified between (1) gender and active information needs and information channels, (2) age and passive information needs and information sources, (3) location and active information needs, passive information needs and information channels and (4) occupation and passive information needs and information sources. Therefore, this paper will discuss demographic factors that have large and medium effects in the context of disaster-related information seeking behaviour instances.

Table 4. Paired samples t-test and effect sizes (d) of demographic factors and disaster-related information seeking behaviour instances

Disaster-related Information seeking behaviour instances	Gender	Age	Location	Occupation
Active Information Needs	$t(505) = -8.537$, $p < 0.001$	$t(505) = 17.954$, $p < 0.001$	$t(505) = 4.347$, $p < 0.001$	$t(505) = 25.699$, $p < 0.001$
d	0.4	0.8	0.2	0.5
Passive Information Needs	$t(977) = -22.544$, $p < 0.001$	$t(977) = 7.029$, $p < 0.001$	$t(977) = -11.601$, $p < 0.001$	$t(977) = 24.046$, $p < 0.001$
d	0.7	0.2	0.4	0.4
Information Sources	$t(832) = -30.238$, $p < 0.001$	$t(832) = -7.017$, $p < 0.001$	$t(832) = -19.881$, $p < 0.001$	$t(832) = 14.228$, $p < 0.001$
d	1.0	0.2	0.7	0.3
Information Channels	$t(760) = -9.088$, $p < 0.001$	$t(760) = 17.945$, $p < 0.001$	$t(760) = 2.677$, $p = 0.008$	$t(760) = 29.273$, $p < 0.001$
d	0.3	0.7	0.1	0.5

4.2 Gender in Disaster-Related Information Seeking Behaviour

Figure 1 shows the gender distribution in the instances of disaster-related information seeking behaviour. Figure 1(a) shows the gender distribution sources between males ($n = 614$) and females ($n = 218$) has a large effect on information sources ($d = 1.0$).

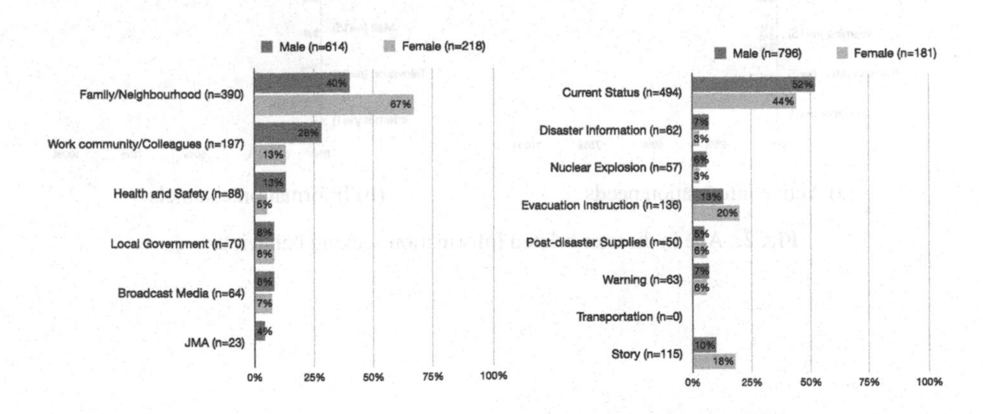

(a) Information sources **(b)** Passive information needs

Fig. 1. Gender in disaster-related information seeking behaviour

The largest proportion of sources belongs to the family/neighbourhood. Work community/colleagues, health and safety and Japan Meteorological Agency (JMA) categories were common among males, and family/neighbourhood was common among females.

Figure 1(b) shows the gender distribution of passive information needs between males ($n = 796$) and females ($n = 181$) has a medium effect ($d = 0.7$). The largest passive needs proportion is the 'current status' segment in gender categories. Disaster information and nuclear explosion categories were common among males, while evacuation instruction and story were common among females.

4.3 Age in Disaster-Related Information Seeking Behaviour

Figure 2 shows the age distribution of instances of disaster-related information seeking behaviour. Figure 2(a) shows age distribution of testimony in active-information-needs categories, in order of size, as adulthood ($n = 311$), maturity ($n = 114$), early adulthood ($n = 76$) and adolescence ($n = 1$) which has a large effect ($d = 0.8$). One common pattern is the large proportion that the 'current status' category represents in early adulthood, adulthood and maturity.

Figure 2(b) shows age distribution for information channels, again organised as adulthood ($n = 443$), maturity ($n = 227$), early adulthood ($n = 82$) and adolescence ($n = 5$) that has a medium effect ($d = 0.5$). The face-to-face category is the largest channel among age categories, and phone and television are common in early adulthood.

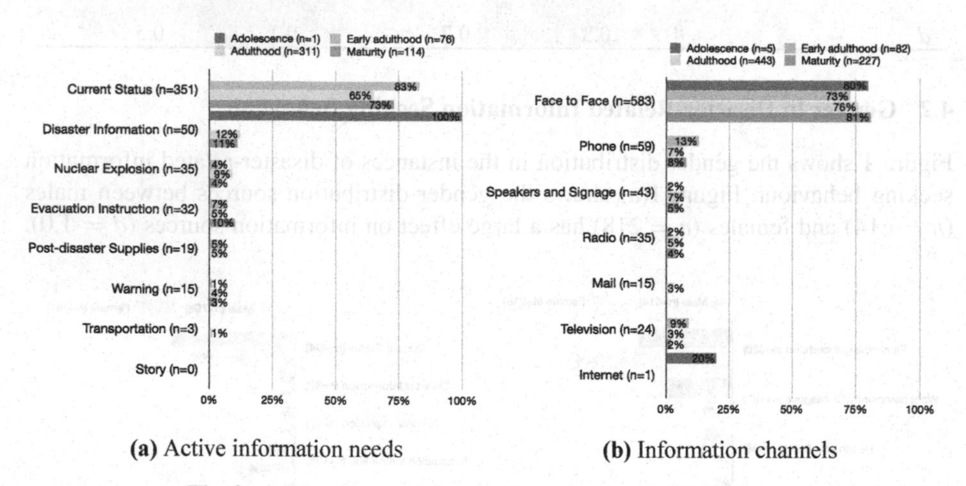

(a) Active information needs **(b)** Information channels

Fig. 2. Age in disaster-related information seeking behaviour

4.4 Location in Disaster-Related Information Seeking Behaviour

Figure 3 shows location distribution for information-sources categories for Fukushima ($n = 335$), Iwate ($n = 287$) and Miyagi ($n = 210$) that has a large effect ($d = 1.0$). Family/neighbourhood was the source with the largest proportion among location categories, followed by work community/colleagues and broadcast media. Information sources from the health-and-safety sector and local government are more common in Fukushima.

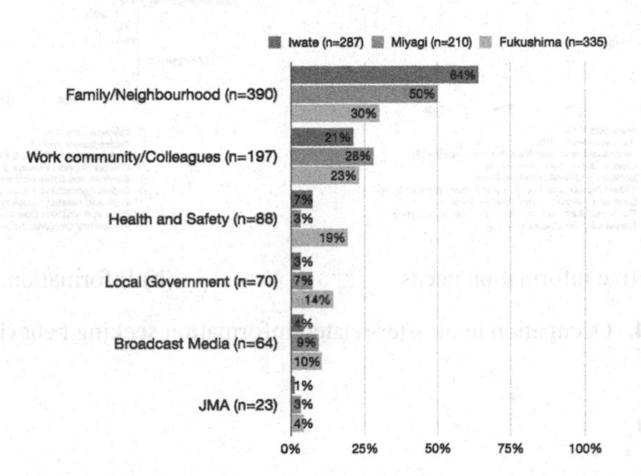

Fig. 3. Location in information sources

4.5 Occupation in Disaster-Related Information Seeking Behaviour

Figure 4 shows the occupation distribution of instances of disaster-related information seeking behaviour. Figure 4(a) shows the occupation distribution for active-information-needs categories that has a medium effect ($d = 0.5$). The current-status category was the largest active-needs proportion among occupation categories, except for 'Technicians and associate professionals' and 'Craft and related trade workers'. 'Professionals' and 'Technicians and associate professionals' actively search for nuclear-explosion information.

Figure 4(b) shows the occupation distribution in information channel categories that has a medium effect ($d = 0.5$). A clear common pattern appears in the large proportion of the face-to-face category among occupation categories. However, 'Technicians and associate professionals' used radio; 'Service and sales workers' used the phone and speakers and signage more than other occupation categories. Patterns in remaining occupations for instances of disaster-related information seeking behaviour varied.

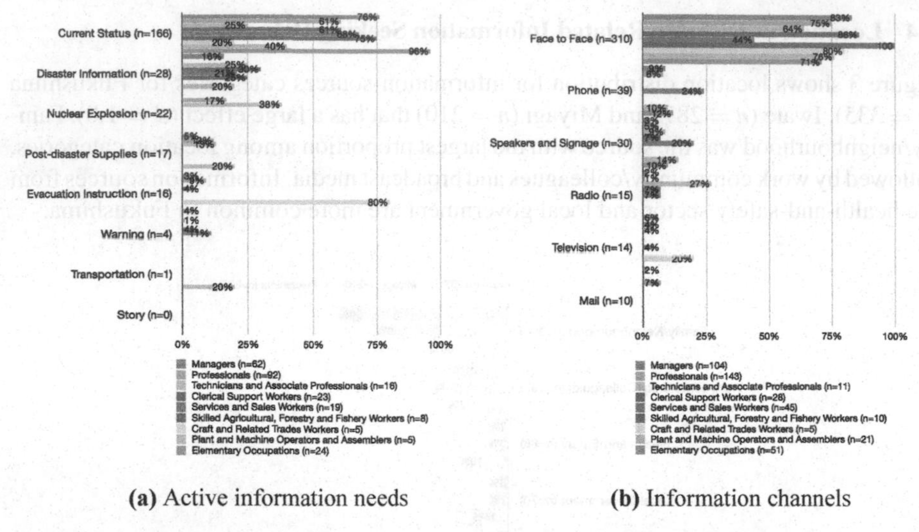

(a) Active information needs **(b)** Information channels

Fig. 4. Occupation in disaster-related information seeking behaviour

5 Discussion

5.1 Demographic Characteristics of Disaster-Related Information Seeking Behaviour

Our findings first showed demographic characteristics, such as gender, age, location and occupation, varied significantly across instances of disaster-related information seeking behaviour, i.e. passive and active information needs, information sources and information channels (see Table 4).

Gender is a key element of human experience that shapes identity and other aspects of social life [3]. In terms of gender (see Fig. 1), our analysis shows that both males and females engage in goal-directed behaviour aimed at ensuring personal and family safety, as well as that of those around them, in the proportion of current status in active and passive information needs. For information sources, females had a higher proportion of using family/neighbourhood, while males had a higher proportion of work community/colleagues, health and safety and JMA. Among information channels, the face-to-face category is the largest channel proportionately in gender categories. Our results echo the findings of existing studies, in that gender differences further influence the creation of disaster-related information seeking behaviour and the practice of disaster management itself [3, 9].

One of the principal receiver demographic characteristics that most studies examine is reflected in age categories (see Fig. 2). Our analysis considered that based on passive information needs, the younger a person is, the more other persons search for information about their current status. As people get older, they search less for information about evacuation instruction and more for information on post-disaster supplies. Regarding active information needs, information channels and information sources, the age proportion was slightly similar and significantly varied across early adulthood, adulthood and maturity. This shows the ageing population in Japan in sharp contrast to the very young populations in most developing countries. This is also important because the very young and the very old disproportionately incur the greatest number of fatalities in disasters [22].

In Fig. 3, location-based capabilities can permeate a multitude of virtual applications and have transformed the ways in which technology could support location-based knowledge generation and decision-making [35]. In terms of location relating to active information needs, 'current status' represents the largest proportion. However, people in Fukushima search for more information about a nuclear explosion and evacuation instruction than in Miyagi and Iwate Prefectures. In Fukushima, a different pattern of information channels and sources appears, perhaps because people in Fukushima experienced an earthquake, a tsunami, and a nuclear explosion. Occupation connects closely with the types of resources people have available for use in crises and the types of public resources available, and they have a strong spatial dimension [40]. Our analysis showed occupation significantly varied across instances of disaster-related information seeking behaviour (see Fig. 4).

Our findings also highlight the effect size (Cohen's *d*) of demographic factors on the instances of disaster-related information seeking behaviour instances. Furthermore, the results were divided into three effects, i.e. large, medium, and small. The large effects were identified between (1) age and active needs and (2) gender and sources. The medium effects were identified between 1) age and channels, (2) gender and passive needs, (3) location and sources and (4) occupation and passive needs and channels. The small effects were identified between (1) gender and active needs and channels, (2) age and passive needs and sources, (3) location and active needs, passive needs and channels and (4) occupation and passive needs and sources.

The potential for feedback mechanisms among these processes is not clear yet from a theoretical perspective, and relatively little empirical work has attempted to examine the interconnections and, thereby, test hypotheses about and provide a better understanding of the disaster-related information seeking behaviour that underlies demographic factors and its processes. Moreover, due to demographers often focusing more on reporting the significant (*p*) value when presenting the results, we emphasise the size of the effect rather than its statistical significance, to promote a more scientific approach to the accumulation of knowledge in reporting and interpreting effectiveness [2, 25]. Policies and programmes related to disaster preparedness and response can affect the outcomes we have considered, as well as their interconnections [8, 20]. Thus, the creation and development of active and passive information needs, sources and channels can alter the immediate consequences of disasters for age, gender, location and occupation as demographic factors.

5.2 Limitation

Since we used published testimonial data, there was little control over how interviews were carried out and their transcripts edited. Although the published data allowed us to access a large collection of testimony, it is possible that the occurrence of some senses was undermined in our datasets. Also, it was difficult to balance data size since this was an underline distribution of data. However, due to the nature of the research methodology adopted, the findings of this study remain at the level of analysing descriptions of self-recollection and self-reporting [23].

6 Conclusion

Disaster-related information seeking behaviour takes on added importance with demographic characteristics, where each instance of that behaviour has come to guide how people's gender, age, location and occupation interplay. We identified the size effects in three categories—large, medium and small effects. The large effects were identified between (1) age and active needs and (2) gender and sources. The medium effects were identified between (1) age and channels, (2) gender and passive needs, (3) location and sources and (4) occupation and passive needs and channels. The small effects were identified between (1) gender and active needs and channels, (2) age and passive needs and sources, (3) location and active needs, passive needs and channels and (4) occupation and passive needs and sources.

The above findings highlight the importance of examining demographic factors in disaster-related information seeking behaviour.This shows the manner of categorising demographics characteristics for information seeking behaviour aligned with active and passive information needs, sources and channels in a disaster. And those demographics—including age, gender, location and occupation—can greatly affect if and how people receive and act upon preparedness or response communications. For example, this research could be a guide to reach particular demographics concerning which medium will be most effective (some technologies or programmes are more likely to reach younger people than older persons) and have a general awareness of how to access them [18]. Also, developing relationships with some occupational groups from various locations can help improve awareness and preparedness while also facilitating more effective communications among people who may not have access to social media, Internet and other modern communication technologies.

Acknowledgments. This work was supported in part by the 2016 iFellows Doctoral Fellowship, the Indonesia Endowment Fund for Education (LPDP), and PUTI Q1 contract NKB-1450/UN2.RST/HKP.05.00/2020. Any opinions, findings, and conclusions described here are the authors' and do not necessarily reflect those of the sponsors.

Appendix A

Disaster-related information seeking behaviour instances by gender, age, and location.

Disaster-related Information Seeking Behaviour Instances	Gender				Age										Location					
	Male	%	Female	%	Adolescence	%	Early adulthood	%	Adulthood	%	Maturity	%	Undefined	%	Miyagi	%	Iwate	%	Fukushima	%
Active Information Needs ($N = 505$)																				
Current Status	278	70.6	73	65.8			63	82.9	202	65	83	72.8	3	100	112	73.2	126	80.8	113	57.7
Disaster Information	47	11.9	3	2.7	1	100			37	11.9	12	10.5			25	16.3	20	12.8	5	2.6
Evacuation Instruction	16	4.1	16	14.4			5	6.6	16	5.1	11	9.6			8	5.2	1	0.6	23	11.7
Nuclear Explosion	27	6.9	8	7.2			3	3.9	27	8.7	5	4.4							35	17.9
Post-disaster Supplies	10	2.5	9	8.1			4	5.3	15	4.8					6	3.9	2	1.3	11	5.6
Transportation	3	0.8							3	1							2	1.3	1	0.5
Warning	13	3.3	2	1.8			1	1.3	11	3.5	3	2.6			2	1.3	5	3.2	8	4.1
***p*-value**	**<0.001**				**0.044**										**<0.001**					
Cohen's *d*	**0.4**				**0.8**										**0.2**					
Passive Information Needs ($N = 977$)																				
Current Status	414	52	80	44.2	17	70.8	45	55.6	318	54.4	108	38.7	6	75	266	67.9	106	38.4	122	39.5
Disaster Information	56	7	6	3.3			8	9.9	32	5.5	22	7.9			18	4.6	34	12.3	10	3.2
Evacuation Instruction	100	12.6	36	19.9			16	19.8	89	15.2	31	11.1			29	7.4	44	15.9	63	20.4
Nuclear Explosion	51	6.4	6	5.8			1	1.2	36	6.2	20	7.2							57	18.4
Post-disaster Supplies	40	5	10	5.5	2	8.3	1	1.2	23	3.9	24	8.6			37	9.4	7	2.5	6	1.9
Story	82	10.3	33	18.2	5	20.8	4	4.9	47	8	57	20.4	2	25	28	7.1	62	22.5	25	8.1
Warning	53	6.7	10	5.5			6	7.4	40	6.8	17	6.1			14	3.6	23	8.3	26	8.4
***p*-value**	**0.002**				**<0.001**										**<0.001**					
Cohen's *d*	**0.7**				**0.2**										**0.4**					
Information Sources ($N = 832$)																				
Broadcast Media	49	8	15	6.9			7	7.4	40	8	17	7.4			19	9	12	4.2	33	9.9
Family/Neigbourhood	244	39.7	146	67	3	75	37	38.9	225	45	124	53.7	1	50	106	50.5	183	63.8	101	30.1
Health and Safety	77	12.5	11	5			8	8.4	54	10.8	26	11.3			7	3.3	19	6.6	62	18.5
JMA	23	3.7			1		5	5.3	10	2	7	3			6	2.9	4	1.4	13	3.9
Local Government	52	8.5	18	8.3		25	4	4.2	47	9.4	18	7.8	1	50	14	6.7	8	2.8	48	14.3
Work community/Colleagues	169	27.5	28	12.8			34	35.8	124	24.8	39	16.9			58	27.6	61	21.3	78	23.3
***p*-value**	**<0.001**				**0.015**										**<0.001**					
Cohen's *d*	**1**				**0.2**										**0.7**					
Information Channels ($N = 760$)																				
Face to Face	392	74.1	191	82.7	4	80	60	73.2	335	75.6	183	80.6	1	33.3	172	78.5	219	92	192	63.4
Internet	1	0.2			1	20									1	0.5				
E-Mail	5	0.9	10	4.3					13	2.9	1	0.4	1	33.3	4	1.8			11	3.6
Phone	46	8.7	13	5.6			11	13.4	30	6.8	18	7.9			12	5.5	2	0.8	45	14.9
Radio	29	5.5	6	2.6			2	2.4	23	5.2	10	4.4			10	4.6	6	2.5	19	6.3
Speakers and Signage	36	6.8	7	3			2	2.4	29	6.5	11	4.8	1	33.3	17	7.8	8	3.4	18	5.9
Television	20	2.6	4	0.5			7	8.5	13	2.9	4	1.8			3	1.4	3	1.3	18	5.9
***p*-value**	**0.001**				**<0.001**										**<0.001**					
Cohen's *d*	**0.3**				**0.7**										**0.1**					

Appendix B

Disaster-related information seeking behaviour instances by occupation.

Disaster-related Information Seeking Behaviour Instances	Occupation																			
	Managers	%	Professionals	%	Technicians and Associate Professionals	%	Clerical Support Workers	%	Services and Sales Workers	%	Skilled Agricultural, Forestry and Fishery Workers	%	Craft and Related Trades Workers	%	Plant and Machine Operators and Assemblers	%	Elementary Occupations	%	Undefined	%
Active Information Needs ($N = 505$)																				
Current Status	47	75.8	56	60.9	4	25	14	60.9	13	68.4	6	75	1	20	2	40	23	95.8	185	73.7
Disaster Information	10	16.1			4	25	7	30.4	4	21.1	2	25			1	20			22	8.8
Evacuation Instruction	2	3.2	8	8.7			1	4.3					4	80			1	4.2	16	6.4
Nuclear Explosion			16	17.4	6	37.5													13	5.2
Post-disaster Supplies	3	4.8	11	12	2	12.5									1	20			2	0.8
Transportation															1	20			2	0.8
Warning			1	1.1			1	4.3	2	10.5									11	4.4
***p*-value**											**<0.001**									
Cohen's *d*											**0.5**									
Passive Information Needs ($N = 977$)																				
Current Status	78	56.1	54	34.2	51	81	12	46.2	65	65.7	25	71.4	6	66.7	11	25.6	41	59.4	151	44.9
Disaster Information	15	10.8	8	5.1	4	6.3	5	19.2			4	11.4	1	11.1	8	18.6	3	4.3	14	4.2
Evacuation Instruction	17	12.2	28	17.7	1	1.6	2	7.7	8	8.1	1	2.9	1	11.1	2	4.7	6	8.7	70	20.8
Nuclear Explosion			29	18.4					1	1							3	4.3	24	7.1
Post-disaster Supplies			15	9.5	4	6.3									17	39.5	2	2.9	12	3.6
Story	19	13.7	17	10.8	1	1.6	3	11.5	17	17.2	5	14.3	1	11.1	2	4.7	8	11.6	42	12.5
Warning	10	7.2	7	4.4	2	3.2	4	15.4	8	8.1					3	7	6	8.7	23	6.8
***p*-value**											**<0.001**									
Cohen's *d*											**0.4**									
Information Sources ($N = 832$)																				
Broadcast Media	8	7.6	11	5.5	4	20	2	7.7	4	10.8			1	20			3	7.1	31	8.6
Family/Neigbourhood	65	61.9	39	19.4	1	5	14	53.8	10	27	12	92.3	4	80	7	31.8	22	52.4	216	59.8
Health and Safety	11	10.5	35	17.4	1	5	1	3.8	8	21.6					3	13.6	2	4.8	27	7.5
JMA	2	1.9	1	0.5	1	5			2	5.4					3	13.6	4	9.5	10	2.8
Local Government	2	1.9	19	9.5	6	30	2	7.7	4	10.8					4	18.2			33	9.1
Work community/Colleagues	17	16.2	96	47.8	7	35	7	26.9	9	24.3	1	7.7			5	22.7	11	26.2	44	12.2
***p*-value**											**<0.001**									
Cohen's *d*											**0.3**									
Information Channels ($N = 760$)																				
Face to Face	86	82.7	107	74.8	7	63.6	24	85.7	20	44.4	10	100	4	80	16	76.2	36	70.5	273	79.8
Internet																			1	0.3
E-Mail			10	7															5	1.5
Phone	9	8.7	11	7.7					11	24.4					2	9.5	6	11.8	20	5.8
Radio	2	1.9	2	1.4	3	27.3	2	7.1	3	6.7							3	5.9	20	5.8
Speakers and Signage	3	2.9	7	4.9	1	9.1	2	7.1	9	20					3	14.3	5	9.8	13	3.8
Television	4	3.8	6	4.2					2	4.4			1	20			1	2	10	2.9
***p*-value**											**<0.001**									
Cohen's *d*											**0.5**									

References

1. Alexander, D., Magni, M.: Mortality in the l'Aquila (central Italy) earthquake of 6 April 2009. PLoS Currents **5** (2013)
2. Coe, R.: It's the effect size, stupid: What effect size is and why it is important (2002)
3. Enarson, E., Fothergill, A., Peek, L.: Gender and disaster: foundations and new directions for research and practice. In: Rodríguez, H., Donner, W., Trainor, J.E. (eds.) Handbook of Disaster Research. HSSR, pp. 205–223. Springer, Cham (2018). https://doi.org/10.1007/978-3-319-63254-4_11
4. Erikson, E.H.: Childhood and Society, WW Norton & Company. Inc., New York (1950)
5. Erikson, E.H.: Identity: Youth and crisis (No. 7). WW Norton & Company, New York (1968)
6. Esnard, A.M., Sapat, A.: Population/community displacement. In: Rodríguez, H., Donner, W., Trainor, J. (eds.) Handbook of Disaster Research. Handbooks of Sociology and Social Research, pp. 431–446. Springer, Cham (2018). https://doi.org/https://doi.org/10.1007/978-3-319-63254-4_21
7. Field, A.: Discovering Statistics Using SPSS (3. baskı). Sage Publications, London (2009)
8. Frankenberg, E., Laurito, M., Thomas, D.: The demography of disasters. Prepared for the International Encyclopedia of the Social and Behavioural Sciences. 2nd edn. (Area), 3, 12 (2014)
9. Gill, D.A., Ritchie, L.A.: Contributions of technological and natech disaster research to the social science disaster paradigm. In: Rodríguez, H., Donner, W., Trainor, J.E. (eds.) Handbook of Disaster Research. HSSR, pp. 39–60. Springer, Cham (2018). https://doi.org/10.1007/978-3-319-63254-4_3
10. Hsieh, H.F., Shannon, S.E.: Three approaches to qualitative content analysis. Qual. Health Res. **15**(9), 1277–1288 (2005)
11. Hyndman, J.: Feminism, conflict and disasters in post-tsunami Sri Lanka. Gend. Technol. Dev. **12**(1), 101–121 (2008)
12. International Labour Office: International Standard Classification of Occupations 2008 (ISCO 2008): Structure, group definitions and correspondence tables. International Labour Office (2012)
13. Kelley, K., Preacher, K.J.: On effect size. Psychol. Methods **17**(2), 137 (2012)
14. Kouadio, I.K., Aljunid, S., Kamigaki, T., Hammad, K., Oshitani, H.: Infectious diseases following natural disasters: prevention and control measures. Expert Rev. Anti Infect. Ther. **10**(1), 95–104 (2012)
15. Lachlan, K., Spence, P., Nelson, L.: Age, gender and information seeking. National Communication Association, San Diego (2008)
16. Lindell, M.K.: Communicating imminent risk. In: Rodríguez, H., Donner, W., Trainor, J.E. (eds.) Handbook of Disaster Research. HSSR, pp. 449–477. Springer, Cham (2018). https://doi.org/10.1007/978-3-319-63254-4_22
17. MacCrimmon, K.R., Wehrung, D.A.: A portfolio of risk measures. Theor. Decis. **19**(1), 1–29 (1985)
18. Monahan, B., Ettinger, M.: News media and disasters: navigating old challenges and new opportunities in the digital age. In: Rodríguez, H., Donner, W., Trainor, J.E. (eds.) Handbook of Disaster Research. HSSR, pp. 479–495. Springer, Cham (2018). https://doi.org/10.1007/978-3-319-63254-4_23
19. Muto, K., Yamamoto, I., Nagasu, M., Tanaka, M., Wada, K.: Japanese citizens' behavioral changes and preparedness against COVID-19: an online survey during the early phase of the pandemic. PLoS ONE **15**(6), e0234292 (2020)
20. Najafi, M., Ardalan, A., Akbarisari, A., Noorbala, A.A., Jabbari, H.: Demographic determinants of disaster preparedness behaviours amongst Tehran inhabitants, Iran. PLoS Currents **7** (2015)

21. Perry, R.W., Lindell, M.K.: Aged citizens in the warning phase of disasters. Int. J. Aging Hum. Dev. **44**(4), 257–267 (1997)
22. Quarantelli, E.L., Boin, A., Lagadec, P.: Studying future disasters and crises: a heuristic approach. In: Rodríguez, H., Donner, W., Trainor, J.E. (eds.) Handbook of Disaster Research. HSSR, pp. 61–83. Springer, Cham (2018). https://doi.org/10.1007/978-3-319-63254-4_4
23. Rahmi, R.: The use of sight, hearing, and touch on information seeking behaviour of the great east Japan earthquake. J. Inf. Media Stud. **18**(Issue 1), 13–27 (2019)
24. Rahmi, R., Joho, H., Shirai, T.: An analysis of natural disaster-related information-seeking behaviour using temporal stages. J. Am. Soc. Inf. Sci. **70**(7), 715–728 (2019)
25. Ray, J.W., Shadish, W.R.: How interchangeable are different estimators of effect size? J. Consult. Clin. Psychol. **64**(6), 1316 (1996)
26. Roszkowski, M.J., Snelbecker, G.E., Leimberg, S.R.: Risk tolerance and risk aversion. Tools Tech. Financial Plann. **4**(1), 213–225 (1993)
27. Ryan, B.: Information seeking in a flood. Disaster Prev. Manage. Int. J. (2013)
28. Sadiq, M.N., Ishaq, H.M.: The effect of demographic factors on the behaviour of investors during the choice of investments: evidence from twin cities of Pakistan. Global J. Manage. Bus. Res. (2014)
29. Schultz, J., Elliott, J.R.: Natural disasters and local demographic change in the United States. Popul. Environ. **34**(3), 293–312 (2013)
30. Shiga, S., Joho, H., Blanco, R., Trippas, J.R., Sanderson, M.: Modelling information needs in collaborative search conversations. In: Proceedings of the 40th International ACM SIGIR Conference on Research and Development in Information Retrieval, pp. 715–724, August 2017
31. Smith, S.K., McCarty, C.: Demographic effects of natural disasters: a case study of Hurricane Andrew. Demography **33**(2), 265–275 (1996)
32. Sonnenwald, D.H.: Evolving perspectives of human information behaviour: contexts, situations, social networks and information horizons. In: Exploring the Contexts of Information Behaviour: Proceedings of the Second International Conference in Information Needs. Taylor Graham (1999)
33. Spence, P.R., Westerman, D., Skalski, P.D., Seeger, M., Sellnow, T.L., Ulmer, R.R.: Gender and age effects on information seeking after 9/11. Commun. Res. Rep. **23**(3), 217–223 (2006)
34. Stough, L.M., Mayhorn, C.B.: Population segments with disabilities. Int. J. Mass Emerg. Disasters **31**(3), 384–402 (2013)
35. Thomas, D.S.: The role of geographic information science & technology in disaster management. In: Rodríguez, H., Donner, W., Trainor, J. (eds.) Handbook of Disaster Research. Handbooks of Sociology and Social Research, pp. 311–330. Springer, Cham (2018). https://doi.org/10.1007/978-3-319-63254-4_16
36. Watson, J.T., Gayer, M., Connolly, M.A.: Epidemics after natural disasters. Emerg. Infect. Dis. **13**(1), 1–5 (2007). https://doi.org/10.3201/eid1301.060779
37. Wei-Haas, M.: What happens when natural disasters strike during a pandemic? 17 April 2020. https://www.nationalgeographic.com/science/2020/04/what-happens-when-natural-disasters-strike-during-coronavirus-pandemic/. Accessed 05 May 2020
38. Xie, B., et al.: Global health crises are also information crises: a call to action. J. Assoc. Inf. Sci. Technol. (2020)
39. Yuen, M.C., King, I., Leung, K.S.: A survey of crowdsourcing systems. In: 2011 IEEE Third International Conference on Privacy, Security, Risk and Trust and 2011 IEEE Third International Conference on Social Computing, pp. 766–773. IEEE, October 2011
40. Bolin, B., Kurtz, L.C.: Race, class, ethnicity, and disaster vulnerability. In: Rodríguez, H., Donner, W., Trainor, J.E. (eds.) Handbook of Disaster Research. HSSR, pp. 181–203. Springer, Cham (2018). https://doi.org/10.1007/978-3-319-63254-4_10

41. Barnewall MacGruder, M.: Examining the psychological traits of passive and active investors. J. Financial Plann. (1988)
42. McKenzie, P.J.: A model of information practices in accounts of everyday-life information seeking. J. Doc. (2003)
43. Wilson, T.D.: Human information behavior. Inf. Sci. **3**(2), 49–56 (2000)
44. Imran, M., Castillo, C., Lucas, J., Meier, P., Rogstadius, J.: Coordinating human and machine intelligence to classify microblog communications in crises. In: ISCRAM, May 2014

Information Practices of French-Speaking Immigrants to Israel

An Exploratory Study

Yohanan Ouaknine(✉)

Bat Hefer, Israel

Abstract. *Purpose:* This preliminary study focuses on information practices of French-speaking immigrants to Israel with three goals in mind: a) understanding their information needs; b) mapping information sources used before and after immigration to cope with these needs, and c) analyzing the information practices related to immigrant's wellbeing. The Israeli law allows immigration and citizenship for any person of Jewish ascendant, and Israel is the fifth country in OECD by immigrants share in its population. Immigration is not only mobility but also a significant transition in life, impacting on immigrant's wellbeing, levels of income, health, and housing conditions. Efficient information practice plays an essential role in coping with these issues.

Methodology: French questionnaires were published on Facebook and LinkedIn groups dealing with immigration to Israel in August and September 2020. Seventy-one responses were collected.

Findings. This preliminary study revealed a shift in Jewish immigrants' information sources to Israel, from familial and organizational to digital information sources. A second finding is a factor analysis of the main topics researched by immigrants, developed in three components: Integration, Short-term settlement, and long-term settlement. The last finding shows how these components are correlated to information sources used after immigration.

Originality. Original aspects of this preliminary study are exploring information sources before and after immigration and their correlation to immigrants' information needs. These findings may also pave the way to include information practices in immigration policy and government agencies' work.

Keywords: Information practice · Information sources · Information needs · Immigrants · Immigration policy

1 Introduction

The world is usually represented nowadays as a "global village" due to increased population mobility and digital connections. International migration is usually defined as "the movement of persons away from their place of usual residence and across an international border " to another country [1]. Faist [2] affirms that the term "migrant refers to

Y. Ouaknine—Independent researcher.

K. Toeppe et al. (Eds.): iConference 2021, LNCS 12646, pp. 66–74, 2021.
https://doi.org/10.1007/978-3-030-71305-8_5

a person who moves from one country to another, intending to take up residence there for a relevant period." OECD reports that more than 120 million people live in other countries than their land of birth [3]. In Israel, 22% of the population was born abroad, making it the fifth OECD country by immigrants share [4].

Immigration is not only mobility but also a significant transition in life, impacting on immigrant's wellbeing, income, healthcare, or housing conditions. Efficient information practice plays an essential role in coping with these issues. Information practice is a concept encompassing information needs, information use, and information sharing [5]. In a study about immigration to Canada, Caidi [6] applies this concept to immigrants to study how they apply their information needs to information sources [6].

In this exploratory research, information practices of French-speaking immigrants to Israel are studied with three goals in mind: a) mapping information sources used before and after immigration, b) understanding their information needs, and c) analyzing the correlation between information sources and information needs.

2 Related Work

2.1 Immigration and Information Practices

Alyiah is the Hebrew term for Jewish immigration to Israel. This unique case of immigration from all Jewish communities worldwide is mainly based on religious and traditional motives [7]. Since 1950 with the Law of Return, the Israeli policy grants immediate citizenship to any Jewish immigrant. After their arrival, new immigrants start their integration under two models. Direct absorption: where the new immigrants learn Hebrew and help to find housing and employment; or Community absorption, where the new immigrants are oriented to a community center delivering their immediate needs until their full integration [8].

Jewish immigrants to Israel are a particular group who decided to voluntarily leave their country to a place they consider their homeland, sometimes at the price of a financial and social loss [9]. However, even if immigration is not forced, it is still a significant disruption in an immigrant's life, a highly exhausting experience usually related to emotional disturbance [10]. This state of mind is usually caused by the lack of basic needs, like work, social life, and housing [10] and the lack of information about these needs.

Julien [11] affirms that information practice was a common term for information search behavior and information literacy, especially among researchers in the social construction stream of social sciences (social constructionism). In Talja & Lloyd's view [12], knowledge and information in information practice are place-based, meaning that the place where the practice occurs directly impacts its performance.

The study of the relationship between immigration and information practices started only twenty years ago with studies related to different countries and types of immigrants [9, 13, 14]. The literature usually shows how information practices are impacted by immigration [15] and that the participants in most studies are usually extreme cases, like refugees and other forced migration. Research about migration between developed countries is limited [15].

In a study focusing on immigration to Canada, Caidi [6] investigated information practices of immigrants in their everyday lives, drawing on the notion of everyday life information seeking (or ELIS) developed by Savolainen [5]. Caidi [6] developed four concepts related to immigrant's needs:

1. Integration: the acculturation and assimilation process of the newcomer.
2. Inclusion: making informed decisions about economic, social, and political issues.
3. Transnationalism: keeping ties with the home country through social networks, languages, and customs.
4. Settlement: adjustment to practical issues in the new country, like housing, banking, schools, healthcare, and language skills.

Caidi [6] uses the concept of information practices, which includes needs, use, and sharing of information [5] to explain how immigrants to Canada use their information sources to cope with information needs.

2.2 Problem Statement

The current research explores the information practices of Jewish immigrants to Israel. Information practices are analyzed by mapping their information needs and the information sources used before and after immigration to cope with these needs.

The shift in the information sources used by immigrants, before and after immigration, and their relationship to information needs are barely studied. These changes may have policy implications, which could impact the way information is presented to new immigrants in Israel. Therefore, this study will focus on the following research questions:

- RQ1: Was there a significant difference in the information sources used by French immigrants to Israel between and after their immigration?
- RQ2: To what extent are information sources related to the French immigrants' information needs to Israel after their immigration?

3 Methodology

3.1 Data Collection

This survey was conducted on Facebook and LinkedIn groups related to French immigrants to Israel. An invitation was sent to 13 public Facebook and LinkedIn groups to participate in the study, and 71 immigrants who arrived in Israel in the last five years answered this survey (in French). The questionnaire was limited to five years before the immigration, as the participants had to rely on memory to answer questions about their pre-immigration information behavior. This limit also allowed the comparison between the social media platforms studied in this paper, which all existed in the last five years.

3.2 Data Analysis

Participants

Of the 71 participants, 60 (84.5%) were born in France, and 11)16.5%) from other French-speaking countries. Fifty-seven women (80.3%) and 14 men (19.7%) answered the questionnaire. Twenty-six participants earned a professional diploma (36.6%) and 25 (35.2%) Master or Ph.D. degrees. Twenty immigrants (28.2%) arrived in Israel with a High school or college level. Most of them were married (80.2%) and with at least one child (76.1%). Their average age was 42.7 years (SD = 12.81).

At present, 12.7% of the immigrants are looking for a job after their immigration, 63.4% are employed, 7% kept their job abroad, and 16.9% are students or retired. 9.9% (n = 7) do not speak Hebrew at all, 28.2% (n = 20) report a beginner level in Hebrew, and 44 immigrants (61.9%) report a language level ranging from elementary to professional.

Measures

Three questionnaires were used to gather data: information sources, information needs, and demographics. The full survey in French is given in Appendix https://forms.gle/8WKZQtmMYXo1T9nZ7. Descriptions of the questionnaires are as follows:

- The *information sources* questionnaire was employed to measure different digital or physical information sources. It consisted of twelve statements rated on a four-point Lickert scale indicating how frequently they used this source (1 = very rarely; 4 = very frequently). The scale also included a "never" answer. This tool, specifically developed for this research, did not include a neutral answer and revealed a high level of internal consistency, as shown by a Cronbach's α of 0.92 before immigration and 0.82 for answers after immigration.
- The *information needs* questionnaire was based on previous works [9] on north-American immigrants. Participants were asked to indicate topics for which they seek information and the frequency of information search. The questionnaire consisted of ten topics, rated on a five-point forced Lickert scale (1 = very rarely; 4 = very frequently). A Cronbach alpha was calculated and showed that the five-item scale's internal consistency of the five-item scale was at a high level (Cronbach's α = .85).
- The *demographics* section included the following variables: age, birthplace, education level, family situation (married or not), Hebrew language level, and employment.

4 Results

4.1 Information Sources

In the first questionnaire, participants indicated the level of use of twelve information sources before and after their immigration. A paired-samples t-test was used to determine whether the mean difference between paired observations was statistically significant before and after their immigration.

The test aimed to investigate whether each information source's level changed after the immigration, as detailed in RQ1. The results for each information source, including means, standard deviations, and effect, are shown in Table 1.

Table 1. Means changes of information sources used before and after immigration

Information sources	Before immigration		After immigration		*t*	Cohen's *d*
	Mean	SD	Mean	SD		
Social networks	2.54	0.98	3.07	0.94	−4.28***	−0.65
Friends	2.22	1.03	2.64	1.08	−3.44**	−0.45
Google	2.32	1.07	2.70	1.09	−2.33**	−0.32
Blogs	1.76	0.97	1.96	0.95	−2.17**	−0.63
Immigration agency	2.14	1.04	1.68	0.84	2.94**	0.57
Books	1.84	0.96	1.74	0.89	1.00	0.19
Family	2.31	1.04	2.39	0.99	− 0.84	−0.11
Professional networks	1.65	1.06	1.94	0.93	−1.40	−0.44
WhatsApp	2.50	1.07	2.92	0.95	−1.49	−0.13
Synagogues	1.69	0.78	2.12	1.01	−1.62	−0.33
YouTube	1.84	1.07	2.18	1.22	−1.73	−0.50
Voluntary organizations	1.95	0.84	2.24	0.87	−1.76	−0.39

Participants used information sources at a higher level after immigration (M = 2.35, SD = 0.72) instead of information sources used before immigration (M = 2.11, SD = 0.65). The level of information sources used show a significant mean increase of 11.4% (t(69) = −3.13, p = .003, d = −0.37).

Five information sources showed significant changes after immigration and are described here as digital and human information sources. Seven information sources showed no significant changes after immigration.

Digital information Sources. Social networks, like Facebook, became a critical source of information after immigration (M = 3.07, SD = 0.94) while they were less used before immigration (M = 2.54, SD = 0.98). Using social networks show a significant mean increase of 20%, (t(42) = −4.28, p < .001, d = −0.65).

Using the Google search engine to find information in Is-rael increased after immigration by 17%, from (M = 2.32, SD = 1.07) before immigration to (M = -2.70, SD = 1.09) after moving to Israel. The change was significant(t(49) = −2.33, p = .024, d = −0.32).

Blogs featured a modest increase from immigration (M = 1.96, SD = 0.97) compared to their use before immigration (M = 1.76, SD = 0.97). The change of 11% was significant (t(11) = −2.17, p = .05, d = −0.63).

Human Information Sources. Friends are a well-known important information source. Using these sources increased after immigration. Friends as information source increased by 19% ($t(57) = -3.44$, $p = .001$, $d = -0.45$). Interestingly, a significant Pearson moderate negative correlation was found between friends as information source and age ($r(66) = -.28^*$).

The immigration agency (known as the Jewish Agency) is an organization with the mission to reinforce the bonds between Jews and Israel and plays a significant role in the immigration process[1]. Participants indicated that the use of the Jewish Agency personnel as an information source decreased by 21% after the immigration ($t(26) = 2.94$, $p = .007$, $d = 0.57$). This finding may be explained by this specific government agency's orientation to help people before their immigration and leave their absorption process to other government agencies [9].

Other Information Sources. Seven information sources showed low levels of use and no significant change before and after immigration: YouTube, WhatsApp, and professional networks like LinkedIn, synagogues, family, voluntary organizations, and books.

4.2 Information Needs

The second questionnaire was related to information needs and their level of importance when searching for information. A principal component analysis (PCA) was run on this questionnaire on how frequently French-speaking immigrants researched these ten topics.

PCA revealed three components that had eigenvalues higher than one, which explained 28.8%, 52.6%, and 68.8% of the total variance.

The three components explained 68.8% of the total variance. The varimax orthogonal rotation was employed to aid interpretability. The data interpretation was consistent with the information needs the questionnaire was designed to measure with strong loadings of basic needs items, Short-term settlement needs, and Long-term settlement needs. Component loadings and communalities of the rotated solution are presented in Table 2.

The first construct, "Integration," consisted of five topics related to the information researched right after immigration. Education and diploma recognition; employment; integration rights of the new immigrants (like tax exemption and allowance); learning Hebrew and recognizing foreign driver's license by Israeli authorities. The scale had an adequate internal consistency level, as determined by a Cronbach's alpha of 0.74.

The second construct, "Short-term settlement," consisted of two questions related to the information researched when preparing children for school or when new immigrants look for healthcare. This scale had a high internal consistency level, as determined by a Cronbach's alpha of 0.88.

The last construct, "Long-term settlement," consisted of three questions related to the information researched when new immigrants deal with banking and legal issues, usually for a house mortgage or rent. This scale had an adequate internal consistency level, as determined by a Cronbach's alpha of 0.76.

[1] https://www.jewishagency.org/.

Table 2. Information needs of French-speaking immigrants

Information needs	Integration	Short term settlement	Long term settlement	Mean	SD	N
1. Education recognition	**.80**			2.32	.99	56
2. Employment	**.72**			2.67	1.08	59
3. Immigrant's rights	**.71**			2.79	.88	69
4. Learning Hebrew	**.65**			2.88	.96	59
5. Driving license	**.48**			2.09	.94	59
6. Healthcare		**.89**		2.61	.90	70
7. Schools		**.81**		2.62	.93	51
8. Legal issues			**.49**	2.72	1.01	62
9. Banking			**.46**	2.80	.88	66
10. Housing			**.90**	2.22	1.05	61

4.3 The Relation Between Information Needs and Information Sources

The second research question dealt with the correlation between information sources and the information needs of the French-speaking immigrants to Israel after their immigration. Pearson's test was run to assess the relationship between information sources and information needs variables. Table 3 shows Pearson correlations for information sources and needs variables.

Table 3. Pearson correlation of information needs and sources variables

Information sources	Integration	Short-term settlement	Long-term settlement
Social networks	.57***	.38**	.49***
WhatsApp	.27	.41**	.33*
Google search engine	.29	.14	.30*
Family	.33*	.27	.45**
Friends	.32*	.15	.26

Note. *** $p < .001$, ** $p < .05$, * $p < .01$. *Only information sources with statistically significant correlations are reported*

Digital Information Sources. Social networks were found as the only information source highly correlated to all information needs. Statistically significant and positive correlations were found between Social networks and integration ($r(34) = .57, p < .001$), Short-term settlement $r((44) = .39, p = .01)$, and Long-term settlement ($r(43) = .49$, $p = .001$) information needs.

The WhatsApp communication service is an information source correlated to Short-term settlement and Long-term settlement information needs. Statistically significant and positive correlations were found between WhatsApp and Short-term settlement ($r(42) = .41, p = .007$), and Long-term settlement information needs ($r(41) = .33, p = .038$).

Surprisingly, the Google search engine was moderately correlated to Long-term settlement information needs only ($r(45) = .30, p = .048$).

Human Information Sources. A statistically significant, moderate positive correlation was found between family as an information source with integration ($r(37) = .33, p = .046$) and Long-term settlement ($r(45) = .45, p = .002$) information needs. Friends as information source were found to be moderately correlated to integration information needs ($r(40) = .32, p = .045$).

No statistically significant correlation was found between other information sources and information needs variables.

5 Discussion and Preliminary Conclusions

This exploratory research dealt with information practices of French-speaking immigrants to Israel, aiming to map information sources used before and after immigration, understand their information needs, and analyze the correlation between information sources and information needs.

The first part of the results indicates a significant change in information sources after immigration, with a significant increase in digital and human information sources. On the other hand, the use of government agencies decreased after immigration.

The second part of the results shows three groups of information needs after immigration: information needs related to integration, short-term settlement, and long-term settlement. It should be noted that the immigrant settlement concept described by Caidi [6] split into short and long-term settlements.

The last part of the findings shows a significant correlation between information needs groups and digital and human information sources. Findings also show that some information sources are more adapted to specific information needs.

These findings may lead to two conclusions.

Information sources of immigrants changes after immigration and move to social networks. These findings need to be addressed by governmental agencies. New immigrants rely heavily on social networks to cope with their information needs rather than immigration agency information. This situation is challenging as information in social networks may be fake, and the level of language knowledge is low.

The second conclusion is that there may be a hierarchy of information needs based on the concepts described by Caidi [6] and information needs described by Shoham and Strauss [9], with information sources used to respond to information needs at different points of the immigration timeline. This mapping is critical for governmental agencies, as their mission is to help the new immigrant in his first years in Israel. The three groups of information needs reflect the new immigrant's complex pathway in need of this help. No components related to the other concepts developed by Caidi [6] were found, especially inclusion and transnationalism, and this should be studied in further research.

The fact that information sources respond to information needs is not new and well discussed in the literature [15]. However, this study is one of the first to address information needs and sources correlation after immigration between two developed countries – France and Israel. This point is essential as economic immigration between undeveloped and developed countries may lead to different results, mainly because of the internet and social network access gaps.

This study is limited by its sample size and the focus on French-speaking immigrants to Israel. A second limitation is that this survey was conducted during the Covid-19 crisis, which might imply confounding variables and the need for a second survey at the end of this crisis. Further research in this field will compare immigrants' information practices by birthplace, language, motivations, and wellbeing.

References

1. IOM: International migration. Key terms (2020)
2. Faist, T.: The Volume and Dynamics of International Migration and Transnational Social Spaces. Oxford University Press, New York (2000)
3. OECD: How's Life? 2017 Life Satisfaction, March 2017
4. OECD: Share of migrants in the population, by gender, Paris (2017)
5. Savolainen, R.: Everyday Information Practices: A Social Phenomenological Perspective. Scarecrow Press, Plymouth (2008)
6. Caidi, N., Allard, D., Quirke, L.: Information practices of immigrants. Annu. Rev. Inf. Sci. Technol. **44**(1), 491–531 (2010)
7. Amit, K.: Determinants of life satisfaction among immigrants from Western countries and from the FSU in Israel. Soc. Indic. Res. **96**(3), 515–534 (2010)
8. Desille, A.: Jewish immigrants in Israel: disintegration within integration?. In: S. Hinger, R. Schweitzer (eds.) The Politics of Disintegration, pp. 129–147. Springer, Osnabrück (2019). https://doi.org/10.1007/978-3-030-25089-8_8
9. Shoham, S., Kaufman-Strauss, S.: Immigrants' information needs: their role in the absorption process. Inf. Res. **13**, 1–4 (2008)
10. Adler, S.: Maslow's Need Hierarchy and the Adjustment of Immigrants. Int. Migrat. Rev. **11**(4), 444–451 (1976)
11. Julien, H., Pecoskie, J.J.L., Reed, K.: Trends in information behavior research, 1999–2008: a content analysis. Libr. Inf. Sci. Res. **33**(1), 19–24 (2011)
12. Talja, S., Lloyd, A.: Challenges for future research on learning, literacies and information practices. In: Pract. Inf. Lit. Bringing Theor. Learn. Pract. Inf. Lit. Together, pp. 357–364, 2010.
13. Caidi, N., MacDonald, S.: Information practices of Canadian Muslims post 9/11. Gov. Inf. Q. **25**(3), 348–378 (2008)
14. Shepherd, J., Petrillo, L., Wilson, A.: Settling in: how newcomers use a public library. Libr. Manag., 39(8–9), 583–596 (2018)
15. Case, D., Given, L.M.: Looking for Information: A Survey of Research on Information Seeking, Needs, and Behavior, 4th edn. Emerald Group Publishing Limited (2016)

How the Intellectually Humble Seek and Use Information

Tim Gorichanaz(✉)

Drexel University, Philadelphia, PA 19104, USA
gorichanaz@drexel.edu

Abstract. Addressing humanity's grand challenges requires productive dialogue. One way to improve dialogue is by helping people build skills in dealing with information using an approach rooted in the intellectual virtues. There are a number of intellectual virtues, but a recent line of multidisciplinary research suggests beginning with intellectual humility (IH) as a cornerstone. At heart, IH is being open to the possibility that one might be mistaken. It has been defined as recognizing one's intellectual limitations, having little concern with intellect-derived social status, and accurately valuing one's beliefs according to the evidence. IH has obvious relevance to the information field, but to date there has been scant mention of IH in the information literature. This short paper reports on the first study connecting measures of IH to information seeking and use, a quantitative online survey with 201 participants. The results of this study suggest that individuals with higher IH are more likely to: be older; favor easily accessible information sources; search in multiple places for information; and find that discovering information ignites further interest in their search. Moreover, those with higher IH are less likely to look upon themselves negatively, or think others would, for using a given information source. These findings suggest avenues for further research on IH and information behavior, literacy and design.

Keywords: Intellectual humility · Intellectual virtues · Information behavior · Information seeking · Information use

1 Introduction

Several observations can be made today that suggest a breakdown of dialogue in many spheres of life. To speak of politics in the United States, for example, political polarization has been on the rise for at least the past few decades [25, 37]. Some possible causes for this include: *ideological sorting*, both geographical and sociotechnical, such that people are less likely to encounter opposing viewpoints; *little incentive* for people to critically examine the evidence for their beliefs; and a growing *culture of disrespect* for differing views. Regarding this latter point, a recent survey found that 42% of respondents in both parties believed the opposition to be "downright evil" [24]. As in politics, so in other domains—from science, such as evidenced in the recent crisis of reproducibility in psychology, to religion, such as evidenced in the problems attending extremism.

K. Toeppe et al. (Eds.): iConference 2021, LNCS 12646, pp. 75–84, 2021.
https://doi.org/10.1007/978-3-030-71305-8_6

Humanity is facing many grand challenges—climate change, race- and class-based justice, migration, and so on. If we are to confront these challenges, productive dialogue is crucial [32, 39]. One approach to improving dialogue is to help people build skills in speaking and listening, broadly construed. This constitutes a virtue-based approach, as it focuses on instilling certain character traits in people, in this case intellectual (or epistemic) virtues. Philosophers have identified a number of intellectual virtues, including curiosity, tenacity, intellectual humility, etc. [8, 40, 46].

Saint Augustine suggested that humility is the basis of all the other virtues. The same could be said of intellectual humility: to learn anything, one must acknowledge that one's perspective is limited. Consonantly, philosophers, psychologists and educators have begun researching intellectual humility (IH) as a cornerstone of the intellectual virtues. At heart, IH is being open to the possibility that one might be mistaken. It has been defined as recognizing one's intellectual limitations [44], having little concern with intellect-derived social status [40], and accurately valuing one's beliefs according to the evidence at hand [11]. To date, researchers have sought to clearly define IH, establish ways to measure it, and create tools for increasing it.

Given the close relationship of IH to information and evidence, the information field has much to contribute to the research on IH. What's more, IH has been described as a component of wisdom [15, 23], and there have been several recent calls to conduct further research on wisdom within information science [e.g., 5, 14, 29], to which work on IH could contribute. To date, though, there has been scant mention of IH in the literature.

As an initial contribution, this short paper reports on a quantitative survey study connecting measures of IH to people's information seeking and use. The survey was intended to explore possible correlations between people's levels of IH and their information behavior with respect to a particular task. Though this is preliminary work, the results suggest that there is much to be gained from further research at the intersection of IH and information.

2 Literature Review

2.1 Defining and Measuring Intellectual Humility

Over the past decade, scholars have worked to define intellectual humility (IH). Extant definitions comprise three basic views:

- owning one's intellectual limitations and valuing others' intellects [44]
- lacking concern with how intellect bears on one's social status [40]
- accurately tracking the force of one's evidence for their beliefs [11]

In addition to defining IH, researchers are working to establish valid and reliable measures of IH. To date, most rely on self-report measures. These have limitations; for instance, a person boasting about their humility may not truly be humble. To address this limitation, researchers have focused their self-report instruments on differences in IH and manifestations thereof that people can assess with reasonable accuracy and without social desirability bias; and these efforts are bearing fruit [4, 17, 26, 39]. The most promising scale for IH is a self-report measure developed by Alfano et al. [4], which uses a 25-item questionnaire to assess IH on four dimensions (open-mindedness, modesty, corrigibility and engagement).

The ability to measure IH has made it possible to explore connections between IH and other behaviors and traits. Recent work shows, for example, that IH predicts intellectual openness and that high IH is associated with more general knowledge [27]. Having high IH is also predictive of mastery behaviors such as seeking out challenges and persisting after setbacks [38]. According to Porter et al. [38], "those higher in intellectual humility will behave in mastery-oriented ways so that they can become more knowledgeable and accurate."

There is evidence that IH can change, opening the door for improvement through education and intervention. While IH does have some trait-like permanence, it also fluctuates across situations [45]. The ways IH fluctuates are still being researched; to date, it seems that a person may manifest IH differently about different beliefs and attitudes [22], and the personal relevance of the topic also seems to matter [30]. Porter and Schumann [39] suggest that IH may be fostered through instilling a growth mindset.

2.2 Intellectual Humility and Information

The internet presents myriad difficulties regarding information seeking and evaluation. Online search, for example, may prioritize misleading information or trap users in filter bubbles [31, 33, 41]. Gunn [16] concludes that online search has made us worse listeners (broadly defined), even while it has the capacity to make us better ones. As a case in point, the Stanford History Education Group [42] conducted a large-scale study of how young people evaluate information in open web searches. "Overall, young people's ability to reason about the information on the Internet can be summed up in one word: bleak." They found, for instance, middle schoolers unable to distinguish ads from news articles and college students who trusted web content simply because it was on a.org domain. In response, the researchers developed curricular materials to benefit future students.

Gunn [16] argues that to become better listeners, we need to foster IH. The materials developed by SHEG [42] may be useful to that end to some extent, but such a large problem is worth addressing from several angles. Porter and Schumann [39] established a connection between IH and openness to hearing opposing political views. According to their findings, higher IH is related to more information seeking, suggesting a role for research and intervention in information behavior and information literacy.

To date, IH has scarcely been mentioned in the information science literature. Extant references are in connection with the Association of College and Research Libraries (ACRL) Framework for Information Literacy [7]. The framework specifies that information seekers should "demonstrate intellectual humility." Commenting on this in light of the fake news phenomenon, Becker [10] correctly surmises that IH moves information literacy from questions of technique and technology to those of logic, reasoning and virtue, and that this means librarians can no longer work on information literacy alone: "Over the course of the next few generations, we need to get this recognized as a basic literacy that's embedded into every level of education and development." Similarly, Cooke and Magee [12] comment on IH in connection with the ACRL framework and metaliteracy. As they write, metaliteracy is a move toward a more holistic and reflexive form of literacy, much-needed in the present day. And likewise, Baer [9] discusses the personal dimensions of information behavior and argues that information literacy today should involve intellectual empathy, a concept akin to IH, as part of realizing more informed societies.

All this suggests it would be fruitful to study IH through the lens of information behavior, though this has not yet been done. In looking for precedents, we should consider the research on personality and information behavior. Personality is a somewhat stable inclination toward certain behaviors and attitudes that can predict individual differences in a variety of contexts. The currently favored model of personality is the "Big Five" model, which defines personality along five dimensions: neuroticism, extraversion, openness to experience, agreeableness, and conscientiousness [36]. Heinström [21] showed that the Big Five traits correlate with different patterns of information behavior among students. For example, preference to retrieve information that confirms previous knowledge was negatively correlated with traits extraversion and conscientiousness; and willingness to expend considerable effort in obtaining information was positively correlated with conscientiousness and openness to experience. These findings were also borne out in other survey studies with students in other national contexts [18, 43]. Further, a recent eye-tracking study shows that individuals highest in conscientiousness, agreeableness and extraversion perform many information-seeking tasks faster than others [3]. IH seems to share some of the same stability as personality, even though it is not reducible to any personality dimension [11]. At the least, this suggests that IH and information behavior can be studied with similar methods to those used to study personality and information behavior.

3 Research Design

3.1 Research Questions and Hypotheses

The overarching research question of this study is: *What, if any, correlations exist between measures of intellectual humility and information behavior?* To be more precise, a number of more specific variables were developed based on the literature reviewed above. The following hypotheses were constructed:

H1. IH is not correlated with age or education level.
H2. Those with higher IH will use more information sources for a given task.
H3. Those with higher IH will be less likely to favor easily accessible sources.
H4. Those with higher IH will put less emphasis on the attractiveness of information sources.
H5. Those with higher IH will be more likely to search multiple places for information.
H6. Those with higher IH will have their interest fueled by the discovery of new information.
H7. Those with higher IH will be less likely to have a negative self-image because of the information sources they used.

3.2 Data Collection

A survey was designed and administered collecting data on participants' IH score, information behavior in context, and demographics (age and education level). To measure IH, Alfano et al.'s [4] scale was used. To collect data on participants' information behavior, a questionnaire was developed. Participants were asked to reflect on a real-life task requiring information seeking that they recently undertook, and the questions examined their seeking and use employing a 7-point Likert-type scale ranging from "Strongly disagree" to "Strongly agree." Items and guidance for developing this questionnaire were drawn from the literature (e.g., 1, 34]; feedback on the questionnaire was solicited from colleagues and tested in a small pilot study.

The survey was implemented on Qualtrics and administered online through Prolific, a research recruitment service, in October 2020. Though it is not yet common in the information field, at present most survey research in psychology and cognitive science is conducted online, the vast majority using Amazon's Mechanical Turk (MTurk) platform [6]. MTurk presents several concerns regarding ethics [19] and data quality [2]. Recently, Prolific has emerged as a rigorous and more ethical and reliable alternative to MTurk. Participants are paid better, and they are prescreened, verified and monitored. As of October 2020, Prolific has nearly 38,000 U.S.-based participants; over 3,000 researchers use the platform. For this study, 201 participants residing in the United States were hired; each participant was paid $1.50 (USD) to take the survey, the equivalent of $12.48 per hour on average.

3.3 Data Analysis

The survey results were analyzed using SPSS 26. To ascertain correlations between participants' IH scores and aspects of their information behavior, Spearman's rank-order correlation was used. This test is appropriate for ordinal data in paired observations with a monotonic relationship.

4 Results

All 201 participants completed the survey with very few missing items. Most participants were in the 25–34 age range, and the mean education level was a two-year college degree (see Table 1). Regarding the recent task that the participants considered, 90% chose a task that had already been completed (rather than still ongoing). The majority selected a task whose duration would be measured in "hours" (41%) or "minutes" (21%). As far as category, 29% chose a work-related task, 24% a home- or family-related task, 23% a hobby-related task, and 21% a school-related task. The vast majority of participants (98%) used a web search in looking for information.

The results of the correlation analysis are presented in Table 2.

H1 (age and education) was partly supported by the data. There was a weak, positive correlation between IH score and age ($r_s = 0.158, p = 0.026$), but none between IH and education level.

H2 (number of information sources) was not supported. Participants were asked to select the different types of information sources they used (person face-to-face, person

Table 1. Overview of the participants

Age	N	Education Level	N
18–24	52 (26%)	Less than high school	3 (1%)
25–34	76 (38%)	High school graduate	24 (12%)
35–44	39 (19%)	Some college	48 (24%)
45–54	22 (11%)	2-year degree	22 (11%)
55–64	9 (4%)	4-year degree	71 (35%)
65–74	2 (1%)	Professional degree	29 (14%)
		Doctorate	4 (2%)

by phone, website, book, etc.), but there was no significant relationship between the number of information source types selected and IH.

H3 (easily accessible sources) was refuted. There was a weak, positive correlation between participants favoring easily accessible sources and their IH score ($r_s = 0.226$, $p = 0.001$).

H4 (attractiveness of information) was not supported. There was no significant relationship between the attractiveness of information favored and IH.

H5 (multiple places) was supported by the data. There was a weak, positive correlation between participants' searching multiple places for information and their IH score ($r_s = 0.172$, $p = 0.015$).

Table 2. Results from correlation analysis with each participant's Intellectual Humility score. * indicates $p < 0.05$, ** indicates $p < 0.01$

Item	Corr. Coeff	Significance	N
Age	0.158*	0.026	200
Education	– 0.010	0.891	201
Number of Types of Sources	– 0.016	0.817	201
Favor Easily Accessible Sources	0.226*	0.001	200
Favor Attractive Information	0.007	0.925	201
Searched Multiple Places	0.172*	0.015	201
Information Fueled Further Interest	0.193*	0.006	201
Negative View of Self	– 0.464**	0.000	201
Perceived Negative View from Others	– 0.466**	0.000	201

H6 (generating further interest) was supported. There was a weak, positive correlation between participants agreeing that the more information they found, the more their interest in searching for information grew, and their IH score ($r_s = 0.193$, $p = 0.006$).

H7 (negative self-image) was supported. There was a moderate negative correlation between participants' IH scores and their agreement with statements about being nervous, embarrassed or uncomfortable using information in this task. These questions were grouped into two analytical categories, one for how the participant regarded themselves ($r_s = -0.464$, $p = 0.000$) and another for how they thought others would perceive them ($r_s = -0.466$, $p = 0.000$).

5 Discussion

This paper presented the first study of the relationship between intellectual humility (IH) and information behavior. In this exploratory, correlational study, participants were asked to consider how they sought and used information during a recent task of their choosing, and their IH was measured using a robust and validated scale from the literature.

The results of this study suggest that individuals with higher IH are more likely to: be older; favor easily accessible sources; search in multiple places for information; and find that discovering information ignites further interest in their search. However, these correlations are weak, and further research should be done to further understand the dynamics at play. Beyond this, the results more strongly indicate that those with higher IH are less likely to look upon themselves negatively, or think others would, for using a given information source.

While in prior research Porter and Schumann [39] suggested that higher IH is correlated with more information seeking, this study casts some doubt on that broad claim. This study found that while those with higher IH are more likely to search in multiple places for information, they are not necessarily more likely to engage with more types of information sources, and they do not favor harder-to-find sources, both of which might be thought to be associated with more information seeking. Again, these issues raise further questions that may be explored in future research.

As a preliminary study, these results are limited in their generalizability. While the study included a sizable 201 participants, it is possible that participants working with the Prolific platform exhibit certain idiosyncrasies of information behavior that are not representative of the broader population.

All in all, this work does demonstrate that further research in this area would be useful. To start, further analysis may be conducted with the data collected for this study. While some correlations were discovered in this preliminary analysis, most of the uncovered correlations are weak. Further analysis may uncover stronger correlations by isolating participant groups within the dataset (e.g., defining groups by the length of the task under consideration, as there may be differing effects based on how long a person was engaged in a task).

Additional analysis of this data will lay the groundwork for further research still at the intersection of IH and information. This research may show how our information infrastructure (e.g., search engine results pages and social media news feeds) might be designed to better support people's IH, as well as techniques for how we might help people build their IH, increasing their information capabilities through information literacy and other educational interventions.

References

1. Agarwal, N.K., Xu, Y., Poo, D.C.: A context-based investigation into source use by information seekers. J. Am. Soc. Inform. Sci. Technol. **62**(6), 1087–1104 (2011)
2. Ahler, D.J., Roush, C.E., Sood, G.: The micro-task market for lemons: Data quality on Amazon's Mechanical Turk. Manuscript under review (2020). https://www.gsood.com/research/papers/turk.pdf. Accessed 19 Sept 2020
3. Al-Samarraie, H., Eldenfria, A., Dawoud, D.: The impact of personality traits on users' information-seeking behavior. Inf. Process. Manage. **53**, 237–247 (2017)
4. Alfano, M., et al.: Development and validation of a multi-dimensional measure of intellectual humility. PLoS ONE **12**(8), e0182950 (2017)
5. Allen, J., Khader, M., Rosellini, A.: Investigating wisdom: call for research. Proc. Assoc. Inf. Sci. Technol. **56**(1), 599–600 (2019)
6. Anderson, C.A., Allen, J.J., Plante, C., Quigley-McBride, A., Lovett, A., Rokkum, J.N.: The MTurkification of social and personality psychology. Pers. Soc. Psychol. Bull. **45**(6), 842–850 (2019)
7. Association of College and Research Libraries, Framework for Information Literacy for Higher Education (2016). https://www.ala.org/acrl/standards/ilframework. Accessed 19 Sept 2020
8. Baehr, J.: Cultivating Good Minds: A Philosophical & Practical Guide to Educating for Intellectual Virtues. Intellectual Virtues Academy, Long Beach, CA (2015)
9. Baer, A.: What intellectual empathy can offer information literacy education. In: Goldstein, S. (ed.) Informed Societies: Why Information Literacy Matters for Citizenship, Participation and Democracy, pp. 47–68. Facet, London (2020)
10. Becker, B.W.: The librarian's information war. Behav. Soc. Sci. Libr. **35**(4), 188–191 (2016)
11. Church, I.M., Samuelson, P.L.: Intellectual Humility: An Introduction to the Philosophy and Science. Bloomsbury, New York (2017)
12. Cooke, N.A., Magee, R.M.: Teaching and learning with metaliterate LIS professionals. In: Mackey, T.P., Jacobson, T.E. (eds.), Metaliterate Learning for the Post-Truth World, pp. 143–157. ALA Neal-Schuman (2019)
13. Ericsson, K., Simon, H.: Protocol Analysis: Verbal Reports as Data, 2nd edn. The MIT Press, Boston (1993)
14. Gorichanaz, T., Latham, K.F.: Contemplative aims for information. Inf. Res. **24**(3), 836 (2019). https://informationr.net/ir/24-3/paper836.html
15. Grossmann, I.: Wisdom in context. Perspect. Psychol. Sci. **12**, 233–257 (2017)
16. Gunn, H.K.: Has Googling made us worse listeners? Contemp. Fr. Francoph. Stud. **23**(4), 512–520 (2019)
17. Haggard, M., et al.: Finding middle ground between intellectual arrogance and intellectual servility: development and assessment of the limitations-owning intellectual humility scale. Personality Individ. Differ. **124**, 184–193 (2018)
18. Halder, S., Roy, A., Chakraborty, P.K.: The influence of personality traits on information seeking behaviour of students. Malays. J. Libr. Inf. Sci. **15**(1), 41–53 (2010)
19. Hara, K., Adams, A., Milland, K., Savage, S., Callison-Burch, C., Bigham, J.P.: A data-driven analysis of workers' earnings on Amazon Mechanical Turk. In: Proceedings of the 2018 CHI Conference on Human Factors in Computing Systems (CHI 2018). Association for Computing Machinery (2018)
20. Heersmink, R.: A virtue epistemology of the internet: search engines, intellectual virtues and education. Soc. Epistemol. **32**(1), 1–12 (2018)
21. Heinstrom, J.: Five personality dimensions and their influence on information behavior. Information Research, 9(1), paper 165 (2003). https://InformationR.net/ir/9-1/paper165.html, last accessed 2020/10/19.

22. Hoyle, R.H., Davisson, E.K., Diebels, K.J., Leary, M.: Holding specific views with humility: conceptualization and measurement of specific intellectual humility. Personality Individ. Differ. **97**, 165–172 (2016)
23. Huynh, A.C., Grossmann, I.: A pathway for wisdom-focused education. J. Moral Educ. **49**(1), 9–29 (2020)
24. Kalmoe, N.P., Mason, L.: Lethal mass partisanship: prevalence, correlates, and electoral contingencies. Paper presented at the NCAPSA American Politics Meeting, College Park, MD (2019)
25. Klein, E.: Why We're Polarized. Avid Reader Press, New York (2020)
26. Krumrei-Mancuso, E.J., Rouse, S.V.: The development and validation of the comprehensive intellectual humility scale. J. Pers. Assess. **98**, 209–221 (2015)
27. Krumrei-Mancuso, E.J., Haggard, M.C., LaBouff, J.P., Rowatt, W.C.: Links between intellectual humility and acquiring knowledge. J. Posit. Psychol. **15**(2), 155–170 (2020)
28. Kuhlthau, C.C.: A principle of uncertainty for information seeking. J. Documentation **49**(4), 339–355 (1993)
29. Latham, K.F., Hartel, J., Gorichanaz, T.: Information and contemplation: a call for reflection and action. J. Documentation, Early View (2020). https://doi.org/10.1108/JD-05-2019-0076
30. Leary, M.R., et al.: Cognitive and interpersonal features of intellectual humility. Person. Soc. Psychol. Bull. **43**(1), 1–21 (2017)
31. Lynch, M.: The Internet of Us: Knowing More and Understanding Less in the Age of Big Data. Norton, New York (2016)
32. Meagher, B.R., Leman, J.C., Heidenga, C.A., Ringquist, M.R., Rowatt, W.C.: Intellectual humility in conversation: distinct behavioral indicators of self and peer ratings. J. Posit. Psychol. Early View (2020). https://doi.org/10.1080/17439760.2020.1738536
33. Miller, B., Record, I.: Justified belief in a digital age: on the epistemic implications of secret internet technologies. Episteme **10**(2), 117–134 (2013)
34. Nardi, P.M.: Doing Survey Research: A Guide to Quantitative Methods, 4th edn. Routledge, New York (2018)
35. Nielsen, J.: Thinking Aloud: The #1 Usability Tool. NNGroup (2012). https://www.nngroup.com/articles/thinking-aloud-the-1-usability-tool/. Accessed 19 Sept 2020
36. Paunonen, S.V., Ashton, M.S.: Big Five factors and facets and the prediction of behavior. J. Pers. Soc. Psychol. **81**(3), 524–539 (2001)
37. Pew Research Center, The partisan divide on political values grows even wider, 5 October 2017. https://www.people-press.org/2017/10/05/the-partisan-divide-on-political-values-grows-even-wider/. Accessed 19 Sept 2020
38. Porter, T., Schumann, K., Selmeczy, D., Trzesniewski, K.: Intellectual humility predicts mastery behaviors when learning. Learn. Individ. Diff. **80**, 101888 (2020). https://doi.org/10.1016/j.lindif.2020.101888,lastaccessed2020/10/19
39. Porter, T., Schumann, K.: Intellectual humility and openness to the opposing view. Self Identity **17**(2), 139–162 (2018)
40. Roberts, R.C., Wood, W.J.: Intellectual Virtues: An Essay in Regulative Epistemology. Oxford University Press, Oxford (2007)
41. Simpson, D.: Evaluating Google as an epistemic tool. Metaphilosophy **43**(4), 426–445 (2012)
42. Stanford History Education Group. Evaluation information: The cornerstone of civic online reasoning (2016). https://sheg.stanford.edu/upload/V3LessonPlans/Executive%20Summary%2011.21.16.pdf. Accessed 19 Sept 2020
43. Tidwell, M., Sias, P.: Personality and information seeking: understanding how traits influence information-seeking behaviors. J. Bus. Commun. **42**, 51–77 (2005)
44. Whitcomb, D., Battaly, H., Baehr, J., Howard-Snyder, D.: Intellectual humility: owning our limitations. Philos. Phenomenol. Res. **94**(3), 509–539 (2015)

45. Zachry, C.E., Phan, L.V., Blackie, L., Jayawickreme, E.: Situation-based contingencies underlying wisdom-content manifestations: examining intellectual humility in daily life. J. Gerontol. Psychol. Sci. **23**(8), 1404–1415 (2018)
46. Zagzebski, L.: Virtues of the Mind. Cambridge University Press, Cambridge (1996)

Gamification as a Way of Facilitating Emotions During Information-Seeking Behaviour: A Systematic Review of Previous Research

Amira Ahmed(✉) and Frances Johnson

Department of Languages, Information & Communication, Manchester Metropolitan iSchool, Manchester Metropolitan University, Manchester, UK

amira.ahmed@stu.mmu.ac.uk

Abstract. Games and game elements have been studied and applied in various domains to encourage sustainable behaviours, often without a clear understanding of how they achieve these effects. One unconventional domain is the management of emotions during information seeking. This paper presents a review of current research literature covering gamification and game elements beyond entertainment and outlines their theoretical practicalities. A total of 40 records from different perspectives were reviewed studying the effects or user perceptions of gamification within motivation, emotions and information-seeking. The results reported in the reviewed records indicate that applying gamification is a promising approach for managing emotions in the information-seeking research process. However, the study identified shortcomings, thematic gaps, and the direction of future research which is discussed in this paper.

Keywords: Gamification · Games with a purpose · Emotions · Information-seeking · Systematic review

1 Introduction

Substantial advances have been made in research on emotions and the affective, not only in disciplines in which that interest might be expected (such as psychology, medicine or neurology) but, more generally, in the realm of the humanities and social sciences [1]. There is an increasing interest in both academia and relevant industries in how motivational technologies, i.e., gamification, serious games, and persuasive technology, can affect human behaviour in various domains, such as information systems, or education, and learning behaviours [2].

One of the reasons for this is the rise of the affective dimension in social life, where relevant transformations in the expression of effect are made in the private and public arenas (i.e., the growth of the vibrant culture in the realms of education, the proliferation of self-narratives and the changes in information seeking practices). In this regard, some of the most fruitful theoretical approaches are developed from psychology (particularly social psychology) and the 'sociology of emotions' [3]. Therefore, various theoretical

K. Toeppe et al. (Eds.): iConference 2021, LNCS 12646, pp. 85–98, 2021.
https://doi.org/10.1007/978-3-030-71305-8_7

frameworks on emotions are conceptualised and explained both from neurobiological and sociocultural standpoints [4]. Moreover, emotions are not merely a subject matter of research but rather the perspective for a new epistemological turn [5]. Conversely, the complex reality of this facet of human nature makes it an interdisciplinary study object, albeit one about which there is still no comprehensive vision, capable of bringing together and integrating all these different disciplines.

The present study sets out to address this by providing a thorough description, via a literature review, how the digital realm is an emotional space, and how the various types of game elements and its gamefulness are activated for expressing emotion. The review focuses on gamification and gamification technologies in the domain of emotion and use of information seeking and retrieval systems to synthesise the literature. Historically, the body of literature has been somewhat fragmented into areas such as motivational technologies [2] that primarily include gamification ([3, 7, 8]), and serious games [9] and persuasive technology [10] that engage individuals with activities that are commonly considered mundane, such as using the internet for buying products. Gamification refers to transforming systems, services, products, organisational structures or practically any activity to afford similar experiences as games do use game design [8].

Serious games commonly represent full games designed for purposes other than pure entertainment [7]. On the other hand, persuasion technology did not emerge from game research and commonly refers to adding a type of design onto an existing practice. A second area or body of literature relates to the study and tools which focus on emotion regulations and expression in various scenarios [6]. The use of technologies can be intricately tied to affective processes, both with emotions altering technology use patterns and technology use changing one's emotional state. A specific context, feelings during information seeking has shown this behaviour to be influenced by emotions and different persuading factors [11, 12].

The third area is the specific application of motivational technologies using gamification to activate emotion within a particular domain of information seeking and retrieval technologies. As a result of this fragmentation, whilst successful applications of motivational technology have been developed, previous research lacks systematic investigations of the mechanisms of gamification [13], and in this review, we aim to address this in synthesising the literature on gamification, and on emotion and its influence in motivation, and specifically in the domain of information-seeking behaviour.

Therefore, the present study can be conceptualised as a comprehensive description of how the digital realm is an emotional space and how gamification mechanisms are activated for facilitating emotion. By focusing on the specific context of information-seeking behaviour, and the influence of emotion, the thematic review of previous research aims to identify information-seeking behaviour as a prominent context for further gamification research.

2 Materials and Methods

To compile a comprehensive scientifically sound literature review about theoretical fundamentals of using games and game elements in emotions and information-seeking, a systematic literature search was conducted on and through available and the most relevant online databases, specifically Scopus, ScienceDirect, IEEE and Springer. The search was conducted from January-July 2020 over records' titles, abstracts and keywords to include only records whose focus was "emotions and information-seeking using gamification". The search terms used are depicted in Table 1 (2010–2020 records).

Table 1. Search terms employed in the keyword search

Focus Area	Search Term(s)
Gamification	"gamifi*" OR "gamify"
	"gamification"
Games with a purpose	"games" AND "purpose"
Serious Games	"serious" AND "games"
	"serious" AND "gaming"
Emotions	"emotions" AND "emoti*"
	"emotions" AND "games"
Information-seeking	"information-seeking" OR "information-behaviour"
	"information-seeking" AND "gamification"

Types of records included in the search were conference papers, articles and book chapters. A total of 504 records were obtained with the user search query. After the search, the titles and abstracts were examined. Fourteen records were excluded for not being available in English. A total of 205 records were identified as not related to gamification, serious games, or persuasive technology, thus excluding further analysis. Of the remaining records, 259 are excluded as they are not focused on emotions or information seeking, or not based on empirical studies. Sixty records were not fully accessible even through inter-library loan. After screening the titles and abstracts, a total of 35 emotion and search behaviour related records were included in the review. Two additional records were identified using a forward search and three using backward search. Thus, a total of 40 records are included in Fig. 1.

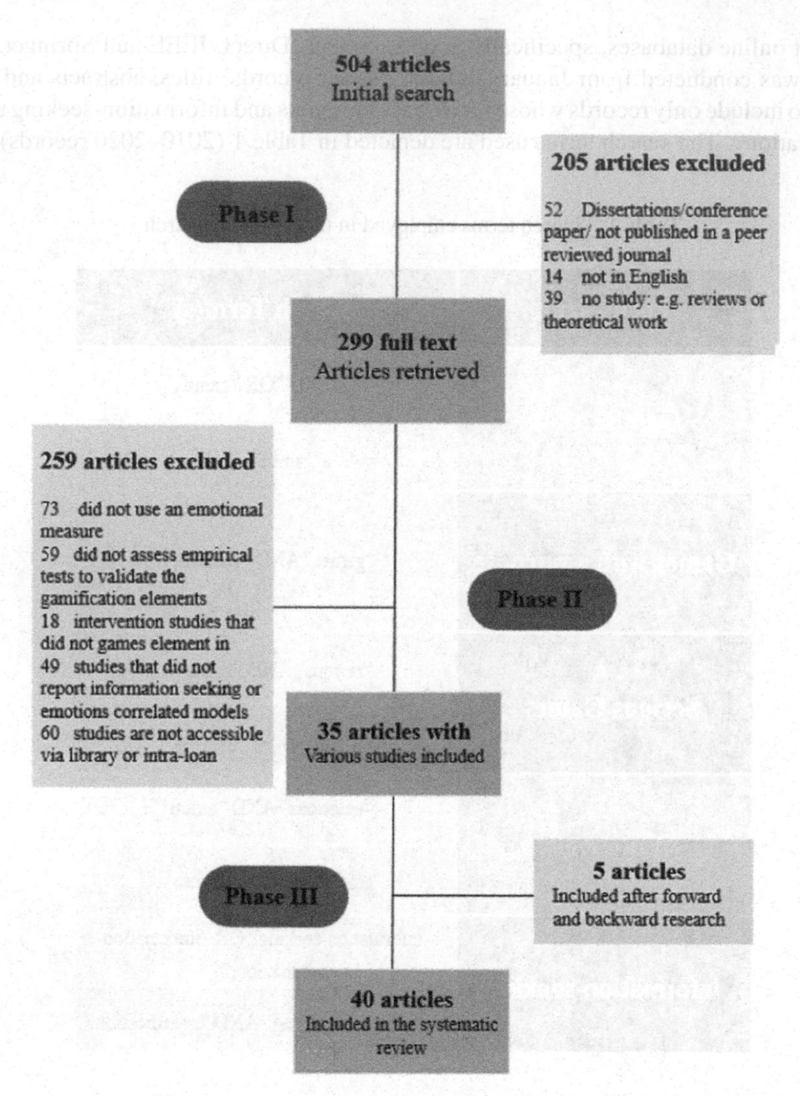

Fig. 1. Procedure and results of the literature search

In reviewing the study employs the review format suggested by Fettke [14]. This categorisation can clarify the characteristics of a review study and is based on several literature recommendations [14]. According to this framework, this study presents a review in natural language that focuses on theory, takes a neutral perspective and aims to highlight central aspects of the literature selected for review. This framework is based.

on characteristics of Type, Application, Perspective, Aim, Structure, Literature, Target Audience and Future Research (Table 2).

Table 2. Characterisation of this review based on Fettke [14]

Characteristics	Classification					
Type	Natural Language	Mathematical-Statistical				
Application	Research Result	Theory	Experience	Research Method		
Perspective	Neutral	Positioned				
Aim	Not Mentioned	Mentioned	Integration	Central Aspects	Criticism	
Structure	Chronologically	Methodologically	Thematically			
Literature	Not Explained	Explained	Keyworks	Representative	Selective	Exhaustive
Target Audience	Common Public	Practitioners	Researchers in General	Specialized Researchers		
Future Research	Mentioned	Not Mentioned				

3 General Description of the Reviewed Research

Although the differentiation between different types of gamification definitions and game elements can be ambiguous, all the records were considered according to whether the technology described in them matched the definitions of the topic gamification, as outlined in the introduction section. Drawing on the terms and labels used by the authors of each manuscript, this process was conducted to ensure that all the papers included were about gamification. It resulted in two additional sub-labels 'serious games' and 'games with a purpose' where this differentiation was needed. Papers that used similar terms to 'serious games' such as 'simulation games', or 'learning games' were categorised as serious games in this study.

Furthermore, gamified simulators, learning platforms and training applications for librarians in information seeking were classified as 'games with purpose'. They focused on providing games, explicitly using empirical data, on completing a task for learning. Of the 40 records included in this review, 32 focused on gamification, six on serious games, and two related 'games with purpose' technology.

Moreover, all 40 records were analysed according to the different search concepts of motivation, emotion, and information seeking. This identified four labels which could differentiate the papers by their focus area (Fig. 2).

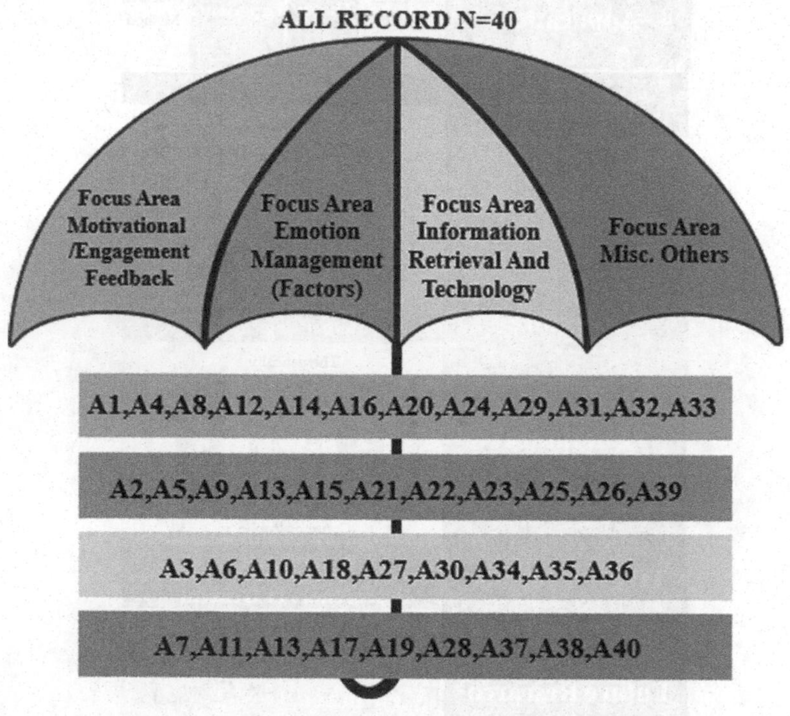

Fig. 2. Contexts of the records

These broad areas identified as a) *Motivational/Engagement Feedback.* Which draw on the notion that emotions triggered in this feedback influence behaviour such as technology use patterns; b) *Emotion Management.* Papers focus on emotional responses and state through using technologies and managing responses people have, e.g. customers behaviour or library researchers and c) *Information Retrieval & Technology* (specifically in eLearning) papers focus on the specific context of information seeking and in which emotions influence and alter technology use patterns. Alternatively, these papers focus on the technology used to conduct information seeking and retrieval tasks that can change one's emotional state. *d) Misc. Others,* which includes all related papers which are advanced research on gamification and its use in different contexts.

3.1 Gamification Affordances and Outcomes in Reviewed Studies

The four broad focus themes emerged while examining the papers, namely *Motivational/Engagement Feedback, Emotion Management, Information Retrieval & Technology and Misc. Others***.** Papers in each focus theme were further explored to identify subcategories relating to the most applied affordances. For example, gamification affordances could include, visual demonstration in-game behaviour consequences, textual feedback, praise or rebuke by in-game characters, forms of ambient audio or optical input and motivational reports. This study of affordances further helped distinguish the research in the main areas of Motivation/Engagement and Emotion Management and Information Retrieval & Technology. The next section details each of the four themes and its subcategories and to emphasise the distinctions made.

A) Motivational/Engagement Feedback. The 11 reviewed papers reported the motivational affordances as forms of motivation and engagement are shown in Table 3. The most applied affordances in most of the studies are forms of *motivation and game engagement*. In paper A4, A8, A16 for *positive intrinsic motivation.*, various motivators techniques are identified such as visual demonstration of in-game behaviour consequences, textual feedback, praise or rebuke by in-game characters, forms of ambient audio or visual motivators.

One paper (A12) applied explicit *performance feedback*. In the form of performance reports. Study A12 and A20 explains *progress visualisation* and reward system. **Simultaneously,** in A31, A32 *technology challenges* and **in A33** *social collaboration* are also identified as gamification affordances and positive outcomes related to motivation and gamification in the domain of information search behaviour.

Table 3. Motivational outcomes

Motivational /Engagement Outcome	Records	Frequency
Motivation & Engagement	A1,A4,A8,A12,A14,A16,A20,A24,A29,A31,A32,A33	12
Intrinsic Motivation	A4,A8,A16	3
Performance Feedback	A12	1
Progress Visualization	A12,A20	2
Technology Challenges	A31,A32	2
Social Networking, Social Collaboration	A33	1

B) Emotion Management. Emotion management in this study refers to regulating emotion in the gamification context where the rule is the effort individuals submit to amplify, maintain, or decrease emotion aspects. It requires a conscious effort to change the emotions following the rules to express emotions specific to the gamification. Eleven of the papers were grouped as focusing on emotion management and were further grouped by one or more emotional outcomes studied (Table 4).

Outcomes related to effects of *behaviour agreement, emotional engagement and perceived enjoyment* of the player (individual user) are the most studied similar context during the emotional engagement. Seven of the records reported (A2, A5, A15, A22, A23, A26, A39) that a gamified system was key in its user's positive attitude, which resulted in effective learning outcome. Six records reported that gamification brings emotional connectivity and engagement when applied to different scenarios, e.g. customer empathy, employee productivity with positive system assignation.

Table 4. Emotional outcome

Emotional Outcome	Records	Frequency
Behaviour Agreement, Attitude	A2,A5,A15,A22,A23,A26, A39	7
Emotional Engagement	A9,A13,A22,A23,A24,A26	6
Enjoyment, Fun, Entertainment, Flow	A2,A5,A15,A39	4
Self-Efficiency	A25,A26	2
Fatigue, Boredom, Arousal	A5	1
User Experience, Ease of use	A2	1
Perceived Vulnerability	A23	1
Perceived Challenge and Gameplay Difficulty	A22	1
Perceived Positivity	A2	1

Four records in A2, A5, A15, and A39 inform that when individuals are pleased with the activities or feel positive emotions while doing gamified tasks, they tend to do a better job. It reflects upon their behaviours (fun, entertainment). In A2, A22, A23 the resulting outcomes are related to user experience, perceived positivity, challenge and vulnerability and occur due to individuals' interaction with their gamified environment. Specifically, this feedback relates to whether their expectations have been met or not. Records A25, A26 specifically identified a positive outcome of users' feeling of self-efficacy, whereas A5 focused on Fatigue or Boredom's negative outcome.

The theme of emotion management relates to individuals' learning as influenced by their emotional states and those game elements make people feel more ownership and purpose when engaging with tasks. The literature reviewed considers gamification's effectiveness as positively impacting the emotions of a 'player'.

C) Information Retrieval & Technology. The third emerging theme while synthesising the literature was *Information retrieval & technology*. In the reviewed manuscripts, Table 5, specified that gamifying Information Retrieval in A3, A10, A18, A27, A35 highlighted gamification phenomenon during system research and human computation games. Outcomes related to e-learning, in A6, A30 & A34 indicated that a gamification is a powerful tool in raising engagement in information system research, e.g. in collaborated systems research, and a bookmarking system. The paper A3 and A36 concluded that gamification elements created a competitive environment and positively influenced the e-learning participation of individuals. Lastly, articles A10 and A18 studied the potential for computation games and gamification in making search tasks fun and that a competitive environment had a positive influence on an online web search.

Table 5. Information retrieval & technology outcome

Information Retrieval & Technology Outcome	Records	Frequency
Information Retrieval	A3,A10,A18,A27,A35	5
Information System Research	A6,A30,A34	3
E-learning	A3,A36	2
Search Tasks	A10,A18	2

D) Misc. Others. The related topics and affordances related to gamification and its ongoing advance research have also been reviewed. The reported manuscripts identified perspectives on the importance of gameful designs and serious games for water sustainability, service marketing, gamification in learning and education. Nine of the records included in systematic review came under this category, as shown in Fig. 2.

4 Discussion and Concluding Remarks

Most of the reviewed studies focusing on the effect and activation of emotion during information seeking approached gamification, by designing with game elements, to enhance serious single-player gaming. The literature grouping of *Emotion Management, Motivational Feedback and Information Retrieval and Technology, identified* gamification affordances. These describe the fragmentation of the literature on gamification identified in the introduction. On the one hand, we see the gamification studies with a core focus of emotions associated with outcomes such as enjoyment and their management in motivational technology/gamification. These studies appear to relate to the deployment of gamification to help complete otherwise mundane or unrewarding tasks. On the other hand, the studies focusing on motivational outcomes, such as feedback on progress or praise which activate emotion associated with influencing and motivating. The papers on information seeking and retrieval and gamification identified how game elements could enhance the search task involved and the emotions experienced and activated by the user(player).

The following key points are identified from this analysis of each of the themes:

- Emotion-based studies focus on how gamification could activate many emotional responses, such as player/user engagement, fun, self-efficiency, perceived challenges and behaviour agreement.
- Motivational studies recognise that by gamifying a task, the flow of emotion activation could be identified. It enhances intrinsic motivation, progress visualisation, motivational feedback and social collaborations.
- Information Seeking/Retrieval studies on gamification branded the affective states of user/player and associated the emotions of task search e.g. in A10 the Library-Tree searching game explained how gameful elements trigger users' intrinsic motivation to increase their interaction with the system. This helps identify the affective states of searching according to [11] such as uncertainty, optimism, confusion/frustration/doubt, clarity, sense of direction/confidence and satisfaction or a sense of disappointment.
- Further analysis of the inter-related gamification fields helps to acknowledge the gamification mechanism, e.g. reward that leads to various enhancement to system and process. The conceptual basis of gamification is exploring new phenomena such as multilevel achievements agreement, virtual reality, and customer enhancements.

In conclusion, research on emotions and motivation in information-seeking tasks provides a bridge for gamification. The affective state in information-seeking tasks is understood to influence technology use and can be activated by gamification outcomes by using gamified technology. All the records that studied gamification in the 'field' were related to the task emotions and information search behaviour. The results implicate that most of the current body of research focuses on emotion frameworks.

Moreover, gamification in information retrieval is mostly aimed at enhancing motivation and individual level (gamifying). The literature reviewed themes and outcome to serve as a foundation for future studies by opening up several research possibilities. This is achieved by identifying several research gaps that need further investigation of the research community. Furthermore, in-depth literary analysis of the exposed under-researched aspects of gamification in the unconventional study of emotions and information-seeking behaviour promises essential insights.

5 Limitations

In this study, the research reviewed various aspects of gamification in the context of emotion and information-seeking behaviour. In the literature search, the critical terms derived from gamification, games with a purpose, serious games, emotions and information seeking are used. Thus, records that have not used the previous terms to describe their research focus were not included even if they studied gamification. Moreover, as the research focused on just emotions-related search terms, records that did not use these terms to describe their focus were not reviewed. The literature search is limited to related databases and although it indexes most relevant databases, using only these databases might have resulted in missing out some relevant records. Moreover, the study analysed the applied affordances based on the descriptions of applied gamification technologies reported by each manuscript's authors. However, it is possible that some of the implemented affordances remained unreported and therefore, not included in this manuscript.

Reviewed Records

A1 Deterding, S.: Gamification: Designing for motivation. Interactions, 19(4), 14- 17 (2012).

A2 S. Deterding, M. Sicart, L. Nacke, K. O'Hara, and D.Dixon, Gamification: using game-design elements in non-gaming contexts. In: 28th Proceedings on Abstracts on Human Factors in Computing Systems, pp. 2425–2428 (2011).

A3 Zhang, P.: Toward a positive design theory: Principles for designing motivating information and communication technology. In: Designing Information and Organisations with a Positive Lens, vol. 2, 0 vols., Emerald Group Publishing Limited, pp. 45–74. (, 2008).

A4 Witt, M., Robra-Bissantz, S.: Sparking Motivation and Creativity with "Online Ideation Games". In: Goltz, U. et al. (eds.) Informatik 2012, Ges. für Informatik, Bonn, p. 1006–1023. (2012).

A5 Jibril, T. A. and Abdullah, M.H.:Relevance of emoticons in computer-mediated communication contexts: An overview, Asian Social Science 9(4): 201–7. (, 2013).

A6 Muntean, C.I.: Raising engagement in e-learning through gamification. In: Proc. 6th International Conference on Virtual Learning ICVL (2011).

A7 Hassan, L.: Governments Should Play Games: Towards a Framework for the Gamification of Civic Engagement Platforms, Simulation & Gaming, vol. 48, no. 2, pp. 249–267 (2017).
A8 Burke, B.: Gamify: How Gamification Motivates People to Do Extraordinary Things, Routledge, Brookline, MA, (2014).
A9 Koivisto, J. and Hamari, J.: Demographic differences in perceived benefits from gamification. In: Computers in Human Behavior, vol. 35, pp. 179–188 (2014).
A10 Broer, J. & Poeppelbuss, J.: Gamification: A new phenomenon in information systems research. In: Proceedings of the 24th Australasian Conference on Information Systems (ACIS) pp. 1–13. RMIT University (2013).
A11 Albertarelli, S. et al.: A Survey on the Design of Gamified Systems for Energy and Water Sustainability, Games, vol. 9, no. 3, pp. 38 (2018).
A12 Gordon, E., Walter, S., & Suarez, P.:Engagement Games: A Case for Designing Games to Facilitate Real-World Action. Boston. (2014).
A13 Oliver, K., Sandra, B., Albrecht, S.: The Effect of Gamification on Emotions. The Potential of Facial Recognition in Work Environments. Lecture Notes in Computer Science, vol. 9169, pp. 489–499, Springer (2015).
A14 Tang, J. and Zhang, P., Exploring the relationships between gamification and motivational needs in technology design, International Journal of Crowd Science, vol. 3, no. 1, pp. 87–103 (2019).
A15 Hamari, J. and Koivisto, J.: Why do people use gamification services?. In: International Journal of Information Management, vol. 35, no. 4, pp. 419–431 (2015).
A16 Hamari, J.: Gamification: Motivations & Effects. Doctoral Dissertation. Aalto University (2015).
A17 Huotari, K. and Hamari, J.: A definition for gamification: anchoring gamification in the service marketing literature, Electron Markets, vol. 27, no. 1, pp. 21–31 (2015).
A18 Pe-Than, E. P. P., Goh, D. H.-L. and Lee, C. S.: Making work fun: Investigating antecedents of perceived enjoyment in human computation games for information sharing.In Proceedings Computers in Human Behavior, vol. 39, pp. 88–99 (2014).
A19 Landers, R. N., Tondello, G. F., Kappen, D. L., Collmus, A. B., Mekler, E. D., and Nacke, L. E.: “Defining gameful experience as a psychological state caused by gameplay: Replacing the term ‘Gamefulness’ with three distinct constructs,” International Journal of Human-Computer Studies, vol. 127, pp. 81–94. (, 2019).
A20 Deterding, S.: The lens of intrinsic skill atoms: A method for gameful design. In: Human-Computer Interaction, vol. 30, no. 3–4, pp. 294–335 (2015).
A21 Tettegah, Y. and Espelage, D.: In Emotions and Technology, Emotions, Technology, and Behaviors, Academic Press, ISBN 9780128018736 (2016).
A22 Hamari, J., & Keronen, L., Why do people play games? A Meta-Analysis. International Journal of Information Management, 7(3), 125–141 (2017).
A23 Chou, Y.: Yu-kai Chou: Gamification & Behavioral Design. https://yukaichou.com/, last accessed 2020/06/20 (2016).
A24 Sailer, M., Hense, J. U., Mayr, S. K., and Mandl, H.: How gamification motivates: An experimental study of the effects of specific game design elements on psychological need satisfaction, Computers in Human Behavior, vol. 69, pp. 371–380 (2017).

A25 Wolf, T., Weiger, W. H. and Hammerschmidt, M.: Gamified services: How gameful experiences drive customer commitment, in GamiFIN Conference, Pori, Finland, p. 75–82 (2018).
A26 Du, J., Xu, J., Liu, F., Huang, B., Li, Z.: Factors Influence Kindergarten Teachers: Emotion Management in Information Technology: A Multilevel Analysis Asia-Pacific Education Researcher, 28 (6), pp. 519–530 (2019).
A27 Kazai, G., Hopfgartner, F., Kr-Auschwitz, U. and Meder, M.: In: E3CIR work-shop on gamification for information retrieval (gamifir'15). SIGIR (2016).
A28 Huotari, K., & Hamari, J.: Defining gamification: A service marketing perspective. In: Proceedings of the 16th International Academic MindTrek Conference, ACM,(pp. 17–22) (2012).
A29 He, J., Bron, M., Azzopardi, L., de Vries, A.: Studying user browsing behaviour through gamified search tasks. In: Proc. GamifIR '14. ACM (2014).
A30 Meder,M.,Plumbaum,T. and Hopf-Gartner, F.: A gamified enterprise bookmarking system. In: Advances in Information Retrieval 36th European Conference on IR Research, ECIR, Amsterdam (2014).
A31 Mekler, E. D., Brühlmann, F., Tuch, A. N. and Opwis, K.: Towards understanding the effects of individual gamification elements on intrinsic motivation and performance," Computers in Human Behavior, vol. 71, pp. 525–534 (2017).
A32 Hassan, L., Dias, A. and Hamari, J.: How motivational feedback increases user's benefits and continued use: A study on gamification, quantified-self and social networking. In: International Journal of Information Management, vol. 46, pp. 151–162 (2019).
A33 Hamari, J., Hassan, L. and Dias, A.: Gamification, quantified-self or social networking? Matching users' goals with motivational technology. In: User Modeling and User-Adapted Interaction, vol. 28, no. 1, pp. 35– 74, Mar. 2018.
A34 Andez-Luna, J.M., Huete, J.F., Rodr__guez-Avila, H., Rodr__guez-Cano, J.C.: Enhancing collaborative search systems engagement through gamification. In: Proc. GamifIR '14. ACM (2014).
A35 Munro, K. and Talbot, C.: Librarygame –evaluating gamification as a means of increasing customer engagement. In38th ELAG Conference (2014).
A36 Liu, D., Santhanam, R. and Webster, J.: Toward Meaningful Engagement: A Framework for Design and Research of Gamified Information Systems. In MIS.Quarterly, vol. 41, no. 4, pp., 1011–1034 (2017).
A37 Hassan, L.: Means to Gameful Ends: How Should Gamification Be Designed? (2018).
A38 Tang, J. & Jia, Y. & Zhang, P.: Using Gamification to Support Users' Adoption of Contextual Achievement Goals. (2020).
A39 Landers, R. N.: Developing a gamified learning theory: Linking serious games and gamification of learning, In Simulation & Gaming, vol. 45, no. 6, pp. 752–768 (2014).
A40 Herzig, P., Ameling, M., Wolf, B., & Schill, A.: Implementing gamification: Requirements and gamification platforms. In: Gamification in Education and Business, pp. 431–450, Cham: Springer International Publishing (2015).

References

1. González, A.M.: The recovery of emotion in philosophy and social sciences. In: Lecture at the Università Cattolica del Sacro Cuore (Scuola di Dottorato in Scienze Sociali), Milan (Italy) (2013)
2. Koivisto, J., Hamari, J.: The rise of motivational information systems. In: Turner, J.H., Stets, J.E. (eds.) A review of gamification: Handbook of the Sociology of Emotions. Springer, New York (2016)
3. Huotari, K., Hamari, J: A definition for gamification: anchoring gamification in the service marketing literature. Electron. Mark. **27**(1), 21–31 (2017)
4. Lewis, M., Havilland-Jones, J., Feldman Barret, L. (eds.): Handbook of Emotions-3rd edition. The Guilford Press, New York (2008)
5. Clough, P.T., Halley, J.: The Affective Turn: Theorising the Social, Durham. Duke University Press, NC (2007)
6. Gross, J.J.: Emotion and emotion regulation: Personality processes and individual differences. In: John, O.P., Robins, R.W., Pervin, L.A. (eds.) Handbook of Personality: Theory and Research, pp.701–724. The Guilford Press (2008)
7. Deterding, S., Dixon, D., Khaled, R., Nacke, L.: From game design elements to gamefulness: defining "gamification". In: Proceedings of the 15th International Academic MindTrek Conference: Envisioning Future Media Environments, MindTrek, pp. 9–15 (2011).
8. Hamari, J.: Gamification, Wiley Blackwell Encyclopedia of Sociology, https://onlinelibrary.wiley.com/doi/abs/10.1002/9781405165518.wbeos1321. Accessed 21 Mar 2020
9. Connolly, T.M., Boyle, E.A., MacArthur, E., Hainey, T., Boyle, J.M.: A systematic literature review of empirical evidence on computer games and serious games. Comput. Educ. **59**(2), 661–686 (2012)
10. Oinas-Kukkonen, H., Harjumaa, M.: Persuasive systems design: Key issues, process model, and system features. Commun. Assoc. Inf. Syst. **24**(1), 485–500 (2012)
11. Kuhlthau, C.C.: Seeking meaning: a process approach to library and information services. 2nd. edn. CTLibraries Unlimited, Westport (2004
12. Nahl, D., Bilal, D.: Information and emotion: the emergent affective paradigm in information behaviour research and theory. Medford, N.J: Information Today.rterly **39**(2), 239–252 (2007)
13. Kankanhalli, A., et al.: Gamification: a new paradigm for online user engagement. In: Huang, M.H., Piccoli, G., Sambamurthy, V. (eds.) Proceedings of the 33rd International Conference on Information Systems (ICIS). Shanghai, p. 10 (2012)
14. Fettke, P.: Wirtschaftsinformatik **48**(4), 257–266 (2006). https://doi.org/10.1007/s11576-006-0057-3

The Dark Side of Personalization Recommendation in Short-Form Video Applications: An Integrated Model from Information Perspective

Jing Li[1], He Zhao[2], Shah Hussain[1], Junren Ming[3], and Jie Wu[1](✉)

[1] Sun Yat-Sen University, Guangzhou, Guangdong, China
18202770400@163.com
[2] Jilin Agricultural Science and Technology University, Jilin, Jilin, China
[3] Wuhan Institute of Technology, Wuhan, Hubei, China

Abstract. Based on the psychological reactance, this study tries to explore the dark side and grey role of the personalization recommendation system of short-form video application in understanding the discontinuance behavior. Specifically, two major depressing consequences of the personalization recommendation system are proposed, namely, privacy concerns and perceived information narrowing. Specifically, personalization recommendation system of short-form video App has significant positive influence on both privacy concern and perceived information narrowing. Besides, the empirical study shows perceived information narrowing is positively related to psychological reactance. However, personalization recommendation system does not lead to discontinuous usage behavior through privacy concerns or perceived information narrowing. Although personalization recommendation has not an indirect effect on discontinuous usage behavior, personalization recommendation has a potential risk to create psychological pressure on users, making personalized recommendations counterproductive. This study renders new insights on the dark side of the personalization recommendation system and provides practical suggestions for short-form video application providers.

Keywords: Personalization recommendation system · Privacy concern · Perceived information narrowing · Psychological reactance · Discontinuous usage behavior

1 Introduction

In recent decades, companies strive to develop and produce various personalization recommendation systems (PRS) and personalized services to their customers, and to overcome the user's needs, the best example is Amazon App's personalized book recommendations [1, 2]. Computer science engineers and scholars are committed to collect users' behavior data, through various kinds of optimized algorithms and advanced analysis techniques, which present more accurate recommendations. However, the personalized recommendation would not be a bonus for the users and may cause them

K. Toeppe et al. (Eds.): iConference 2021, LNCS 12646, pp. 99–113, 2021.
https://doi.org/10.1007/978-3-030-71305-8_8

unfavorable beliefs to the contrary [1, 3–7]. For instance, as PRS hugely relied on the data collection of users' personal information (e.g. personal identifiers and biographical information), privacy concern may be the foremost reason for consumer reactance [3, 6–8]. Besides, Pariser also warns that high personalization levels limit the diversity of material one sees and resulting in reduced creativity [9]. Although there are potential negative effects of PRS, scant studies have systematically investigated the dark side of PRS from information perspective by integrating the privacy concern and psychological reactance theory. Furthermore, it is unknown whether a negative impact caused by PRS can become a progressively spreading problem or even a phenomenon of user's decline. It may be interesting to find if there is a relationship between users' discontinuance behavior related to the negative impact of PRS.

In the current decade, PRS is widely applied in short-form video application (App). The short-form video gives provision to the users for quickly and easily creating and uploading 15-s videos to share with their friends and family or the whole universe of the internet. The App developers estimate the user's preference through algorithm by tracking the videos click and display those videos that match the user's preference of watching. This article chooses TikTok as the research object, because the APP has a very prominent personalized recommendation function. In the previous interviews, many users talked about being affected by the personalized recommendation function. The characteristics of the personalized recommendation function can be clearly perceived by the users, so it can be regarded as "Personalized recommendation added short video App". High personalization level, especially customize contents and special effect filters, reduces the users' choice cost and information load, but consequently has aroused public concern over the risk of excessive use [10]. Preliminary studies highlighted over-customized content of the App may result in negative feelings and behavior [1, 2]. Like the Chinese short-form video App, namely, TikTok for example, the statistics show that TikTok has been downloaded nearly 800 million times worldwide but approximately 22% of TikTok users have used for some time, then they deleted the App and doesn't use it anymore [11]. Inevitably, this discontinuous usage behavior of users will generate challenges for the App providers because of the decline in consumer's interest and high cost of developing and presenting Apps on mobile devices, thus the roller coaster phenomenon reflects users' negative consequences. That's what considerable attention is required for the production of mechanism, that what are the negative consequences which can be derived from the usage of the PRS.

Therefore, this study conducts an empirical study to examine this issue. Specifically, two research questions are addressed in this study. First, from the perspective of psychological reactance theory, we would like to know how PRS leads to users' perceived information narrowing, which causes an influence on users' discontinuous usage behavior. We will measure the real usage behavior rather than the usage intention or attitude. In the second one, we would like to dig out how PRS influenced privacy concern, which leads to influence users' discontinuous usage behavior. Our study gives theoretical implications for the discipline as well as new insights about the personalized recommendation for practitioners. To the best of our knowledge, this will be the first study not only to propose an integrated model but also systematically investigate the dark side of personalized recommendation from both the privacy concern and perceived information narrowing angles in Apps simultaneously.

The rest of this study is organized as follows. First, we present an overview of PRS and the theory of psychological reactance and then develop hypotheses. Next, we delineate the research methodology, carry out the data analysis, and conclude the results. Finally, we exhibit our findings and implications for both researchers and professionals.

2 Literature Review and Theoretical Background

2.1 Overview of PRS

To solve the problem of information flood and data overload, system developers introduced a solution which named as PRS [12]. This solution can proactively and smartly suggest items of interest for the consumers based on their objective behavior or explicitly searched preferences [13]. PRS, depends on various types of algorithms that can be classified into four categories: content-based filtering, collaborative filtering, knowledge-based filtering, and hybrid filtering [14–17]. The Amazon website utilizes collaborative filtering, in which the list of recommended items would include a set of books and music that are supposed to be of interest to the user who can browse them anytime anywhere using a smartphone. Furthermore, current PRS is in App generally calculate on collaborative filtering systems [17].

2.2 Privacy Concern

Various scholars defined privacy as a process of anonymity preservation, which is strongly connected with control over one's information [18, 19]. Although PRS helps provide tailored services, the process of collecting users' data and the use of algorithm or techniques to analyze would make users uncomfortable in the way they presumed the invasion of their data by an unauthorized handling of people or unknown third-party organizations, and thus loss over control on their data [17]. The higher degree of seriousness the privacy violation, the stronger would be the users' anxiety. According to the Protection motivation theory, the awareness of "my personal information was threatened and was in the risk" can motivate individuals' protection response: to bound others from accessing personal information [18, 20, 21]. As matter of fact to decrease the use of PRS of App means to abandon the use of App, it refers to individuals' protection response. The previous studies also reported that the higher level of privacy concern more likely motivates individuals to adopt actions to reduce risks [22–24].

2.3 Perceived Information Narrowing

Perceived information narrowing is firstly proposed by Sunstein [25]. In Republic. Com, he described that people tend to selectively associate with individuals of similar viewpoints and consume information appealing to their perspectives, thus resulting in a homogeneous material one sees and information cocoons [25]. Sunstein also presented empirical studies and demonstrated that PRS of the system of communications on one hand has greatly increased people's ability to filter what they want to read, see, and hear, but on the other hand, such unlimited filtering and years of information assimilation would largely decrease the range of possible choices [25]. The information cocoon room increases the difficulty and trouble of communication between different groups, limits the scope for individuals to receive specific information, thereby continuously enhancing current interests and opinions, and hindering the possibility of accepting other opinions [26]. Due to easier selectivity online, individuals prefer to intake attitude-consistent messages and thus exhibit a confirmation bias [27]. According to thc present investigations, Sude considers when the time allocated to attitude-consistent messages exceeds time allocated to attitude-discrepant messages, even though both messages types are available to the same extent, this pattern indicates a confirmation bias [28]. When people are in the information cocoon that they construct, the information society and environment will become more closed, just like in the "echo chamber". Hundt used the word "echo chamber effect" to indicate that in a closed system environment, repeated and recurring dissemination of information can lead to people's ideas to constantly magnifying and consolidated, the absence of critical discussions and a controlled set of opinions about a topic [29]. Negroponte also points out that PRS successfully helps users to constitute "the daily me", but leads them to lose the freedom for information encountering and serendipity [30]. Other similar studies highlighted this phenomenon and named it "filter bubbles". While PRS in mobile Apps could help to filter information and provide the accurate information congruent with users, it could also foster the users' proneness to tunnel vision when they search and process information. In this paper, we use the concept of perceived information narrowing to describe the psychological state after a long time of used PRS.

2.4 Psychological Reactance Theory

Psychological reactance theory (PRT) concerned with how individuals react when their liberty to choose is limited [31, 32]. The magnitude of reactance is posited to be a direct result of how much individuals are aware that they have the freedom to engage in that particular behavior [31, 32]. Previous research has shown that reactance is positively correlated with consumers' contractor behavior, such as the increased attraction for non-recommended products [33], to ignore recommendations [4] or limit the intention to usage service [3]. PRT has been applied to a diverse array of research fields including online advertising [34, 35], the persuasive communication [36, 37], and product or service recommendations [33, 38, 39]. The current article invokes psychological reactance theory to explain the effects that PRS have on the discontinuous usage behavior.

3 Research Hypotheses

PRS is a technique and methodology to solve the problem of information overload in the internet and cyber environment, short-form video App engineers generate a list of recommendations for users to help them find similar short-form videos of their interest. However, personalization has a mixed effect on users, and that personalized services may lead users to negative reactions [40]. Some empirical studies show that users may presume this personalized recommendation phenomenon negatively as an invasion of privacy [41], and users' disapproval may become a major obstacle for the continuous use of short-form video Apps. Other studies show that personalized recommendation technology reduces the possibility of the occurrence of diverse information, which will lead users to receive a large number of homogeneous information and the user's antipathy [42]. Summing up, this study makes the following hypotheses.

H1. PRS of short-form video App is positively related to privacy concerns.
H2. PRS of short-form video App is positively related to perceived information narrowing.

Although information is recommended and referred according to the user needs, inductive information without user authorization and permission can obstruct the user's normal decision-making and information search process. Taking a business scenario for an example, consumers do not like to buy recommended commodities, which may lead to them feeling anger [43]. High volumes of the same short-form videos may also make users feel invariable and boring. If the methodology of pushing this personalized recommendation information is obvious and single, other sounds have actively deteriorated [44]. The users may think that the application is trying to influence the strategies and methods of their decision which will create psychological reactance. Therefore, this study makes the following hypothesis.

H3. Perceived information narrowing is positively related to psychological reactance.

Recommendation Systems rely on users' cognitive aspects included personality, behavior, and attitude [45]. Through a psychological lens, excessive inducement and persuasion can threaten or limit people's freedom of decision-making, and lead to psychological reactance [3]. The performance of users' psychological reactance to personalized recommendation information is mainly the discontinuous usage behavior. The concept of discontinuous usage was originated from the field of medicine, it was originally used to study the discontinuous usage of drugs [46]. Later on, some scholars introduced it into the field of information systems to study the termination of information systems. For example, researchers studied that, the final stages of the information system life cycle, exploring the discontinuity of information systems from a deeper and digger perspective, using semi-structured interviewing methods to summarize and determine the significant impact on the intent of information system disruption [47]. In addition to that, scholars have also studied discontinuous usage in other modes, such as the discontinuous usage of consumer products [48], the discontinuous usage of voice-activated Intelligent Personal Assistants [49], the users' discontinuous usage in social network context [50], and users' cessation or reduction of Facebook usage due to psychological and behavioral consequences [51]. When also know that the unsustainable usage caused by psychological reactance is worthy of attention. Therefore, this study proposes another hypothesis.

H4. Psychological reactance is positively related to discontinuous usage behavior.

For the purpose to improve the accuracy and precision of recommendation, short-form video App providers collect more personalized information from users when providing a personalized recommendation. Concern over privacy matters, the user's curiosity increases more and more when they are learning to take care of whether their personal information is being collected in extraordinary ways. With the advent of internet technology, the unrestricted nature of the cybersecurity has made privacy a major concern for all online activities [52]. Studies show that present concern about information privacy can affect the usage of applications. When customers become aware of privacy concerns, the click-through ratio drops dramatically [6]. User's anxiety about personal privacy leads them to more and more cautious privacy disclosure behavior, which shows a further reduction in usage behavior [53]. Therefore, this study generates another hypothesis.

H5. Privacy concern is positively related to discontinuous usage behavior.

Summarizing the above literature review and hypotheses, we develop and construct our conceptual model as shown in Fig. 1. Based on psychological reactance and privacy concerns, we discover the impacts of personalized recommendations enabled by short-form video Apps on users' discontinuous usage behavior.

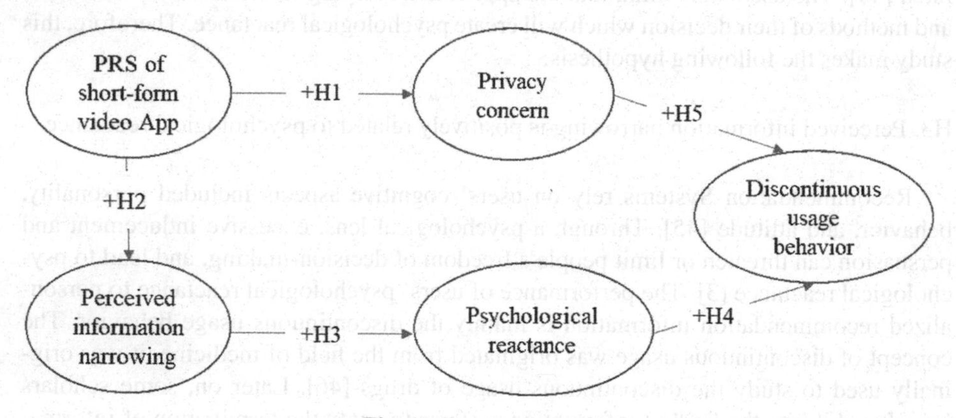

Fig. 1. Structural model

4 Research Methodology

4.1 Research Design

All the items were adapted from the existing literature. Specifically, item of PRS of short-form video App was adapted from Pappas et al. [54], item of Privacy concern was adapted from Huang et al. [55], item of Psychological reactance was adapted from Chen et al. [7], item of Discontinuance usage behavior was adapted from Maier [56], item of Perceived information narrowing was adapted from Du et al. [57]. As the sample comes from China, we translated the questions into Chinese. All the items were measured using

Table 1. Measures of construct

Construct	Items
PRS of short-form video App	**PRS1.** Short-form video Apps can provide and show me with personalized videos tailored to my activity context
	PRS2. Short-form video Apps can provide me with more relevant promotional information tailored to my preferences or personal interests
	PRS3. Short-form video Apps can provide me with the kind of videos that I might like
Privacy concern	**PC1.** I am concerned that too much personal information is collected when I use short-form video Apps
	PC2. I have doubts over how sound my privacy is protected when I use short-form video Apps
	PC3. My personal information could be subject to misuse and unauthorized access when transacting through short-form video Apps
Perceived information narrowing	**PIN1.** The type of information that short-form video Apps recommend to me is relatively single, which makes me miss other types of information
	PIN2. It is more like similar information to me when I use short-form video Apps
	PIN3. Long-term use of the personalized information recommended by short-form video Apps makes my options limit and limit
Psychological reactance	**PR1.** The usage of personalized recommendations on short-form video Apps is forced upon me
	PR2. The usage of personalized recommendations on short-form video Apps is unwelcomed
	PR3. The usage of personalized recommendations on short-form video Apps is interfering
	PR4. The usage of personalized recommendations on short-form video Apps is intrusive
Discontinuous usage behavior	**DIS1.** I use other alternatives to short-form video Apps
	DIS2. I use the alternatives to short-form video Apps for ore than two weeks
	DIS3. I already have one or more alternatives to short-form video Apps

a five-point Likert scale, ranging from (1) strongly disagree to (5) strongly agree. In the first part of the survey, we construct a screening option to ask respondents if they still use the App (the valid answer is the "No"). The content, terminology and language of the survey were evaluated with the help of two experts. Table 1 shows the items and constructs.

4.2 Sample and Data Collection

In order to measure the real usage behavior and access to the target sample, we contacted with teachers from different university departments and requested them to distribute the link for the online survey to students in their classes. Students were assured about the confidentiality of their personal information and were requested by their teachers to volunteer in this study by completing a questionnaire that explored their short-form video Apps usage. To increase the response rate, the researchers conducted a second round of follow-up with their colleagues. The questionnaire was distributed in 2019 for over two months. The link to the questionnaire was distributed among 300 students, out of the 275 who returned the questionnaires. A total of 27 responses were discarded and deleted due to incomplete information and answered "Yes" to the question of still use the App, resulting in a final sample consisting of 248 valid responses.

5 Data Analysis

For data analysis, we applied partial least squares (PLS) as the structural equation modeling approach with SmartPLS 3.0 software [58]. PLS was chosen to validate the conceptual model for it, as it doesn't require multivariate normality of the data [59] and was fit for the exploration study [58].

5.1 Measurement Model

At first, we assessed the reliability of the measurement items by examining the values of Cronbach's α and composite reliability (CR). As presented in Table 2, the scores of Cronbach's α and CR for all the constructs are higher than the criterion values of 0.7 suggested by Fornell and Larcker (1981) [60], which indicates that all the constructs have good reliabilities. Next, we evaluated convergent validity and discriminant validity. The average variance extracted (AVE) values of each construct are higher than 0.5 and the loading values of all items are above 0.7, suggesting that all the constructs have good convergent validity [60]. Table 3 shows the correlations between constructs and square roots of AVEs. We can see that the square root of each construct's AVE is larger than its correlations with other constructs, suggesting sufficient discriminant validity [61].

Table 2. Overview of the measurement model

Construct	Item	Outer loading	Cronbach's Alpha	CR	AVE
PRS of short-form video App	PRS 1	0.925	0.944	0.965	0.902
	PRS 2	0.963			
	PRS 3	0.960			
Privacy concern	PC 1	0.953	0.939	0.961	0.892
	PC 2	0.930			
	PC 3	0.950			
Perceived information narrowing	PIN 1	0.932	0.927	0.954	0.873
	PIN 2	0.919			
	PIN 3	0.951			
Psychological reactance	PR 1	0.887	0.937	0.956	0.843
	PR 2	0.920			
	PR 3	0.927			
	PR 4	0.939			
Discontinuous usage behavior	DIS 1	0.921	0.912	0.945	0.851
	DIS 2	0.940			
	DIS 3	0.906			

Table 3. Correlations between constructs and square roots of AVEs

Construct	DIS	PRS	PR	PIN	PC
DIS	**0.923**				
PRS	0.427	**0.950**			
PR	0.263	0.250	**0.918**		
PIN	0.246	0.524	0.632	**0.934**	
PC	0.240	0.526	0.475	0.711	**0.944**

Note: DIS = Discontinuous usage behavior; PRS = PRS of short-form video App; PR = Psychological reactance; PIN = Perceived information narrowing; PC = Privacy concern. The diagonal elements represent the square root of AVE

5.2 Structure Model

Figure 2 displays standardized path coefficients. Overall, this study examines five hypotheses, and the results support most of them. Specifically, PRS of short-form video App had significant positive influences on both privacy concern ($\beta = 0.526$, $p < 0.001$)

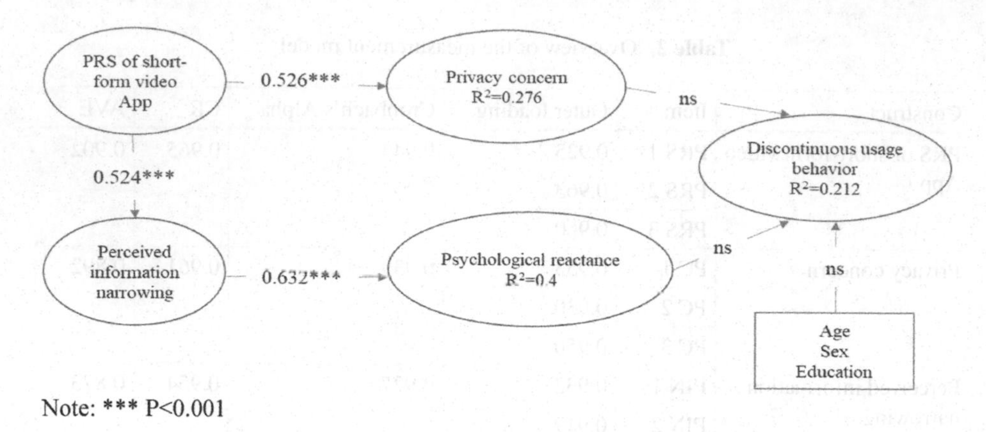

Fig. 2. Results of the structure model

and perceived information narrowing ($\beta = 0.524, p < 0.001$), thus H1 and H2 were supported. Perceived information narrowing was positively related to psychological reactance ($\beta = 0.632, p < 0.001$), but not related to discontinuous usage behavior ($\beta = 0.186$, $p > 0.05$). Thus, H3 was supported, but H4 was not. To our surprise, although privacy concern has a negative but non-significant relationship with discontinuous usage behavior ($\beta = -0.06, p > 0.05$), suggesting H5 was not supported. Besides, we also check the influence of control variables, including age, sex and education, but none of them has a significant effect. Furthermore, we calculated the explained variances. Specifically, it shows discontinuous usage behavior is 0.212, and privacy concern is 0.276. We give more detailed discussions in the following section.

6 Discussion and Conclusion

6.1 Discussion

The aim of this study is to explore and dig out the negative effects of the PRS of short-form video App. This study is particularly important in the way because it is one of the first studies to take the PRS of smartphone App as the research object. Specifically, this study supplements the existing research on information cocoon room based on content intelligent distribution platform from the perspective of PRS of short-form video App, to determine and analyze the impact of PRS on users' discontinuous usage behavior [62]. To the best of our knowledge, this is the first study to propose and empirically two major depressing consequences of the personalization recommendation system. Although prior research acknowledges that long time of PRS could make people boring, neither it is the role of PRS that not enough to lead to the discontinuous usage behavior (indirect effect), nor the direct influence of privacy concern and psychological reactance expressed.

Based on the information cocoons theory, this study highlighted on the construct of Perceived information narrowing, and proved how the PRS of short-form video App induces the perceived information narrowing which in turn causes users' psychological reactance. The study strongly supported the hypotheses, that is a negative psychological state (such as perceived information narrowing) could generate after a long time of used PRS.

Based on the Privacy concern theory, this study reconfirmed PRS of short-form video App induces and highlight users' privacy concerns. It is confirmed with the previous studies [63]. Surprisingly, privacy concern is not predicted by the discontinuous usage behavior, which is possibly explained by the privacy paradox. According to Norberg et al., the privacy paradox is the relationship between individuals' intentions and disclosure behavior either to disclose fake personal information or their actual personal information. In the Internet environment, users want to know their privacy will be revealed, but would rather sacrifice privacy in exchange for services. In this context, users are aware of the risk of information leakage when the use of short-form video App, but haven't a strong willingness to stop usage behavior. We can see privacy paradox plays its pivotal role. From the empirical study, it shows both privacy concerns and psychological reactance have no significant effect on discontinuous usage behavior, which clearly indicated PRS of short-form video App has no indirect effect on discontinuous usage behavior. There are two possible explanations. One is that the perceived information narrowing is an unconscious state of mind, thus the negative effect of personalization recommendation system in short-form App is not fully perceived by users and does not necessarily lead to users' discontinuous usage behavior. The other possible explanation is that PRS in short-form App is not enough to cause users' discontinuous usage behavior, there are other more critical factors, such as exhaustion or technostress [64].

6.2 Theoretical Implications

This research has several theoretical contributions. First, instead of focusing on the positive side of PRS, our study investigated the users of short-form video Apps who have a suspension behavior and research on the dark side of using PRS, which was previously been ignored. With the increasing App developers are crazy about PRS and the recommended algorithm, the dark side of PRS becomes more and more salient and quashed, and also has not raised enough attention in existed literature. However, in this study, we verified two types of negative consequences that resulted from PRS of short-form video Apps, through which, this research complements previous studies, and contributes to a more comprehensive understanding of the dark side of PRS developer.

Second, based on the information cocoons theory, we contributed to provide an understanding of how the PRS narrowing the user's vision which further leads to the users' negative psychological reactions. Previous studies also highlighted such negative psychological state [25, 29], but lack of a deep understanding of the concept, as well as the analysis of antecedents and outcomes.

Third, this study investigated discontinuance usage behavior rather than attitude or intention, thus complements previous studies that have mainly focused on the behavior intention [65].

6.3 Practical Implications

This study indicated that the App engineers should consider the disadvantages of personalized recommendation, pay strong attention to users' psychological reactions, and provide personalized recommendation services appropriately. As we pointed out, while personalized services are usually convenient, easy to use, and accurate, users may be psychologically resistant to them, making personalized recommendations counterproductive. Besides, with the use of short-form video more and more in people's daily life, perceived privacy concern will be increasing. It is also a potential risk to cause users' reject using [63], as a direct influence between PRS and discontinuous usage behavior could be verified in the future study. Apps providers thus should prevent the centralization of information and increase the diversification of recommendation information. Personalization would not be a bonus but the fuse of the abandoned behavior, which should be taken into account by App designers and their companies. Therefore, these Apps need to be rigorously reviewed and continuously controlled.

6.4 Limitation and Future Research

This study has some limitations that provide direction for future research. Firstly, this study does not discuss the effect of types of privacy on psychological reactance and discontinued usage. Future studies may go into deep the impact of privacy types. Secondly, the study collects data only in China, and its results should be studied and tested in other cultures. Thirdly, the samples are focused on the student user groups, thus the impact of other types of user groups is not discussed. To overcome these shortcomings, future research should focus on and consider the moderate impact of different occupations and experience groups.

References

1. Nunes, P.F., Kambil, A.: Personalization? No thanks. Harv. Bus. Rev. **79**(4), 32–33 (2001). https://doi.org/10.1111/1468-0440.00114
2. Piggot, J.: Micro-Segmentation and Personalization in Information Systems in the Financial Service Industry. (2015). https://doi.org/10.13140/RG.2.1.3073.9366
3. Lee, G., Lee, W.J.: Psychological reactance to online recommendation services. Inf. Manage. **46**(8), 448–452 (2009). https://doi.org/10.1016/j.im.2009.07.005
4. Fitzsimons, G., Lehmann, D.: Reactance to recommendations: when unsolicited advice yields contrary responses. Market. Sci. **23**(1), 82–94 (2004). https://doi.org/10.1287/mksc.1030.0033
5. Berk, M., Blank, J., Daniels, D., Schatsky, D.: Beyond the personalization myth: cost-effective alternatives to influence intent. Jupiter Research Site Technologies and Operations (2003). 2
6. Aguirre, E., Mahr, D., Grewal, D., Ruyter, K.D., Wetzels, M.: Unraveling the personalization paradox: the effect of information collection and trust-building strategies on online advertisement effectiveness. J. Retail. **91**(1), 34–49 (2015). https://doi.org/10.1016/j.jretai.2014.09.005
7. Chen, Q., Feng, Y.Q., Liu, L.N., Tian, X.: Understanding consumers' reactance of online personalized advertising: a new scheme of rational choice from a perspective of negative effects. Int. J. Inf. Manage. 44(FEB), 53–64(2019). https://doi.org/10.1016/j.ijinfomgt.2018.09.001.
8. Newell, S., Marabelli, M.: Strategic opportunities (and challenges) of algorithmic decision-making: a call for action on the long-term societal effects of 'datification.' J. Strat. Inf. Syst. **24**(1), 3–14 (2015). https://doi.org/10.1016/j.jsis.2015.02.001

9. Pariser, E.: The Filter Bubble: How the New Personalized Web Is Changing What We Read and How We Think. Penguin, London (2011)
10. Zhang, X., Wu, Y., Liu, S.: Exploring short-form video application addiction: socio-technical and attachment perspectives. Telemat. Inform. **44**(SEP), 101–121 (2019). https://doi.org/10.1016/j.tele.2019.101243.
11. O'Connell, C.: 24% of Users Abandon an App After One Use (2017). https://info.localytics.com/blog/24-of-users-abandon-an-app-after-one-use.
12. Hossein, A., Rafsanjani, N., Salim, N., Aghdam, A.R., Fard, K.B.: Recommend. Syst. Rev. **3**(5), 47–52 (2013)
13. Pu, P., Chen, L., Hu, R.: Evaluating recommender systems from the user's perspective: survey of the state of the art. User Model. User-Adap. Interact. **22**(4–5), 317–355 (2012). https://doi.org/10.1007/s11257-011-9115-7
14. Jannach, D., Zanker, M., Felfernig, A., Friedrich, G.: Recommender Systems: An Introduction. Cambridge University Press, Cambridge (2011)
15. Yue, S., Larson, M., Hanjalic, A.: Collaborative filtering beyond the user-item matrix: a survey of the state of the art and future challenges. ACM Comput. Surv. **47**(1), 1–45 (2014). https://doi.org/10.1145/2556270
16. Su, X., Khoshgoftaar, T.: A survey of collaborative filtering techniques. Adv. Artif. Intell. **2009**(12), 1–9 (2009). https://doi.org/10.1155/2009/421425
17. Pimenidis, E., Polatidis, N.,Mouratidis, H.: Mobile recommender systems: identifying the major concepts. J. Inf. Sci. 165–176 (2018). https://doi.org/10.13140/RG.2.2.24011.08488
18. Portilla, I.: Privacy concerns about information sharing as trade-off for personalized news. El profesional de la informacíón **27**(1), 19–26 (2018). https://doi.org/10.3145/epi.2018.ene.02
19. Taddei, S., Contena, B.: Privacy, trust and control: Which relationships with online self-disclosure? Comput. Hum. Behav. **29**(3), 821–826 (2013). https://doi.org/10.1016/j.chb.2012.11.022
20. Posner, R.: The economics of privacy. Am. Econ. Rev. **71**(2), 405–409 (19 81). https://doi.org/10.2139/ssrn.2580411.
21. Mai, J.: Three models of privacy: new perspectives on informational privacy. Nordicom Rev. **37**, 171–175 (2016). https://doi.org/10.1515/nor-2016-0031
22. Youn, S.: Determinants of online privacy concern and its influence on privacy protection behaviors among young adolescents. J. Consum. Aff. **43**(3), 389–418 (2009). https://doi.org/10.1111/j.1745-6606.2009.01146.x
23. Mohamed, N., Ahmad, I.: Information privacy concerns, antecedents and privacy measure use in social networking sites: evidence from Malaysia. Comput. Hum. Behav. **28**(6), 2366–2375 (2012). https://doi.org/10.1016/j.chb.2012.07.008
24. Li, Y.: Theories in online information privacy research: a critical review and an integrated framework. Decis. Support Syst. **54**(1), 471–481 (2012). https://doi.org/10.1016/j.dss.2012.06.010
25. Sunstein, C.R.: Republic. com. Princeton University Press. Princeton (2002)
26. Chen, T.: Opportunities, anomalies and governance of new media under capital logic: case study of wechat subscription "Mi Meng". In: 2nd International Symposium on Social Science and Management In-novation (SSMI 2019). Atlantis Press. https://doi.org/10.26914/c.cnkihy.2019.048893
27. Bennett, W.L., Lyengar, S.: A new era of minimal effects? The changing foundations of political communication. J. Commun. **58**(4), 707–731 (2010). https://doi.org/10.1111/j.1460-2466.2008.00410.x
28. Sude, D.J., Pearson, G.D.H., Knobloch-Westerwick, S.: Journal pre-proof Self-expression just a click away source interactivity impacts on confirmation bias and political attitudes. Comput. Hum. Behav. **114**, 2020. https://doi.org/10.1016/j.chb.2020.106571.

29. Hundt, M., Schneider, B., Mennatallah, E.F., Daniel, A.K., Alexandra, D.: Visual analysis of geolocated echo chambers in social media. In: EuroVis 2017 Eurographics/IEEE VGTC Conference on Visualization 2017, pp. 125–128 (2017). https://doi.org/10.2312/eurp.20171185.
30. Negroponte, N.: Being Digital. Alfred A. Knopf, New York(1995)
31. Chandler, T.: Why discipline strategies are bound to fail. Clear. House J. Educ. Strat. Issues Ideas **64**(2), 124–126 (1990). https://doi.org/10.1080/00098655.1990.9955826
32. Brehm, J.W.: A Theory of Psychological Reactance. Academic Press, New York (1966)
33. Kwon, S., Chung, N.: The moderating effects of psychological reactance and product involvement on online shopping recommendation mechanisms based on a causal map. Electron. Commerce Res. Appl. **9**(6), 522–536 (2010). https://doi.org/10.1016/j.elerap.2010.04.004
34. Edwards, S., Li, H., Lee, J.: Forced exposure and psychological reactance: antecedents and consequences of the perceived intrusiveness of pop-up ads. J. Advert. **31**(3), 83–95 (2002). https://doi.org/10.1080/00913367.2002.10673678
35. Youn, S., Kim, S.: Understanding ad avoidance on Facebook: antecedents and outcomes of psychological reactance. Comput. Hum. Behav. **98**, 232–244 (2019). https://doi.org/10.1016/j.chb.2019.04.025
36. Dillard, J., Shen, L.: On the nature of reactance and its role in persuasive health communication. Commun. Monogr. **72**(2), 144–168 (2005). https://doi.org/10.1080/03637750500111815
37. Moyer-Gusé, E.: Toward a theory of entertainment persuasion: explaining the persuasive effects of entertainment-education messages. Commun. Theory **18**(3), 407–425 (2008). https://doi.org/10.1111/j.1468-2885.2008.00328.x
38. Han, K., Kim, S.: Toward more persuasive diabetes messages: effects of personal value orientation and freedom threat on psychological reactance and behavioral intention. J. Health Commun. **24**(2), 95–110 (2019). https://doi.org/10.1080/10810730.2019.1581304
39. Lee, G., Lee, J., Sanford, C.: The roles of self-concept clarity and psychological reactance in compliance with product and service recommendations. Comput. Hum. Behav. **26**(6), 1481–1487 (2010). https://doi.org/10.1016/j.chb.2010.05.001
40. Samah, N., Ali, M.: Individual differences in online personalized learning environment. Educ. Res. Rev. **6**(7), 516–521 (2011). https://doi.org/10.5897/ERR.9000199
41. Lee, S., Lee, Y., Lee, J., Park, J.: Personalized e-services: consumer privacy concern and information sharing. Soc. Behav. Pers. Int. J. **43**(5), 729–740 (2015). https://doi.org/10.2224/sbp.2015.43.5.729
42. Guo, X., Gan, X.Y.: Burst your bubbles: reflection on the formation and resolution of filter bubbles in an era of recommendation algorithm. Global Media J. **5**(2), 76–90 (2018)
43. Rajat, K.B., Pradip, K.B., Rashmi, J.: A rule-based automated machine learning approach in the evaluation of recommender engine. Benchmark. Int. J. **27**(10), 2721–2757(2020). https://doi.org/10.1108/BIJ-01-2020-0051.
44. Nguyen, C.T.: Echo Chambers and Epistemic Bubbles. Cambridge Unniversity Press **17**(2), 141–161 (2020). https://doi.org/10.1017/epi.2018.32
45. Beheshti, A., Yakhchi, S., Mousaeirad, S., Ghafari, S.M., Goluguri, S.R., Edrisi, M.A.: Towards cognitive recommender systems. Algorithms **13**(8), 12–13 (2020). https://doi.org/10.3390/a13080176
46. Belknap, J.K., Ondrusek, G., Berg, J., Waddingham, S.: Barbiturate dependence in mice: effects of continuous vs. discontinuous drug administration. Psychopharmacology **51**(2), 195–198 (1977). https://doi.org/10.1007/BF00431740.
47. Furneaux, B., Wade, M.R.: An exploration of organizational level information systems discontinuance intentions. MIS Q. **35**(3), 573–598 (2011). https://doi.org/10.1016/j.biopsycho.2011.11.004

48. Lakshmanan, A., Krishnan, H.S.: The aha! experience: Insight and discontinuous learning in product usage. J. Market. **75**(6), 105–123 (2011). https://doi.org/10.2307/41406862
49. Zhao, L., Lu, X., Hu, Y.: a proposed theoretical model of discontinuous usage of voice-activated intelligent personal assistants (IPAs). In: PACIS (2018)
50. Zhao, L., Lu, Y.B., Yang, J., Zhang, S.W.: Get tired of socializing as social animal? An empirical explanation on discontinuous usage behavior in social network services. In PACIS (2015)
51. Luqman, A., Cao, X., Ali, A., Masood, A., Yu, L.: Empirical investigation of Facebook discontinues usage intentions based on SOR paradigm. Comput. Hum. Behav. **70**(5), 544–555 (2017). https://doi.org/10.1016/j.chb.2017.01.020
52. Wu, K.W., Huang, S.Y., Yen, D.C., Popova, I.: The effect of online privacy policy on consumer privacy concern and trust. Comput. Hum. Behav. **28**(3), 889–897 (2012). https://doi.org/10.1016/j.chb.2011.12.008
53. Naveen, F.A., Krishnan, M.S.: The personalization privacy paradox: an empirical evaluation of information transparency and the willingness to be profiled online for personalization. MIS Q. **30**(1), 13–28 (2006). https://doi.org/10.2307/25148715
54. Pappas, I.O., Kourouthanassis, P.E., Giannakos, M.N., Chrissikopoulos, V.: Sense and sensibility in personalized e-commerce: how emotions rebalance the purchase intentions of persuaded customers. Psychol. Market. **34**(10), 972–986 (2017). https://doi.org/10.1002/mar.21036
55. Huang, C.D., Goo, J., Nam, K., Yoo, C.W.: Smart tourism technologies in travel planning: the role of exploration and exploitation. Inf. Manage. **54**(6), 757–770 (2017). https://doi.org/10.1016/j.im.2016.11.010
56. Maier, C., Laumer, S., Weinert, C., Weitzel, T.: The effects of technostress and switching stress on discontinued use of social networking services: a study of Facebook use. Inf. Syst. J. **25**(3), 275–308 (2015). https://doi.org/10.1111/isj.12068
57. Du, J., You, J.: Consumers' willingness to adopt personalized push under the effect of "information cocoon house": the perspective of psychological resistance. Enterpr. Econ., 103–110(2019).
58. Ringle, C.M., Wende, S., Becker, J.M.: SmartPLS 3. SmartPLS GmbH, Boenningstedt (2015)
59. Teo, H.H., Wei, K.K., Benbasat, I.: Predicting intention to adopt interorganizational linkages: an institutional perspective. Society for Information Management and The Management Information Systems Research Center (2003). https://doi.org/10.2307/30036518.
60. Fornell, C., Larcker, D.F.: Structural equation models with unobservable variables and measurement error: algebra and statistics. J. Market. Res., 382–388 (1981). dhttps://doi.org/10.2307/3150980.
61. Straub, D., Gefen, D.: Validation guidelines for IS positivist research. Commun. Assoc. Inf. Syst. **24** (2004). 10.17705/1CAIS.01324.
62. Wang, Y., Wang, P., Zhang, L., Zhang, W.: Research on the "Information Cocoons" of content intelligent distribution platform from the perspective of network information ecological chain. Res. Library Sci. **2** (2018)
63. Alyson, L.Y., Anabel, Q.H.: Information revelation and internet privacy concerns on social network sites: a case study of Facebook. Communities Technol. (2009). https://doi.org/10.1145/1556460.1556499
64. Chen, J.V., Tran, A., Nguyen, T.: Understanding the discontinuance behavior of mobile shoppers as a consequence of technostress: an application of the stress-coping theory. Comput. Human Behav. 83–93 (2019). https://doi.org/10.1016/j.chb.2019.01.022.
65. Huang, C.K., Chen, S.H., Tang, C.P., Huang, H.Y.: A trade-off dual-factor model to investigate discontinuous intention of health app users: fFrom the perspective of information disclosure. J. Biomed. Inform. 2–10 (2019). https://doi.org/10.1016/j.jbi.2019.103302

Effects of Question Type Presentation on Raised Questions in a Video Learning Framework

Hinako Izumi[1](✉), Masaki Matsubara[2], Chiemi Watanabe[3], and Atsuyuki Morishima[2]

[1] Master's Programs in Informatics, Graduate School of Comprehensive Human Sciences, University of Tsukuba, 1-2 Kasuga, Tsukuba-shi, Ibaraki 305-8550, Japan
hinako.izumi.2019b@mlab.info

[2] Faculty of Library, Information and Media Science, University of Tsukuba, 1-2 Kasuga, Tsukuba-shi, Ibaraki 305-8550, Japan
{masaki,mori}@slis.tsukuba.ac.jp

[3] Faculty of Industrial Technology, Tsukuba Univesity of Technology, Amakubo 4-3-15, Tsukuba, Ibaraki 305-8520, Japan
chiemi@a.tsukuba-tech.ac.jp

Abstract. This paper examines the effect of explicitly presenting types of questions to students in an online learning environment. In our setting, people are shown online videos for lectures and then asked to raise questions. We compare the collected questions with those collected in the same situation but without presenting types of questions. We find that presenting types of questions improves the quality of questions, but does not increase the number of questions raised.

Keywords: Learning framework · Question prompt · Facilitation

1 Introduction

Understanding what we do not understand is an important step for learning, and one of the most difficult things for beginners [10,14]. One technique for helping students understand what they do not understand is presenting types of questions to prompt the students to raise questions. In offline classes, presenting types of questions was shown to be effective for learners to raise questions [9]. In their settings, facilitators present types of questions when they prompt questions from students.

This paper examines the effect of explicitly presenting types of questions to students in an online video learning environment.

Our challenge is to develop a *fully automated workflow* to prompt questions with presenting types of questions for online classes. The focus of this paper is whether presenting types of questions to students is still effective in such a setting. In our setting, students watch online videos for lectures (Fig. 1). The

K. Toeppe et al. (Eds.): iConference 2021, LNCS 12646, pp. 114–126, 2021.
https://doi.org/10.1007/978-3-030-71305-8_9

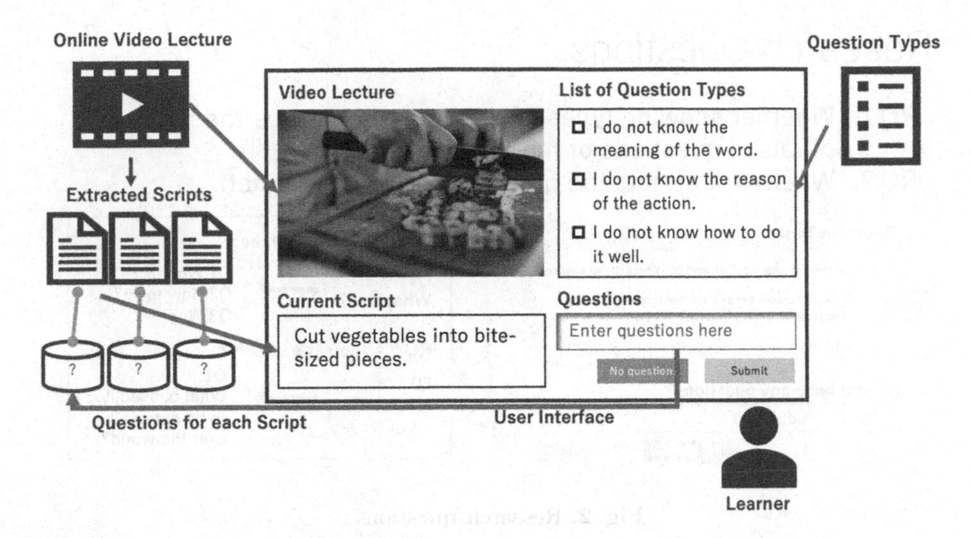

Fig. 1. An automatic framework for prompting questions with presenting types of questions for online classes

process of prompting questions is fully automated as follows. First, the script of the video lecture is extracted and divided into chunks of sentences in advance[1]. Second, the video lecture is played in the screen, and scripts are shown in synchronization with the video. A list of question types is shown with checkbox to help the learner come up with their question. An example of question type is "I do not know the meaning of the word." There is a text field for typing a question with two buttons "submit" and "no question." The video does not move to the next chunk if none of the two buttons are pressed before the current chunk ends.

This paper focuses on the following two research questions (Fig. 2). The first question is whether showing types of questions increases the number of raised questions or not, and the second one is whether it affects the quality of questions or not.

To obtain the answers to the two questions, we conduct an experiment in a crowdsourcing setting. We hire crowd workers and ask them to raise questions on one of two online lectures. We compare the collected questions with those collected in the same situation but without presenting types of questions. We find that presenting types of questions improves the quality of questions.

The contributions of this paper are as follows.

(1) It shows that presenting types of questions is effective in terms of the quality of raised questions, in an automatic framework that does not require human facilitators' intervention, for prompting questions on online lectures from learners.
(2) It gives the result of a large scale experiment on the framework with hundreds of people recruited from a crowdsourcing service.

[1] In the simplest way, each chunk is composed of one sentence.

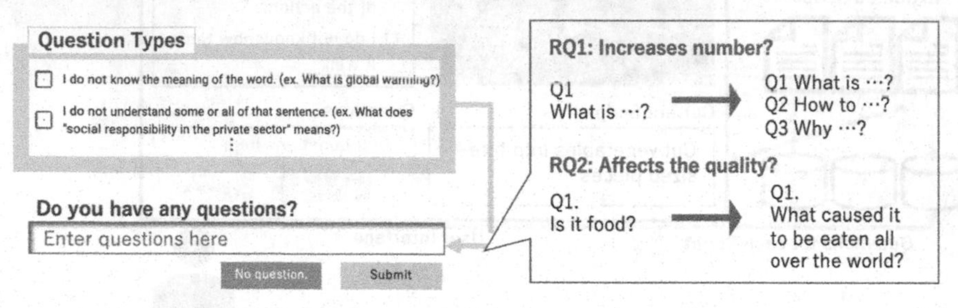

Fig. 2. Research questions

The remainder of this paper is as follows. Section 2 introduces related work. Section 3 explains the experiment and its result. Section 4 gives a discussion. Section 5 concludes.

2 Related Work

Our challenge is to develop an automatic and effective way to prompt questions in an online learning environment. However, effective ways to prompting questions is not straightforward. Miyake et al. [13] showed one of the reasons students have difficulty eliciting questions is that they do not understand what they do not understand. Therefore, we have been exploring an approach to promote metacognition of students. In our previous work [8], we found that when eliciting questions, having them select parts of the script that they do not understand elicits more specific questions. We adopted the task design used in [8] that we found effective.

Existing approaches to facilitating learning include peer review of writing [11], having students solve problems created by each other [6,7,12], and supporting the process of recognizing and solving complex problems [1,4]. In particular, studies that focus specifically on eliciting questions from students include King [9] and Endo [3]. Both studies showed an improvement in performance as a result of prompting question types and other students' existing questions. In addition, [9] showed an increase in the number of some types of questions, too. Since it has been shown that promoting questioning is an effective approach to aid learning, we try to incorporate it into an automatic online workflow.

CoNet-C [2] is similar to our approach in that it is an online scaffolding tool for discussion that gives students several types of questions in each of three patterns of clarification, rebuttal, and viewpoint change so that they can ask questions about each student's opinions. CoNet-C allows students to individually ask their classmates. Since its focus was on a discussion situation, it was assumed that the questions would be about the opinions of each student. Our study differs from the CoNet-C in that ours aims to promote questions on the content of the class in a general classroom situation where the teacher gives a lecture to students.

3 Experiment

We conducted an experiment in which students were asked questions about a video lecture's content to examine the following research questions (Fig. 2).

RQ1: Whether showing types of questions increases the number of raised questions or not.
RQ2: Whether it affects the quality of questions or not.

In the experiment, we provided the online video and its script to the participants, and asked them to list up pairs of their question and the place (chunk) to ask their question.

For investigating the research questions, we applied the following two types of question collecting processes to compare the number and the qualities of questions.

- When collecting the questions, we show the types of general questions.
- When collecting the questions, we do not show the question types.

The detailed procedure is described in Sect. 3.3. The IRB approved the experiment at the Faculty of Library, Information, and Media Science at the University of Tsukuba.

Summary of Results: RQ1 did not support, and RQ2 did support.

3.1 Teaching Materials

The topic of teaching materials were a cooking recipe video[2] (COOK) and TED Talks[3] (social issue, SI). Table 1 shows title, video length, and number of chunks of each video. We extracted a script from the subtitles in the video, and divided it into chunks according to the timing of the screen transitions in the video. We used that video because the script of the video content was available and that it was the right time length for the crowd workers to engage.

[2] "How to make boiled and soaked aubergines" https://www.youtube.com/watch?v=27sVEYPlomw.
[3] "The evolution of the coffee cup lid" https://www.ted.com/talks/a_j_jacobs_the_evolution_of_the_coffee_cup_lid.

Table 1. Details of teaching materials

Domain	Title	Video length	Number of chunks
COOK	How to make boiled and soaked aubergines	00:03:42	17
SI	The evolution of the coffee cup lid	00:03:02	8

3.2 Participants

A total of 400 participants (100 in each condition and domain) were recruited via Yahoo! Crowdsourcing. We didn't set any requirements for the age or gender, but the workers can work on a PC for the environment.

To investigate effectiveness of presenting question type, the participants were divided into two groups: an experimental group and a control group. In the experiment group, the question type was shown when the participants asked the questions, whereas it was not in the control group.)

Before watching the online video, participants were shown a table (Table 2) to explain the skill levels for each domain and asked to assess and declare their skills based on the table.

Table 2. Criteria for self assessment of skill level in each domain

Level	COOK	SI[a]
1	I've barely ever cooked	I watch or read the news at least once a month for a total of 5 to 30 min
2	I may cook a simple meal once a week to once a month	I watch or read the news at least once a month for a total of 30 min or more
3	I may cook an elaborate meal once a week to once a month	I watch or read the news at least once a week for a total of 5 to 30 min
4	Cook a simple meal almost every day	I watch or read the news at least once a week for a total of 30 min or more
5	I cook an elaborate meal almost every day	I watch or read the news for a total of 5 to 30 min almost every day
6	I have worked in a cooking job for more than a month	I watch and read the news for a total of 30 min or more almost every day

[a] The content of TED talks is one of social issues. Since many news and newspapers also deal with social issues, people who read news frequently are likely to be familiar with understanding social issues. For this reason, we used this index.

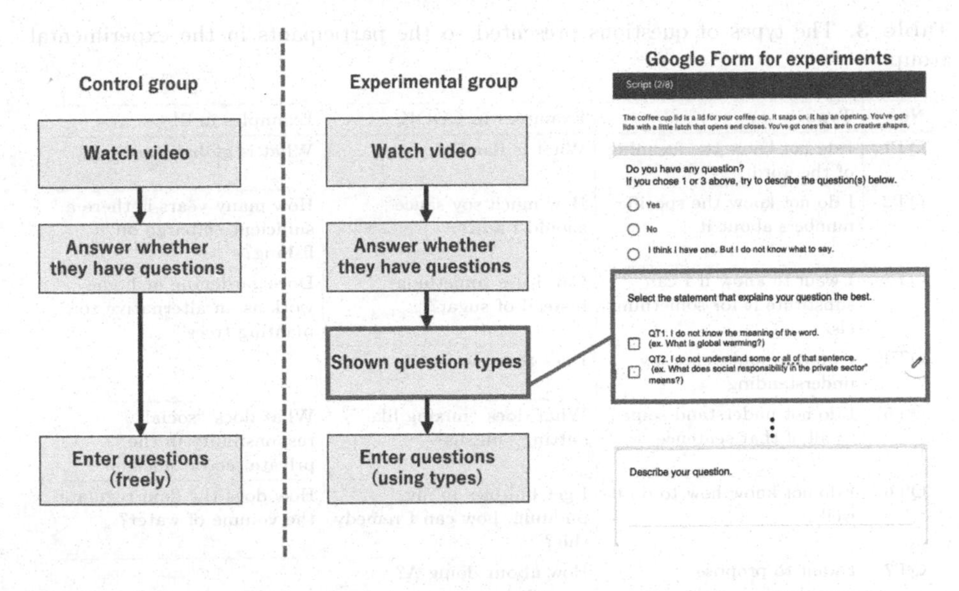

Fig. 3. Experimental procedure: First, participants watch a video. Second, they are asked whether they have questions. If they have a question, they enter it. In the experimental group, with type presentation only in the experimental group.

3.3 Procedure

Figure 3 shows the experimental procedure. First, the participants were asked to watch the video, then the chunk of segmented scripts was presented to them in turn. They were instructed to ask any questions if they have at each chunk. In asking questions, we used Google Form. Each section has one chunk and a text area to write questions down if they have.

While the control group was asked questions without presenting types, the experimental group was presented with types in each section of the form. It was asked to choose the one that matches the content of the question they wanted to ask and report it to the participants. The purpose of presenting the types is to have the participants use the presented types as hints to raise their own questions.

Question Types. The question types were designed to present questions with roughly the same meanings in COOK and SI, based on the questions on the mathematics domain collected in [8]. Table 3 shows question types and example questions in two domains. In the table, the "type" column explain what are asked in question types, and "Examples in COOK" and "Examples in SI" columns show examples of the question types in the cooking and social issues domain we used for our two videos COOK and SI, respectively.

Table 3. The types of questions presented to the participants in the experimental group

No	Type	Examples in COOK	Examples in SI
QT1	I do not know the meaning of the word	What is flambé?	What is global warming?
QT2	I do not know the specific numbers about it	How much soy sauce should I add?	How many years is there a sufficient embargo on fishing?
QT3	I want to know if I can substitute it for something else	Can I use something instead of sugar?	Does gardening at home work as an alternative to planting trees?
QT4	I want to make sure my understanding	Does that mean A?	
QT5	I do not understand some or all of that sentence	What does "mixing like cutting" means?	What does "social responsibility in the private sector" means?
QT6	I do not know how to do it well	I get bubbles in my pudding, how can I remedy this?	How does the dam regulate the volume of water?
QT7	I want to propose something	How about doing A?	
QT8	I do not know why you do it	Why do you let that food sit in the fridge?	Why do we need to separate our waste?
QT9	I do not know why you say "A, so B."	"Radishes can fall apart in boiling, so cut the corners." Why do you say so?	"Sea surface temperatures will rise, so dwellings could be flooded." Why do you say so?
QT10	A question that is none of the above	–	

Table 4. Number of subjects and questions in each conditions. In both domains, the number of questions did not increase or decrease despite the type presentation.

	COOK		SI	
	Control	Experimental	Control	Experimental
Valid answers	98	98	100	100
Questions	289	289	222	218

3.4 Result

Table 4 shows the numbers of valid answers (participants) and the numbers of questions raised by participants in each domain. In the COOK domains, the number of raised questions was the same (289). In SI, 222 questions were raised by participants in the control group and 218 in the experimental group.

Table 5 shows the distributions of skill levels of participants. In COOK, the majority of participants were either beginners (Levels 1, 2) or moderate (4), while in SI, they were mostly skilled workers (5, 6).

Table 5. Self-reported skill of subjects. Many participants reported skill level 1 or 2 or 4 in COOK domains and skill level 5 or 6 in SI domains.

Skill level	COOK		SI	
	Control	Experimental	Control	Experimental
1	19	11	6	3
2	32	37	1	0
3	6	5	3	3
4	32	35	4	2
5	6	7	42	51
6	3	3	44	41

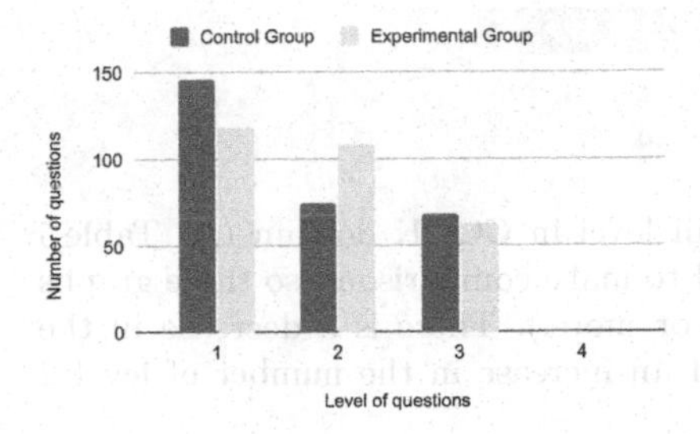

Fig. 4. Level of questions in COOK domain. The presentation of the type reduced the number of skill level 1 questions and increased the number of skill level 2 questions (p = 0.01468).

Fig. 5. Level of questions in SI domain. The presentation of the type reduced the number of level 1 questions and increased the number of level 2 questions (p = 0.00019).

Figures 4 and 5 show the distributions of *questions levels* raised by participants in the COOK and SI domains, respectively. Here, we adopted the question levels defined in [5]:

Level 1 Questions that can be answered with Yes/No or a few words.
Level 2 Questions that can be answered with a simple explanation at the general theory level.
Level 3 Those that require a reasoned and complex explanation.
Level 4 Those that require a reasoned explanation of the causal relationships between multiple concepts.

Levels of the raised questions were determined by a different set of five evaluators, one of which is an author of this paper. They give the levels independently (Cronbach's alpha = 0.99), and we computed the level by majority voting of the results.

For both COOK and SI, the number of Level 1 questions decreased and that of Level 2 questions increased significantly in the experimental group. The -square test results showed a significant difference in the distribution of question levels in both domains (COOK: p = 0.01468, SI: p = 0.00019).

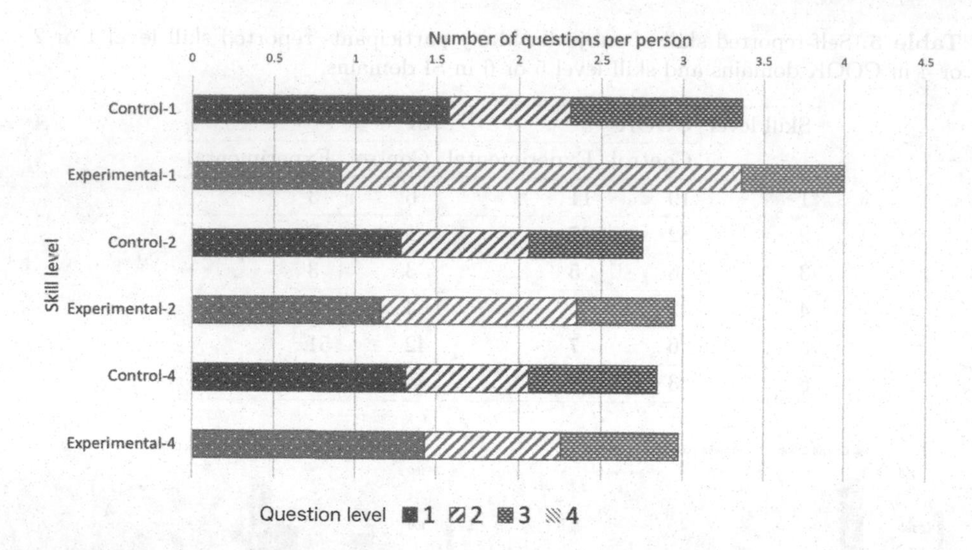

Fig. 6. Number of questions per person by skill level in COOK domain (As Table 5 shows, at some levels the numbers are too small to make comparisons, so these graphs show only skill levels with a population of 10 or more.). There is a decrease in the number of level 1 questions at skill level 1 and an increase in the number of level 2 questions at skill levels 1 and 2.

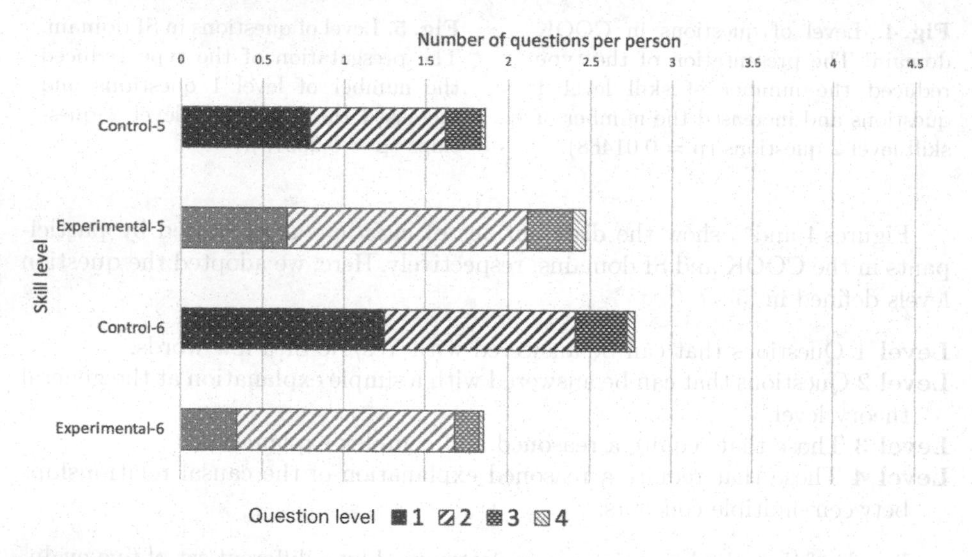

Fig. 7. Number of questions per person by skill level in SI domain (As Table 5 shows, at some levels the numbers are too small to make comparisons, so these graphs show only skill levels with a population of 10 or more.). There is a decrease in the number of level 1 questions and an increase in the number of level 2 questions at skill levels 5 and 6.

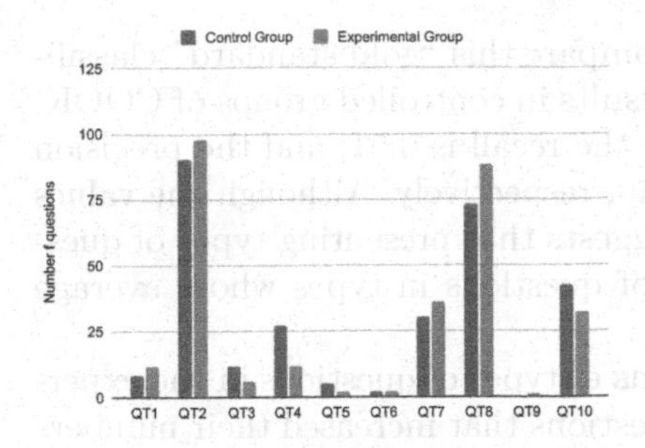

Fig. 8. Type of questions in COOK domain. In the experimental group, there are some noticeable different such as a decrease in the number of QT4 questions with a mean question level of around 1 (in Table 6), and an increase in the number of QT8 questions with a mean question level of around 3.

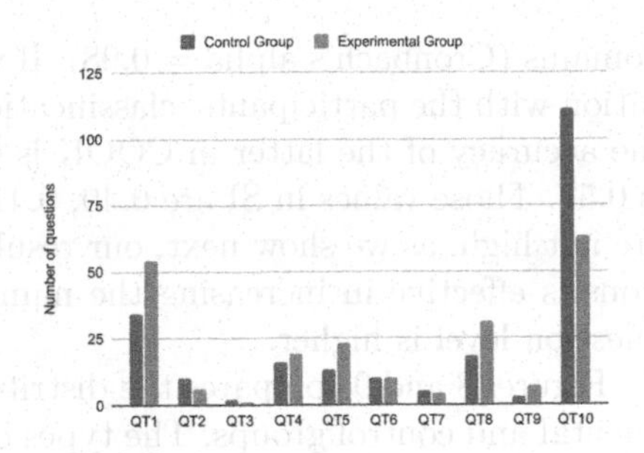

Fig. 9. Type of questions in SI domain. In the experimental group, there are some noticeable different such as an increase in the number of QT1 and QT8 questions with a mean question level of around 2 (in Table 6).

Figures 6 and 7 show the results of the analysis for each skill level. The horizontal axis is the number of questions at each question level per person on average. There is a significant increase in level 2 questions at skill levels 1 and 2 in COOK and at skill levels 5 and 6 in SI within the skill levels where there are enough people to make comparisons.

4 Discussion

As the result shows, presenting a list of question types did not affect the number of questions at all, although we did not force participants to raise questions. Note that the experiment setting was different from our framework's actual deployment, in that we asked crowd workers to raise questions. However, we did not find any cause that enforces the number of questions to be almost the same. Therefore, the result gave a negative answer to our first research question. Identifying factors to raise the number of questions is an interesting future work.

On the other hand, the result clearly showed that presenting a list of question types increased the number of questions at higher levels out of those raised by participants, and we can conclude that our second hypothesis, whether it affects the quality of questions, is true.

We assumed that presenting a list of question types reminded participants of questions they had not noticed so that they consider a more variety of questions. In order to see what happened in this aspect, we analyzed the types of questions raised by the participants.

We re-classified the collected questions because we found that some of the questions were labeled with wrong question types by participants. The five evaluators classified the questions raised by participants in both groups of both domains (Cronbach's alpha = 0.98). If we compare this "gold standard" classification with the participants' classification results in controlled groups of COOK, the accuracy of the latter in COOK is 0.57, the recall is 0.51, and the precision is 0.57. Those values in SI are 0.49, 0.41, 0.40, respectively. Although the values are not high, as we show next, our result suggests that presenting types of questions is effective in increasing the number of questions in types whose average question level is higher.

Figures 8 and 9 compares the distributions of types of questions in the experimental and control groups. The types of questions that increased their numbers in the experimental groups of both domains are QT1, QT8, QT9. Those that decreased their numbers are QT3, QT10. Although some showed different behaviors (QT2, QT4, QT5, QT6, QT7), the result shows that presenting question types may affect the distribution of types of questions raised by participants. Table 6 shows the relationship between question types and question levels. The result shows that presenting question types increased the average of question levels because it increased the number of question types that result in higher question levels.

It has not been verified yet whether the types used in this study are effective for this system, or which types are more effective. These issues will be included in the future work.

In this experiment, the distribution of skills was skewed, especially in SI. In order to improve this, it is necessary to obtain the distribution of skills beforehand and measure the ability to understand the course content appropriately.

Table 6. Average level of questions in each types (In Table 6, the part that cannot be calculated because there is no question is left blank.). This table shows that there are several types of questions that tend to be at a higher level.

	COOK		SI	
	Control	Experimental	Control	Experimental
QT1	2.00	1.92	1.85	1.96
QT2	1.52	1.49	1.60	1.67
QT3	1.00	1.33	1.50	1.00
QT4	1.00	1.23	1.25	1.32
QT5	1.80	2.00	1.92	1.91
QT6	2.00	2.50	2.20	1.90
QT7	1.27	1.11	2.00	1.50
QT8	2.81	2.61	2.33	2.39
QT9	–	3.00	3.00	3.00
QT10	1.29	1.53	1.56	1.76

5 Conclusion

We examined the effect of explicitly presenting types of questions to students in an online learning environment. In our setting, people were shown online videos for lectures and then asked to raise questions. We compared the collected questions with those collected in the same situation but without presenting types of questions. We found that presenting types of questions improves the quality of questions, but did not increase the number of questions raised. Future work includes a more detailed exploration of the relationship between posed question types and collected questions, investigation of factors to increase the number of raised questions, and in-the-wild studies with deployed systems.

Acknowledgments. This work was partially supported by JST CREST Grant Number JPMJCR16E3, JSPS KAKENHI Grant Number JP19K11978, and JST-Mirai Program Grant Number JPMJMI19G8, Japan.

References

1. Chen, C.H., Bradshaw, A.C.: The effect of web-based question prompts on scaffolding knowledge integration and ill-structured problem solving. J. Res. Technol. Educ. **39**(4), 359–375 (2007)
2. Choi, I., Land, S.M., Turgeon, A.J.: Scaffolding peer-questioning strategies to facilitate metacognition during online small group discussion. Instr. Sci. **33**(5–6), 483–511 (2005)
3. Endo, K.: A pilot study on the facilitating effect of the questioning behavior of students through the internet. J. Educ. Appl. Inf. Commun. Technol. **14**(1), 11–15 (2011)
4. Ge, X., Chen, C.H., Davis, K.A.: Scaffolding novice instructional designers' problem-solving processes using question prompts in a web-based learning environment. J. Educ. Comput. Res. **33**(2), 219–248 (2005)
5. Guthrie, J.T., et al.: Increasing reading comprehension and engagement through concept-oriented reading instruction. J. Educ. Psychol. **96**(3), 403 (2004)
6. Hirai, Y., Hazeyama, A.: A learning support system based on question-posing and its evaluation. In: Fifth International Conference on Creating, Connecting and Collaborating through Computing (C5 2007), pp. 178–184. IEEE (2007)
7. Hirai, Y., Hazeyama, A., Inoue, T.: Assessment of learning in concerto III: a collaborative learning support system based on question-posing. In: Proceedings of Computers and Advanced Technology in Education, pp. 36–43. ACTA Press, Calgary (2009)
8. Izumi, H., Matsubara, M., Watanabe, C., Morishima, A.: A microtask approach to identifying incomprehension for facilitating peer learning. In: 2019 IEEE International Conference on Big Data (Big Data), pp. 4624–4627. IEEE (2019)
9. King, A., Rosenshine, B.: Effects of guided cooperative questioning on children's knowledge construction. J. Exp. Educ. **61**(2), 127–148 (1993)
10. Kruger, J., Dunning, D.: Unskilled and unaware of it: how difficulties in recognizing one's own incompetence lead to inflated self-assessments. J. Pers. Soc. Psychol. **77**(6), 1121 (1999)

11. Lin, S.S.P., Samuel, M.: Scaffolding during peer response sessions. Proc. Soc. Behav. Sci. **90**, 737–744 (2013)
12. Luxton-Reilly, A., Bertinshaw, D., Denny, P., Plimmer, B., Sheehan, R.: The impact of question generation activities on performance. In: Proceedings of the 43rd ACM Technical Symposium on Computer Science Education, pp. 391–396 (2012)
13. Miyake, N., Norman, D.A.: To ask a question, one must know enough to know what is not known. J. Verbal Learn. Verbal Behav. **18**(3), 357–364 (1979)
14. White, R.T., Gunstone, R.F.: Metalearning and conceptual change. Int. J. Sci. Educ. **11**(5), 577–586 (1989)

The Impact of Question Type and Topic on Misinformation and Trolling on Yahoo! Answers

Pnina Fichman(✉) and Rachel Brill

Indiana University, Bloomington, IN 47405, USA
Fichman@indiana.edu, rabrill@iu.edu

Abstract. Trolling and misinformation are ubiquitous on social media platforms, such as Yahoo! Answers. Yet, little is known about the impact of question type and topic on the extent of trolling and misinformation in answers on these platforms. We address this gap by analyzing 120 transactions with 2000 answers from two Yahoo! Answers categories: Politics & Government and Society & Culture. We found that trolling and misinformation are widespread on Yahoo! Answers. In most cases, trolling in questions was echoed by more trolling in answers, and misinformation in questions with more misinformation in answers. We also found that 1) more misinformation and more trolling were found in answers to conversational questions than to informational questions; 2) more misinformation occurred in answers to questions in politics than answers to questions in culture; and 3) trolling significantly differed between politics and culture.

Keywords: Trolling · Misinformation · Question answering sites

1 Introduction

Question answering sites are online communities in which users interact with one another to ask and answer questions. On Yahoo! Answers, for example, users can ask questions on a variety of topics and provide answers to other users' questions. Question answering sites allow anyone to contribute, and thus they can abound in misinformation (inaccurate information) and disinformation (*intentionally* inaccurate information); answer accuracy is an important aspect of the site and misinformation impedes accuracy. While trolling is often found on these sites, little research on question answering sites has aimed to examine the impact of trolling on the quality of information shared on these sites and the well-being of the users and administrators of these communities [11, 13]. This lacuna might be due to the fact that trolling is perceived to have little impact on answer quality [19], or because it is not as evident on question answering sites as other concerning behaviors [11]. It is also possible that it is because trolling is notoriously difficult to define, but typically a troll is "a person who intentionally antagonizes others online by posting inflammatory, irrelevant, or offensive comments or other disruptive content" (Merriam-Webster, n.d.). Trolling can range from light-hearted and humorous to offensive and threatening, and can take different forms on different Internet platforms

K. Toeppe et al. (Eds.): iConference 2021, LNCS 12646, pp. 127–140, 2021.
https://doi.org/10.1007/978-3-030-71305-8_10

[e.g. 15, 26]. Trolls often aim at eliciting an emotional reaction from other Internet users; the troll may do this purely for their own enjoyment, or to spread political or ideological beliefs. Trolling involves deception, misinformation and disinformation [14]. There is much confusion over the meaning and use of the concepts of misinformation and disinformation; at times, these are used interchangeably. In general, misinformation seems to refer more generally to incorrect information, whether it is malicious in nature or simply a mistake of some sort, whereas disinformation carries with it an implication of intent and malice; in other words, misinformation is created on accident, while disinformation is created with the intention to harm people [18]. However, because it can often be difficult to discern intent, especially online, misinformation and disinformation are commonly confused and often used as synonyms, and labeling a piece of information as misinformation or disinformation can depend on the actor's intentions as much as on the standards of the one evaluating it. Intentionally false or misrepresentative information that succeeds in misleading the recipient involves deception. As Søe [33] explains, deception is a "success term," while the term disinformation does not imply success. Caddell [4] says that there are two forms of deception, fabrication and manipulation. He describes fabrication as "false information [created] and presented as true … for the purpose of disinformation," and manipulation as "the use of information which is technically true, but is being presented out of context in order to create a false implication" (p. 1); thus, deception does not always imply the presence of disinformation. Trolls often act under hidden identity in deceiving others [31].

Scholars that focused attention on trolling on question answering sites report that trolling was one of seven content-bearing terms in user-reported behavior on the site Answerbag, but other issues, such as users creating multiple profiles to manipulate voting, were of equal or more concern [11]. Others found that trolling was mentioned only once in a study on how users judge answer quality on Yahoo! Answers, when a participant in the study did not think trolling occurred on the site [19]. Nonetheless, Guy and Shapira [13] conducted a large survey of troll questions on Yahoo! Answers and proposed a seven-point classification system to separate "troll questions" from "clean questions." They identified characteristics that distinguish troll questions from legitimate questions and found that 1) questions categories most prone to trolling are conversational rather than informational in nature and include: society & culture, sports, social science, food & drink, and politics & government; 2) the average length of trolling questions is longer and the average answer to these questions is shorter; and that 3) trolling questions attract more user activity, provoke similar answers, and elicit more negative feedback. Trolls often aim at eliciting an emotional reaction from other Internet users; the troll may do this purely for their own enjoyment, or to spread political or ideological beliefs; trolling can range from light-hearted and humorous to offensive and threatening, and can take different forms on different Internet platforms [e.g., 15, 26]. Attention to trolling from both scholars and the media is quickly growing, perhaps because it has become a ubiquitous part of our daily life online. With these few studies and their mixed findings, there is clearly a need to examine if and to what extent trolling impacts questions and answers on question answering sites; it is also critical to gain a better understanding of the relationships between trolling and misinformation.

Furthermore, of specific interest is the possible impact of question topic and question type on the extent of trolling and misinformation; questions about certain topics may lead to more trolling and misinformation, and Yahoo! Answers categories might predict levels of trolling and misinformation in answers [13]. Then, certain trolling behaviors may differ across categories because each category attracts its own community of users forming different norms of behaviors [29]. For example, questions with overly polite tones on Yahoo! Answers or Stack Overflow communities were less likely to be answered [5]. Conversely, others argue that linguistic indications of gratitude can increase the likelihood of success [2], and that politeness is a key factor in assessing the quality of answers on social Q&A sites [37]. Clearly, question answering differs on various platforms and on various categories, and these differences may impact the extent of trolling and misinformation. We thus propose that:

H1: Question category will impact the extent of trolling and misinformation.

Question type may also impact the extent of trolling and misinformation. Troll questions [13] may have unique characteristics that may lead to more misinformation or disinformation, and more trolling in their answers. Some questions never even receive an answer, partially depending on the level of details, specificity, clarity, accuracy, and socio-emotional value of the questions [6]. Because not all questions are alike, it is possible that conversational questions, which "are asked with the intent of stimulating discussion… [and are aimed] at getting opinions" [17] may lead to more trolling than other questions. At the same time it is possible that informational questions, which "are asked with the intent of getting information that the asker hopes to learn or use via fact- or advice-oriented answers" [17], may lead to more misinformation than trolling. Thus we propose that:

H2: Question type will impact the extent of trolling and misinformation.

We designed a study with 120 transactions (and 2000 answers) from two categories on Yahoo! Answers to address test the two hypotheses and answer the following research questions: 1) What is the extent of trolling and misinformation on Yahoo! Answers?; and 2) Does trolling and misinformation vary based on question topic (politics vs culture) and type (conversational vs informational)?

2 Background

Scholars have focused attention on SQA communities [28], trying to understand the motivation to answer questions, for example [e.g., 3, 7, 22]. They found that users are motivated by their identification with and joy of helping the community [3, 7], and they continue to contribute when they feel that the other members of the community treat them fairly, appreciate their contributions, and in general meet their expectations [22].

Others focused on the quality of answers on various SQAs, mainly in an effort to re-use high quality answers, showing that answer quality varies between questions, topics, and sites [e.g., 5, 8, 27, 34]. Site popularity does not always correspond to answer quality [30]. Answer quality varies widely on different platforms, with Wikipedia Reference Desk having the highest quality answers [30], and Yahoo! Answers, which is a community based SQA, having the lowest [8]. The quality of answers on Yahoo! Answers was commonly assessed through the platform's best answer (the answer selected by the question-asker as being the most helpful; this feature exists on Yahoo! Answers), and occasionally other answer quality dimensions have been used [e.g., 1]. John, Goh and Chau [21] posit that the quality of answers as ranked by users on Yahoo! Answers does not correlate with answer quality ranked by experts, and Fichman [8] suggests that user rankings are subjective and therefore problematic. In line with this approach Kim and Oh [23] found that in 29.8% of cases where users chose "best answers" in Yahoo! Answers, their selections were based on socio-emotional criteria rather than on the content or utility of the answer. Studies found that better answers are longer [1, 16, 20], and include references to external sources [12]. John, Goh, and Chua [21] further proposed that quality of answers should be measured by social (user interaction and feedback) and content features (intrinsic and extrinsic content quality); other scholars agree that content features are critical in assessing answer quality [8, 16, 20, 30]. Fichman [8] used three content measures: accuracy (whether the answer to the question is correct), completeness (whether the answer thoroughly responds to all parts of the question), and verifiability (whether the answer provides sources). Ong, Day, and Hsu [25] also consider format (whether the answer is presented well) and currency (whether the answer is up to date) in measuring quality as determined based on the user's perception. Fichman [8] also examined if the "whole" answer, which is the composite of all the answers provided by users to the question, forms a higher quality answer than the first answer or the "best" answer, and found that the whole answer is more complete and more verifiable but not more accurate than the first or best answer. Best answers generally are of the same quality as whole answers; both are of higher quality than first answers [30]. High-quality answers include, besides positive votes, completeness, clear presentation, and reliability and accuracy of information; answer length is a feature weakly associated with high-quality answers [20]. On average, seven answers to a question provides the highest quality whole answer on the collaborative question answer site Yahoo! Answers [30].

Harper et al. [16] propose that there are two types of questions on social question answer sites: informational, which are asked with the intent of getting information to learn or use via fact- or advice-oriented answers; and conversational, which are asked with the intent of stimulating discussion, aiming at getting opinions. Some questions are not answered, especially if they are very short [1, 16], or when they are unclear or overly polite [6, 36]. Clearly, question types and attributes impact the extent and quality of answers [e.g. 6]. It is likewise possible that they impact the extent of trolling and misinformation.

3 Method

Using data from Yahoo! Answers we designed a study to address the research questions: 1) What is the extent of trolling and misinformation on Yahoo! Answers?; and 2) Does trolling and misinformation vary based on question topic (politics vs culture) and type (conversational vs informational)?

3.1 Data Collection

We collected data at midnight (EST) on November 9, 2016 from two Yahoo! Answers categories: "Politics & Government" (=politics) and "Society & Culture" (=culture), scrapping all the data from the previous day, which included all the questions belonging to each category and their corresponding answers. This resulted in 163 transactions[1], with 4,668 total answers, out of which we chose to analyze a sample of 2,000 answers (1,000 answers from each category), along with the 120 questions associated with these answers (65 questions from culture and 55 questions from politics). We uploaded the sampled data into Nvivo 12. While the number of questions varied between the two categories (55 in politics and 65 in culture), the number of answers did not (1,000 answers per category). These data, taken from two categories on one question-answering site on one day, the day of the 2016 American presidential election, are necessarily limited.

3.2 Data Analysis

We developed a coding scheme based on prior research on trolling [29, 32], and social question answering [8, 17] and modified it in an iterative process of coding, discussion among the two authors, and revisions; the final coding scheme included 16 codes (Appendix I). We coded the data at the individual post. To assure coding reliability, one coder coded the entire data set and a second coder coded a sample of the data; intercoder reliability was 89% simple agreement.

Using Nvivo matrix queries we were able to generate frequency tables, and using SPSS 21 we tested for statistical significance. Pattern coding and code-co-occurrence [24], using Nvivo matrix query, helped us identify the topic and type of questions that are associated frequently with misinformation and trolling. We then created a subset of all the transitions with questions that included misinformation, disinformation, or deception in at least one of the corresponding answers for further analysis; we compared code-co-occurrences in that subset of questions with the total set of questions.

4 Findings

We report our findings in two sections, each addressing one of the two research questions.

4.1 Trolling, Answer Quality, and Misinformation on Yahoo! Answers

Answer Quality and Misinformation on Yahoo! Answers. We found inaccurate information frequently, with disinformation in 6% of the posts (132), misinformation in 8% of the posts (171), and deception in 4 posts (Table 1). Most of the inaccurate information appeared in answers, while in questions it appeared infrequently.

Answer quality in our data is low - accuracy (24) and completeness (19) levels around 1%, verifiability (93) at 5%, and relevance (393) at 20% (Table 1). The low levels of answer accuracy and completeness are in sharp contrast to some prior research that reports much higher levels of answers quality. There are several factors that may

[1] A transaction includes a question and all of its corresponding answers.

contribute to these low levels. First, while these other studies examined the quality of informational questions only [8, 30], we included conversational questions, and informational questions account only for about 25% of the questions; since opinionated answers to conversational questions cannot be defined as accurate or inaccurate, this results in a lower overall accuracy level. Second, while other studies have looked at the transaction level or at the best answer post [8, 30] we have analyzed quality at the individual post level.

Table 1. Code frequencies by category in questions and answers

Code Category	Code	Culture				Politics				Total	
		Qs		As		Qs		As			
		#	%	#	%	#	%	#	%	#	%
Answer Quality	Accuracy	0	0	17	1.7	0	0	7	.7	24	1
	Completeness	0	0	12	1	0	0	7	.7	19	1
	Relevance	0	0	299	30	0	0	94	9	393	20.
	Verifiability	0	0	77	8	0	0	15	1.5	92	5
Question Type	Conversational	49	75	0	0	42	76	0	0	91	76
	Informational	16	25	0	0	13	24	0	0	29	24
Inaccurate Information	Misinformation	2	.3	51	5	7	13	111	11	171	8
	Deception	0	0	1	.1	0	0	3	.3	4	.2
	Disinformation	2	.3	31	3	0	0	99	10	132	6
Trolling	Derailment	0	0	118	12	0	0	289	29	407	19
	Emotional Display	2	6	60	6	10	18	85	8.5	157	7
	Insulting	7	11	184	18	10	18	316	32	517	24
	Personal Attacks	0	0	83	8	0	0	139	14	222	10
	Provocation	28	43	514	51	36	65	459	46	1,037	49
	Sarcasm	6	9	408	41	10	18	276	28	700	33
	Swearing	1	3	35	3.5	2	4	68	7	106	5
Total		113		1890		130		1968		4101	

Trolling on Yahoo! Answers. We found trolling behaviors frequently, with more trolling behaviors in answers, compared with questions, regardless of category. As can be seen in Table 1, the most frequent trolling behaviors involve provocation, sarcasm, insulting, and derailment. We also found that the tone set up by the questions is most of the time echoed in answers; when certain trolling behaviors appeared frequently in questions, they appeared more frequently in answers. For example, insulting appears in 18% of the questions on politics and then even more frequently in 32% of the answers on that category. A similar increase is evident in culture when 11% of the questions included insulting language, and then 18% of the answers did. It is possible that each of these trolling behaviors serve as effective trolling bait that triggers others to react.

4.2 Differences in Trolling and Misinformation Based on Question Topic and Type

The average number of answers per question was higher in the political category (18.1) than in culture (15.38), but the proportions between informational and conversational questions did not significantly differ between the categories (conversational questions in politics 75.4% and in culture 76.4%).

Differences Between Politics and Culture in Answer Quality, Trolling and Misinformation. We found that more answers on culture, compared with politics, were relevant, accurate, and verifiable (Table 1). We found that question topic impacts the extent of misinformation; there were more instances of misinformation and disinformation in politics compared with culture (111 vs. 51 and 99 vs. 31 respectively) (Table 1). These differences were statistically significant and hypothesis H1 was partially supported.

Table 2. Differences between politics and culture in answer quality, trolling and misinformation

Category	Code	χ^2
Answer quality	Accuracy	4.217**
	Completeness	1.32
	Verifiability	43.797***
	Relevance	133.085***
Trolling	Derailment	92.201***
	Emotional display	4.647**
	Insulting	46.464***
	Personal attacks	15.89***
	Provocation	6.389*
	Sarcasm	38.714***
	Swearing	11.147***
Misinformation	Deception - N/A	
	Disinformation	35.332***
	Misinformation	24.181***

*p < .01, **p < .05, ***p < .001

Differences in Trolling and Misinformation Based on Question Type. As can be seen in Table 3, misinformation appeared in 63.3% of the answers regardless of the type of questions; yet, misinformation was found slightly more frequently in answers in conversational questions (63.7%) than in informational questions (62.1%). The differences in level of misinformation between the two types of questions was not significant ($\chi^2(1, N = 200) = 1.49, P = .699$), and hypothesis H2 was partially not supported. future research may try to use other question typologies to identify a more nuanced understanding of the impact of question type on misinformation.

Table 3. Question type, misinformation and trolling

Category	Code	Conversational		Informational		All questions	
		#	%	#	%	#	%
Question type	Conversational	91	100.0	0	0.0	91	75.8
	Informational	0	0.0	29	100.0	29	24.2
Trolling	Emotional display	9	9.9	2	6.9	11	9.17
	Insulting	13	14.3	2	6.9	15	12.5
	Provocation	51	56.0	11	37.9	62	51.7
	Sarcasm	10	11.0	4	13.8	14	11.7
	Swearing	2	2.2	0	0.0	2	1.7
Misinformation		6	6.6	3	10.3	9	7.5
Questions that led to misinformation		58	63.7	18	62.1	76	63.3

We found that the differences in the extent of trolling based on question type were statistically significant ($\chi^2(1, n = 200) = 6.503, p = .011$); for example, we found significantly more provocation in conversational questions (56%) than informational questions (37.9%) (Table 3), and Hypothesis H2 was partially supported.

It is perhaps not surprising that we found more provocation in conversational questions, because using provocation is one way to assure responses from other users. Provocation on its own cannot be a proxy for trolling, because only when it appears along with other trolling behaviors does it count as trolling [29]. In fact, in our study, provocation co-occurred frequently with each of the other trolling behaviors (Table 4), and mainly with insulting, derailment and sarcasm.

Table 4. Trolling behaviors code co-occurrence

	Derailment	Emotional display	Insulting	Personal attacks	Provocation	Sarcasm	Swearing
Derailment	407	36	151	64	280	168	29
Emotional display	36	157	45	19	86	37	15
Insulting	151	45	517	94	401	223	65
Personal attacks	64	19	94	222	147	101	34
Provocation	280	86	401	147	1037	474	79
Sarcasm	168	37	223	101	474	700	38

5 Discussion and Conclusion

Our findings demonstrate variation in information quality across subject domains, similar to findings on other platforms, such as Wikipedia [33], or across subject categories on Yahoo! Answers [1]. Wilkinson and Huberman [32] note that Wikipedia articles in more "popular" subject domains will have more edits, and that in general, a greater number of edits is correlated with higher quality. Others argued that seven answers provide the optimal answer quality on Yahoo! Answers [9]. However, we found that more answers per question (in politics) actually increased the level of misinformation and disinformation and reduced the quality of answers in terms of accuracy and completeness; still more answers per questions increased answer quality in terms of relevance and verifiability in politics, compared to culture. Regardless of the direction of the change, question category impacted all quality measures and misinformation. Thus the relationships between number of answers, quality measures, and misinformation is complex and multidimensional, and future research may focus on the relationships between the various measures of quality with misinformation.

We also found that there are significant differences in trolling between the two categories, supporting hypothesis H1 (Table 2). However, while we found in the politics category significantly more derailment, emotional display, insulting, and swearing than in the culture category, we found significantly more sarcasm and provocation in culture than in politics (Table 1). As such, we conclude that question category significantly impacts the level of trolling, but specific trolling behaviors are more common in the culture category while others are more common in the politics category. Others found more trolling in politics compared to health, entertainment, and religion, in a study that compared trolling on subreddits [10]. Future research should further examine and explain the more nuanced differences across topics.

Our findings are in line with prior research that suggests that trolling is context-dependent and varies by context [29]; however, we provide a more nuanced understanding of these variations, by using the same socio-technical platform, Yahoo! Answers, in comparing the two categories. Interestingly, in the politics we found more trolling tactics that are associated with malevolent anti-social trolling, such as insulting and swearing, but in the culture, we identified more tactics that are often associated with humorous and light-hearted trolling, such as provocation and sarcasm. Expanding on Sanfilippo, Fichman and Yang's [29] work that linked seven behavioral dimensions with four types of trolling, we found that the topic impacts the extent of some trolling behaviors and tactics, within one sociotechnical environment.

In this study we demonstrate that trolling is an integral part of social question answering sites. More than half of the posts involved provocation and about one-third involved sarcasm. We also provide evidence to show that the extent of trolling and misinformation significantly vary based on question topic and question type. Misinformation appeared in questions regardless of the nature of the question or the topic, but more misinformation appeared in politics than culture and more trolling appeared in conversational questions than informational questions. We found no significant differences in levels of misinformation based on question type, and while we found that question topic impacts the extent of trolling, we report mixed findings in terms of the specific trolling behaviors. In the politics category, more posts contained malevolent trolling (swearing, insulting, and personal attacks), while in the culture category, more posts contained sarcasm. We also provide evidence that trolling, as a socio-technical concept, varies not only across socio-technical platforms, but also between categories within the same platform.

Appendix - Coding Scheme

Code	Description	Example
Trolling/Swearing	Using vulgar language, usually to elicit a reaction	This is because CATHOLICS ARE PEDOPHILLIE CULT that rapes innocent children. F*uck catholicism!!!
Trolling/Insulting	Making intentional statement to insult an individual or group of people	Because atheists have integrity and self respect. Christians are sinners anyway, so raping and killing are OK so long as they ask for forgiveness
Trolling/Sarcasm	Using humor, hyperbole, and other rhetorical devices to convey a nuanced public opinion	Sure, she will put that at the top of her list. Forget trade agreements, national debt, war in middle east, and national health insurance. Lets get right to the important issues: hamburgers
Trolling/Provocation	Making intentional claims to elicit a specific reaction	Gods are imaginary creatures, and they don't exist. But if one did exist, and it chose a piece of siht like Trump, then everything I've ever thought about the BuyBull god is accurate…

(continued)

(*continued*)

Code	Description	Example
Trolling/Derailment	Purposely leading a conversation off course	Never mind triune, first you have to prove that gods exist…but you can't so why be concerned with interpretation of mythical stories?
Trolling/Emotional display	Emotional displays in reference to a subject/their behaviors (e.g., all caps writing, multiple exclamation points). This deals with the message's form	BYE BYE LIBTARDS! GET OUT!
Inaccurate information/Deception	Intentionally making someone believe something that is not true	The idea is to be far away on Dec 17th when Russia uses conventional weapons to bomb that missile base in Romania that violates the INF treaty
Inaccurate information/Misinformation	Making an untrue or inaccurate statement unintentionally	There are many good muslins, so hating all of them is not right, but the muslins have to realize that 95% of all terrorist activity is related to people of the muslin faith. so all muslins unfortunately have to bear that burden
Inaccurate information/Disinformation	Intentionally making an inaccurate (fabricated, manipulated, or simply false) statement	He [Obama] can return to Kenya where he was born and run there
Question/Informational	*Informational questions* are asked with the intent of getting information that the asker hopes to learn or use via fact- or advice-oriented answers	Which 'specific' denomination of Christianity did Jesus Himself start? Does that 'specific' denomination still exist today?
Question/Conversational	*Conversational questions* are asked with the intent of stimulating discussion. They may be aimed at getting opinions, or they may be acts of self-expression	Are you hopeful about the future of the United States given the outcome of this election?
Quality/Accuracy	Accuracy of an answer refers to a correct response	Nope. Unless the 22nd ammendent is repealed he is term limited

(*continued*)

(*continued*)

Code	Description	Example
Quality/Completeness	Completeness of an answer refers to an answer that is thorough, provides enough information, and answers all parts of a multi-part question	(In response to a question asking what people think Trump will do as president) Repeal and replace Obamacare. Enforce immigration laws. Reverse Obama orders and rules that are stifling the economy. End wasteful spending. Strengthen the military. Restore US image as world leader. Defeat ISIS. And that's just in the first year
Quality/Relevance	Answer is relevant to the question	(In response to a question asking how the Holy Trinity is composed) God is revealed in three persons--God the Father, God the Son and God the Holy Spirit
Quality/Verifiability	Verifiability of an answer refers to an answer that provides a link or a reference to another source where the information can be found	The Bible tells us, that the Father is Jehovah, Jesus is his 'Firstborn' Son. And the holy spirit, is not a person like God. Rather, it is God's active force. Psalm 104:30

References

1. Adamic, L.A., Zhang, J., Bakshy, E., Ackerman, M.S.: Knowledge sharing and Yahoo! Answers: everyone knows something. In: Proceedings of the International World Wide Web Conference, pp. 665–667. ACM, Beijing (2008)
2. Althoff, T., Danescu-Niculescu-Mizil, C., Jurafsky, D.: How to ask for a favor: a case study on the success of altruistic requests. In: Proceeding of AAAI International Conference on Weblon and Social Media. Association for the Advancement of Artificial Intelligence, Palo Alto (2014)
3. Bao, Z., Han, Z.: What drives users' participation in online social Q&A communities? An empirical study based on social cognitive theory. Aslib J. Inf. Manag. **71**, 637–656 (2019)
4. Caddell, J.: Deception 101 – primer on deception. Strategic Studies Institute, US Army War College (2004). https://fas.org/irp/eprint/deception.pdf
5. Chua, A.Y.K., Balkunje, R.S.: Comparative evaluation of community question answering websites. In: Chen, H.-H., Chowdhury, G. (eds.) ICADL 2012. LNCS, vol. 7634, pp. 209–218. Springer, Heidelberg (2012). https://doi.org/10.1007/978-3-642-34752-8_27
6. Chua, A.Y.K., Banerjee, S.: Measuring the effectiveness of answers in Yahoo! Answers. Online Inf. Rev. **3**, 104–118 (2015)
7. Fang, C., Zhang, J.: Users' continued participation behavior in social Q&A communities: a motivation perspective. Comput. Hum. Behav. **92**, 87–109 (2019)

8. Fichman, P.: A comparative assessment of answer quality on four question answering sites. J. Inf. Sci. **37**, 476–486 (2011)
9. Fichman, P.: Information quality on Yahoo! Answers. In: Tsiaakis, T., Kargidis, T., Katsaros, P. (eds.) Approaches and Processes for Managing the Economics of Information Systems, pp. 295–307. Idea Group Publishing Inc., Hershey (2014)
10. Fichman, P., Sharp, S.: Successful trolling on Reddit: a comparison across subreddits in entertainment, health, politics, and religion. In: Proceedings of the 83rd Annual Conference of the American Society of Information Science and Technology, vol. 57, p. e333 (2020). https://doi.org/10.1002/pra2.333
11. Gazan, R.: Seven words you can't say on Answerbag: contested terms and conflict in a social Q&A community. In: Proceedings of the 27th ACM Conference on Hypertext and Social Media, pp. 27–36. ACM, New York (2016)
12. Gazan, R.: Specialists and synthesists in a question answering community. In: Proceedings of the American Society for Information Science & Technology Annual Meeting, p. 10. ASIST, Austin (2006)
13. Guy, I., Shapira, B.: From royals to vegans: characterizing question trolling on a community question answering website. In: Proceedings of the 41st International ACM SIGIR Conference on Research & Development in Information Retrieval (SIGIR 2018), pp. 835–844. ACM, New York (2018).
14. Hara, N., Fichman, P., Meyer, E.T., Chen, Y., Rieh, S.: Deception in online trolling. A social informatics perspective on misinformation, disinformation, deception, and conflict. In: Proceedings of the 82nd Annual Conference of the American Society of Information Science and Technology, Melbourne, Australia (2019).
15. Hardaker, C., McGlashan, M.: "Real men don't hate women": Twitter rape threats and group identity. J. Pragmat. **91**, 80–93 (2016)
16. Harper, F.M., Raban, D., Rafaeli, S., Konstan, J.A.: Predictors of answer quality in online Q&A sites. In: Proceedings of the Conference on Human Factors in Computing Systems, pp. 865–874. ACM, Florence (2008)
17. Harper, F.M., Moy, D., Konstan, J.A.: Facts or friends? Distinguishing informational and conversational questions in social Q&A sites. In: CHI 2009, Boston, Massachusetts, pp. 759–768 (2009)
18. Ireton, C., Posseti, J.: Journalism, 'Fake News' and Disinformation: A Handbook for Journalism Education and Training. UNESCO, France (2019). https://en.unesco.org/sites/default/files/journalism_fake_news_disinformation_print_friendly_0_0.pdf
19. Jeon, G.Y., Rieh, S.Y.: Answers from the crowd: how credible are strangers in social Q&A? In: iConference 2014 Proceedings, pp. 663–668, Berlin (2014)
20. John, B.M., Chua, A.Y., Goh, D.H.: A predictive framework for retrieving the best answer. In: Proceedings of the 2008 ACM Symposium on Applied Computing, pp. 1107–1111. ACM, Fortaleza (2008)
21. John, B.M., Goh, D.H., Chua, A.Y.: Predictors of high-quality answers. Online Inf. Rev. **36**, 383–400 (2012)
22. Kang, M.: Active users' knowledge-sharing continuance on social Q&A sites: motivators and hygiene factors. Aslib J. Inf. Manag. **70**, 214–232 (2018)
23. Kim, S., Oh, S.: Users' relevance criteria for evaluating answers in social Q&A site. J. Am. Soc. Inf. Sci. Technol. **60**, 716–727 (2009)
24. Miles, M.B., Huberman, A.M.: Qualitative Data Analysis: An Expanded Sourcebook, 2nd edn. Sage Publications Inc., Thousand Oaks (1994)
25. Ong, C., Day, M., Hsu, M.: The measurement of user satisfaction with question answering systems. Inf. Manag. **46**, 397–403 (2009)
26. Philips, W.: This is Why We Can't Have Nice Things: Mapping the Relationship Between Online Trolling and Mainstream Culture. MIT Press, Cambridge (2015)

27. Rehavi, A., Refaeli, S.: Knowledge and social networks in Yahoo! Answers. In: 45th Hawaii International Conference on System Science (HICSS), Maui, Hawaii, pp. 781–789 (2012)
28. Rosenbaum, H., Shachaf, P.: A structuration approach to online communities of practice: the case of Q&A communities. J. Am. Soc. Inf. Sci. Technol. **61**, 1933–1944 (2010)
29. Sanfilippo, M., Yang, S., Fichman, P.: Managing online trolling: from deviant to social and political trolls. In: Proceedings of the 50th Hawaii International Conference on System Sciences, Hawaii, pp. 1802–1811 (2017)
30. Shachaf, P.: The paradox of expertise: is the Wikipedia reference desk as good as your library? J. Doc. **65**, 977–996 (2009)
31. Shachaf, P., Hara, N.: Beyond vandalism: Wikipedia trolls. J. Inf. Sci. **36**(3), 357–370 (2010)
32. Sobieraj, S., Berry, J.M.: From incivility to outrage: political discourse in blogs, talk radio, and cable news. Polit. Commun. **28**, 19–41 (2011)
33. Søe, S.O.: Algorithmic detection of misinformation and disinformation: Gricean perspectives. J. Doc. **74**, 309–332 (2017)
34. Wu, P.F., Korfiatis, N.: You scratch someone's back and we'll scratch yours: collective reciprocity in social Q&A communities. J. Am. Soc. Inf. Sci. Technol. **64**, 2069–2077 (2013)
35. Wilkinson, D., Huberman, B.: Cooperation and quality in Wikipedia. In: WikiSym 2007, pp. 157–163, Montreal (2007)
36. Yang, L., Bao, S., Lin, Q., et al.: Analyzing and predicting not-answered questions in community-based question answering services. In: Proceedings of the Association for the Advancement of Artificial Intelligence Conference, pp. 1273–1278. AAAI, Palo Alto (2011)
37. Zhu, Z., Bernhard, D., Gurevych, I.: A multi-dimensional model for assessing the quality of answers in social Q&A sites. In: Proceedings of 14th International Conference on Information Quality, vol. 1, pp. 264–265, Potsdam (2009). https://tuprints.ulb.tu-darmstadt.de/1940/1/TR_dimension_model.pdf

Counteracting Misinformation in Quotidian Settings

Abdul Rohman(✉)

RMIT University, 521 Kim Ma Street, Ba Dinh, Hanoi 100000, Vietnam
abdul.rohman@rmit.edu.vn

Abstract. Recent studies investigating misinformation spread have been situated within political contexts and have used psychological and technological approaches. In response, this study illuminates everyday life situations where people discover misinformation. Based on interviews conducted in Vietnam, it found that people's decision to counteract misinformation in part links to their existent relationship with its sharer. People tend to counteract misinformation shared by significant others rather than by strangers. The need to adhere to norms in order to keep the relationships harmonious and to avoid embarrassing the sharer shapes what methods are used to counteract misinformation. The findings demonstrate the role of maintaining relationships in choosing appropriate ways of counteracting misinformation, offering insights for reconciling ideological polarizations in everyday life.

Keywords: Misinformation · Fake news · Disinformation · Information sharing · Vietnam

1 Introduction

People encounter misinformation through quotidian interactions with others in online and offline settings. This phenomenon has principally appeared in extant studies on how confirmation biases and groupthinks expedite the spread of misinformation [1]. Despite being aware of the falsehood, the ideology promulgated by the source, people still seem to share misinformation because it serves their needs [2]. Consequently, having adequate media literacy skills is insufficient to deal with misinformation without the awareness to interrogate the biases that people hold when evaluating the quality of information circulating in everyday life [3]. In this vein, the decision whether to share, ignore, or counteract misinformation is cognitive in nature. The reaction to misinformation depends on the individual's assessment about how their action will directly affect their interest at a point in time [4, 5]. This approach to why people share misinformation pushes the fundamental understanding that people contextually interact with information, regardless of its truth or falsehood, to the periphery. In fact, misinformation and its subsequent interpretation are embedded in social interactions [6].

The interactions shape what action people deem appropriate when encounter any information they consider false. Studies have demonstrated that people tend to believe

K. Toeppe et al. (Eds.): iConference 2021, LNCS 12646, pp. 141–155, 2021.
https://doi.org/10.1007/978-3-030-71305-8_11

and share misinformation because it is aligned with their predispositions [7], resulting in a strong belief that information shared by others from different perspective is false and irrelevant [8]. This psychological approach to understanding misinformation spread underestimates that such a belief indeed is a byproduct of interactions with others [9]. In everyday life, people may cordially counteract misinformation shared by others, allowing for exchanging correct information with one another [10]. In this sense, the potential for misinformation to be corrected or shared is situated within mundane interactions that motivate people from different backgrounds to engage in information sharing activities. These interactions allow corrections to misinformation to be presented organically, as well as the falsehood in the misinformation to be solidified [11]. People ascribe meanings to interactions as they consider maintaining relationships with others involved in the interactions is important to gather, use, and share information impacting their life [12]. Hence, counteracting misinformation that others share can affect the continuity of the relationships. In this sense, preserving the relationships is preferred, leading people to seek appropriate methods for counteracting the misinformation. Like the act of sharing information, the counteraction is a form of social performances, in which people choose to act in accordance to what is suitable for themselves and the situation they are in [13]. Counteracting can risk both the people's and the sharer's social situations, as the need to maintain the relationships out-weighs the intention to provide correct information. Essentially, the meaning that people put into the relationship with the sharer affects the evaluation of whether to counteract misinformation is appropriate [14].

Building on the above understanding, this study aims to illuminate mundane situations where people discover misinformation and how their relationships with the sharer lead to different ways of counteracting. This endeavor potentially provides an understanding that any form of information, regardless of its truthfulness, circulates within complex social interactions, in which people assess their positions, their relationships with others involved in the interactions, the norms that shape what actions are appropriate to perform, and the consequences to the relationships [15].

2 Literature

2.1 Quotidian Interactions as a Context

Understanding what lies beneath quotidian interactions allows for a closer proximity to the dynamics of the encounters in everyday situations [16]. Observing the everyday unfolds the opportunity to illuminate the insights that are less apparent in research employing experimental or survey methods [17]. The everyday represents the actions and behaviors that people perform when naturally interacting with others, providing an organic understanding of what it is that underlies the interactions [18]. Contextualizing a study in a quotidian setting potentially enables us to untangle the complexity and subtlety that bind social interactions. Like information, misinformation is shared in situations where people interact with others, demonstrating that its spread and consequences are a byproduct of the interactions rather than strictly situated within the people's individual psychologies [19].

Recent studies on misinformation have largely used politics as a context. These studies, mostly situated in democratic countries, have documented the mechanisms, repercussions, and political ramifications of misinformation spread [20]. While portraying the current social dynamic affecting the political arena, such studies seem to assume that most people are able to identify the falsehood in the misinformation they encounter and to freely converse about political subjects. The falsehood in fact is often thin, hard to identify by non-experts, easily luring ordinary people to make false decisions [21]. In an environment where political content is sensitive, principally because of limited freedom of speech and expression, most people tend to concern themselves with non-political misinformation impacting their everyday lives. In this sense, misinformation is embedded in a nexus of mundane interactions, in which a delicate strategy to protect existing relationships with others characterizes appropriate ways of counteracting it.

In response, this study is situated within an environment where political discussions are scant. Vietnam, the most populous Socialist-Communist country in Southeast Asia, is chosen as a locus of data collection. The country has transitioned to an open policy-oriented economy since it became a member of World Trade Organization in 2007 [22]. Over 70% of approximately 97 million Vietnamese are connected to the Internet and 50% of these are social media users.[1] Other forms of information and communication technologies rapidly grow, enabling initiatives to mitigate the potential of the technologies to become tools for spreading misinformation. The government imposes formal approaches in the forms of laws and sanctions to combat it, as its spread has become a mutual concern among countries in the region [23].

The everyday interactions of the ordinary Vietnamese people are marked by the tendency to avoid conflicts in order to maintain social harmony.[2] Preserving relationships is key to interacting with others, perpetuating the inclination to conceal different opinions and disagreements. Correcting others sharing misinformation thus can be seen as challenging the extant norm that requires people to maintain harmonious relationships with others by avoiding sharing information that potentially disrupts regular interactions [25]. In a collectivistic society such as Vietnam, the relationships with others lie in a complex, delicate social hierarchy. Making others look good in the public eye is considered a social obligation while everyday interactions between different ages are often formal and less equal. Considering such characteristics, Vietnam offers promising insights to unravel the complexity of situations when ordinary people living in a non-democratic, collectivistic setting discover information they deem false in everyday life.

2.2 Discovering Misinformation

To unfold the counteracting misinformation as an everyday practice, this study builds on the concept of discovering information in context [26]. This concept emphasizes the understanding that people use and share information in social settings. Thus, people interact with information, together with others situated within a certain time and space, and this interaction is not entirely driven by individual needs. Rather, people's situations shape the appropriate actions, allowing for the continuous construction of the meanings that emerge from the interactions [27]. Hence, the actions reflect the people's ability to examine the consequences of the actions to themselves, their situations, and others

[1] www.internetworldstats.com.

[2] www.hosftedeinsights.com.

involved in the situations [28]. With that in mind, the concept of discovering information is a response to a strong tendency to understand information from a merely cognitive and psychological approach. Information indeed is a byproduct of human social interactions with others, which can affect the depth and breadth of the interactions over time [29].

Solomon's concept of discovering information implies that people use and share the information they discover without caution, giving inadequate attention to the fact that information may contain false or true content. In response, this study attempts to highlight situations where ordinary people discover misinformation, a form of information that can be true or false [30]. Misinformation potentially triggers uncertainty and polarization, and, if it remains unverified, can lead to casualties and other undesirable outcomes [31].

People respond to misinformation differently. They may correct its sharer immediately [32], authenticate it with their inner circle [33], simply ignore it because of viewing it as irrelevant to their circumstances [34], or refuse to accept the falsehood it contains [11]. Documenting different actions people performed when discovering misinformation, these studies have provided a limited understanding of the situation in which people decide to counteract misinformation in quotidian interactions.

The situation comprises people, norms, actions, and spaces in which social interactions occur [35]. Considering that misinformation is a critical incident that happens in social settings [36], a focus on situations has the potential to identify how people counteract misinformation they discover in mundane interactions, contributing to extant cognitive and psychological approaches to misinformation spread. Counteracting misinformation is an unexpected occurrence but calls for further reflection and examination to prevent unwanted events from happening, affecting people's relationships with others or harming their own situations. Given that, this study pays attention to everyday interactions in a collectivistic society and the norms that govern the interactions, shaping the decision whether counteracting misinformation is appropriate.

The presence of others, together with information exchanged, shape what form of counteractions are appropriate, as people reflect on what is the proper action to react to the situation faced [37]. Characteristics of other people involved in the situation when people discover misinformation, such as age and social positions, inform whether counteracting will threaten the existing relationships. The closeness between people and the misinformation sharer plays a role in whether or not counteracting the misinformation will disrupt the relationship. As such, there is a possibility that people will choose to ignore the misinformation when perceiving that correcting it will be detrimental to their relationship with the sharer. In other words, the perceived risk of counteracting outweighs the benefit, hence leaving the misinformation uncorrected is preferred [38].

The perceived risks stem from people's preconceptions about whether the spaces where the interactions take place are safe [39]. Sharing different content and information will likely occur in a certain space where people view it safe [40]. A perception of safety shapes the decision of whether to share or conceal information, where assessments of the benefits and costs for either decision take place during the span of the interactions [41]. Thus, people may refrain from sharing different information if doing otherwise will jeopardize their own safety in the space where they interact. A perceived safe space creates an inclination to share an array of information, allowing people from different backgrounds to discuss information impacting their lives [42]. Hence, counteracting false information seems to be unlikely to happen in a space where people feel that providing the sharer with corrections is risky.

The perceived risks can also come from the norms that govern social interactions. The norms provide guidance to appropriate responses to the situation that people are part of. Hence, punishments and rewards are an integral part of the norms, requiring people to adhere to them in order to maintain relationships with others and society in general [43]. In other words, a set of normative behaviors is expected to be present in order to keep different forms of social interaction in place, leading to negotiations of what information is acceptable to share and a continuous examination of appropriate actions in a certain situation. The norms, in this sense, can therefore be constraining when the need to counteract misinformation emerges [44]. Counteracting misinformation shared by others can be seen as inappropriate, because the impact of the norm of conformity is to maintain social harmony. People may prefer not to counteract mis-information, considering that it might disrupt the relationships.

Thus, people search for appropriate ways to counteract misinformation, when believing that leaving it uncorrected will potentially bring damage to the society. In this sense, people tend to choose a communication channel that is unlikely to embarrass the misinformation sharer, taking the norm into considerations for navigating between making the decision to counteract misinformation and to protect existing relationships with the sharer. Using a private communication channel can become an appropriate way of preventing misinformation from spreading without compromising the existing relationship with the sharer. In such a channel, people can safely address the falsehood in the misinformation while assuring the sharer that the correction provided comes from a good intention. This way of counteracting indicates an incentive to sow a mutual understanding despite differences in the information people share with one another [45].

Having said that, counteracting the misinformation sharer publicly remains feasible, provided that the people trying to counteract are confident with their preemptive knowledge and the sharer's ability to accept corrections [46]. In this sense, self-efficacy and the perceived outcome lead to the decision to counteract misinformation. The stock of knowledge that people have about a subject matter generates the intention to counteract or believe misinformation, together whether its spread will directly impact them or others who matter to them [47]. Such an individual approach however undermines the complexity of misinformation spread through mundane interactions, resulting in a simplification of the social dynamics that gradually develop as people interact with others, in a particular space, and with different types of information [48].

Broadly, the presence of different types of people shapes the ways people interact with information in general [49]. Materialized in different characteristics and meanings, these social types influence whether it is appropriate to counteract misinformation and what types of communication channels are aligned with the expected norms. Counteracting misinformation, embodying an act of sharing different information, thus is embedded in social situations. It involves foreseeing the impact of taking a certain action on existing relationships with others and the situations that people faced [50]. The decision to counteract misinformation lies in a constant examination of potential damages that it may cause and the relationships between people and the sharer, in which a certain norm shapes what counteracting methods are considered appropriate.

With the foregoing discussion in mind, this study asks:

a. What are situational factors that shape people's decisions to whether counteract misinformation?
b. How do such situational factors affect the methods that people choose to counteract misinformation in everyday life?

3 Method

To answer the above questions, we conducted interviews with 36 participants (28 women). Among these, 24 had college degrees, nine had vocational diplomas, two graduated from high schools and one graduated from a primary school. The participant's average age was 34 years. These participants were recruited via social media and the researchers' connections with the local communities. To be eligible to participate in the interviews, the participants had to have at least once corrected any information they considered false online.

The data collection period was from May to September 2019. Most interviews lasted between 60 and 75 min. Two Vietnamese native speakers conducted the interviews, maintaining the cultural and contextual elements of the situations that the participants experienced when counteracting misinformation. At the end of the interviews, the participants were asked to nominate their associates to participate in this study [51]. Most participants nominated female associates for convenient reasons, dis proportionally skewing the sample towards female participants. Gender was not the focus of the present study, this limitation to data collection however should be acknowledged.

Data analysis began in October 2019. We used NVivo for data analysis, after the interviews were transcribed and translated into English. On a weekly basis, we discussed some of the themes and patterns regarding the situations where the participants encountered information they deemed false as well as their considerations to whether counteract. This later process, guided by the proposed research questions, helped us develop a holistic understanding of the data. Thus, the findings reported in the following sections were able to capture the participants' experiences, while addressing the existing theoretical gap asserted in the literature [52].

4 Findings and Discussion

Discovering misinformation, people took into account whether the sharer was a stranger or significant other. These different social types reflected the degree of closeness between people and the sharer, resulting in a decision to whether counteract the misinformation or not. People tended not to counteract misinformation shared by strangers if the perceived risks of counteracting were high. In comparison, when discovering significant others shared misinformation, people tended to counteract it. Guided by the norm governing everyday interactions, the age of the sharer and the existing relationships affected the methods that people used to counteract misinformation. People preferred an indirect method to counteract misinformation shared by older people. Conversely, if the sharer was younger, or at the same age, people seemed to choose a direct method to counteract misinformation. Despite employing different methods, people preferred private communication channels for counteracting misinformation. Figure 1 visualizes these findings. The sections that follow illuminate some of the situations where people discovered misinformation and demonstrate the manner of counteraction according to the norms applicable to their everyday lives.

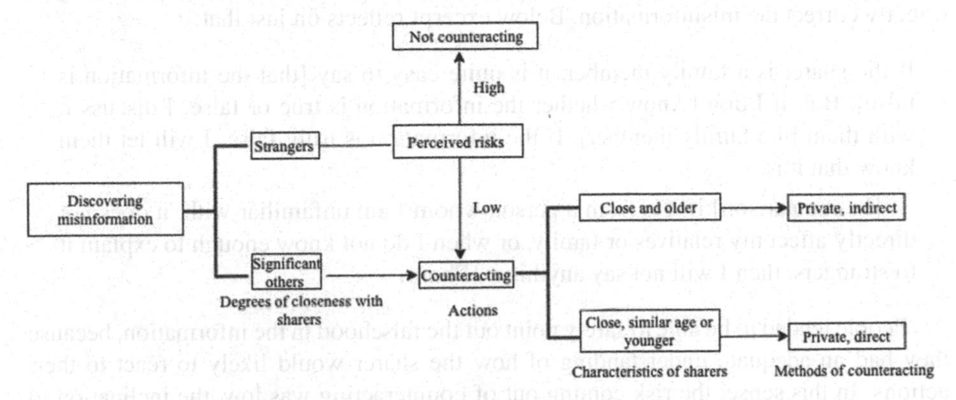

Fig. 1. Counteracting misinformation in quotidian settings.

4.1 How Close Was the Sharer with Me?

The degree of closeness with the sharer affected whether people decided to counteract misinformation. The closeness ranged from significant others to strangers, facilitating the inclination to address the falsehood in the information the sharer intentionally or unintentionally spread through different occasions. Family members and friends were some of the examples that the participants reported as significant others, to whom counteracting misinformation was considered valuable in order to protect themselves and others from harm. A participant described different actions she took based on the closeness of her relationship with a significant other who happened to share misinformation:

> There are some cases when I am quiet [did not counteract misinformation]. I feel uncomfortable of directly correcting [false] information from strangers. For example, when I saw someone shared an article falsely claiming that two [famous Chinese] singers were truly in love with each other, I shared that article on my Facebook and added my own opinion on it instead of directly confronting the sharer. (Anh)

The decision to counteract misinformation appeared to be related to whether people had meaningful relationships with the sharer or not. The excerpt shows that she tended to ignore strangers who shared information she considered false regarding some of her favorite singers. This decision linked to the feeling of discomfort for correcting strangers. While believing that the information shared on social media (i.e. Facebook) was false, she chose not to directly correct the sharer. She, instead, corrected the falsehood by re-posting the article and adding some corrections and contexts to the content [53], providing clarifications to others who may have believed the false content that the article contained. She believed that such a way of counteracting was safer because a direct counteraction on Facebook could have risked her of being attacked by other Facebook users.

However, if the sharer was a significant other, then people seemed to be willing to directly correct the misinformation. Below excerpt reflects on just that:

> If the sharer is a family member, it is quite easy to say [that the information is false]. But, if I don't know whether the information is true or false, I discuss it with them [the family member]. If the information is truly false, I will let them know that it is.
>
> … [In comparison] if it is from a person whom I am unfamiliar with, it does not directly affect my relatives or family, or when I do not know enough to explain it to strangers, then I will not say anything. (Pham)

People tended to be able to safely point out the falsehood in the information, because they had an adequate understanding of how the sharer would likely to react to their actions. In this sense, the risk coming out of counteracting was low, the inclination to address the falsehood in the information was high. People believed that the correction provided to the misinformation would not disrupt their existing relationships, because they knew that the sharer would be able to accept the correction [54]. In this vein, the sharer had the belief that the people correcting the misinformation had no negative intention to embarrass them nor to impose their beliefs and ideologies.

In relation to such good faith, people counteracted misinformation shared by significant others was in part to protect them from harm. This altruistic intention was rooted in the belief that the significant others would endanger themselves or others if the mis-information left uncorrected. A participant commented:

> If I correct people that I'm not close with, I don't know what they will think about me afterwards. Besides, I think, they will try hard to defend the information they shared, which, [if I correct it], would quickly turn into an unnecessary argument. That would never end well. … Most time I will correct [misinformation shared by] my family members.
>
> There was one time my sister told me information I consider incorrect regarding a concept taught in Vietnam for students grade 2–3. Since I was in the education field, I sent her a link of an article to correct her. The concept was very old. And, she might not have enough information to fairly comprehend it. Since she was my sister, and I saw that she was lacking such information and would share it with many people, so I corrected her. …
>
> My mother-in-law has diabetes so I look for her information on a low sugar diet. Thus, every time I hear false information that eating a certain food will reduce sugar level, I will correct it. Similar to that, my dad is quite overweight. He said that eating an unripe banana before dinner is good for losing weight, whereas I read

that eating it will badly affect health. I correct such false information immediately because it might endanger my family members. (Hung)

The comment demonstrates that correcting misinformation stemmed from an altruistic intention to prevent significant others from harming themselves and others. This act of altruism embodied a sense of civic duty that revolved around protecting the ones within the persons' social circle while at the same time mitigating the risk that the misinformation potentially brought to a larger social circle [45]. People believed counteracting misinformation shared by significant others could have the potential to prevent it from wildly spreading. In this sense, the agency that people exercised to deal with misinformation showed a need to protect the significant others, while indirectly preventing the misinformation from reaching the public. Put simply, protecting members of the smaller social circle embodied the people's intention to protect the society from damages that the misinformation could bring.

In broad terms, counteracting misinformation in everyday life seemed be selfish in intent as reflected in the people's inclination to counteract misinformation only if they knew the sharer, resulting in a tendency to ignore misinformation shared by strangers. Paired with that, counteracting misinformation was social in intent, as people believed that counteracting it would have affected the larger social circle. People thought of counteracting misinformation that circulated within their social circle as a way of protecting society and preventing a wider audience from basing their actions on the misinformation. Counteracting misinformation began from individual interests, as people tended to care about the wellbeing of others that mattered to them rather than strangers. Within that however there was also a collectivistic intention to keep others from making decisions based on misinformation. Hence, mitigating the potential damages that it could possibly bring was necessary.

4.2 How Old Was the Sharer?

The social norms provided guidance to interactions, shaping the decision on whether countering misinformation was appropriate in a given situation. These norms helped inform what actions were acceptable as people from different backgrounds interacted, allowing for the maintenance of social harmony [9]. The excerpt below illustrates this, principally related to the need of respecting elders, shaped the methods that Vietnamese people chose when deciding to counteract misinformation without disrupting their existing relationships with the sharer:

> Whether counteracting misinformation or not depends on the relationship between an older person and the younger one. For examples, if the relationship is close, then I can correct the misinformation directly. Otherwise, I have to examine whether it is true or not. If I think it is false, I am not close to the sharer, I have to double-check it and correct the sharer via [Facebook] Messenger or chat later. In short, I am not brave enough to speak the truth in such a situation. …
>
> [With an older person] I will correct it [misinformation], but with a much softer style. "I hear about it differently …. I hear it from a legitimate source of information and so on" The elderly is keen to listen to stories from people and believe them.

> It's like their hobby to listen to rumors, to sources of false information that only want to attract the attention of the receivers …. For example, when my parents give wrong information, I react immediately. "Where do you hear that from?" When something potentially goes wrong, I will usually counteract misinformation. In a family setting, I say it softly; the response is usually less severe. For people like my parents, I still have to care [about their feelings] a little bit. For my brother, no need. (Linh)

The excerpt illuminates a norm that guided ways of counteracting misinformation. The age of the sharer appeared to shape appropriate ways of counteracting, based on the norm requiring the younger to respect the elder. In this circumstance, providing corrections could be seen as embarrassing the elder or challenging their positions embedded in extant social hierarchies [24]. In the above excerpt, she carefully corrected her elder who happened to spread misinformation within her social circle. The fact that she was younger than the sharer conditioned her to look for a way of counteracting that would not be considered as disrespectful. Rather than directly addressing the falsehood in the information, she pointed her elder to other information that could motivate them to rectify the misinformation they already shared voluntarily. Indirectly talking about related content surrounding the misinformation with the elder seemed to be more appropriate than directly correcting it. She believed doing so would mitigate potential conflicts that might follow direct corrections, which could have disrupted existing relationships with the elder.

In some events, people preferred not to counteract misinformation because doing so would be seen as offensive. People chose to avoid interpersonal conflicts over correcting the falsehood in the misinformation. Staying silent was a way to keep the relationships normal [13]. Doing otherwise would trigger the impression that the younger was challenging the elder, fraying the relationships as unnecessary arguments ensued. Thus, the decision not to counteract misinformation that the elder shared was a result of a continuous examination to its ramifications to the relationships. Believing that keeping a good relationship was more important than providing the sharer with correct information, people were disinclined to counteract misinformation. If, on the other hand, they were so inclined, methods appropriate to the extant norms were employed in order for preventing the act of correcting from disrupting the relationships.

By comparison, when knowing that the sharer was younger or at the same age, people counteracted misinformation in a direct way. In the above excerpt, discovering that her younger brother shared misinformation on Facebook, she directly corrected it in person, as perceiving that their age positioned them in the same social level. Interacting with the elder, as prescribed by extant norms governing social interactions, required subtle and soft methods of counteracting. On the other hand, with others whom people considered within the same position, such an expectation seemed to be absent. The risk that correcting the misinformation would break the existing relationship was therefore low, making it feasible to directly counteract the misinformation.

The findings suggest that the norms that guided how people should behave within their social circles shaped what methods were appropriate to counteract misinformation. The norms provided a set of ideas about the consequences of choosing a counteracting method that would not disrupt the existing relationships with the sharer. In quotidian settings, considering that correcting the elder who shared misinformation could potentially harm the existing relationship, people used a soft, indirect method to counteract misinformation. People believed that this way was viable and would not compromise the existing relationship with the sharer, especially if the sharer was normatively perceived as located in a higher social position. The indirect counteracting method stemmed from a continuous assessment of the need to counteract misinformation and to keep the relationship with the sharer in harmony. This normative consideration then informed the type of communication channels was suitable for use.

4.3 Private or Public?

People preferred private communication channels for counteracting misinformation shared by significant others. A one-on-one messaging app or personal conversation was considered more appropriate than public communication channels such as social media. People believed that a use of private communication channels minimized the risk of the sharer being embarrassed by being corrected, which could be detrimental to interpersonal relationships. Correcting misinformation shared by significant others seemed to be delicate; people did that in private as they considered it more acceptable in the public eye. The excerpt below points out reasons for choosing a private over public communication channel to counteract misinformation:

> If someone I know posts false information, I will choose to talk to them in person rather than directly correcting them on social media. I feel like it is unnecessary to leave a comment on their post. Maybe, after correcting the information personally, they will correct it themselves. If I correct it directly on social media, what might other people think. They would say, "oh they both know each other but choose to criticize each other on Facebook rather than talking in private." The good intention [to counteract misinformation] will become a backlash. (Kieu)

The decision to counteract misinformation through private channels reflected the norm that bound the relationship between people and the sharer. In this case, protecting the sharer's public reputation was deemed necessary while correcting the falsehood in the misinformation they shared. The decision to use one-on-one communication channels linked to the concern to what other people might think about one's relationship with the sharer. The excerpt demonstrates that using public communication channels such as social media platforms to counteract misinformation was inappropriate. To do so potentially created a misperception that both parties had a bad relationship, which seemed to be unaligned with the social expectation to keep existing relationships harmonious. Correcting the sharer on Facebook potentially made other Facebook friends develop a view that both parties were in a dispute. As a result, the good intention to counteract misinformation would become a source of misunderstandings, which could lead to interpersonal conflicts if left unresolved.

The norm discouraged people from embarrassing others in public, making the use of private communication channels appropriate. The comment below shows how the norm affected people's decision as to which communication channel was appropriate to use when deciding to counteract misinformation:

> I won't correct any false information most time, if my friends shared it. I don't really want to get involved in any arguments with them .... There was information that I knew was incorrect about a fire near my friend's workplace. She had told me everything about the accident. But when some other friends shared information about it that I thought false [on Facebook], I told them about what I heard from her without saying that these friends were wrong I knew there would be others who invested time to find information and be ready to use it to argue anytime. Generally, if misinformation is from my close friends and I feel like I know the correct information, I will tell them in private. I don't comment online because I don't want to embarrass them in public. (Phuong)

The comment illuminates why people considered using a private communication channel more appropriate for correcting misinformation shared by significant others. The private channel prevented the sharer from feeling embarrassed when people provided the correction. Such a consideration implied that people aimed at balancing their intention to share correct information and to protect the sharer from being seen as untrustworthy or incapable of identifying the falsehood. This altruistic intention demonstrated that correcting misinformation remained important and that the method used to correct the sharer however should be appropriate. People had a reason to believe that using private communication channels would mitigate the risk of unnecessary arguments, as the sharer would not feel embarrassed. Besides, private communication channels opened the chance to interpersonally share the intention to prevent the sharer from harming themselves and others.

Broadly, the present finding suggests that people deemed using private communication channels to be more appropriate to counteract misinformation that significant others shared. The appropriateness stemmed from the norm that expected people to maintain the reputation of the other and keep relationships with them harmonious. The private communication channels seemed to be able to minimize the risk that the sharer would feel embarrassed, compared to when the sharer received the corrections publicly. Paired with that, counteracting misinformation in private was linked to the act of protecting the sharer from harm and bringing damages to other people. In the context of this study, a private communication channel such as a one-on-one messaging app seemed to be able to offer civility in counteracting misinformation. Although potential disruptions to the existing relationship with the sharer could still occur, people perceived that using such a communication channel was more appropriate than confronting the sharer publicly.

5 Conclusion

This study has illuminated the situations where people discover misinformation in everyday life. The decision to counteract misinformation in part links to existent relationships with its sharer. People tend to counteract misinformation shared by significant others rather than by strangers. This tendency stems from the intention to prevent the significant others from being seen as untrustworthy or bringing damages to others if the misinformation is left unaddressed. The need to keep the relationships harmonious and to avoid embarrassing the sharer shape what methods are appropriate to counteract misinformation in the situations that people face. If the sharer is older, people prefer an indirect counteracting method. Believing that a public, direct counteracting method is inappropriate and can potentially disrupt their relationships with the sharer, people are inclined to choose private communication channels when deciding to counteract misinformation, regardless of the age of sharer.

As such, the finding expands the current understanding of the concept of discovering information. It sheds light on nuances in which people discover misinformation as well as the appropriate counteracting methods. Discovering misinformation invites the need to address the falsehood in the misinformation, in which assessments to the situation where it circulates affects the evaluation of whether counteracting it will harm existing relationships with the sharer. In this sense, the spread of misinformation, together with the appropriate counteracting methods, lies within a delicate matter that influences the continuity of the relationship with the sharer and the intention to prevent misinformation from spreading widely. Choosing an inappropriate method will likely harm the relationship, as counteracting can disrupt normal interactions with the sharer. On the other hand, leaving misinformation uncorrected can potentially bring damage to the sharer and others. In short, the spread of misinformation and the methods that people choose to counteract it with are embedded in complex social interactions, rather than purely situated within cognitive and technologically deterministic situations.

References

1. Gorman, S.E., Gorman, J.M.: Denying to the Grave. Why we Ignore the Facts that will Save Us. Oxford University Press, . New York (2017)
2. Sunstein, C.: #republic. Divided Democracy in the Age of Social Media. Princeton University Press, Princeton (2017)
3. Boyd, D.: You think you want media literacy… Do you? (2018). https://points.datasociety.net/you-think-you-want-media-literacy-do-you-7cad6af18ec2. Accessed 15 Oct 2018
4. Nickerson, R.S.: Confirmation bias: a ubiqutis phenomenon in many guises. Rev. Gen. Psychol. **2**(2), 175–220 (1998)
5. Walter, N., Murphy, S.T.: How to unring the bell: a meta-analytic approach to correction of misinformation. Commun. Monogr. **85**(3), 423–441 (2018)
6. Fallis, D.: What is disinformation? Libr. Trends **63**(3), 401–426 (2015)
7. Thorson, E.: Belief Echoes: the persistent effects of corrected misinformation. Polit. Commun. **33**(3), 460–480 (2016)
8. Ecker, U.K.H.: Why rebuttals may not work: the psychology of misinformation. Media Asia **44**(2), 79–87 (2017)
9. Berger, P., Luckmann, T.: Social Construction of Reality: Treatise in the Sociology of Knowledge. Penguin Books, London (1966)
10. Fetzer, J.H.: Information: does it have to be true? Minds Mach. **14**(2), 223–229 (2004)
11. Mercier, H., Sperber, D.: The Enduring Enigma of Reason. Harvard University Press, Cambridge (2017)
12. Widen-Wulff, G., Ek, S., Ginman, M., Perttila, R., Sodergard, P., Totterman, A.-K.: Information behaviour meets social capital: a conceptual model. J. Inf. Sci. **34**(3), 346–355 (2008)
13. Chatman, E.A.: A theory of life in the round. J. Am. Soc. Inf. Sci. **50**(3), 207–217 (1999)
14. Chatman, E.A.: Framing social life in theory and research. New Rev. Inf. Behav. Res. **1**, 3–17 (2000)

15. Solomon, P.: Information mosaics: patterns of action that structure. In: Exploring the Contexts of Information Behaviour, pp. 150–175 (1999)
16. Goffman, E.: Encounters. Two Studies in the Sociology of Interaction. Penguin University Books, Middlesex (1961)
17. Pollio, H.R., Henley, T.B., Thompson, C.J.: The Phenomenology of Everyday Life. Cambridge University Press, Cambridge (1997)
18. Scott, S.: Making Sense of Everyday Life. Polity Press, Cambridge (2009)
19. Karlova, N.A., Fisher, K.E.: A social diffusion model of misinformation and disinformation for understanding human information behaviour. Inf. Res. **18**(1), 1–17 (2013)
20. Bennett, W.L., Livingston, S.: The disinformation order: disruptive communication and the decline of democratic institutions. Eur. J. Commun. **33**(2), 122–139 (2018)
21. Helfand, D.J.: A Survival Guide to the Misinformation Age: Scientific Habits of Mind. Columbia University Press, New York (2016)
22. Sakata, S. (ed.): Vietnam's Economic Entities in Transition. Palgrave Macmillan, New York (2013)
23. Victor, P.: The rise of 'fake news' and disinformation in ASEAN | The ASEAN Post. The ASEAN Post (2017). https://theaseanpost.com/article/rise-fake-news-and-disinformation-asean. Accessed 30 Oct 2018
24. Hofstede, G.: The cultural relativity of organizational practices and theories. J. Int. Bus. Stud. **14**(2), 75–89 (1983)
25. Burnett, G.: Colliding norms, community, and the place of online information: the case of archive.org. Libr. Trends **57**(4), 694–710 (2009)
26. Solomon, P.: Discovering information in context. Annu. Rev. Inf. Sci. Technol. **36**(1), 229–264 (2005)
27. Solomon, P.: Discovering information behavior in sense making. III. The person. J. Am. Soc. Inf. Sci. **48**(12), 1127–1138 (1997)
28. Savolainen, R.: Everyday Information Practices. A Social Phenomenological Perspective. The Scarecrow Press Inc, Lanham (2008)
29. Solomon, P.: Discovering information behavior in sense making. I. Time and timing. J. Am. Soc. Inf. Sci. **48**(12), 1127–1138 (1997)
30. Fallis, D.: A functional analysis of disinformation. In: iConference 2014 Proceedings (2014)
31. O'Connor, C., Weatherall, J.O.: The Misinformation Age. How False Beliefs Spread. Yale University Press , New Haven & London (2019)
32. Vraga, E.K., Bode, L.: I do not believe you: how providing a source corrects health misperceptions across social media platforms. Inf. Commun. Soc. **21**(10), 1337–1353 (2018)
33. Tandoc, E.C., et al.: Audiences' acts of authentication in the age of fake news: a conceptual framework. New Media Soc. **20**(8), 2745–2763 (2018)
34. Marwick, A.E.: Why do people share fake news? A sociotechnical model of media effects. Geo. L. Tech. Rev. **2**(474), 1–52 (2018)
35. Clarke, A.E.: Situational Analysis. Grounded Theory after the Postmodern Turn. Sage Publications , Thousand Oaks (2005)
36. Tandoc, E.C., Jenkins, J., Craft, S.: Fake news as a critical incident in journalism. J. Pract. 2786 (2018)
37. Chatman, E.A.: Channels to a larger social world: older women staying in contact with the great society. Libr. Inf. Sci. Res. **13**, 281–300 (1991)
38. Wilson, T.D.: Information sharing: an exploration of the literature and some propositions. Inf. Res. **15**(4) (2010)
39. Lingel, J., Boyd, D.: 'Keep it secret, keep it safe': information poverty, information norms, and stigma. J. Am. Soc. Inf. Sci. Technol. **64**(5), 9891–9981 (2013)
40. Widen-Wulff, G.: The Challenges of Knowledge Sharing in Practice: A Social Approach. Chandos Publishing, Oxford (2007)

41. Williamson, K., Kennan, M.A., Johanson, G., Weckert, J.: Data sharing for the advancement of science: overcoming barriers for citizen scientists. J. Assoc. Inf. Sci. Technol. **67**(10), 2392–2403 (2016)
42. Fisher, K.E., Naumer, C.M.: Information grounds: theoretical basis and empirical findings on information flow in social settings. In: Spink, A., Cole, C. (eds.) New Directions in Human Information Behavior, pp. 93–111. Springer, Dordrecht (2006)
43. Collins, R.: Interaction Ritual Chains, 2nd edn. Princeton University Press, Princeton (2005)
44. Giddens, A.: The Constitution of Society. Polity Press, Cambridge (1984)
45. Rohman, A., Ang, P.H.: Truth, not fear: countering false information in a conflict. Int. J. Commun. **13**, 4586–4601 (2019)
46. Cook, J., Lewandowsky, S., Ecker, U.K.H.: Neutralizing misinformation through inoculation: exposing misleading argumentation techniques reduces their influence. PLoS One **12**(5) (2017)
47. Young, D.G., Jamieson, K.H., Poulsen, S., Goldring, A.: Fact-checking effectiveness as a function of format and tone: evaluating FactCheck.org and FlackCheck.org. J. Mass Commun. Q. **95**(1), 49–75 (2018)
48. Tabak, E., Willson, M.: A non-linear model of information sharing practices in academic communities. Libr. Inf. Sci. Res. **34**(2), 110–116 (2012)
49. Turner, T., Boyd, D., Burnett, G., Fisher, K.E., Adlin, T.: Social types and personas: typologies of persons on the web and designing for predictable behaviors. Proc. Am. Soc. Inf. Sci. Technol. **44**(1), 1–6 (2008)
50. Solomon, P.: Discovering information behavior in sense making. II. . Soc. J. Am. Soc. Inf. Sci. **48**(12), 1127–1138 (1997)
51. Patton, M.Q.: Qualitative Research & Evaluation Methods, 3rd edn. Sage Publications Inc., California (2002)
52. Charmaz, K.: Constructing Grounded Theory: A Practical Guide Through Qualitative Analysis. Sage Publication Ltd., London (2006)
53. Amazeen, M.A., Thorson, E., Muddiman, A., Graves, L.: Correcting political and consumer misperceptions: the effectiveness and effects of rating scale versus contextual correction formats. J. Mass Commun. Q. **95**(1), 28–48 (2018)
54. Margolin, D.B., Hannak, A., Weber, I.: Political fact-checking on Twitter: when do corrections have an effect? Polit. Commun. **35**(2), 196–219 (2018)

Towards Target-Dependent Sentiment Classification in News Articles

Felix Hamborg[1,2](✉), Karsten Donnay[2,3], and Bela Gipp[2,4]

[1] Department of Computer Science, University of Konstanz, Konstanz, Germany
felix.hamborg@uni-konstanz.de

[2] Heidelberg Academy of Sciences and Humanities, Heidelberg, Germany

[3] Department of Political Science, University of Zurich, Zurich, Switzerland

[4] Data and Knowledge Engineering, University of Wuppertal, Wuppertal, Germany

Abstract. Extensive research on target-dependent sentiment classification (TSC) has led to strong classification performances in domains where authors tend to explicitly express sentiment about specific entities or topics, such as in reviews or on social media. We investigate TSC in news articles, a much less researched domain, despite the importance of news as an essential information source in individual and societal decision making. This article introduces NewsTSC, a manually annotated dataset to explore TSC on news articles. Investigating characteristics of sentiment in news and contrasting them to popular TSC domains, we find that sentiment in the news is expressed less explicitly, is more dependent on context and readership, and requires a greater degree of interpretation. In an extensive evaluation, we find that the current state-of-the-art in TSC performs worse on news articles than on other domains (average recall $AvgRec = 69.8$ on NewsTSC compared to $AvgRev = [75.6, 82.2]$ on established TSC datasets). Reasons include incorrectly resolved relation of target and sentiment-bearing phrases and off-context dependence. As a major improvement over previous news TSC, we find that BERT's natural language understanding capabilities capture the less explicit sentiment used in news articles.

Keywords: Sentiment classification · Stance detection · News bias · Media bias

1 Introduction

Target-dependent sentiment classification (TSC) is a sub-task of sentiment analysis that aims to identify the sentiment of a text, usually on sentence-level, towards a given target, such as named entities (NEs) or other semantic concepts [19]. Aspect-based sentiment classification (ABSC) [29], a closely related task, defines such targets as aspects of a given topic, e.g., "service" and "food" may be aspects of the topic "restaurant." Previous research on TSC and ABSC (due to their technical similarity we will refer to both as TSC) has focused mostly

K. Toeppe et al. (Eds.): iConference 2021, LNCS 12646, pp. 156–166, 2021.
https://doi.org/10.1007/978-3-030-71305-8_12

on domains in which authors tend to express their opinions explicitly, such as reviews, surveys, and social media [7,24,26,29].

In this paper, we investigate TSC in the domain of news articles – a much less researched domain that is of critical relevance, especially in times of "fake news," echo chambers, and news owner centralization [15]. How persons and other entities are portrayed in articles on political topics is, e.g., very relevant for individual and societal opinion formation [3,14,17].

The main contributions of this paper are: (1) We introduce *NewsTSC*, a manually annotated dataset for the exploration of TSC in political news articles. (2) We discuss similarities and differences between political news and established TSC domains. (3) We perform an extensive evaluation of state-of-the-art TSC approaches on NewsTSC. To improve classification performance, we also fine-tune a BERT language model [6] on a large news dataset, thereby establishing the current state-of-the-art in TSC on political news.

We provide the dataset including code book, code to reproduce our experiments, and the fine-tuned BERT at: https://github.com/fhamborg/newstsc.

2 Related Work

Most TSC-related research uses three annotated datasets: *Restaurant* and *Laptop*, containing reviews on restaurants and laptops [26], and *Twitter*, consisting of tweets [7]. Each example in these datasets consists of a *target*, *context* (often a single sentence or tweet), and the target's *sentiment* within its context.

The advent of word embeddings and deep learning including neural language models, such as BERT [6], has led to a performance leap in many natural language processing (NLP) disciplines including TSC, where, e.g., macro F1 gained from $F1_m = 63.3$ [21] to $F1_m = 75.8$ on the Twitter set [38]. Whereas traditional TSC research focused on careful feature engineering and dictionary creation (cf. [21]), researchers now focus on designing neural architectures suited to catch the relation between target and context [32,38,39]. By fine-tuning the underlying language model for the particular classification domain, performance can be improved further [28].

Text in news articles differs from reviews and social media in that news authors typically do not express sentiment towards a target explicitly (exceptions include opinion pieces and columns). Instead, they implicitly or indirectly express sentiment because language in news is expected to be neutral and journalists to be objective [1,10,15]. For example, news texts express sentiment by describing actions performed by a target, or by including and highlighting information in favor or against a target (or omitting and downplaying such information, respectively) [31]. Adding to the difficulty of news TSC, different readers may assess an article's sentiment towards a target differently [1], depending on their own political or ideological views (we discuss real-world examples in Sect. 3.3). Previous news TSC approaches mostly employ manually created [1] or semi-automatically extended [10] sentiment dictionaries. To our knowledge, there exist one dataset for evaluation of news TSC methods [34], which – perhaps due to its small size ($N = 1274$) – has not been used or tested in recent TSC literature. Another dataset contains quotes extracted from news articles, since quotes more likely contain explicit sentiment ($N = 1592$) [1].

To our knowledge, no suitable datasets for news TSC exist nor have news TSC approaches been proposed that exploit recent advances in NLP.

3 Dataset

We describe how we create the news TSC dataset, including the collection of articles and the annotation procedure. Afterward, we discuss the characteristics of the dataset.

3.1 Data Collection and Example Extraction

We create a base set of articles of high diversity in topics covered and writing styles, e.g., whether emotional or factual words are used (cf. [8]). Using a news extractor [16], we collect news articles from the Common Crawl news crawl (CCNC, also known as CC-NEWS), consisting of over 250M articles until August 2019 [23]. To ensure diversity in writing styles, we select 14 US news outlets,[1] which are mostly major outlets that represent the political spectrum from left to right, based on selections by [2,4,13]. We cannot simply select the whole corpus, because CCNC lacks articles for some outlets and time frames. By selecting articles published between August 2017 and July 2019, we minimize such gaps while covering a time frame of two years, which is sufficiently large to include many diverse news topics. To facilitate the balanced contribution of each outlet and time-range we perform binning: we create 336 bins, one for each outlet and month, and randomly draw 10 articles reporting on politics for each bin, resulting in 3360 articles in total.[2] During binning, we remove any article duplicates by text equivalence.

To create examples for annotation, we select all mentions of NEs recognized as PERSON, NROP, or ORG for each article [37].[3] We discard NE mentions in sentences shorter than 50 characters. For each NE mention, we create an example by using the mention as the target and its surrounding sentence as its context. We remove any example duplicates. Afterward, to ensure diversity in writing styles and topics, we use the outlet-month binning described previously and randomly draw examples from each bin.

---

[1] BBC, Breitbart, Chicago Tribune, CNN, LA Daily News, Fox News, HuffPost, LA Times, NBC, NY Times, Reuters, USA Today, Washington Post, and Wall Street Journal.

[2] To classify whether an article reports on politics, we use a DistilBERT-based [30] classifier with a single dense layer and softmax trained on the HuffPost [22] and BBC datasets [12]. During the subsequent manual annotation, coders discard remaining, non-political articles.

[3] For this task, we use spaCy v2.1.

Different means may be used to address expected class imbalance, e.g., for the Twitter set, only examples that contained at least one word from a sentiment dictionary were annotated [24,25]. While doing so yields high frequencies of classes that are infrequent in real-world distribution, it also causes dataset shift and selection bias [27]. Thus, we instead investigate the effectiveness of different means to address class imbalance during training and evaluation (see Sect. 4).

3.2 Annotation

We set up an annotation process following best practices from the TSC literature [24,26,29,34]. For each example, we asked three coders to read the context, in which we visually highlighted the target and assess the target's sentiment. Examples were shown in random order to each coder. Coders could choose from *positive*, *neutral*, and *negative* polarity, whereby they were allowed to choose positive and negative polarity at the same time. Coders were asked to *reject* an example, e.g., if it was not political or a meaningless text fragment. Before, coders read a code book that included instructions on how to code and already annotated examples. Five coders, students, aged between 24 and 32, participated in the process.

In total, 3288 examples were annotated, from which we discard 125 (3.8%) that were rejected by at least one coder, resulting in 3163 non-rejected examples. From these, we discard 3.3% that lacked a majority class, i.e., examples where each coder assigned a different sentiment class, and 1.8% that were annotated as positive and negative sentiment at the same time, to allow for better comparison with previous TSC datasets and methods (see Sect. 2). Lastly, we split the remaining 3002 examples into 2301 training and 701 test examples. Table 1 shows class frequencies of the sets.

We use the full set of 3163 non-rejected examples to illustrate the degree of agreement between coders: 3.3% lack a majority class, for 62.7%, two coders assigned the same sentiment, and for 33.9% all coders agreed. On average, the accuracy of individual coders is $acc_h = 72.9\%$. We calculate two intercoder reliability (ICR) measures. For completeness, Cohen's Kappa is $\kappa = 25.1$, but it is unreliable in our case due to Kappa's sensitivity to class imbalance [5]. The mean pairwise observed agreement over all coders is 72.5.

Table 1. Class frequencies of NewsTSC sets.

	Negative	Neutral	Positive	Total
Training	530	1600	171	2301
Test	167	487	47	701
Total	697	2087	218	3002

3.3 Characteristics of Sentiment in News Articles

In a manual, qualitative analysis of NewsTSC, we find two key differences of news compared to established domains: first, we confirm that news contains mostly implicit and indirect sentiment (see Sect. 2). Second, determining the sentiment in news articles typically requires a greater degree of interpretation (cf. [34]). The second difference is caused by multiple factors, particularly the implicitness of sentiment (mentioned as the first difference) and that sentiment in news articles is more often dependent on non-local, i.e., off-sentence, context. In the following, we discuss annotated examples (part of the dataset and discarded examples) to understand the characteristics of target-dependent sentiment in news texts.

We find that in news articles, a key means to express targeted sentiment is to describe actions performed by the target. This is in contrast, e.g., to product reviews where more often a target's feature, e.g., "high resolution", or the mention of the target itself, e.g., "the camera is awesome," express sentiment. For example, in "The Trump administration has worked tirelessly to impede a transition to a green economy with actions ranging from opening the long-protected Arctic National Wildlife Refuge to drilling, [...]." the target (underlined) was assigned negative sentiment due to its actions.

We find sentiment in ≈3% of the examples to be strongly reader-dependent (cf. [1]).[4] In the previous example, the perceived sentiment may, in part, depend on the reader's own ideological or political stance, e.g., readers focusing on economic growth could perceive the described action positively whereas those concerned with environmental issues would perceive it negatively.

In some examples, targeted sentiment expressions can be interpreted differently due to ambiguity. As a consequence, we mostly find such examples in the discarded examples and thus they are not contained in NewsTSC. While this can be true for any domain (cf. "polarity ambiguity" in [26]), we think it is especially characteristic for news articles, which are lengthier than tweets and reviews, giving authors more ways to refer to non-local statements and to embed their arguments in larger argumentative structures. For instance, in "And it is true that even when using similar tactics, President Trump and President Obama have expressed very different attitudes toward immigration and espoused different goals." the target was assigned neutral sentiment. However, when considering this sentence in the context of its article [36], the target's sentiment may be shifted (slightly) negatively.

From a practical perspective, considering more context than only the current sentence seems to be an effective means to determine otherwise ambiguous sentiment expressions. By considering a broader context, e.g., the current sentence and previous sentences, annotators can get a more comprehensive understanding of the author's intention and the sentiment the author may have wanted to communicate. The greater degree of interpretation required to determine non-explicit sentiment expressions may naturally lead to a higher degree of subjectivity. Due

[4] We drew a random sample of 300 examples and concluded in a two-person discussion that the sentiment in 8 examples could be perceived differently.

to our majority-based consolidation method (see Sect. 3.2), examples with non-explicit or apparently ambiguous sentiment expressions are not contained in NewsTSC.

4 Experiments and Discussion

We evaluate three TSC methods that define the state-of-the-art on the established TSC datasets Laptop, Restaurant, and Twitter: AEN-BERT [32], BERT-SPC [6], and LCF-BERT [38]. Additionally, we test the methods using a domain-adapted language model, which we created by fine-tuning BERT (base, uncased) for 3 epochs on 10M English sentences sampled from CCNC (cf. [28]). For all methods, we test hyperparameter ranges suggested by their respective authors.[5] Additionally, we investigate the effects of two common measures to address class imbalance: weighted cross-entropy loss (using inverse class frequencies as weights) and oversampling of the training set. Of the training set, we use 2001 examples for training and 300 for validation.

We use average recall (*AvgRec*) as our primary measure, which was also chosen as the primary measure in the TSC task of the latest SemEval series, due to its robustness against class imbalance [29]. We also measure accuracy (*acc*), macro F1 ($F1_m$), and average F1 on positive and negative classes ($F1_{pn}$) to allow comparison to previous works [24].

Table 2 shows that LCF-BERT performs best (*AvgRec* = 67.3 using BERT and 69.8 using our news-adapted language model).[6] Class-weighted cross-entropy loss helps best to address class imbalance (*AvgRec* = 69.8 compared to 67.2 using oversampling and 64.6 without any measure).

Performance in news articles is significantly lower than in established domains, where the top model (LCF-BERT) yields in our experiments *AvgRev* = 78.0 (Laptop), 82.2 (Restaurant), and 75.6 (Twitter). For Laptop and Restaurant, we used domain-adapted language models [28]. News TSC accuracy *acc* = 66.0 is lower than single-human-level $acc_h = 72.9$ (see Sect. 3.3).

We carry out a manual error analysis (up to 30 randomly sampled examples for each true class). We find *target misassociation* as the most common error cause: in 40%, sentences express the predicted sentiment towards a different target. In 30%, we cannot find any apparent cause. The remaining cases contain various potential causes, including usage of euphemisms or sayings (12% of examples with negative sentiment). Infrequently, we find that sentiment is expressed by rare words or figurative speech, or is reader-dependent (the latter in 2%, approximately matching the 3% of reader-dependent examples reported in Sect. 3.3).

[5] Epochs $\in \{3, 4\}$; batch size $\in \{16, 32\}$; learning rate $\in \{2e-5, 3e-5, 5e-5\}$; label smoothing regularization (LSR) [35]: $\epsilon \in \{0, 0.2\}$; dropout rate: 0.1; $\mathcal{L}_2$ regularization: $\lambda = 10^{-5}$. We use Adam optimization [20], Xaviar uniform initialization [9], and cross-entropy loss [11]. Where multiple values for a hyperparameter are given, we test all their combinations in an exhaustive search.

[6] Each row in Table 2 shows the results of the hyperparameters that performed best on the validation set.

Table 2. Experiment results. *LM* refers to the language model used, where *base* is BERT (base, uncased) and *news* is our fine-tuned BERT model.

LM	Method	AvgRec	acc	$F1_m$	$F1_{pn}$
Base	AEN-BERT	59.7	62.9	55.0	47.3
	BERT-SPC	62.1	62.1	53.3	44.9
	LCF-BERT	**67.3**	61.3	54.4	46.5
News	AEN-BERT	59.8	62.9	54.5	46.2
	BERT-SPC	66.7	63.5	55.0	45.8
	LCF-BERT	**69.8**	66.0	58.8	51.4

Previous news TSC approaches, mostly dictionary-based, could not reliably classify implicit or indirect sentiment expressions (see Sect. 2). In contrast, our experiments indicate that BERT's language understanding suffices to interpret implicitly expressed sentiment correctly (cf. [1,6,10]). NewsTSC does not contain instances in which the broader context defines sentiment, since human coders could not classify them correctly in the first place. Our experiments therefore cannot elucidate this particular characteristic discussed in Sect. 3.3.

5 Conclusion and Future Work

We explore how target-dependent sentiment classification (TSC) can be applied to political news articles. Our main contributions are as follows: first, we introduce NewsTSC, a dataset to explore target-dependent sentiment classification (TSC) in political news articles, consisting of over 3000 manually annotated sentences.

Second, in a qualitative analysis, we find notable differences concerning how authors express sentiment towards targets as compared to other well-researched domains of TSC, such as product review or posts on social media. In these domains, authors tend to explicitly express their opinions. In contrast, in news articles, we find dominant use of implicit or indirect sentiment expressions, e.g., by describing actions, which were performed by a given target, and their consequences. Thus, sentiment expressions may be more ambiguous, and determining their polarity requires a greater degree of interpretation.

Third, in a quantitative evaluation, we find that state-of-the-art TSC methods perform lower on the news domain (average recall *AvgRec* = 69.8 using our news-adapted BERT model, *AvgRec* = 67.3 without) than on popular TSC domains (*AvgRec* = [75.6, 82.2]).

We identify multiple future research directions for news TSC. While NewsTSC contains clear sentiment expressions, it lacks other sentiment types that occur in real-world news coverage. For example, sentences that express sentiment more implicitly or ambiguously. To create a labeled TSC dataset that better reflects real-world news coverage, we suggest to adjust annotation instructions to raise annotators' awareness of these sentiment types and clearly define how they should be labeled. Technically, apparently ambiguous sentiment expressions might be easier to label when considering a broader context, e.g., not only the current sentence but also previous sentences. Considering more context might also help to improve a classifier's performance.

We envision to integrate TSC methods into a system that identifies slanted news coverage [18,33]. For example, given a set of articles reporting on the same topic, a system could identify articles that similarly frame the actors involved in the event. To do so, the system would analyze frequently mentioned persons' polarities in each article. Then, it would group articles that similarly portray these persons.

Acknowledgements. The authors thank the students who participated in the manual annotation as well as the anonymous reviewers for their valuable comments.

References

1. Balahur, A., et al.: Sentiment analysis in the news. In: Proceedings of the Seventh International Conference on Language Resources and Evaluation (LREC 2010). European Language Resources Association (ELRA), Valletta, Malta (2010)
2. Baum, M.A., Groeling, T.: New media and the polarization of American political discourse. Polit. Commun. **25**(4), 345–365 (2008). https://doi.org/10.1080/10584600802426965
3. Bernhardt, D., Krasa, S., Polborn, M.: Political polarization and the electoral effects of media bias. J. Public Econ. **92**(5), 1092–1104 (2008)
4. Budak, C., Goel, S., Rao, J.M.: Fair and balanced? Quantifying media bias through crowdsourced content analysis. Public Opin. Q. **80**(S1), 250–271 (2016). https://doi.org/10.1093/poq/nfw007
5. Cicchetti, D.V., Feinstein, A.R.: High agreement but low kappa: II. Resolving the paradoxes. J. Clin. Epidemiol. **43**(6), 551–558 (1990). https://doi.org/10.1016/0895-4356(90)90159-M, https://linkinghub.elsevier.com/retrieve/pii/089543569090159M
6. Devlin, J., Chang, M.W., Lee, K., Toutanova, K.: BERT: pre-training of deep bidirectional transformers for language understanding. In: Proceedings of the 2019 Conference of the North, pp. 4171–4186. Association for Computational Linguistics, Stroudsburg (2019). https://doi.org/10.18653/v1/N19-1423, http://aclweb.org/anthology/N19-1423
7. Dong, L., Wei, F., Tan, C., Tang, D., Zhou, M., Xu, K.: Adaptive recursive neural network for target-dependent twitter sentiment classification. In: 52nd Annual Meeting of the Association for Computational Linguistics, ACL 2014, pp. 49–54, Baltimore (2014). https://doi.org/10.3115/v1/p14-2009
8. Gebhard, L., Hamborg, F.: The POLUSA dataset: 0.9M political news articles balanced by time and outlet popularity. In: Proceedings of the ACM/IEEE Joint Conference on Digital Libraries in 2020, pp. 467–468. ACM, New York (August 2020). https://doi.org/10.1145/3383583.3398567, https://dl.acm.org/doi/10.1145/3383583.3398567
9. Glorot, X., Bengio, Y.: Understanding the difficulty of training deep feedforward neural networks. In: Journal of Machine Learning Research, pp. 249–256 (2010)

10. Godbole, N., Srinivasaiah, M., Skiena, S.: Large-scale sentiment analysis for news and blogs. In: Proceedings of the International Conference on Weblogs and Social Media (ICWSM), vol. 7, pp. 219–222, Boulder (2007)
11. Goodfellow, I., Bengio, Y., Courville, A.: Deep Learning. MIT Press (2016). http://www.deeplearningbook.org
12. Greene, D., Cunningham, P.: Practical solutions to the problem of diagonal dominance in kernel document clustering. In: Proceedings of the 23rd International Conference on Machine Learning, pp. 377–384. ACM Press (2006)
13. Groseclose, T., Milyo, J.: A measure of media bias. Q. J. Econ. **120**(4), 1191–1237 (2005). https://doi.org/10.1162/003355305775097542, http://dx.doi.org/10.1162/003355305775097542
14. Hamborg, F.: Media bias, the social sciences, and NLP: automating frame analyses to identify bias by word choice and labeling. In: Proceedings of the 58th Annual Meeting of the Association for Computational Linguistics: Student Research Workshop, pp. 79–87. Association for Computational Linguistics, Stroudsburg (2020). https://doi.org/10.18653/v1/2020.acl-srw.12, https://www.aclweb.org/anthology/2020.acl-srw.12
15. Hamborg, F., Donnay, K., Gipp, B.: Automated identification of media bias in news articles: an interdisciplinary literature review. Int. J. Digit. Libr. **20**(4), 391–415 (2019). https://doi.org/10.1007/s00799-018-0261-y, http://link.springer.com/10.1007/s00799-018-0261-y
16. Hamborg, F., Meuschke, N., Breitinger, C., Gipp, B.: news-please: a generic news crawler and extractor. In: Proceedings of the 15th International Symposium of Information Science, pp. 218–223. Verlag Werner Hülsbusch (2017)
17. Hamborg, F., Zhukova, A., Donnay, K., Gipp, B.: Newsalyze: enabling news consumers to understand media bias. In: Proceedings of the ACM/IEEE Joint Conference on Digital Libraries in 2020, pp. 455–456. ACM, New York (August 2020). https://doi.org/10.1145/3383583.3398561, https://dl.acm.org/doi/10.1145/3383583.3398561
18. Hamborg, F., Zhukova, A., Gipp, B.: Automated identification of media bias by word choice and labeling in news articles. In: 2019 ACM/IEEE Joint Conference on Digital Libraries (JCDL), pp. 196–205. IEEE, Champaign (June 2019). https://doi.org/10.1109/JCDL.2019.00036, https://ieeexplore.ieee.org/document/8791197/
19. Jiang, L., Yu, M., Zhou, M., Liu, X., Zhao, T.: Target-dependent Twitter sentiment classification. In: Proceedings of the 49th Annual Meeting of the Association for Computational Linguistics, pp. 151–160, Portland (2011)
20. Kingma, D.P., Ba, J.: Adam: a method for stochastic optimization. arXiv preprint arXiv: 1412.6980 (2014). http://arxiv.org/abs/1412.6980
21. Kiritchenko, S., Zhu, X., Cherry, C., Mohammad, S.: NRC-Canada-2014: detecting aspects and sentiment in customer reviews. In: Proceedings of the 8th International Workshop on Semantic Evaluation (SemEval 2014), pp. 437–442. Association for Computational Linguistics, Dublin (2014). https://doi.org/10.3115/v1/s14-2076
22. Misra, R.: News category dataset (2018). https://www.kaggle.com/rmisra/news-category-dataset
23. Nagel, S.: Common crawl: news crawl (2016). https://web.archive.org/web/20191118111519/commoncrawl.org/2016/10/news-dataset-available/
24. Nakov, P., Ritter, A., Rosenthal, S., Sebastiani, F., Stoyanov, V.: SemEval-2016 Task 4: sentiment analysis in Twitter. In: Proceedings of the 10th International Workshop on Semantic Evaluation (SemEval-2016), pp. 1–18. Association for Computational Linguistics, San Diego (2016). https://doi.org/10.18653/v1/S16-1001

25. Nakov, P., Rosenthal, S., Kozareva, Z., Stoyanov, V., Ritter, A., Wilson, T.: SemEval-2013 Task 2: sentiment analysis in Twitter. In: Second Joint Conference on Lexical and Computational Semantics (SEM). Volume 2: Proceedings of the Seventh International Workshop on Semantic Evaluation (SemEval 2013), pp. 312–320. Association for Computational Linguistics, Atlanta (2013)
26. Pontiki, M., Galanis, D., Papageorgiou, H., Manandhar, S., Androutsopoulos, I.: SemEval-2015 Task 12: aspect based sentiment analysis. In: Proceedings of the 9th International Workshop on Semantic Evaluation (SemEval 2015), pp. 486–495. Association for Computational Linguistics, Stroudsburg (2015). https://doi.org/10.18653/v1/S15-2082, http://aclweb.org/anthology/S15-2082
27. Quionero-Candela, J., Sugiyama, M., Schwaighofer, A., Lawrence, N.D.: Dataset Shift in Machine Learning. The MIT Press, Cambridge (2009)
28. Rietzler, A., Stabinger, S., Opitz, P., Engl, S.: Adapt or get left behind: domain adaptation through BERT language model finetuning for aspect-target sentiment classification. arXiv preprint arXiv:1908.11860 (2019). http://arxiv.org/abs/1908.11860
29. Rosenthal, S., Farra, N., Nakov, P.: SemEval-2017 Task 4: sentiment analysis in Twitter. In: Proceedings of the 11th International Workshop on Semantic Evaluation (SemEval-2017), pp. 502–518. Association for Computational Linguistics, Vancouver (2017). https://doi.org/10.18653/v1/s17-2088
30. Sanh, V., Debut, L., Chaumond, J., Wolf, T.: DistilBERT, a distilled version of BERT: smaller, faster, cheaper and lighter. arXiv preprint arXiv: 1910.01108 (2019). http://arxiv.org/abs/1910.01108
31. Schreier, M.: Qualitative Content Analysis in Practice. SAGE Publications, Thousand Oaks (2012)
32. Song, Y., Wang, J., Jiang, T., Liu, Z., Rao, Y.: Targeted sentiment classification with attentional encoder network. In: Artificial Neural Networks and Machine Learning - ICANN 2019: Text and Time Series, pp. 93–103. Springer, Cham (2019). https://doi.org/10.1007/978-3-030-30490-4_9, http://link.springer.com/10.1007/978-3-030-30490-4_9
33. Spinde, T., Hamborg, F., Donnay, K., Becerra, A., Gipp, B.: Enabling news consumers to view and understand biased news coverage: a study on the perception and visualization of media bias. In: Proceedings of the ACM/IEEE Joint Conference on Digital Libraries in 2020, pp. 389–392. ACM, New York (August 2020). https://doi.org/10.1145/3383583.3398619, https://dl.acm.org/doi/10.1145/3383583.3398619
34. Steinberger, R., Hegele, S., Tanev, H., Della Rocca, L.: Large-scale news entity sentiment analysis. In: RANLP 2017 - Recent Advances in Natural Language Processing Meet Deep Learning, pp. 707–715. Incoma Ltd., Shoumen (November 2017). https://doi.org/10.26615/978-954-452-049-6_091, http://www.acl-bg.org/proceedings/2017/RANLP2017/pdf/RANLP091.pdf
35. Szegedy, C., Vanhoucke, V., Ioffe, S., Shlens, J., Wojna, Z.: Rethinking the inception architecture for computer vision. In: Proceedings of the IEEE Conference on Computer Vision and Pattern Recognition, pp. 2818–2826. IEEE (2016). https://doi.org/10.1109/CVPR.2016.308, http://arxiv.org/abs/1512.00567
36. Taub, A.: How liberals got lost on the story of missing children at the border (2018). https://web.archive.org/web/20191120151037/www.nytimes.com/2018/05/31/upshot/liberals-immigration-children-border-misinformation.html
37. Weischedel, R., et al.: OntoNotes 5.0 (2013). https://web.archive.org/web/20190705173013/catalog.ldc.upenn.edu/LDC2013T19

38. Zeng, B., Yang, H., Xu, R., Zhou, W., Han, X.: LCF: a local context focus mechanism for aspect-based sentiment classification. Appl. Sci. **9**(16), 1–22 (2019). https://doi.org/10.3390/app9163389, https://www.mdpi.com/2076-3417/9/16/3389
39. Zhaoa, P., Houb, L., Wua, O.: Modeling sentiment dependencies with graph convolutional networks for aspect-level sentiment classification. arXiv preprint arXiv:1906.04501 (2019). http://arxiv.org/abs/1906.04501

What Are Researchers' Concerns: Submitting a Manuscript to an Unfamiliar Journal

Chang-Huei Lin(✉) and Jeong-Yeou Chiu

Graduate Institute of Library, Information and Archival Studies, National Chengchi University, Taipei 11605, Taiwan
107155501@nccu.edu.tw

Abstract. Publishing academic works can be regarded as an indispensable extension of researchers' academic career. To find suitable journals for manuscript submission and to avoid the trap of predatory/questionable publishers, researchers need a large amount of information about the facts of publishing and conduct a comprehensive evaluation before submission. All information that the publisher discloses should address any concerns raised by researchers. As the information and level of journal disclosures affect researchers' decision of submission, this paper takes their information needs on submitting manuscripts to publishers as the starting point, compares and contrasts the third edition of *Principles of Transparency and Best Practice in Scholarly Publishing*, and explores the differences between the two. The results can serve as future guidelines for publishing information disclosure by scholarly publishers in the course of a paper's submission.

Keywords: *Principles of Transparency and Best Practice in Scholarly Publishing* · Publishing information disclosure · Researchers' information needs

1 Introduction

To disseminate research works to academic peers and exert scholarly influence, most researchers choose journals as the platform for publishing their results. The target journal for submission is considered as the heart of the problem from every aspect, such as the purpose of receiving the manuscript, the publication time, the nature of readers, and its presentation method. Nevertheless, to maintain a high academic reputation, it is also necessary to avoid submitting manuscripts to a predatory/questionable publisher. Therefore, before submitting research results to journal publishers, many factors of assessment need to be examined in more detail to find a quality journal [1].

Many predatory/questionable publishers in recent years have tried to use improper editorial measures to publish articles without real academic value or offer a rapid track of publication such as non-peer review procedures in order to attract papers and increase profitability of the Article Processing Charge (APC). This improper academic publishing behavior not only contradicts academic ethics, but also puts researchers in the predicament of academic misconduct [2]. To prevent scholarly publishers from accidentally

K. Toeppe et al. (Eds.): iConference 2021, LNCS 12646, pp. 167–175, 2021.
https://doi.org/10.1007/978-3-030-71305-8_13

entering such forbidden territory, several academic publishing professional organizations jointly formulated the *Principles of Transparency and Best Practice in Scholarly Publishing* (referred to as the *Principles*) in 2013 and officially released it in 2014. They strongly urge the Principles as publishers' administration guidelines, so as to assist publishers in meeting quality journal assessment [3].

From the perspective of researchers, it is necessary to collect public information from scholarly publishers and evaluate them before submitting papers. This can prevent future journals from being identified as having academic misconduct publishing behaviors, which affect scholarly works and the prestige of those researchers who are impacted [1]. Publishers should present a firm understanding of scholarly publishing operations and fully expose all necessary information on their official website for institutions and contributors to evaluate their publishing quality. Based on the above arguments, this article compares the information needs of researchers who are looking to submit papers with the content of the *Principles* and further discusses the similarities and differences between the two.

2 Literature Review

Accessed on October 6, 2020, Ulrich's Web Directory listed 140,573 active academic/scholarly periodicals (including "Journal," "Magazine," "Bulletin," and "Proceedings"), of which 67,028 are online periodicals and 21,541 OA journals are available online. The above statistical datapoints illustrate that the number of journals make up a very large base. In addition to finding an appropriate publishing platform from this pool, researchers must distinguish the publication quality and academic value of a target journal to avoid predatory/questionable publishers abducting their academic achievements.

Scholarly publishers need to stand out from their competitors. Whether their disclosed journal information can meet the needs of researchers and attract contributions runs in line with complying with the norms of academic ethics. The following shall discuss the information needs of researchers at the time of submission and the development and standardization of the *Principles.*

2.1 The Information Needs of Researchers' Submission

Based on the earliest documents available for this article, Oster believes that most scholars consider "prestige", "familiarity", "mean waiting time", and "acceptance probability" as four of the key journal factors when submitting articles [4]. Now twenty years into the 21st century, researchers not only continue to consider the "reputation of the journal" and "estimated length of time to article publication", but also are concerned about "the readership of the journal" and "recommendations from colleagues about the journal". Because of the development of the Internet and information technology, researchers now include copyright restrictions and open access among the criteria affecting their willingness to submit a paper [5].

Wijewickrema and Petras therefore in 2017 compiled a number of related studies on researcher submission factors and sorted out a total of 16 selection factors (see Table 1).

They classified the selection factors into three categories: a journal's scientific reputation; the performance or production issues of a journal; the reliability of and the demand for a journal [6].

Table 1. Selection factors of journals for researchers [6]

Wijewickrema and Petras' classification	Selection factors
Journal's scientific reputation	peer reviewed
	IF
	journal's prestige
	abstracting and indexing
	publisher's prestige
Performance or production issues of a journal	time from submission to first online appearance
	acceptance rate
	online submission with tracking facility
	no author charges
	number of journal issues per year
	number of papers published per year
Reliability of and the demand for a journal	journal represents an institution or society
	number of subscribers per year
	author contributions from different countries
	availability of a persistent article identifier
	age of journal
	publisher's prestige
	number of journal issues per year
	number of papers published per year

Note: Gray backgrounds are repeated classifications

Although the selection factors of journals for researchers in Table 1 are roughly classified, they are diverse and scattered. To help researchers save time in finding a suitable submission platform, some academic groups listed journals they publish such as *Journal Statistics and Operations Data* of the American Psychological Association. Large scholarly publishers have designed online journal recommendation systems to provide researchers with online retrieval of publication information about their journals, such as Elsevier Journal Finder, Springer Journal Suggester, IEEE Publication Recommender™ and Wiley Journal Finder, etc. Researchers can use these journal selection tools to search and filter, and the results of a search present several considerations for journal selection, such as the subject of the manuscript received, rejection rate, publication lag time (embargo), impact factor, publishing cost, frequency of publication, and other journal statistics.

Cabell's Scholarly Analytics, Endnote Manuscript Matcher, Edanz Journal Selector, JournalGuide, SciRev, and other comprehensive journal recommendation systems also help researchers to select journals. The problem of predatory journals for researchers, such as through Cabell's Scholarly Analytics, is that there are two sets of subsystems, "Journalytics" and "Predatory Reports", for users to input their specific selection criteria to search for academic journals [7]. On the other hand, SciRev and MedSci databases offer, basic information on a journal such as review time, rejection rate, and an acceptance rate of manuscripts created by members of the system instead of any official information established by the publisher of a journal.

Even though these journal systems mentioned above are designed to allow users to save time for information acquisition and retrieval, it should be noted that users must rely on the number of samples and their knowledge of the journal so as to judge its content. Before this, the level of journal information disclosure determines that the various data in the journal system can be verified again. Moreover, researchers still need to fully understand the concepts of various journal selection indicators to have the ability to submit their work appropriately.

2.2 Development and Regulation of the *Principles*

Due to problems of predatory/questionable publishers, international academic and professional publishing organizations such as the Open Access Scholarly Publishers Association (OASPA), the Open Access Journal Guide (DOAJ), the Committee on Publication Ethics (COPE), and the World Association of Medical Editors (WAME) jointly formulated the *Principles* in December 2013 and officially released it in January 2014. The goal is that the *Principles* can be used to provide scholarly publishers with a guideline for operations of publishing and to give academic institutions and contributors a reference when evaluating journal quality [3].

The four organizations above incorporate the *Principles* into the evaluation standards of their journal databases and regularly review their members or journals to ensure their good quality [8]. To highlight journals that adhere to the *Principles*, DOAJ created a journal qualification certification, "DOAJ Seal for Open Access Journals," to encourage strict compliance with the *Principles* and to review the relevant regulations and operational strategies of a publisher [9]. However, there scant studies on the *Principles* in the literature. At present, only Hyung Wook Choi et al. classified the 33 items under the 16 norms of the *Principles* and then compared them with the official website contents of 781 journals indexed in Science Citation Index Expanded [10].

The content of the *Principles* is mainly based on the operations of scholarly publishers. It is now in its third edition and has been translated into 18 languages. However, to follow the formal publishing rules, scholarly publishers must consider the information needs of researchers when submitting manuscripts. If the information disclosed by the publisher can fit the needs of researchers, then it will attract researchers to submit works and allow the publisher to enrich its manuscript sources. Therefore, this present paper expands further to ensure that the details of the international standards of the *Principles* are more in line with the needs of researchers.

3 Method and Analysis

This paper re-consolidates the information needs of researchers who are submitting their works from Oster [4], Regazzi and Aytac [5], and Wijewickrema and Petras [6] and uses the third edition of the *Principles* as the comparative basis.

3.1 Comparison of the *Principles* and Information Needs of Researchers Submitting Their Manuscripts

Items that are the Same or Similar to Each Other. Among the information needs of researchers when submitting their manuscript, items that are the same or similar to those in the third edition of the *Principles* include readership of the journal, peer reviewed, online submission with tracking facility, journal represents an institution or society, copyright restrictions, no author charges, number of journal issues per year, open access, number of subscribers per year, abstracting, and indexing (see Table 2).

Table 2. Items that are the same or similar to each other

The *Principles*	Information needs for researchers' submission	Reference sources
1. Website: An 'Aims & Scope' statement should be included on the website and the readership clearly defined	The readership of the journal	Regazzi and Aytac [5]
3. Peer review process: Journal content must be clearly marked as whether peer reviewed or not. …This process, as well as any policies related to the journal's peer review procedures, shall be clearly described on the journal website, including the method of peer review used	Peer reviewed	Wijewickrema and Petras [6]
	Online submission with tracking facility	Wijewickrema and Petras [6]
4. Ownership and management: Information about the ownership and/or management of a journal shall be clearly indicated on the journal's website	Journal represents an institution or society	Wijewickrema and Petras [6]
7. Copyright and Licensing: The policy for copyright shall be clearly stated in the author guidelines and the copyright holder named on all published articles	Copyright restrictions	Regazzi and Aytac [5]
8. Author fees: Any fees or charges that are required for manuscript processing and/or publishing materials in the journal shall be clearly stated …	No author charges	Wijewickrema and Petras [6]

(*continued*)

Table 2. (*continued*)

The *Principles*	Information needs for researchers' submission	Reference sources
11. Publishing schedule: The periodicity at which a journal publishes shall be clearly indicated	Number of journal issues per year	Wijewickrema and Petras [6]
12. Access: The way(s) in which the journal and individual articles are available to readers and whether there are associated subscription or pay per view fees shall be stated	Open access	Regazzi and Aytac [5]
	Number of subscribers per year	Wijewickrema and Petras [6]
13. Archiving: A journal's plan for electronic backup and preservation of access to the journal content...	Abstracting and indexing	Wijewickrema and Petras [6]

Items of Conflict to Each Other. Among the three groups of scholars, the information needs for submission of a manuscript by researchers all mention the relevant description of a manuscript's publication time (listed in Table 3). However, this information conflicts with the third peer review process in the *Principles*, which describes that "journal websites should not guarantee manuscript acceptance or very short manuscript acceptance or very short peer review times".

Table 3. Items of conflict between information needs of researchers' submission and the *Principles*

The *Principles*	Information needs of researchers' submission	Reference sources
3. Peer review process: … Journal websites should not guarantee manuscript acceptance or very short peer review times	Mean waiting time	Oster [4]
	Estimated length of time to article publication	Regazzi and Aytac [5]
	Time from submission to first online appearance	Wijewickrema and Petras [6]

Items of Information Needed for Submission Not Included in the *Principles*. From the information needs of submission by researchers listed in the three groups of scholars, after deducting overlapping items, those not included in the *Principles* are: familiarity, acceptance probability/rate, reputation/prestige of the journal, recommendations from colleagues about the journal, IF, publisher's prestige, number of papers published per year, author contributions from different countries, availability of a persistent article identifier, and age of journal (see Table 4 listed).

Table 4. Items of information needed for submission not included in the *Principles*

Information needs for researchers' submission	Reference sources
Prestige Familiarity Acceptance probability	Oster [6]
The reputation of the journal Recommendations from colleagues about the journal	Regazzi and Aytac [7]
IF Journal's prestige Publisher's prestige Acceptance rate Number of papers published per year Author contributions from different countries Availability of a persistent article identifier Age of journal	Wijewickrema and Petras [8]

Note: Gray backgrounds are repeated classifications

3.2 Discussion

The information needs for submission by researchers listed in Table 2 are also in some of the rules of the third edition of the *Principles.* This covers whether scholarly publishers comply with the *Principles* and detailed descriptions of the above points. When researchers directly obtain the required information from a publisher's official website, it will significantly increase researchers' willingness to submit their work.

Only the information about "publishing time" conflicts with the *Principles.* For researchers, academic research results relate to timeliness and priority of publication. They also closely relate to the researchers' academic performance and job promotion, pushing them to consider timeliness during the submission process. However, some scholarly publishers are recognized as predatory/questionable, as they emphasize the ability to quickly review a work and can shorten the publishing schedule, thus attracting researchers into the trap of giving away their academic achievements.

Although the impact factor (IF), number of papers published per year, author contributions from different countries, and age of journal are not included in the *Principles,* before researchers submit a paper, they can obtain this information indirectly from third-party institutions or academic databases that index journals or directly obtain the information from a publisher's website. In addition, the acceptance probability/rate must be calculated based on the publisher's internal submission statistics and the number of articles published. Although the *Principles* does not regulate the above factors, and the above factors do not conflict with existing provisions, the disclosure of the above information still depends on the wishes of the journal publisher.

The two items of information in "the reputation/prestige of the journal" and "publisher's prestige" are also indicators for researchers to measure a publisher's quality, influence, and importance. However, they are also based on the researcher's subjective view of the scholarly publishers under various conditions. This requires the researcher to rely on the "familiarity" of the publisher or recommendations from colleagues about the journal. Other external information can also be used as reference materials for selecting journals.

A description on the availability of a persistent article identifier for the information needs of researchers is not regulated in the *Principles*. The use of persistent article identifiers not only belongs to the category of digital scholarship applications, but is also a part of the value-added of academic publishing services. However, it can only depend on the current operating direction and publishing mode of scholarly publishers.

As the contents regulated by the *Principles* are quite diverse, some of them are not required by researchers when submitting articles. However, this does not mean that scholarly publishers should neglect the handling of such information. Because these contents relate to the overall operations and future development of publishers, they must be treated with caution, and their details should be made public for future reference and use by journal evaluating institutions or researchers.

4 Conclusion and Further Research

With the advent of the information society, most required information that researchers need can be obtained from the Internet. From the current relationship between scholarly publishers and researchers, it is not difficult to find that the level of publishing information disclosure strongly relates to the willingness of researchers' submission. Therefore, the information released by scholarly publishers must comply not only with the norms of academic publishing ethics, but they must also reasonably disclose relevant information according to the needs of researchers.

In the past most scholarly publishers' information disclosures or information required by researchers for submitting papers have been treated as a starting point for their own research. At present, no scholar has been able to combine all the information disclosure of scholarly publishers with the information required by researchers and with an in-depth review of the contents on a journal's official website.

The academic publishing community in recent years has been broadening the focus of the *Principles* and the information needs of researchers who are submitting their works. Therefore, when the *Principles* is revised in the future, this research suggests that the information needs of researchers should be taken into consideration. Moreover these two principles can be combined together for discussing and conducting empirical research. Through all these efforts, it is hoped that a win-win situation between scholarly publishers and researchers can be eventually created.

References

1. Babor, T.F., Morisano, D., Stenius, K., Ward, J.H.: How to choose a journal: scientific and practical considerations. In: Babor, T.F., Stenius, K., Pates, R., Miovský, M., O'Reilly, J., Candon, P. (eds.) Publishing Addiction Science: A Guide for the Perplexed, pp. 37-70. Ubiquity Press, London (2017). https://doi.org/10.5334/bbd.c
2. Beall, J.: Medical publishing triage – chronicling predatory open access publishers. Ann. Med. Surg. **2**(2), 47–49 (2013). https://doi.org/10.1016/S2049-0801(13)70035-9

3. Open Access Scholarly Publishers Association: Principles of transparency and best practice in scholarly publishing (2018). https://oaspa.org/information-resources/principles-of-transparency-and-best-practice-in-scholarly-publishing/, Accessed 10 Oct 2020
4. Oster, S.: The optimal order for submitting manuscripts. Am. Econ. Rev. **70**(3), 444–448 (1980). https://www.jstor.org/stable/1805232
5. Regazzi, J.J., Aytac, S.: Author perceptions of journal quality. Learn. Publ. **21**(3), 225–235 (2008). https://doi.org/10.1087/095315108X288938
6. Wijewickrema, M., Petras, V.: Journal selection criteria in an open access environment: a comparison between the medicine and social sciences. Learn. Publ. **30**(4), 289–300 (2017). https://doi.org/10.1002/leap.1113
7. Cabell's International: Journal selection policy. https://www2.cabells.com/selection-policy, Accessed 10 Sep 2020
8. DOAJ: Principles of transparency and best practice in scholarly publishing, version 3 (2018). https://blog.doaj.org/2018/01/15/principles-of-transparency-and-best-practice-in-scholarly-publishing-version-3, Accessed 12 Jul 2020
9. DOAJ: What is the DOAJ seal of approval for open access journals (the DOAJ Seal)? https://doaj.org/publishers#seal, Accessed 12 Jul 2020
10. Choi, H.W., Choi, Y.J., Kim, S.: Compliance of "principles of transparency and best practice in scholarly publishing" in academic society published journals. Sci. Ed. **6**(2), 112–121 (2019). https://doi.org/10.6087/kcse.171

German Art History Students' Use of Digital Repositories: An Insight

Cindy Kröber(✉)

Media Center, TU Dresden, 01062 Dresden, Germany
cindy.kroeber@tu-dresden.de

Abstract. The paper describes a study on art history students' research behavior and needs connected to digital resources and repositories. It tries to identify aspects of and approaches to improving and developing these repositories. These students make up a large proportion of the users of digital libraries and their content; their supposedly distinct attitude and skill level concerning technology renders them an important group to observe. Qualitative data derives from three focus groups with 25 students from two German universities. Thematic analysis is based on questions concerning research approaches, curriculum, and the students' connected desires as avid users of technology in everyday life.

Keywords: Digital libraries · Art history · Human information behavior · Qualitative research · User study

1 Introduction

Innovation occurs in research support structures like archives, libraries, and museums [1]. Recent developments have forced many libraries and facilities to close down and only operate online, making the need to offer users digital access more obvious. While digitization is still very much on the minds of most libraries and repositories, they are aware that needs are changing. Wanting to update their interfaces and make their digital supply more accessible and usable, they are turning their attention toward the users. This paper examines the idea that digital libraries and especially digital image repositories can support and improve art history research in many ways, particularly for its students. On another note, it might uncover ways for repositories to reach students and therefore grow their user group.

Previous research concerning developments and improvements in libraries and repositories has focused on evaluation and reviews [2–8], looked at technological advancements in general and their potential for development in repositories [9–11] and the discipline of art history itself [12–17], or examined the behavior and needs of advanced scholars in art history and the humanities [18–27]. Findings are always very similar regarding improvements (metadata, usability of interfaces and functions, information on digitization, amount and quality of data available) but seeking and handling images poses additional challenges that have received less attention.

K. Toeppe et al. (Eds.): iConference 2021, LNCS 12646, pp. 176–192, 2021.
https://doi.org/10.1007/978-3-030-71305-8_14

Art historians are hesitant to adopt new technologies [28], so art history students as a user group with similar information and especially image needs [29] seems a promising line of enquiry. To date, art history students have only been included in studies providing instructions for their online research [30–32] or for advice on how to deal with teaching in the digital age [33]. Studies on visual [34] and information literacy [35] provide interesting insights into student skills but do not consider art historians. Students are more comfortable with using social media. So far, social media has only been considered as a means of communication and exchange with peers [1] but it is thought to have greater potential and impact on art history that needs to be exposed [3].

Many studies comprise international views and findings and do not pay specific attention to the situation and issues in Germany. The German art history community is still concerned with debating pros and cons of digital art history [36]. This paper focuses primarily on art history students as users of digital image repositories. It is expected that the German situation differs from the international one because of the way students are taught at university and their use of local solutions for research.

Regarding differences between students and established scholars of art history, experts are able to define their information needs and suitable search approaches well due to their prior knowledge of the subject and sources [26, 37]. Students have not gained a thorough insight into possible research areas and all the available sources. But their experience and attitude toward digital content and new technologies make them an interesting group to investigate.

To gain insight into how art history students use digital repositories, their investigative practices within the research setting need to be explored. Information and image seeking is the active and intentional behavior to satisfy the information need connected to the research for assignments [38] and includes planning, executing and evaluating a search [39]. Browsing is another way to satisfy a very loosely defined information need [40]. Information seeking has changed somewhat with the development of digital content and online solutions. Before digitization, finding an image involved just three tasks: describe, search, and interact [41]. Now, it requires understanding digitization, online search features, presentation software, and manipulation software [1].

2 Intention and Research Questions

This is a study of students who are connected to the academic environment and obliged to meet scientific standards for their seminar assignments, which places them in the position of an apprentice investigator [22, 42]. Their motivations and approaches when seeking information and images differ from those of the general public but they are not as experienced as advanced scholars. Their assignments pose as traditional research interest in art history on a smaller scale. But students might have a different attitude to new technologies and opportunities for research and a higher skill level in this due to their private use of media and technology.

The main intention is to identify aspects and approaches of how digital repositories can be improved to address the needs of art history students. The connected research questions are:

- How do students approach and structure their research for assignments and when do they rely on online resources and digital image repositories during the process?
- Do their private interactions with digital technologies influence their approaches and desires for new resources and features?
- How do general habits of art history and the curriculum impact their research actions?

3 Research Design

The findings presented here are derived from three focus groups held by the author in September 2016, June 2019 and January 2020[1] (see Table 1).

The focus group method was chosen because the topic of students' research is very broad, linking many complex aspects, and a qualitative approach is suitable. It is a way of exploring the students' opinions and experiences; statements can be discussed and clarified on the spot [43]. The use of different image repositories and platforms is very much connected to individual preferences and research routines but students may also agree on certain standards and habits. The sampling was designed to investigate their approaches, perceptions, and thoughts with regard to their different levels of experience. The first focus group helped to get data from students at the very beginning of their academic career. The purpose of the second cohort was to collect data from slightly more advanced students at a different university. The last group provided an insight into changes and ideas related to working on bigger assignments like a thesis, because the participants were further into their studies. During the first two focus groups new behaviors and questions had emerged. With the third focus group, saturation was reached.

Using interview guidelines, focus groups provided a semi-structured way of collecting data. Multiple rounds focused on the same topics but different reactions and discussion arose; direct feedback by peers offered a broader perspective [44]. Having groups with the same level of experience ensured that participants could discuss experiences and challenges without being intimidated by more advanced students. The small group size helped every participant to engage.

An initial survey with the first focus group in 2016 focused on the step-by-step process of research for a traditional art history assignment [3]. This was done so the process could be classified based on information behavior models and helped to identify stages when new approaches and features become relevant. The focus here is on the use and perception of online platforms and repositories for information and image seeking.

3.1 Focus Groups and Participants

The first focus group was carried out in September 2016 at the University of Würzburg, Germany and involved 15 students from the art and architectural history program. The majority of them were just a few semesters into their bachelor's program. The students were recruited during a seminar on certain architecture and buildings of the city of Dresden. The main aim was to investigate how students deal with seeking images for

[1] The two-and-a-half year gap between the 1st and 2nd cohort was due to the author's extended parental leave.

Table 1. Overview of focus groups

Number of participants	Length of focus group	University	Details of participants
1st cohort in September of 2016			
15	1 h 20 min	University of Würzburg, Germany	Sex: 4 males, 11 females Age: 20s to 70s Experience: 2nd and 4th semester of bachelor's program, 2x third age learning program
2nd cohort in June of 2019			
5	3 groups: 30 min, 23 min, 15 min	Technische Universität (TU) Dresden, Germany	Sex: 5 females Age: 20s Experience: 6th semester of bachelor's program and 2nd semester of master's program
3rd cohort in January of 2020			
5	30 min	University of Würzburg, Germany	Sex: 4 females, 1 male Age: 20s and 60s Experience: 1st semester of master's program; 1x third age learning program

a seminar assignment. The moderator prepared a guideline with questions in advance, which was slightly altered or extended for the following focus groups (see Table 2).

The second focus group was carried out in June 2019 at the TU Dresden, Germany. At the beginning of different lectures on the art history program, students were recruited. Five students agreed to take part. The participants named art history as their major and either English studies, Romance studies, or philosophy as their minor. Due to timetabling, the group was split up for the focus group. Three consecutive rounds (2 participants, 2 participants, 1 participant) were carried out which lasted 30 min, 23 min, and 15 min.

The third round in January 2020 was a focus group with five students from the University of Würzburg. Again, the participants were recruited during a seminar on architectural structures and developments in Dresden. Four students had completed their bachelor's degree (including a thesis) at three different German universities and just started the master's program in Würzburg. One had chosen two majors, art history and English studies, the other three had abandoned their minors. All participants took part in the same 30-min focus group.

Table 2. Overview of questions in the interview guideline for each of the three focus groups

Focus group 2016	Focus group 2019	Focus group 2020
How did you approach your research work?		
x	x	x
What type of source material do you search for?		
x	x	x
What criteria are important to you when selecting pictures?		
x	x	x
Where and how do you search? Which databases, repositories, and sources do you know and use?		
x	x	x
How do you know the repositories you mentioned?		
x	x	x
What are your alternative sources when the results that the repositories provide are not satisfactory? Would you turn to social media platforms for images?		
	x	x
Does your university offer a seminar or any material on scholarly practices within art history?		
	x	x
Do you miss any functionalities for repositories or software solutions for data analysis during your research process?		
	x	x
How do you come up with and develop search terms for the (digital) repositories and platforms you use?		
		x
Are you aware of any current trends in the research landscape or any advances concerning digital art history? Does this topic come up in any of your lectures or seminars?		
	x	

3.2 Data Preparation and Analysis

Thematic analysis was applied to identify, analyze and report themes within the data [45]. The focus groups were recorded using handwritten and audio recordings with the consent of the participants. The audio was transcribed using AmberScript.[2] The automatically generated transcripts relied on speech recognition and needed some editing to ensure correct, clean transcripts of the focus groups. The transcripts were analyzed through coding using the software MaxQDA.[3] Coding helped to identify patterns in the students'

[2] AmberScript is a software that automatically transcribes audio into text using speech recognition. See https://www.amberscript.com.

[3] MaxQDA is a software program for computer-assisted qualitative data analysis. See https://www.maxqda.com.

responses [46–48]. The open and inductive coding allowed for multiple codes [49] which led to the themes discussed here.

The study has several limitations: 1) the small number of participants from just two German universities, and 2) the presence of the lecturer (during the focus group in 2016) and of fellow students may induce social desirability bias. The focus group styles were slightly different each time, which of course affects the outcome and to some extent the comparability. However, the mentioned topics and responses were of interest and not necessarily directly comparable between the three groups.

4 Findings

All students stated clearly that they strictly distinguish between a) phases and b) sources dedicated to gathering textual information in the beginning and visual data later on. The findings which will be discussed in this paper are related to 1) information seeking, 2) image seeking and 3) curriculum.

4.1 Information Seeking

Prior to picking a topic for an assignment, the students usually make use of a preliminary search and evaluation of authoritative writing on the designated object or subject. To explore the topic first, they often turn to an online library catalog, and some also use Google to identify new keywords.

Seeking profound textual information is part of the very beginning of research work even before students may come up with a hypothesis and arguments. Hence, their further actions are very much influenced by what they come across. Before browsing the library shelves, students use the online library website to identify and locate suitable sources. During the focus groups, the students named ways of gathering additional information mainly through references in books and other people's work, citations, and indexes, as well as looking at nearby shelves.

Although all students named the library as their first choice for textual information, they did turn to other online resources when they were desperate for material.

It is important to point out that they are not at all inexperienced with the use of online platforms for more recent articles due to assignments for their minor subjects:

> *As I said, essays online, there is just not much available for art history. In English studies I often do research digitally at home and can download the essays directly as PDF documents. [Question by moderator: So there are other solutions you use for your minor?] JSTOR most of the time. They don't have everything, but you can find a lot more. And most importantly, a lot more recent articles from the last few years. It is hard to get those through other ways than digital platforms.*[4] *(January 2020, 1st semester of master's)*

JSTOR, a digital library for academic journals, was named as a solution for minor subjects a few times.

[4] Any quotes from the focus groups are the author's direct translations from the German transcript.

4.2 Image Seeking

When searching for images, the students usually look for either digital or digitized photographs or digital images of art pieces such as paintings or sculptures. The only visualizations potentially enriched with additional information that were mentioned were maps:

The only other thing I have used for art history is actually Google Maps. When it comes to architecture, for example, to show geographical distances. This building is far away from the city center or any other building. (June 2019, 6th semester of bachelor's)

All participants preferred to use digital images instead of digitizing images themselves. The resolution was the main quality criteria for a 'good image.' Quantifiable resolution was not important, only the need to be 'not blurry.' Correct color display was also mentioned:

[Question: How do you cope with possible color deviations?] Compare [many images] to estimate the average colors. But it is almost standard for a good reproduction to have this color strip included. If I find an image with it, then this is my preferred choice. (January 2020, 1st semester of master's)

The students are very aware of time-consuming tasks—e.g., digitizing images and creating visualizations—and many only consider making this much effort when they are in desperate need or the assignment is of greater significance, like a thesis. They are aware that it is accepted in art history to use one's own camera for digitization.

If you want to put more effort into it, for example for a bachelor's thesis, you can also photograph it [the object] yourself. (January 2020, 1st semester of master's)

Students prefer to get as many search results as possible and are frequently afraid that relevant images are left out due to restricting search terms.

I think I always approach it relatively inconveniently, because often, if your search terms are too specific, many images are not included in the results. Therefore you tend to enter less and then have to search through more. (January 2020, 1st semester of master's)

It seems metadata are in general less important for students then they are for advanced scholars.

This is also the question of how many pictures to choose from? If the [additional] information is not there and I have only a few images I don't care. But if I have information, then I would rather choose an image where I know there is more available. Because it comes across as more scientific or certified. Whether this is true, I don't know. (June 2019, 6th semester of bachelor's)

Repositories and Sources for Images. Table 3 provides an overview of all sources for images mentioned by the students.

Table 3. Overview of mentioned repositories and sources used for image access

Repository/Source	Comment
The Prometheus Bildarchiv[a]	Known to all participants and almost always mentioned instantaneously; recommended by university staff during early semesters; usually university staff assists with setting up access
Image database of the university's art history department[b]	Usually set up with easyDB,[c] and referred to by many students as easyDB of [university]
Website of a museum	Digitized images the museum provides
Foto Marburg[d]	Mentioned several times
Artstor[e]	Mentioned once
Deutsche Fotothek[f]	Mentioned once
(Google Images)	Students are aware that Google redirects to the actual resource
Wikipedia	Mentioned several times, especially by younger students
Instagram	Mentioned several times and for different scenarios
Flickr[g]	Mentioned several times during the 2016 focus group as a way to get images with transparent copyright handling; never mentioned during the later focus groups
Travel-related sites and blogs	Typical vacation pictures and selfies are not desired

[a]The Prometheus Bildarchiv is a distributed digital image archive for research and education mainly focused on art and cultural sciences. It belongs to the department of art history of the University of Cologne, Germany. See https://www.prometheus-bildarchiv.de

[b]Before the rise of digital images, art history departments had their own slide collections for students and staff to use for lectures and research work. Now, many departments have digitized their stock and make it available to their associates through their own online solutions.

[c]easyDB is a Digital Asset Management System developed in Germany. See https://www.programmfabrik.de

[d]The Deutsche Dokumentationszentrum für Kunstgeschichte – Bildarchiv Foto Marburg (German documentation center for art history) is an image archive with a focus on art and architecture belonging to the Philipps University of Marburg, Germany. Their online portal for search is called Bildindex der Kunst & Architektur. Foto Marburg and Bildindex are used interchangeably. See https://www.bildindex.de

[e]The Artstor Digital Library provides access to curated images from reliable sources for education and research. See https://www.artstor.org

[f]The Deutsche Fotothek is a picture library and universal archive dedicated to art and cultural history. See https://www.slub-dresden.de/sammlungen/deutsche-fotothek

[g]Flickr is an American image hosting service that was especially popular while it was owned by Yahoo! (2005–2017). In the past years its popularity has significantly decreased.

Most students were completely satisfied with their small selection and did not see the need for many other resources. Apparently, there is a lack of awareness concerning resources. However, it does not seem to pose an issue for the students:

[Access to Prometheus] was actually set up at the beginning. I didn't even look at Google anymore to see if there were any pictures. I only searched there for images for my essays. (June 2019, 6th semester of bachelor's)

Several students stated that they have always found what they needed:

[Question: What if you don't find it on Prometheus?] This has never happened before. Sometimes it was my fault that I didn't enter the search term correctly or it was too specific and the website was a bit overwhelmed. But actually it has never happened that something was not there. (June 2019, 2nd semester of master's)

Now, as far as architecture is concerned, there may not be one photograph from every angle or view. When you are looking for a specific display, or you need a color photograph and there are only black and white ones or something. But apart from that, I have actually always found everything. (June 2019, 6th semester of bachelor's)

Furthermore, students stick to familiar approaches and sources and do not change their ways as long as they do not encounter difficulties.

[Question: Why don't more students use e.g., the Deutsche Fotothek?] Because nobody ever talks about it. That's also the case during the seminar in the first semester, there's never a word about the Deutsche Fotothek. And then it doesn't get picked up later, when you have developed a routine, because you have never heard about it. And you don't start looking for new resources in the fifth semester. Usually. Well, not for pictures. You take the five or six you have or know, and you don't start looking for new databases. Not for an ordinary essay. (June 2019, 2nd semester of master's)

Once the students encounter difficulties in seeking digital visual content and connected information, they do turn to Google.

Well, I think it has happened to me quite often, that you search for an image on Google, because you don't know where the object is hung [at which museum it can be found] or if it is privately owned. (January 2020, 1st semester of master's)

The students are very aware of which resources are good and which are not; they pay attention to stating the source, since this is essential to maintain a good scientific standard.

[Question: Pinterest offers many images. Would it be an alternative for images?] [It is possible] that you find a good image on Pinterest. But most of the time you have to provide the image credits […] and then you would probably get your head torn off for that [using Pinterest]. That's why you don't use them. Only for a quick look maybe. (January 2020, 1st semester of master's)

Instagram was named several times during the focus groups with regard to different scenarios.

- For providing images:

[… I] follow accounts of art galleries [on Instagram] […]. And I actually find it quite exciting when this is offered. They also post images you could use. They write texts for it, too. […] but also the background, who took it, and so on. You can actually use that quite well. I like it. (June 2019, 6th semester of bachelor's)

- To build on knowledge in art history which can be beneficial in the long run:

*I think it's cool that so many public institutions go online, and I always think it's a really good way to incorporate your studies into everyday life without having to sit down and read a scholarly text for three hours. Instead, it's just a very relaxed way of getting further information. [*Certain museum] has a kind of small series: Mythology Wednesday or something. Every Wednesday they post a picture from [*a certain collection], which represents a mythological story from Greek or Roman antiquity and then provide the story briefly. I think this is a great way to learn something. (June 2019, 6th semester of bachelor's)*

- As a research topic:

A fellow student, for example, has done a piece on Instagram as a showroom. […] It's the platform where the artists can present themselves and we as art historians can look at it and analyze it. (June 2019, 6th semester of bachelor's)

- As a research tool during an internship:

[…] to compile a catalog for an artist's work. Just to collect material on him and try to trace where his works are located today. And it was all about the collection of material, and in some cases, simply for documentation purposes, Instagram images and posts were screenshotted and documented. To simply trace back who owns it today. Where and in which exhibition was it displayed when. (June 2019, 6th semester of bachelor's)

4.3 Curriculum

Most art history programs include some kind of training, either in form of a seminar or as handouts for self-study. Topics included in these training materials were: Citations, references and bibliographies, image captions, repositories and resources, use of the online library catalog and search on the spot, approaching and structuring essays and PowerPoint presentations. The students were mostly satisfied with the content and found it helpful. Some said that the issue of developing research questions was neglected:

[Question: What was missing from the seminar?] Maybe developing research questions. But it was just at the beginning of the studies, and it wasn't so important to ask research questions back then. In the beginning it was more about describing images. (June 2019, 6th semester of bachelor's)

Another comment raises the question of how much the assignments help to prepare undergraduates for an active research routine.

[...] it's noticeable that the focus of the work is really not so much on research itself. In art history it's more like, look and read what he and she said and somehow summarize it. So that one's own research part is neglected. (June 2019, 6th semester of bachelor's)

Information on current trends and developments in art history was not part of the curriculum. However, one student noted that it is possible to get a glimpse into advances in the field through guest lecturers. Currently, assignments fit the established ways that university staff has been working for possibly the last decade. The courses and tasks have not changed much and do not rely on the use of new technologies.

The way I have experienced the [university's art] department over the last five years, 'old school' is quite appropriate. They have their common research techniques and their procedures and methods, how it is done, and also their areas of expertise. And that's perfectly fine. But it does not go beyond that. They won't give a lecture on a topic they haven't really dealt with for 20 years. (June 2019, 2nd semester of master's)

5 Discussion

The students' initial approach to research shows that online libraries should consider that users in the early stages of their research need a broad overview of relevant information, more guidance discovering connections, and help recognizing additional keywords. It might be possible to use technological advancements to provide suggestions based on approaches and selections of other users.

Bibliographies, footnotes, citations, and indexes from books have been recommended to find additional literature [18, 31]. Unfortunately, those options are not available as easily accessible links online. When books have been scanned or are available as PDF documents, their citations and indexes are usually not included for keyword search. With all necessary data available, digital libraries could improve by interconnecting texts or at least allowing users to tag and link certain resources.

The lack of access to more recent resources for art history has been mentioned various times by Beaudoin [26], stating that material might be overlooked or outdated [50]. This will only change once the field of art history turns away from 'only print counts' [1] to fully embrace digital and open access publications. However, older materials not available in digital form do not lose relevance [51] and still provide necessary information and ideas for comparison [22]. Currently, digitization of older material is an ongoing process and the selection will certainly grow.

The value of the library shelves as discovery source is undiminished; the readability and browsability of books and their tables of contents is not easily achieved through online search [40]. The issue of finding sources when semantically related material is shelved at distant locations, so important 30 years ago [52], was not specifically mentioned. It is not clear if this problem was not revealed because students do not strive to find every possible source and accept a decent number of sources as sufficient. Several statements indicate that they are willing to put in more effort for assignments with greater significance, like a thesis. Nevertheless, the use of online library websites has made it significantly easier to locate material. This experience can be further enhanced by ensuring usability standards for the library websites.

Art history lives on images and they are needed to comprehend, visualize, and verify arguments, and to emphasize findings in essays. However, students do not actively look for visualizations and do not expect to find them in an image repository. Rather, they create them themselves using resources like Google Maps and image editing tools.

A color strip to estimate color rendition and suitable resolution were the mentioned criteria for selecting a digital image. Any influence by screens or actual analysis of the colors was not addressed. However, this emphasizes the need for image repositories to provide high-resolution images including an indicator for the color rendition and for universities to raise awareness how digitization and display alter images.

Licensing and copyright for images was not part of the study and the topic was only briefly addressed during the 2016 focus group, when the students praised Flickr for its transparent copyright labeling.

The students' answers about which repository or sources they turn to for images were not as diverse or numerous as expected after reviewing previously published articles [3, 50]. It is also very interesting that local or German solutions were so commonly used; bigger, internationally known repositories like the ones suggested by Chen were hardly used at all [31].

The fortunate lack of problems finding images might be because the students' assignments are very well-aligned with the recommended sources, like Prometheus.

All participants that had switched university for their master's program agreed that they still preferred the familiar database of the old university and had not bothered to look at the new one, confirming Beaudoin's findings [25]. Problems may occur later on in the academic career when students have to develop their own research projects or during a thesis where they are suddenly confronted with more rare or 'exotic' pieces of art.

The students' mention of Instagram and its many connected possibilities for use in art history research are very interesting. Advanced scholars in art history do occasionally turn to social media when material is otherwise difficult to access [3]. A shift toward recognizing or trusting social media channels of well-known institutions is not unlikely. This may point institutions and digital libraries in the direction of considering social media to raise awareness of stock and foster serendipitous findings for followers.

Art historians usually look for objects of art, their reproductions, related objects, and any written accounts on them [18]. These categories are not sufficient for today's online image search. The following three use cases are helpful when seeking images:

1. Looking for a specific, well-defined image,
2. Looking for a loosely defined image,
3. Looking for items that can be grouped by characteristics.

When looking for a specific image, a clear name or description (usually by the object's title and artist) is needed for a keyword search [53] and the goal and criteria are well-defined. Even for comparing several items, the literature usually provides objects that help to clarify the description.

Finding a loosely defined image is much harder, because while the goal is clear, the description might not be. Sometimes an idea of what is depicted needs to be put into words [25]. The description needs to correspond with the metadata of the desired image; the user needs to be proficient in describing the image and repositories must maintain metadata well.

The students very much struggle with or avoid constructing search terms. It is not clear if they are not confident enough with their own abilities to construct search terms or if they do not trust the search features or metadata of the repositories. What is clear is that they prefer to look through more irrelevant images during their search. As McKay et al. [40] have put it, the needs are better met by recognition than specification.

Most online search algorithms rely on possibly flawed metadata [25, 54–56] which might make it harder to find images. Crowdsourcing is an option to improve metadata quality [57], which some repositories are embracing [58]. Advances in the field of artificial intelligence, particularly automatic object and context recognition,[5] can be used to improve and standardize metadata entries. Several more recent approaches try to overcome possible flaws in metadata using artificial intelligence by providing advanced keyword tags,[6] object recognition, and georeferencing images [59]. Students appreciate metadata, but clearly value quantity over quality.

Looking for images for e.g., inspiration can be grouped by characteristics and is even more loosely defined. Users prefer to browse through a longer list and see what comes up [40]. Different repositories offer faceted search to tackle this issue which might help students to identify new relevant keywords.

For this study it was assumed that students have similar requirements to scholars because their assignments function as traditional research interest in art history, although on a smaller scale. Students stated that their work was more about summarizing and describing than developing their own research questions; thus, especially when starting their studies, their work differs significantly from that of advanced scholars. Therefore, students need to be discussed as an independent user group within academia.

Many articles highlight the need to prepare art history students for a future that relies more on technology [1, 60]. Higher education should certainly prepare students to use up-to-date concepts and methodology, considering technological advances. Some have argued that it would be better to train postgraduates than undergraduates because training is endless and technology changes rapidly. But including innovations in student training

5 See Artificial intelligence as a bridge for art and reality, by J. H. Dobrzynski, in The New York Times, October 25, 2016.

6 Exploring art with open access and AI: What's next? in The Met Museum. See: https://www.metmuseum.org/blogs/now-at-the-met/2019/met-microsoft-mit-exploring-art-open-access-ai-whats-next September 9, 2019.

will help the whole discipline to move forward. The push for innovation already comes from younger scholars who are more adept in the digital realm [1]. Increasing awareness of technological advances needs to be complemented by training in source criticism, however, concerning the digital or digitized nature of images and data, as well as the reliability of sources.

6 Conclusion

The study provided several clues on topics both familiar (e.g., metadata) and lesser-known (e.g., social media) to archivists and librarians to improve digital repositories based on art history students' approaches to research and their discipline-specific habits.

The potential students saw in Instagram is noteworthy, and sets them apart from many career researchers. Students felt their experience with research was limited; hence, their assignments might not pose as traditional research in art history and their needs might differ significantly from those of advanced scholars. These aspects emphasize that art history students are a user group with distinct requirements which image repositories should consider more closely.

Not all issues raised by the study can be tackled by the repositories. The opportunities presented by new technologies and approaches need to be better communicated to the art history community. Universities should introduce more recent topics, approaches and technologies into their curricula to equip the younger generation and broaden the scope for changes and advances in the field.

The focus groups were among the first efforts to observe art historians' information behavior. In combination with expert interviews, this may supply more details on improvements for digital repositories to meet the needs of art history scholars. A quantitative study could follow, to show the significance of the derived implications. The academic landscape is changing and more art history students are venturing into newly established digital humanities programs, which may have an impact on their future research topics and methodologies.

Acknowledgments. The research upon which this paper is based is part of the activities of the junior research group Urban History 4D, which has received funding from the German Federal Ministry of Education and Research under grant agreement No 01UG1630.

References

1. Zorich, D.M.: Transitioning to a Digital World: Art History, Its Research Centers, and Digital Scholarship. The Samuel H. Kress Foundation (2012)
2. Zorich, D.M.: A Survey of Digital Humanities Centers in the United States. Council on Library and Information Resources Washington, DC (2008)
3. Münster, S., Kamposiori, C., Friedrichs, K., Kröber, C.: Image libraries and their scholarly use in the field of art and architectural history. Int. J. Digit. Libr. **19**(4), 367–383 (2018). https://doi.org/10.1007/s00799-018-0250-1

4. Liang, S., He, D., Wu, D., Hu, H.: Challenges and opportunities of ACM digital library: a preliminary survey on different users. In: Sundqvist, A., Berget, G., Nolin, J., Skjerdingstad, K.I. (eds.) iConference 2020. LNCS, vol. 12051, pp. 278–287. Springer, Cham (2020). https://doi.org/10.1007/978-3-030-43687-2_22
5. Saracevic, T.: Evaluation of digital libraries: an overview. In: Notes of the DELOS WP7 Workshop on the Evaluation of Digital Libraries, Padua, Italy (2004)
6. Kous, K., et al.: Usability evaluation of a library website with different end user groups. J. Libra. Inf. Sc. **52**(1), 75–90 (2020). https://doi.org/10.1177/0961000618773133
7. Hee Kim, H., Ho Kim, Y.: Usability study of digital institutional repositories. Electron. Libr. **26**(6), 863–881 (2008). https://doi.org/10.1108/02640470810921637
8. Zimmerman, D., Paschal, D.B.: An exploratory usability evaluation of Colorado State University libraries' digital collections and the Western Waters digital library web sites. J. Acad. Librariansh. **35**(3), 227–240 (2009). https://doi.org/10.1016/j.acalib.2009.03.011
9. Shiri, A.: Digital library research: current developments and trends. Libr. Rev. **52**(5), 198–202 (2003). https://doi.org/10.1108/00242530310476689
10. Rose, T.: Technology's impact on the information-seeking behavior of art historians. Art Docum. J. Art Libr. Soc. North America **21**(2), 35–42 (2002). https://doi.org/10.1086/adx.21.2.27949206
11. Zimmerman, T., Chang, H.-C.: Getting smarter: definition, scope, and implications of smart libraries. In: Proceedings of the 18th ACM/IEEE on Joint Conference on Digital Libraries (2018). https://doi.org/10.1145/3197026.3203906
12. Lincoln, M.: Predicting the Past: Digital Art History, Modeling, and Machine Learning (2017)
13. Klinke, H., Surkemper, L., Underhill, J.: Editorial. Creating new spaces in art history. Int. J. Digit. Art Hist: Digital Space and Archit. **3**, (2018). https://doi.org/10.11588/dah.2018.3.49921
14. Bruzelius, C.: Digital technologies and new evidence in architectural history. J. Soc. Archit. Histor. **76**(4), 436–439 (2017). https://doi.org/10.1525/jsah.2017.76.4.436
15. Hatchwell, S., Insh, F., Leaper, H.: Born digital: early career researchers shaping digital art history. Vis. Resour. **35**(1–2), 171–179 (2019). https://doi.org/10.1080/01973762.2019.1553448
16. Kohl, A.T.: Revisioning art history: how a century of change in imaging technologies helped to shape a discipline. Vis. Resour. Assoc. Bull. **39**(1) (2012)
17. Gagliardi, S.E., Gardner-Huggett, J.: Introduction to the special issue: spatial art history in the digital realm. Histor. Geogr. **45**(1), 17–36 (2017)
18. Stam, D.C.: How art historians look for information. Art Docum. J. Art Libr. Soc. North Am. **16**(2), 27–30 (1997). https://doi.org/10.1086/adx.16.2.27948896
19. Long, M.P., Schonfeld, R.C.: Supporting the Changing Research Practices of Art Historians. Ithaka S+ R , New York (2014)
20. Kamposiori, C.: Digital infrastructure for art historical research: thinking about user needs. In: Electronic Visualisation and the Arts (EVA 2012), pp. 245–252 (2012). https://doi.org/10.14236/ewic/EVA2012.41
21. Kamposiori, C.: The impact of digitization and digital resource design on the scholarly workflow in art history. Int. J. Digit. Art Hist. (4), 3.11–3.27 (2019)
22. Brilliant, R.: How an art historian connects art objects and information. Libr. Trends **37**(2), 120–129 (1988)
23. Kamposiori, C., Warwick, C., Mahony, S.: Accessing and using digital libraries in art history. In: Münster, S., Friedrichs, K., Niebling, F., Seidel-Grzesinska, A. (eds.) UHDL/DECH - 2017. CCIS, vol. 817, pp. 83–101. Springer, Cham (2018). https://doi.org/10.1007/978-3-319-76992-9_6
24. Beaudoin, J.E.: A framework of image use among archaeologists, architects, art historians and artists. J. Doc. **70**(1), 119–147 (2014). https://doi.org/10.1108/JD-12-2012-0157

25. Beaudoin, J.E.: An investigation of image users across professions: a framework of their image needs, retrieval and use. Drexel University (2009)
26. Beaudoin, J.: Image and text: a review of the literature concerning the information needs and research behaviors of art historians. Art Doc. J. Art Libr. Soc. North Am. **24**(2), 34–37 (2005). https://doi.org/10.1086/adx.24.2.27949373
27. Kamposiori, C.: Personal Research Collections: Examining Research Practices and User Needs in Art Historical Research. UCL (University College London) (2008)
28. Zorich, D.M.: Digital art history: a community assessment. Vis. Resour. **29**(1–2), 14–21 (2013). https://doi.org/10.1080/01973762.2013.761108
29. Layne, S.S.: Artists, art historians, and visual art information. Ref. Libr. **22**(47), 23–36 (1994). https://doi.org/10.1300/J120v22n47_03
30. Chen, C.-J., Dhawan, A.: Online Instruction for Art History Research (2011)
31. Chen, C.-J.: Art history: a guide to basic research resources. Collect. Build. **28**(3), 122–125 (2009). https://doi.org/10.1108/01604950910971152
32. du Toit, G.E., Meyer, H., du Preez, M.: Factors influencing undergraduate art students' information behaviour: a case study of the art students of the open window school of visual communication. Mousaion **36**(1), 1–17 (2018). https://doi.org/10.25159/0027-2639/2489
33. Klusik-Eckert, J., et al.: Digitale Lehre in der Kunstgeschichte. Eine Handreichung (2020). https://doi.org/10.11588/artdok.00006806
34. Beaudoin, J.E.: Describing images: a case study of visual literacy among library and information science students. Coll. Res. Libr. **77**(3), 376–392 (2016). https://doi.org/10.5860/crl.77.3.376
35. Henkel, M., Grafmüller, S., Gros, D.: Comparing information literacy levels of Canadian and German university students. In: Chowdhury, G., McLeod, J., Gillet, V., Willett, P. (eds.) iConference 2018. LNCS, vol. 10766, pp. 464–475. Springer, Cham (2018). https://doi.org/10.1007/978-3-319-78105-1_51
36. Schelbert, G.: Art history in the world of digital humanities. Aspects of a difficult relationship (2017). https://doi.org/10.18452/18694
37. Belkin, N.J.: Anomalous states of knowledge as a basis for information retrieval. Can. J. Inf. Sci. **5**(1), 133–143 (1980)
38. Case, D.O.: Looking for information: a survey of research on information seeking, needs, and behavior. In: Boyce, B.R. (ed.) Library and Information Science. Emerald Group Publishing, London (2007)
39. Rather, M.K., Ganaie, S.A.: Information seeking models in the digital age. In: Advanced Methodologies and Technologies in Library Science, Information Management, and Scholarly Inquiry, pp. 279–294. IGI Global (2019). https://doi.org/10.4018/978-1-5225-7659-4.ch022
40. McKay, D., et al.: The things we talk about when we talk about browsing: an empirical typology of library browsing behavior. J. Assoc. Inf. Sci. Technol. **70**(12), 1383–1394 (2019). https://doi.org/10.1002/asi.24200
41. Jorgensen, C.: Image attributes: An investigation, School of Information Studies (1995)
42. Sweetnam, M.S., et al.: User needs for enhanced engagement with cultural heritage collections. In: Zaphiris, P., Buchanan, G., Rasmussen, E., Loizides, F. (eds.) TPDL 2012. LNCS, vol. 7489, pp. 64–75. Springer, Heidelberg (2012). https://doi.org/10.1007/978-3-642-33290-6_8
43. Stewart, D.W., Shamdasani, P.N.: Focus Groups: Theory and Practice, vol. 20. Sage Publications, New York (2014)
44. Kitzinger, J.: Qualitative research. introducing focus groups. BMJ: Br. Med. J. **311**(7000), 299–302 (1995). https://doi.org/10.1136/bmj.311.7000.299
45. Braun, V., Clarke, V.: Using thematic analysis in psychology. Qual. Res. Psychol. **3**(2), 77–101 (2006). https://doi.org/10.1191/1478088706qp063oa

46. Aronson, J.: A pragmatic view of thematic analysis. Qual. Rep. **2**(1), 1–3 (1995). https://doi.org/10.46743/2160-3715/1995.2069
47. Kuckartz, U., Rädiker, S.: Analyzing qualitative data with MAXQDA. Springer, Cham (2019). https://doi.org/10.1007/978-3-030-15671-8
48. Vaismoradi, M., Turunen, H., Bondas, T.: Content analysis and thematic analysis: implications for conducting a qualitative descriptive study. Nurs. Health Sci. **15**(3), 398–405 (2013). https://doi.org/10.1111/nhs.12048
49. Saldaña, J.: The Coding Manual for Qualitative Researchers. Sage, London (2015)
50. Beaudoin, J.E., Brady, J.E.: Finding visual information: a study of image resources used by archaeologists, architects, art historians, and artists. Art Doc. J. Art Libr. Soc. North Am. **30**(2), 24–36 (2011). https://doi.org/10.1086/adx.30.2.41244062
51. Van Zijl, C., Gericke, E.M.: Information-seeking patterns of artists and art scholars at the vaal triangle technikon. South African J. Libr. Inf. Sci. **66**(1), 23–33 (1998)
52. Losee, R.M.: The relative shelf location of circulated books: a study of classification, users, and browsing. Libr. Resour. Tech. Serv. **37**(2), 197–209 (1993)
53. Matusiak, K.K.: Towards user-centered indexing in digital image collections. OCLC Syst. Serv. Int. Digit. Libr. Perspect. **22**(4), 283–298 (2006). https://doi.org/10.1108/10650750610706998
54. Lopatin, L.: Library digitization projects, issues and guidelines: a survey of the literature. Library hi tech **24**(2), 273–289 (2006). https://doi.org/10.1108/07378830610669637
55. Giral, A.: Digital image libraries and the teaching of art and architectural history. Art Libr. J. **23**(4), 18–25 (1998). https://doi.org/10.1017/S0307472200011251
56. Ellis, D.: A behavioural model for information retrieval system design. J. Inf. Sci. **15**(4–5), 237–247 (1989). https://doi.org/10.1177/016555158901500406
57. Nowak, S., Rüger, S.: How reliable are annotations via crowdsourcing: a study about inter-annotator agreement for multi-label image annotation. In: Proceedings of the International Conference on Multimedia Information Retrieval (2010). https://doi.org/10.1145/1743384.1743478
58. Bunge, E.: Citizen Science in der Bibliotheksarbeit: Möglichkeiten und Chancen, vol. 63. bit online Verlag (2017)
59. Maiwald, F., et al.: Geo-information technologies for a multimodal access on historical photographs and maps for research and communication in urban history. Int. Arch. Photogramm. Remote Sens. Spatial Inf. Sci. **XLII-2/W11**, 763–769 (2019). https://doi.org/10.5194/isprs-archives-XLII-2-W11-763-2019
60. Hoppe, S., Schelbert, G.: Für ein verstärktes engagement in den digital humanities der arbeitskreis digitale kunstgeschichte. AKMB-news **19**(2), 40–42 (2013)

Information Governance and Ethics

Encouraging Diversity of Dialogue as Part of the iSchools Agenda

Simon Mahony[1] and Yaming Fu[2](✉)

[1] Research Centre for Digital Publishing and Digital Humanities, Beijing Normal University at Zhuhai, Zhuhai 519087, China
simon.mahony@gmail.com

[2] Department of Information Studies, University College London, London WC1E 6BT, UK
yaming.fu.17@ucl.ac.uk

Abstract. This paper takes the conference themes of Diversity, Divergence, Dialogue and applies them to an analysis of the published topic headings and keywords from previous iConferences to determine the extent to which diversity is an important aspect within the iSchools community. It follows previous research from 2016 where Bogers and Greifeneder conducted a quantitative analysis of the metrics for submission and acceptance of papers for the 2014 iConference in Berlin. Their interest was in the potential for bias resulting from a lack of diversity in the established review process. We look at topic headings, language and country of presenters as a sub-set of diversity and how we might move away from the Anglophone dominance towards more demographic diversity and in doing so widen the channels for scholarly communication and dialogue. The move to a virtual conference removes any geolocational difficulties and competition for limited travel budgets. The 2021 Chinese track accepts submissions in Chinese, removing the difficulties of the English-language requirement for scholars of the host nation. Language, publication and travel are determining factors for encouraging and facilitating diversity; these should be reinforced within the iSchools movement to develop a sense of community with members as stakeholders so that they feel that they are part of a diverse but inclusive community. This Short Paper is the first stage in a wider study looking at the changes that the innovations for the 2021 iConference have on diversity, divergence, and dialogue for papers and published proceedings.

Keywords: Diversity · Community · Globalisation

1 Introduction[1]

The themes of iConference 2021, Diversity, Divergence, and Dialogue are very pertinent in these challenging times. How do we bring different (diverse) groups

[1] The authors would like to thank the anonymous reviewers for their helpful comments which have allowed us to strengthen this short paper as well as given us a clearer focus for our future work on this topic.

K. Toeppe et al. (Eds.): iConference 2021, LNCS 12646, pp. 195–206, 2021.
https://doi.org/10.1007/978-3-030-71305-8_15

of people with dissimilar attitudes and opinions (divergence) together to facilitate communication (dialogue) that is equally shared and accessible to us all? Our paper looks at the challenge that faces us in a global community, such as the iSchools movement, and how we, as a community, might take a lead in demonstrating ways in which researchers, academics and practitioners can advance an agenda for diversity and divergence in our scholarly dialogue.

We refer to the iSchools movement as a community as that is what it represents with members in an association, a *Gesellschaft*, based on occupation and common interest [5]. As a scholarly and professional community there is the shared interest where we might communicate and collaborate but there are also the shared goals and values that we adhere to. Most of the 'association' takes place online but, importantly, there is the annual iConference which brings people together from the virtual to the physical. This is the place where networks are developed, and casual conversations may result in future collaborations; it brings the community of iSchool members together. A significant determinant of the iConference is its international nature and global reach. Looking back through the history of the iSchools movement, it is clear that the epicentre was North America. All iConferences before 2014 were held in the USA, with the exception of 2012 in Toronto; then came Europe in 2014 (Berlin), UK (Sheffield) in 2018, and Sweden (Boras) in 2020 [16]. The iConference became truly global in 2017 when it was hosted at Wuhan and now in 2021 at Renmin. Following the conference move beyond North America, Bogers and Greifeneder [1] conducted a quantitative analysis of the metrics for submission and acceptance of papers, mostly concerned with review balance and how that might be corrected.

When we look at the regional distribution of iSchool members the North American Directory lists 53 institutions, the European Directory 33, and Asia Pacific 31 [13]. The latter is the largest region by land mass, rather than population or representation in the iSchools, including Australia and East Asia with 13 iSchools in mainland China. The Board of Directors includes members representing each of these regions [6]. The iSchools, then, is a truly global community but, nevertheless, a mainly virtual one with dedicated initiatives and committees all conducted online.

A community is mostly identified by what it does and if it is to flourish and grow, particularly a virtual one, its members need to interact with other members to give them a sense of belonging [17]; and have an emphasis on iSchool research connectivity. This is the importance of the iConference and the associated proceedings which connects members of the global community to give a point of reference for members to come together in person to share research output under the common banner and aegis of iSchools. This strengthens the weaker community ties of *Gesellschaft* to give focus and cohesion by giving a locus and annual physical point of contact for members; it is especially important for new researchers in promoting communications, making connections, and strengthening collaborations. As a community at a distance this temporary locus of the iConference places us somewhere between *Gesellschaft* and *Gemeinschaft*[2], serving to strengthen and consolidate the ties between members. The 2020 and 2021 iConferences are exceptions with the pandemic related travel restrictions; they break the geolocation limitations thus increasing the 'association' with enhanced possibilities for communication and a globalised research agenda. The iConference has a global reach but does that lead to diversity and divergence of dialogue as evidenced by the papers and publications?

2 Methodology

This paper is an initial study into aspects of diversity within the iSchools community to address questions of inclusion; to what extent do our members have an equal voice and opportunity to be heard? The iSchools movement certainly crosses continents and from the 'About' statement envisions 'a future in which the iSchool Movement has spread around the world' [12]. The question is whether or not this global reach does indeed lead to 'diversity and divergence of dialogue' and, if not, how might this be encouraged and facilitated. The majority of the iSchool community are found outside its North American roots (Asia Pacific 31, Europe 33, North America 53) [14] but is this globalisation matched by the iConference papers and publications?

From Bogers and Greifeneder [1], overall, '[t]he 2014 European iConference saw 109.8% more submissions from Europe relative to 2015, whereas the 2015 North American iConference saw 19.5% more submissions from North America than in 2014.' They discussed the cultural bias in the review process following an analysis of the review data from 2014 (Berlin) and looking ahead to Wuhan the following year [1].

Diversity is a wide-ranging term, including gender, race, orientation, as well as intellectual content, but analysing these requires more data than can be found in the public domain and hence, as our starting point, we focus on language and country. As a Short Paper the scope is necessarily limited and our data is taken from the published iConference Past Proceedings, the Papers Proceedings (2018–2020) and conference summaries from the inaugural event (2005) up to and including 2020 [16]. Additional material published in IDEALS, other than 2009, and ADC has not been included as the format did not match the conference summaries; nevertheless, we intend to include these in our next phase to expand the corpus and deepen the analysis. Our emphasis is on the presenters and publications, along with stated conference themes and keywords (where included). The topic titles were collected and run through a simple concordance programme (MonoConc Pro), to identify commonly occurring words and themes; as titles rather than natural language extracted from abstracts or content, there

[2] The former being understood here as the weaker bonds of the post industrialisation, urban or capitalistic society and the latter being the closer ties of family and kinship related to Confucian principles; derived from Tönnies (edited by Harris and translated by Hollis) [5].

was no need to exclude stop words and simple frequency lists, concordances and collocations were generated.

The Springer published proceedings only cover the last three years but, nevertheless, give a good indication of topical interests and concerns; considering the presenters and topics shows movement of both over time. A more exhaustive study would need to include an analysis of the representation within the membership itself as well as the executive postholders and represents a useful follow up project which would also benefit from cross-tabulation of author nationality with keywords.

3 Results

Table 1. iConference participants and host country by year

iConference participants by year (no conference is recorded for 2007)		
Year	Participants	Host country
2005	No data	USA
2006	317	USA
2008	277	USA
2009	305	USA
2010	346	USA
2011	538	USA
2012	486	Canada
2013	512	USA
2014	450	Germany
2015	531	USA
2016	467	USA
2017	482	China
2018	468	UK
2019	539	USA
2020	390 (virtual)	Sweden

Table 1 lists the iConference host country and number of participants by year. The information on the published conference summaries is not consistent and has developed and expanded along with the conference. The Conference Summary for 2014, the first to be held outside North America (Berlin), gives a breakdown of the countries represented as well as the total number of participants (450) [8]. The following year (USA) has similar data on the public page but this is not available publicly for other iConferences [9].

Table 2. Geographic breakdown of participants for 2014 and 2015

Geographic breakdown of participants for the years 2014 and 2015 (data from conference summaries [1,10])			
Year	**2014**		**2015**
iConference location	Berlin, Germany		Newport Beach, California, USA
Host institution	Humboldt University, Berlin		University of California, Irvine
Total number participants	450		531
Countries represented	30		25
Top 10 countries by number of participants	**Only eight listed**		**Ten listed**
USA	242	USA	404
UK	26	Canada	18
Denmark	26	China	15
Canada	22	Germany	14
Germany	22	UK	10
Japan	12	South Korea	9
China	10	Denmark	7
Sweden	8	Japan	7
		Spain	5
		Sweden	5

Table 2 shows an increase in non-North American participants with the iConference held in Europe. This is as expected with easier travel for Europeans but, although it is not possible to make strong claims based on data from two years, nevertheless, it illustrates a turn that is worth further investigation. The overall number of participants is higher in 2015 but with fewer countries represented and following this there seems to be interest in the demographic of participants [1]. A more complete data set could be determined for the presenters by correlating names and affiliations from the conference programmes, although, that would not return the global representation of the 'participants' as non-presenters would not be included. An additional issue is that declared affiliation would be for participants' current institution rather than indicative of their nationality or first language.

For individual conferences, the programmes do not show themed sessions that could be used for analysis, although from 2017 (Wuhan) the Conference Summary lists a series of organisational headings in the Supporting Materials [10]. The Springer proceedings are edited volumes with the content seemingly broken down according to the preferences of the editors rather than any fixed format. Each volume has local editors (different US states 2019) but the number and wording of the section headings is very different. The number of articles published are 75 in 2020, 77 in 2019 and 82 in 2018 and hence fairly consistent and match the number of accepted papers with a stated acceptance rate of 30%, 33%, 30% respectively. The subject headings are, however, very different with the 2019 editors taking a much more granular and verbose approach to the organisation of the material (see Table 3).

Generating a simple concordance (Fig. 1) of the topic headings (Table 3) shows the most frequent (and popular) terms in the titles of the published articles and hence the dominant terms in the accepted conference presentations: 'information', 'data' and 'communities'.

Table 3. Topic Headings taken from the published proceedings, Springer LNCS (2018–2020)

Topic headings in the published proceedings: Springer LNCS		
2020 Sustainable Digital Communities	2019 Information in Contemporary Society	2018 Transforming Digital Worlds
Sustainable Communities	Scientific Work and Data Practices	Social media
Social media	Methodological Concerns in (Big) Data Research	Communication Studies and Online Communities
Information Behavior	Concerns About "Smart" Interactions and Privacy	Mobile Information and Cloud Computing
Information Literacy	Identity Questions in Online Communities	Data Mining and Data Analytics
User Experience	Measuring and Tracking Scientific Literature	Information Retrieval
Inclusion	Limits and Affordances of Automation	Information Behaviour and Digital Literacy
Education	Collecting Data about Vulnerable Populations	Digital Curation
Public Libraries	Supporting Communities Through Public Libraries and Infrastructure	Information Education and Libraries
Archives and Records	Information Behaviors in Academic Environments	
Future of work	Data-Driven Storytelling and Modeling	
Open Data	Online Activism	
Scientometrics	Digital Libraries, Curation and Preservation	
AI and Machine Learning	Social-Media Text Mining and Sentiment Analysis	
Methodological Innovation	Data and Information in the Public Sphere	
	Engaging with Multi-media Content	
	Understanding Online Behaviors and Experiences	
	Algorithms at Work	
	Innovation and Professionalization in Technology Communities	
	Information Behaviors on Twitter	
	Data Mining and NLP	
	Informing Technology Design Through Offline Experiences	
	Digital Tools for Health Management	
	Environmental and Visual Literacy	
	Addressing Social Problems in iSchools Research	

Corpus Frequency List

Count	Pct	Word
20	11.1732%	and
9	5.0279%	information
8	4.4693%	data
6	3.3520%	in
5	2.7933%	communities
4	2.2346%	libraries
4	2.2346%	digital
4	2.2346%	online
3	1.6760%	public
3	1.6760%	work
3	1.6760%	literacy
3	1.6760%	mining
3	1.6760%	social
3	1.6760%	behaviors

179 words, 102 types

Fig. 1. Word frequency of topic headings

Combining the most frequent term with its collocating words (Fig. 2), gives 'information' and 'information behavio(u)r(s)', to be expected at an iConference, as the most frequent heading terms. Of particular interest for our study is that 'communities' is the third most common term and hence an important theme.

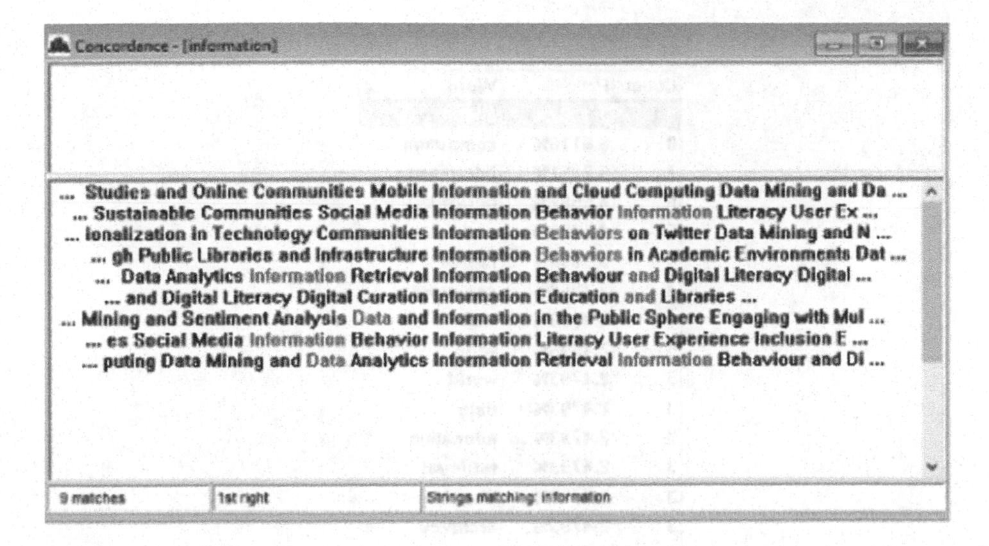

Fig. 2. Key Word in Context concordance for 'information' within topic headings

The second iConference held outside North America was at Wuhan in 2017 – the first in the iSchools Asia Pacific region – with another category added to 'By the numbers': Chinese Papers (45) with more than either Completed Research Papers (30) or Preliminary Results Papers (36). Nevertheless, despite this special track and that the 2017 Supporting Materials and 2017 Proceedings IDEALS show many Asian names as presenters, 'all papers were written and presented in English' [10].

Other changes in 2017 saw the introduction of a series of headings in the Supporting Materials, presumably, to group together the proposals by topics: 'Workshop Proposals and Results; Sessions for Interaction and Engagement Proposals; Special Panel Proposals; iSchool Best Practices Proposals; iSchools and Industry Partnership Presentations and Proposals' [10]. These, however, reduced in number over the following years.

A brief analysis of the published keywords for all conferences (where given) shows the most frequent term to be 'social' (Fig. 3) and when put into its immediate linguistic context (with associated words sorted) it links most often with 'networks' and 'networking' (Fig. 4). This indicates the importance of both social and networking aspects to iSchool concerns and again emphasises the 'community' aspect, a focus of this paper.

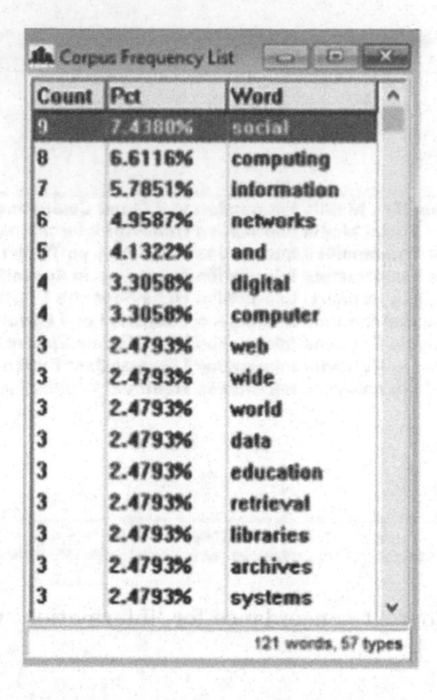

Fig. 3. Word frequency of keywords

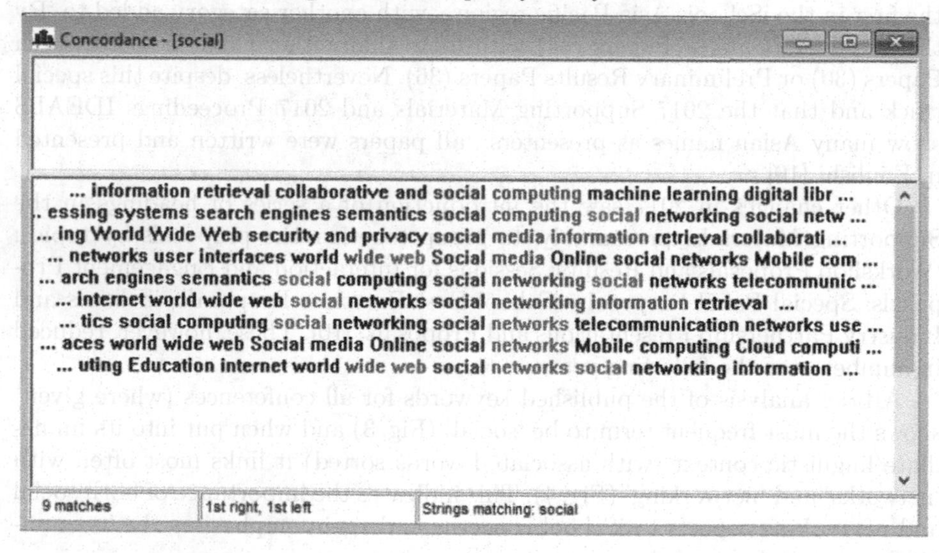

Fig. 4. Key Word in Context concordance for 'social' within keywords

4 Discussion

The analysis above is limited to the terms used in the titles of the published articles and the conference topic headings, rather than the content of abstracts or papers, but nevertheless they are indicative of clear trends.

The iConference has attracted growing numbers since the early days with a dip in 2020, the year of the global pandemic. With the move to Berlin (2014) the countries of participants were published and regions in the following year; an awareness seemed to be growing about the need to consider demographics with the move from North America. Wuhan (2017) saw the addition of a dedicated local Chinese track although proposals and presentations still needed to be in English. The recommendations of Bogers and Greifeneder [1, p. 10] were that reviewers should be more representative of the overall iSchools community by increasing the number of female and Asian reviewers to allow for more representative diversity of accepted papers.

Another significant issue impacting on 'diversity of dialogue' is that of language. With its North American roots, the iSchools movement developed in an environment dominated by English as the language of the Internet (with ICANN) and the *lingua franca* of the Web (with the W3C Consortium). The medium in which we work and correspond has a bias towards the English language leading to linguistic differences and regional inequalities [3]. This is also true of publishing where to have your work widely circulated and read, leading to more citations to support academic advancement and promotion, results in a distortion of the publication metrics [18]. This has been corroborated by studies on the metrics of publication in the cognate disciplines of the Arts and Humanities (as counted in major indices such as Scopus and Web of Science) and how that along with citation counts has a clear Anglophone-bias, resulting not only in incentives for advancement but also for successful funding applications [4]. Hence, there is pressure to publish in English, regardless of native language.

The track for Chinese papers in Wuhan 2017 still required papers and presentations to be in English. This was restrictive and particularly so with the difficulties of Chinese scholars to have papers accepted under the strict language requirements. The 2021 iConference at Renmin similarly has a special track for Chinese papers, but this time submissions 'are exempted from the English-language requirement and may be in Chinese' [7]. Looking at overall submissions and acceptances, tracking the data following Renmin would enable us to see whether dropping the English-language requirement would create a more equitable field.

5 Conclusions

Pulling all this together, what is it that facilitates dialogue and would help to encourage diversity within the iSchools movement? It is argued here that we should develop inclusiveness, a sense of community and ownership of the movement by its members. Members need to be stakeholders and feel that they are part of the community; that they have a voice, and most importantly that they are able to have a positive and valuable interaction with other members of the community; that the community is more than just a symbolic and intellectual construct but one that they can engage with.

Our membership is global and so the iConference needs to move away from Anglophone dominance. There seems to be a correlation between the location and the demographic of conference presenters with a greater number of non-North American attendees when held in Europe or Asia (see Table 2 and Bogers and Greifeneder [1]). Travel seems an obvious restriction and, in competition for funds, institutional support is also fundamental for ensuing a diversity of participants, particularly for graduate students who may lack research budgets. This ability to make connections, establish relationships, and create networks is, to a great extent, dependent on institutional and financial constraints. Nevertheless, without this interconnection we limit the essential exchange of ideas that potentially lead to collaboration; faculties and funders need to be persuaded of this.

Consider the difficulties in getting a proposal accepted in a language other than your own. Allowing more diversity of languages for conference proposals, at a minimum that of the host nation, would go some way to increasing the possibility of more diverse dialogue. Without doing so, we are restricting our participants and our audience, and so limiting the reach and 'impact' of our research.

If we restrict our cultural perspective, we also restrict our field as we all learn from each other; inclusion benefits sustainable discussions among us. Without this, it is those English speakers who have no other language, and no incentive to engage outside the Anglophone sphere, that stand to lose the most.

> "Now, English has emerged as a de facto lingua franca – with of those of us who grew up speaking English losing the most, insofar as the widespread use of English makes it easy for us to ignore the importance of language and to avoid the challenge of mastering languages other than our own. No one would benefit more from a commitment to linguistic diversity than speakers of English [2]."

We need to be willing to engage with researchers and practitioners outside of our linguistic comfort zones, to reach out more widely to new audiences and to engage beyond our limited echo-chamber. Otherwise, we are destined to discuss our research interests only with those that we already know. Language is also an issue for conference organisation, with translation facilities to be considered as part of the package for conference funding. In addition, there are significant benefits for multi-lingual published proceedings to widen their circulation that could also be part of long-term strategic vision to create a truly global iSchools community.

The wish for this global iSchools community is clearly stated in 2020 by both Professor Sam Oh, the first Chair from the Asia-Pacific region: 'I knew from the start that I wanted to do whatever I could to further promote and develop the iSchool movement on a global scale. [...] It is my dearest wish that the iSchools will become truly international in every sense of the word' [15]. And our incoming Chair, Professor Gobinda Chowdhury, 'Inclusiveness and diversity are key attributes of the iSchools community [...]' [11]. There is still some way to go to achieve this but movement with regards to language and concerns about demographics of conference participants are beginning to be addressed.

The 2020 iConference was moved online in response to the pandemic and so the expectation for those submitting proposals was for physical attendance. The announcement that the 2021 iConference would be a virtual one was accompanied by a three-week extension to the deadline for proposals. It will be interesting to see how this affects the demographic diversity of participants, the divergence of published papers, and opportunities for dialogue once the need to travel is removed; this is the topic for our iConference paper in preparation for 2022.

There is a general movement towards, at least, an interest in diversity but to enable any effective examination, the iConference organization needs to make demographic data of presenters and attendees available at a more granular level. This will enable a cross-tabulation and other analyses of language and country before moving into the wider diversity landscape. Data on submissions versus acceptance (and reason for nonacceptance) is needed to assess whether allowing non-English papers at Renmin has led to more equity.

The organisers should be more flexible about languages for submission and presentation (at a minimum those of the host nation); encourage multi-lingual presentation of work (e.g. slides in two or multiple languages) to increase dialogue; consider multi-lingual publications of proceedings; having hybrid online/in-person formats to remove travel and funding related limitations; actively encourage wider participation and community engagement by funding bursaries and workshops.

References

1. Bogers, T., Greifeneder, E.: The ischool community: a case study of iconference reviews. In: IConference 2016 Proceedings (2016). www.ideals.illinois.edu/handle/2142/89312
2. Crane, G.: The big humanities, national identity and the digital humanities in germany (2015). http://www.dh.uni-leipzig.de/wo/the-big-humanities-national-identity-and-the-digital-humanities-in-germany
3. Fiormonte, D.: Towards a cultural critique of the digital humanities. Historical Social Research/Historische Sozialforschung, pp. 59–76 (2012)
4. Fiormonte, D.: Towards monocultural (digital) humanities? Text. Infolet 12 (2015). https://infolet.it/2015/07/12/monocultural-humanities
5. Harris, J., Tönnies, F.: Community and Civil Society. Cambridge University Press, Cambridge (2001)
6. iSchool: ischools leadership (board of directors). https://ischools.org/Meet-Our-People#iSchools_Leadership. Accessed on 7 Oct 2020
7. iSchools: 2021 chinese papers. https://ischools.org/Chinese-Papers. Accessed on 7 Oct 2020
8. iSchools: iconference 2014 summary. https://ischools.org/iConference-2014-Summary/. Accessed on 7 Oct 2020
9. iSchools: iconference 2015 summary. https://ischools.org/iConference-2015-Summary/. Accessed on 7 Oct 2020
10. iSchools: iconference 2017 summary. https://ischools.org/iConference-2017-Summary/. Accessed on 6 Oct 2020
11. iSchools: Incoming chair 2020. https://ischools.org/resources/Documents/iconf%202020/iSchools%20Chair%20message.pdf. Accessed on 7 Oct 2020

12. iSchools: ischools about. https://ischools.org/About. Accessed on 16 Sept 2020
13. iSchools: ischools directory. https://ischools.org/Directory. Accessed on 14 Sept 2020
14. iSchools: ischools regions. https://ischools.org/Regions. Accessed on 16 Sept 2020
15. iSchools: Outgoing chair 2020. https://ischools.org/resources/Documents/iconf%202020/iSchools-Outgoing-Chair-Farewell-Message.pdf. Accessed on 07 Oct 2020
16. iSchools: Past proceedings. https://ischools.org/Past-Proceedings. Accessed on 14 Sept 2020
17. Mahony, S.: The digital classicist: building a digital humanities community. Digital Hum. Q. **11**(3), 264–275 (2017)
18. Mahony, S.: Cultural diversity and the digital humanities. Fudan J. Hum. Soc. Sci. **11**(3), 371–388 (2018)

No Longer "Neutral Among Ends" – Liberal Versus Communitarian Ethics in Library and Information Science

David McMenemy(✉)

University of Strathclyde, Glasgow, Scotland
d.mcmenemy@strath.ac.uk

Abstract. As the concept of neutrality is under significant challenge as an ethic in LIS in the past decade from more critical approaches to social justice, the paper argues that in such a polarized world, we must seek to consider ethical approaches that do not divide us but instead have the capacity to bring people together around a common good, while also respecting aspects of group identity. Communitarianism is presented as a potential solution to that dilemma. The paper begins by exploring the philosophical debate between liberalism and communitarianism in political philosophy, and how concepts like neutrality figure in that debate. It presents the philosophies of both liberalism and communitarianism to encourage debate among the LIS community as to the potential for a communitarian ethic to develop in LIS as an alternative to the one based on liberalism. In doing so it considers what a communitarian ethic might look like for library and information science, and considers that ethical approach in contrast with both individual rights, and group-rights based philosophies. The paper adds to the wider debates within LIS related to the ethic of neutrality and its fit for modern practice and presents an alternative to liberalism that is, nevertheless, still grounded in a liberal tradition.

Keywords: Liberalism · Neutrality · Social justice · Ethics · Communitarianism

1 A Contemporary Concern in LIS

Neutrality is a core facet of liberalism, and it is also a "core – yet controversial – value" in library and information ethics [1]. A search of Library and Information Science Abstracts indicates that in only the first 9 months of 2020 alone, there are almost three times as many references in the LIS literature to neutrality as there was in the entire 1980s (Table 1). The topic has been exercising the discourse of the profession more in the modern era, and this paper seeks to add to that debate by unpacking neutrality from its first principles as a core tenet of liberalism, and considers how a communitarian approach might counter the criticism that liberal neutrality attracts.

This growing focus on the concept of neutrality within LIS reflects wider societal concerns related to social justice that consider neutrality as a value that, rather than guaranteeing detached subjectivity and equity, actually reflects the worldview of one particular group in society over others:

K. Toeppe et al. (Eds.): iConference 2021, LNCS 12646, pp. 207–214, 2021.
https://doi.org/10.1007/978-3-030-71305-8_16

Table 1. Frequency of term "neutral" in Library and Information Science Abstracts – 1980s-2020s ("net neutrality" removed) – September 2020

Decade	Number of references
1980s	14
1990s	127
2000s	487
2010s	755
2020s	36

Universalistic moral theories in the Western tradition from Hobbes to Rawls are substitutionalist, in the sense that the universalism they defend is defined surreptitiously by identifying the experiences of a specific group of subjects as the paradigmatic case of the human as such. These subjects are invariably white, male adults who are propertied or at least professional" [2, p.181].

This debate reflects the wider one that is happening in Western societies related to social justice. Pejoratively dubbed, "culture wars" by some commentators, there has been an emerging emphasis on group rights where previously disenfranchised sections of society seek to assert their positions from the point of view of equity [3].

2 The Tenets of Liberalism

Since neutrality has become such a controversial issue in LIS ethics, it is important to understand the concept from first principles. To effectively critique an idea, we must understand it. In this part of the paper, then, we explore the key tenets of liberalism, and consider neutrality as a fundamental cornerstone of the liberal ethic.

2.1 The Triumph of Autonomy

The emphasis of liberalism on the autonomy of the person has important strands that are of concern for the debate with communitarianism. Primarily in this context liberalism can be defined as an "approach to political power and social justice that determines principles of right (justice) prior to, and largely independent of, determination of conceptions of the good" [4]. This autonomous individual has no theoretical connection with others or the community, it is her own rational processes of decision making that determine her own good. Stemming in large part from the theories of Kant, who placed the emphasis on his rational autonomous agent as they key arbiter of moral duty, the main proponent of liberalism in this vein in the modern era was John Rawls. For Rawls the Kantian inspiration on his notion of justice as fairness was profound; he devotes an entire section (s40) of his *A Theory of Justice* to the Kantian influence [5]. As much as the concept of autonomy is important to Rawls, a crucial component of the Kantian influence on Rawls is an expansion of the notion of the rational choice of the individual. For Rawls, the justice is in the right to be the autonomous rational agent. In other words the right comes before the good and "no particular conception of the good may define or take priority over the principles of justice" [6]. The self is defined as a "chooser of ends", separated from the conception of the good, and exists as a moral being regardless of the good [6].

2.2 Societal Neutrality – The Right Before the Good

This autonomous individual that is the cornerstone of modern liberalism is ideally free to choose her own path in life, and the state should not attempt to impose on her any form of the good life. She must be free to select her own path, and as long as that path does not negatively impact on the rights of other autonomous selves, then no interference is to be supported from the state. This is core of the neutrality that liberalism supports; no one idea of the good is to be considered above any other that is preferred by an autonomous person. Applied to LIS practice then, in theory, this means no one users' preference is to be preferred over any others'.

Griffin identifies three distinct periods of human rights that brought us to where we are today:

> The first generation consists of the classic liberty rights of the seventeenth and eighteenth centuries—freedom of expression, of assembly, of worship, and the like. The second generation is made up of the welfare rights widely supposed to be of the mid-twentieth century though actually first asserted in the late Middle Ages—positive rights to aid, in contrast, it is thought, to the purely negative rights of the first generation. The third generation, the rights of our time, of the last twenty-five years or so, consists of 'solidarity' rights, including, most prominently, group rights [7, p. 256].

The current controversies dubbed "culture wars" seem largely to be a clash between those on the right-wing of politics who espouse the first generation classic liberty rights versus those more left of center who espouse the third generation solidarity or group rights. Like many battles within liberalism, the issue is with competing rights and how society resolves them. First generation liberals may believe their traditional rights like freedom of speech trump the rights of groups not to be offended. Yet supporters of group rights might counter that social cohesion and equity rely on autonomous individuals within previously under-served or ill-served groups being able to have that group identity respected and even lauded. Griffin clarifies this viewpoint:

> Membership of certain groups, especially cultural groups, is of great importance to their members. A good life depends importantly upon the successful pursuit of worthwhile goals and relationships, and they, in turn, are culturally determined. So the 'pragmatic' and 'instrumental' case, as Raz describes it, for the existence of a group's right to self-determination would go along these lines [7, p. 262].

Within this clarification of Raz's argument for the group right we can delineate the beginnings of the concept of what has come to be known as *identity politics*.

The question for us is whether liberal neutrality, *or* a more group-rights focus are the best fit for a library and information science ethic. More broadly, Griffin argues that the "fairly recent appearance of group rights is part of a widespread modern movement to make the discourse of rights do most of the important work in ethics, which it neither was designed to do nor, to my mind, should now be made to do" [7, p. 256]. Put simply, if society becomes merely a collection of groups fighting for their own group rights, can this be done so in a way that preserves respectful civic discourse and promotes a just society? Even more simply, how can the LIS profession respond to such potential civic dysfunction?

3 The Communitarian Critique

The most influential critique of liberalism in modern times has come from the communitarian critics. In this section we explore some of the key critics of liberalism in the modern era from the point of view of communitarian philosophy.

3.1 The Communitarian Critique of Liberalism

It is rather difficult to summarize the communitarian critics in one sentence, as they espouse differing aspects of critiques on liberalism, however Parvin and Chambers neatly do so by defining communitarianism as, "the view that, by emphasizing individual freedom, liberalism and libertarianism [liberals] undermine the shared sense of identity which people need in order to function as a society" [6, p. 205]. As Voice has observed, "for some communitarians, the ideal of autonomy is not only a philosophical error but it is also a practical evil" [8, Kindle Location 1893]. As we will see in our discussion of some of the communitarian critiques that will follow, the philosophical error in question is that the individual autonomy espoused by Rawls and other liberals is a flawed concept of personhood; a metaphysical error. The practical evil in question is that a system of justice based around concepts of autonomy and individualism is argued to be detrimental to society. An individualistic society puts each and every person out on their own, with no good specified but that each individual sets themselves. This mitigates against a common good, of any kind, be it one based on a historical concept of the good, or a religious one. Communitarians believe that such a society can lead to "selfishness and anti-social behavior" [9, p. 12].

One of the most important and influential communitarian critiques of liberalism is Michael Sandel, who in *Liberalism and the Limits of Justice* (LLJ) provides a critique of Rawls' notion of the "unencumbered self". Sandel challenged Rawls' *Theory of Justice* in important areas. He argued that the idea of the freely choosing autonomous individual, rationally choosing the distribution of social goods behind a veil of ignorance, was extremely flawed:

> At issue is not whether individual or communal claims should carry greater weight but whether the principles of justice that govern the basic structure of society can be neutral with respect to the competing moral and religious convictions its citizens espouse. The fundamental question, in other words, is whether the right is prior to the good [10, Kindle location 92].

As well as the state neutrality implied in Sandel's critique above, he also considered the subsequent influence of moral and religious encumbrances on individuals as equally important. He captures this critique that relates to the influence of community/culture on the individual well in the following passage from LLJ: "Construing all religious convictions as products of choice may miss the role that religion plays in the lives of those for whom the observance of religious duties is a constitutive end, essential to their good and indispensable to their identity" [10, Kindle location 134]. In other words, to many the right does not come before the good, the right is an intrinsic aspect of the self. If Sandel's point is proven, we can easily move from here to an argument that suggests any attacks on that religion, or other cultural identities, should potentially be considered attacks on the self. In this idea we can see the core of some of the attacks on free speech espoused by group rights advocates, where the argument is that speech that targets someone based on group characteristics is a harm in that it attacks a fundamental aspect of a person's sense of self, and thus their autonomy.

Another key text in the communitarian critique is Alasdair MacIntyre's, *After Virtue*. MacIntyre's critique goes beyond the notion of Rawls' version of liberalism and attacks the very foundations of liberalism that emerged from the Enlightenment. For MacIntyre, since all liberal ideas are essentially based on the rational voice that emerged during the Enlightenment with its focus on the universality of persons rather than community, history and culture, he believed liberalism as an entire concept was fundamentally flawed. A much quoted passage from *After Virtue* summarizes his thesis:

> I inherit from the past of my family, my city, my tribe, my nation, a variety of debts, inheritances, rightful expectations, and obligations. These constitute the given of my life, my moral starting point. This is in part what gives my life its own moral particularity [11, p. 220].

Fawcett neatly summarizes MacIntyre's thesis as follows:

> Liberals assumed that what people happened to want fixed their values and ideals, whereas in truth, values and ideals fixed what people ought to want. Values and ideals, in addition, grew out of shared practices in society, which alone gave people a purpose in life. Liberal modernity had dislocated society and shattered its practices [12, pp. 351-352].

For MacIntyre, humans are story-telling beings; we exist as products of the legacies we are a part of, and we inherit from this legacy our cultures, mores and values, and these are fundamental parts of our identity. For him, then, the notion of the rational agent, atomized within society is a philosophical error. Again, we can suggest that it is not a large step from MacIntyre's thesis, to consideration of how integral our culture, religion and identity is to our sense of being.

Bell sets out the 3 main communitarian critiques of liberalism as follows:

- The Liberal Self – the liberal, autonomous individual is defined as overly-individualistic. An unencumbered self.
- Liberal Universalism – the liberal, autonomous individual is insufficiently rooted in community and its influences.

- Liberal Atomism – the liberal, autonomous individual is a lonely, isolated figure who is unaware of society and the people in it [13, pp. 4–6].

The key theme, then, is that liberalism fails to consider communities and their needs and influences, and the impact such links and ties have on us as individuals. It fails to consider how important notions of community, culture, group identity, and religion are to the identities of people.

How does then manifest in the professional concerns of library and information science? From the point of view of free speech, for example, we can identify some fundamental issues here. We can see here the core of the argument that if the situated self is a product of their culture or religion, then such attachments form a crucial part of their identity. It is no great leap from there, then, to argue that any attacks on those elements of identity would constitute a potential harm to the person and should therefore be considered for remedy. Attacks on these elements from the point of view of freedom of speech become a potential harm. Sandel further revisits this concept as it relates to state neutrality later in the LLJ when he summarizes the approach a liberal state would normally take towards offensive speech: "To ban offensive or unpopular speech imposes on some the values of others and so fails to respect each citizen's capacity to choose and express his or her own opinions" [10, Kindle location 169]. The state, then, cannot impose any version of the good life on its citizens in a liberal democracy, it must respect the ability of citizens to rationally and autonomously choose their own way. It is clear that such an approach can cause societal tension.

4 What Can Replace the Ethic of Neutrality?

Sandel has written extensively about what he calls the hollowing out of the public sphere, suggesting that it, "makes it difficult to cultivate the solidarity and sense of community on which democratic citizenship depends" [14, p. 267]. Like other communitarians, he has "argued not only for stronger notions of community and solidarity but also for a more robust public engagement with moral and religious questions" [14, p. 247]. The current times seems to many to feel bleak with polarization of society along political, racial, and religious grounds. To arrive at solutions, different groups in society need to engage, but when peoples are poles apart, the challenge is what can bring us together.

Identifying with one group over a wider societal ethic poses challenges for a cohesive society unless we are able to embrace respectful civic discourse. An LIS ethic built around communitarian values may well contribute to the solution. Table 2 illustrates what some aspects of an LIS ethic of communitarianism might look like when compared to the liberal ethos that currently governs much ethical thinking by professions:

A debate around what could constitute a new LIS ethic that is inclusive, but that protects ideas and knowledge from censorship, is not an easy one to have. However, by applying a communitarian ethic, one that favors respectful discourse, one that celebrates diversity of groups and peoples, while recognizing our own individual rights within our own groups, we stand the chance of building a progressive future for our profession that supports a cohesive and respectful society.

Table 2. How might a communitarian LIS ethic differ from a liberal LIS ethic?

Liberal ethos impact on LIS	Communitarian ethos impact on LIS
The profession should be neutral in matters of morals, individual identity, and religious belief	The profession should promote and facilitate civic discourse on matters of moral, group, and religious issues that impact the wider common good
The profession should not limit individual rights even if it is detrimental to a common good	The profession should promote individual rights as a backbone of a shared, agreed common good
The profession should support the individual within society to prioritize and achieve their own goals without consideration of wider societal needs	The profession should support the development of active citizens with an emphasis to enable them to take their place as part of a civic populace able to take part in a shared democratic project

5 Conclusion

This short paper has sought to contribute to the important conversation that is being undertaken within the profession around library and information ethics, and the place of neutrality within the debate. It has explored neutrality as part of a wider liberal ethic related to individualism and autonomy, and presented the communitarian critique that seeks to take on the atomization the critics perceive to be damaging to society. It has also considered how a communitarian approach to social justice might offer a potentially inclusive ethic that respects diversity within the public discourse.

Moral psychologist Jonathan Haidt sums up the key dilemma for us quite well:

> We're not always selfish hypocrites. We also have the ability, under special circumstances, to shut down our petty selves and become like cells in a larger body, or like bees in a hive, working for the good of the group [15].

It has been argued that "liberalism advocates for the principle of priority of individual rights and freedoms over the common good" [16]. The "common good" should be the core concern of the LIS profession, and a communitarian ethic may well help achieve that.

References

1. Macdonald, S., Birdi, B.: The concept of neutrality: a new approach. J. Document. **76**(1), 333–353 (2020)
2. Benhabid, S.: The generalized and the concrete other. In: Benhabib, S., Cornell, D. (eds.), Feminism as Critique, pp. 77–95. Polity Press, London (1987)
3. Chandler, D., Munday, R.: Culture Wars a Dictionary of Media and Communication. Oxford University Press (2020)

4. Christman, J.: Autonomy in Moral and Political Philosophy. The Stanford Encyclopedia of Philosophy (Spring 2011 Edition), Edward N. Zalta (ed.) https://plato.stanford.edu/archives/spr2011/entries/autonomy-moral/. Accessed 30 Sept 20
5. Rawls, J.: A Theory of Justice Mass. Harvard University Press, Cambridge (1971)
6. Parvin, P., Chambers, C.: Political Philosophy - A Complete Introduction: Teach Yourself. Hodder & Stoughton, London (2012)
7. Griffin, J.: On Human Rights. Oxford University Press, Oxford (2008)
8. Voice, P.: Rawls Explained: From Fairness To Utopia (Ideas Explained). Open Court. Kindle Edition. 2011.
9. Campbell, T. Rights: A Critical Introduction. Routledge, London (2006)
10. Sandel, M.J.: Liberalism and the Limits of Justice, p. 1998. Cambridge University Press, Cambridge (1982)
11. MacIntyre, A. After Virtue. 3rd edn., Bloomsbury Publishing, London (1981). 2007
12. Fawcett, E.: Liberalism: The Life of an Idea. Princeton University Press, Princeton (2014)
13. Bell, D.: Communitarianism and its Critics. Oxford University Press, Oxford (1993)
14. Sandel, M.J.: Justice: What's the Right Thing To Do? Allen Lane, London (2009)
15. Haidt, J.: The Righteous Mind: Why Good People are Divided by Politics and Religion. Kindle Edition edn., Penguin Books Ltd., London (2012)
16. Zeidler, K., Łągiewska, M.: Liberalism versus communitarianism in cultural heritage law. Int. J. Semiotics Law (2020). https://doi.org/10.1007/s11196-020-09792-9

Identification of Biased Terms in News Articles by Comparison of Outlet-Specific Word Embeddings

Timo Spinde[1,2](✉), Lada Rudnitckaia[1], Felix Hamborg[1,3], and Bela Gipp[2,3]

[1] University of Konstanz, Konstanz, Germany
{timo.spinde,lada.rudnitckaia,felix.hamborg}@uni-konstanz.de
[2] University of Wuppertal, Wuppertal, Germany
gipp@uni-wuppertal.de
[3] Heidelberg Academy of Sciences and Humanities, Heidelberg, Germany

Abstract. Slanted news coverage, also called media bias, can heavily influence how news consumers interpret and react to the news. To automatically identify biased language, we present an exploratory approach that compares the context of related words. We train two word embedding models, one on texts of left-wing, the other on right-wing news outlets. Our hypothesis is that a word's representations in both word embedding spaces are more similar for non-biased words than biased words. The underlying idea is that the context of biased words in different news outlets varies more strongly than the one of non-biased words, since the perception of a word as being biased differs depending on its context. While we do not find statistical significance to accept the hypothesis, the results show the effectiveness of the approach. For example, after a linear mapping of both word embeddings spaces, 31% of the words with the largest distances potentially induce bias. To improve the results, we find that the dataset needs to be significantly larger, and we derive further methodology as future research direction. To our knowledge, this paper presents the first in-depth look at the context of bias words measured by word embeddings.

Keywords: Media bias · News slant · Context analysis · Word embeddings

1 Introduction

News coverage is not just the communication of facts; it puts facts into context and transports specific opinions. The way how "the news cover a topic or issue can decisively impact public debates and affect our collective decision making" [12], slanted news can heavily influence the public opinion [11]. However, only a few research projects yet focus on automated methods to identify such bias.

One of the reasons that make the creation of automated methods difficult is the complexity of the problem: How we perceive bias is not only dependent on words, but also their context, the medium, and readers' background. While many current research projects focus on collecting linguistic features to describe media bias, we present an

K. Toeppe et al. (Eds.): iConference 2021, LNCS 12646, pp. 215–224, 2021.
https://doi.org/10.1007/978-3-030-71305-8_17

implicit approach to the issue. The main question we want to answer is: Comparing biased words among word embeddings created from different news outlets, are they more distant (or close) to each other than non-biased words?

To answer this question, we measure any word's context by word embeddings, which reflect the specific usage of a word in a particular medium [15]. We focus on the definition of language bias given by Recasens et al. [10], describing biased words as subjective and linked to a particular point of view. Such words can also change the believability of a statement [10].

Overall, our objectives are to:

1) Analyse and compare the word embeddings of potential bias inducing words trained on different news outlets.
2) Test the assumption that distances between vectors of similar bias words trained on different corpora are larger than between neutral words due to usage in a specific context.

2 Related Work

While some scholars propose methods to create bias lexica automatically, none of them is in the domain of news articles. Recasens et al. [10] create a static bias lexicon based on Wikipedia bias-driven edits, which they combine with a set of various linguistic features. Ultimately, they aim to classify words as being biased or not. Hube & Fetahu [5] extend this approach by manually selecting bias-inducing words from a highly biased source (Conservapedia) and retrieving semantically close words in a Wikipedia word embedding space.

Since there is no large-scale dataset from which initial knowledge about biased language can be derived, implementation and extending of the approaches of Recasens et al. and Hube & Fetahu may be relevant for news data. However, in the context of media bias identification, creating static bias lexica is inefficient because the interpretation of language and wording strongly depends on its context [3].

It is therefore desirable to either evaluate every word independent of pre-defined lexica or, even more, enable existing biased lexica to be context-aware. In this regard, exploiting the properties of word embeddings is especially interesting [15]. Word embeddings are highly dependent on training corpora they are obtained from and accurately reflect biases and stereotypes in the training corpora [15]. Kozlowski et al. [6] use word embeddings trained on literature from different decades to estimate the evolution of social class markers over the 20th century.

Mikolov et al. [9] compare word embeddings obtained from different languages and show that similar words have minimal cosine similarity. Tan et al. [15] analyze the usage of the same words in Twitter and Wikipedia by comparing their different word representations – one trained on Twitter data and another on Wikipedia.

3 Methodology

We seek to devise an automated method that ultimately finds biased words by comparing two (or more) word embeddings spaces, each trained on a differently slanted group of text documents. In this exploratory study, we devise a one-time process that consists of four tasks: selection of a word embedding model, selection of biased words, data processing and analysis, and linear mapping of the word embedding spaces.

3.1 Word Embeddings and Parameter Selection

To calculate our embeddings, we use Word2Vec [8] with the Continuous Skip-gram (SG) architecture, which achieves better semantic accuracy and slightly better overall accuracy than the Bag-of-Words architecture [9]. We evaluated our word representations via an estimation of word semantic similarity on two datasets – WordSim-353 [2] and MEN [1] and the Google analogy test set [8].

WordSim-353 consists of 353 pairs assessed by semantic similarity with a scale from 0 to 10. MEN consists of 3,000 pairs assessed by semantic relatedness and scaled from 0 to 50. We use these datasets since they focus on topicality and semantics. The Google analogy test set consists of 8,869 semantic and 10,675 syntactic questions. Generally, for our task, the data sets are not ideal, which we discuss in Sect. 5.

We summarize our hyper-parameters in Table 1 and the summary evaluation of our word embeddings in Table 2. We train the word embeddings on the data preprocessed with Genism simple preprocessing and n-grams generated within two passes.

Table 1. Hyper-parameters for training the word embeddings

Hyper-parameter	Value	Hyper-parameter	Value
Dimensionality	300	Maximum token length	28
Window size	8	n-grams threshold (1st pass)	90
Subsampling rate	10^{-5}	n-grams threshold (2nd pass)	120
# of iterations	10	Articles titles	Included
Minimum frequency	25	Training sentence	The whole article
Function	Hierarchical softmax		

Table 2. Evaluation of the word embeddings

Corpora	# articles	# tokens	Vocabulary size	Semantic similarity		Analogy
				WordSim-353	MEN	Google
HuffPost	101K	68M	53K	0.65	0.71	0.50
Breitbart	81K	39M	37K	0.57	0.59	0.38

3.2 Manual Selection of Bias Inducing Words

We follow the approach proposed by Hube and Fetahu [5] and manually select a small set of "seed" words that are very likely to be related to controversial opinions. They also have a high density of bias-inducing words surrounding them in the embedding space. The 87 seed words are selected based on the description of controversial left and right topics on Allsides.com (Table 3, see also https://www.allsides.com/media-bias/left,.../right). From the list of the closest twenty words to each seed word, we manually extracted words that convey a strong opinion [5]. As the identification of bias is not trivial for humans [10], we validated the extended seed words by four student volunteers, age 23–27, who labeled each word as being biased or not. We discarded any words where less than three students agreed on.

Table 3. The seed words that are likely to be related to controversial opinions and to have a high density of bias-inducing words surrounding them in the word embedding space

Divisive issue	Seed words
The role of the government	Regulation(s), involvement, control, unregulated, government, centralization, law
Economics	Tax(es), taxation, funding, spending, corporation(s), business(es), economy
Equality	Equality, inequality, rights, equal_rights, wealth, living_wage, welfare, welfare_state
Social services	Services, government_services, social_security, benefit(s), help, student(s), loan(s), student_loan(s), education, healthcare, individual, personal_responsibility, collective
Security	Security, military, military_force, defense, intervention, protect, protection, border, border_security, migration, migrant(s), immigration, immigrant(s), terror, terrorist(s)
Traditions, religion, and culture	Tradition, norms, cultural_norms, progress, change(s), race, racism, gender, sexual, orientation, sexual_orientation, identity, religion, Islam, tolerance, multiculturalism, values, family_values, bible, constitution
Miscellaneous	Freedom, speech, freedom_of_speech, free_speech, hate_speech, gun(s), gun_owner(s), abortion, environment, media

3.3 Data

We choose two news outlets that are known to take different views and potentially use different words to describe the same phenomena. We based the choice of news outlets for analysis on the media bias ratings provided by Allsides.com. The news aggregator aims to estimate the overall slant of an article and a news outlet by combining users' feedback and expert knowledge [3, 13]. We choose The HuffPost as a left-wing news outlet and Breitbart News as right-wing. We scraped articles from both news outlets, published in the last decade, from 2010 to 2020, from Common Crawl [4]. For preprocessing, we use Genism simple preprocessing and generate n-grams.

3.4 Linear Mapping Between Vector Spaces

Since the goal is to compare word vectors between two different word embedding spaces, it is necessary to make sure that these two word embedding spaces have similar dimensionality. We use the approach proposed by Mikolov et al. [15] and Tan et al. [20].

The results of training two different mapping matrices – trained on 3,000 most frequent words and on the whole common vocabulary – are presented in Table 4. The only metric to evaluate mapping quality is the number of distant words. Ideally, similar words should be close to each other after linear mapping. A large number of very distant words can be an indicator of a poorly trained matrix.

We assessed the distance between words in an embedding space with cosine similarity. In case the distance between similar words after mapping from one source to another depends on the frequency of the word in either a left- or right-wing source, according to Tan et al. [15], adjusted distances should be compared. The larger an adjusted distance, the less similar the word is between the two sources since positive adjusted distance values belong to words that are less similar than at least half of the words in their frequency bucket.

Table 4. Comparison of two linear mappings

Mapping matrix trained on	# tokens in common vocab	Median cos. sim	# distant words		# close words	Correlation of cos. Sim. With freq. in	
			cos. sim ≤ 0.4	adj. Cos. sim. ≥ 0.1	cos. sim ≥ 0.6	HuffPost	HuffPost
3K	30K	0.48	8K (25%)	5K (17%)	6K (20%)	0.14	0.13
whole vocab.		0.56	2K (6%)	4K (14%)	10K (35%)	0.12	0.12

For both variants, we still obtained many distant words after linear mapping, i.e., median cosine similarity is 0.48 and 0.56 for the first and the second variant, respectively. Ideally, similar words should be close to each other after linear mapping, except those used in different contexts. Possible reasons for having many distant words are:

- low quality of trained word embeddings and, thus non-stable word vectors,
- low quality of mapping matrix, possibly nonlinear transformation is needed,

- a high number of "bad" n-grams,
- a high number of noisy words.

We tried to address the first two causes by training different word embeddings models and different mappings. Since Tan et al. [15] did not discuss the threshold for the definition of distant words, we choose the thresholds to define distant words as lower than 0.4 and higher than 0.1 for pure and adjusted cosine similarities, respectively. Comparing n-grams and unigrams based on their cosine similarity statistics, we conclude that

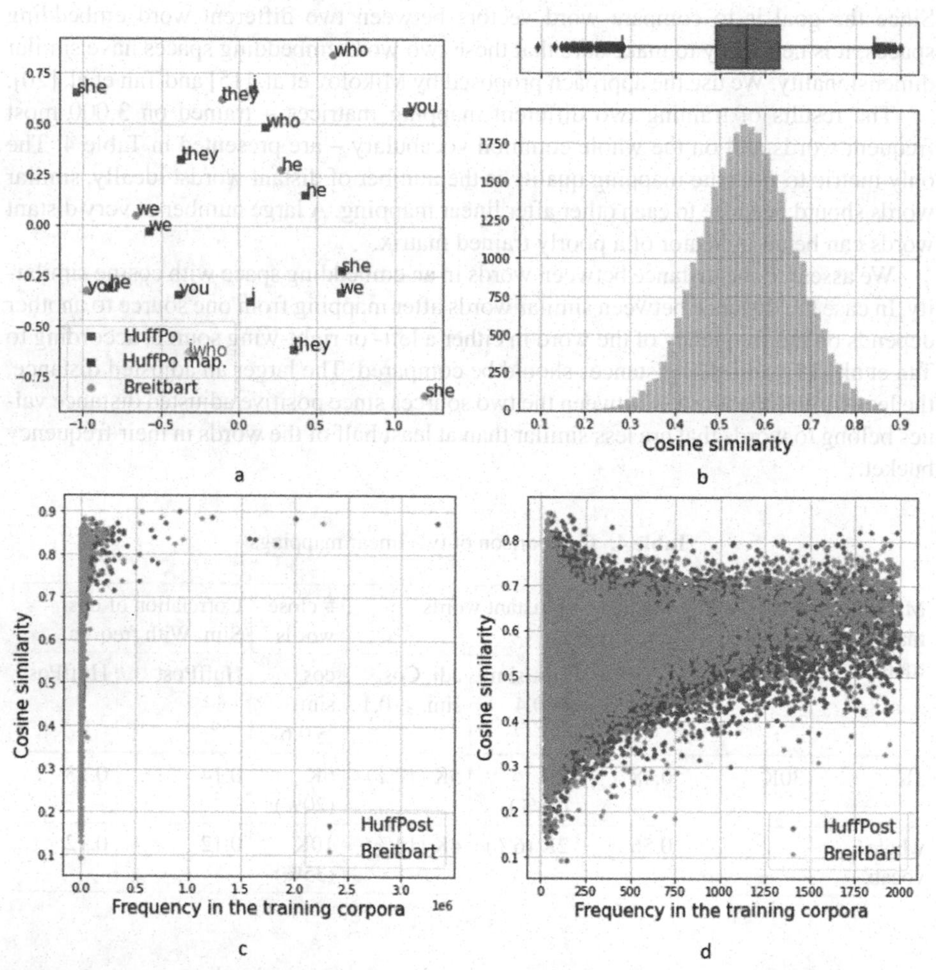

Fig. 1. Linear mapping from HuffPost to Breitbart trained on the whole vocabulary: a) high-frequency word vectors before and after mapping, b) distribution of cosine similarities after mapping, c) dependency of frequency and cosine similarities, d) dependency of frequency and cosine similarities for the words less frequent than 2K, e) median cosine similarities per frequency bucket, f) distribution of adjusted cosine similarities after mapping, g) dependency of frequency and adjusted cosine similarities, h) dependency of frequency and adjusted cosine similarities for the words less frequent than 2K

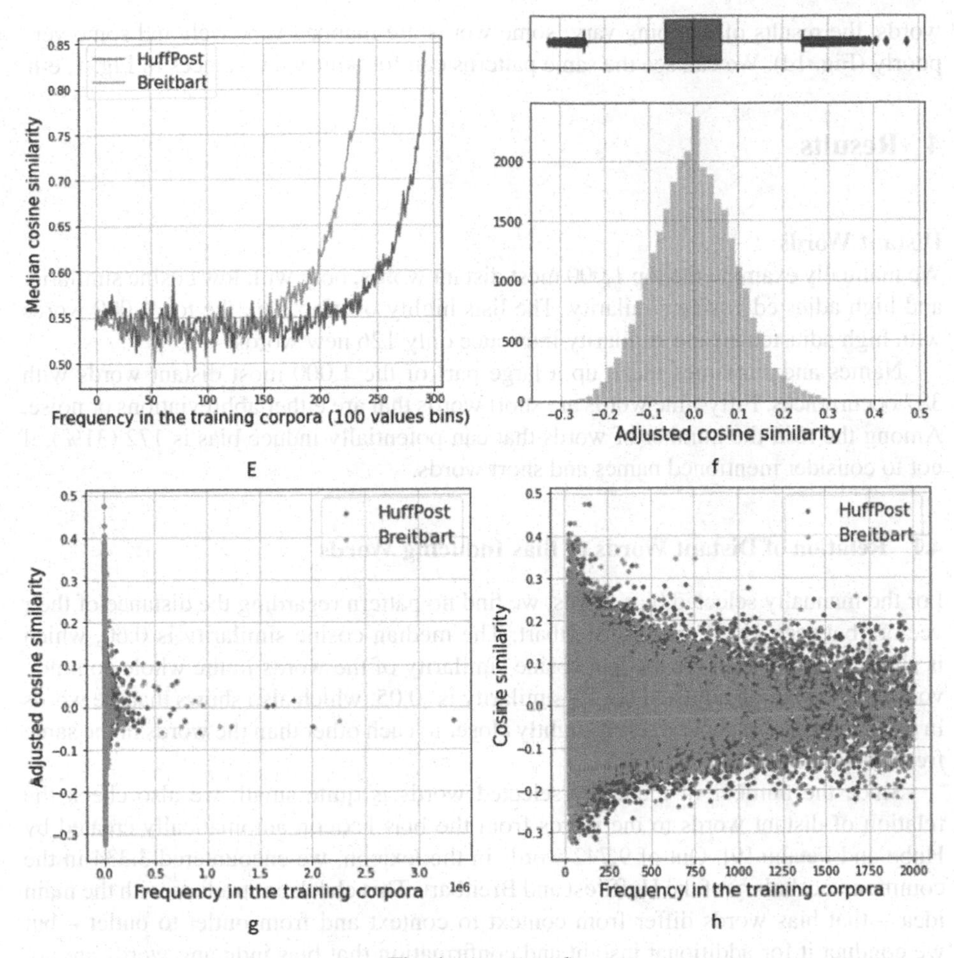

Fig. 1. (*continued*)

there is no apparent reason to think that the generated n-grams are more distant than the unigrams: median cosine similarity for n-grams is 0.62, whereas for unigrams it is 0.55. We manually inspected the distant words to estimate the possible influence of flaws in preprocessing and connection to bias words (Sect. 4.1).

The matrix trained on the whole vocabulary maps similar words better since there are fewer very distant words, and the median cosine similarity for the words is higher. Therefore, we used this mapping for further analysis.

At the two-dimensional graph obtained by reducing the dimensions of the word vectors from 300 to 2 with PCA, it can be seen that the mapping works quite well for the most frequent words, here, the pronouns: the word vectors mapped from HuffPost to Breitbart are indeed closer to the vectors from Breitbart than the initial vectors from the HuffPost (Fig. 1a). However, we also get a high number of distant words (Fig. 1b). We can also see that the higher the frequency, the higher the chance that the words are better mapped from one source to another (Fig. 1c). Simultaneously, for less frequent words, the results of mapping vary: some words are mapped very well and some very poorly (Fig. 1d). We can see the same patterns can for adjusted distances in Fig. 1, e-h.

4 Results

Distant Words

We manually examine the top 1,000 most distant words, both with low cosine similarity and high adjusted cosine similarity. The lists highly overlap, i.e., the top 1,000 words with high adjusted cosine similarity introduce only 126 new words.

Names and surnames make up a large part of the 1,000 most distant words with 387 occurrences. Fifty-nine words are short words that are either abbreviations or noise. Among the rest, the number of words that can potentially induce bias is 172 (31%), if not to consider mentioned names and short words.

4.1 Relation of Distant Words to Bias Inducing Words

For the manually selected bias words, we find no pattern regarding the distance of their vectors between HuffPost and Breitbart. The median cosine similarity is 0.60, which is slightly higher than the median cosine similarity of the words in the whole common vocabulary. Median adjusted cosine similarity is -0.05, which also shows that the words in this group are, in general, even slightly closer to each other than the words in the same frequency buckets.

Since the number of manually selected words is quite small, we also check the relation of distant words to the words from the bias lexicon automatically created by Hube and Fetahu [9]. Out of 9,742 words in the lexicon, we encountered 3,334 in the common vocabulary of the HuffPost and Breitbart. This check contradicts with the main idea – that bias words differ from context to context and from outlet to outlet – but we conduct it for additional insight and confirmation that bias inducing words are not directly related to distant ones.

Similarly to the manually selected bias words, for the words from the bias lexicon, we found no pattern regarding the distance of their vectors between different outlets. The median cosine similarity is 0.52, slightly lower than the median for all the words in the common vocabulary. The Median adjusted distance is 0.03, which means that the words in this group are, in general, just slightly more distant than other words with the same frequency. Therefore, this finding does not allow to claim that bias words are in general more distant than other words but rather corroborates that bias inducing words are not directly connected with distant words.

Overall, there are no salient differences when comparing the context of biased words between HuffPost and Breitbart. The most noticeable differences are between the context of the words "regulations," "welfare," "security," "border," "immigration," "immigrants," "hate_speech," and "abortion". We also notice the differences in the context of the words that have more than one meaning, e.g., the word "nut" is surrounded by the words describing food in the word embeddings trained on the HuffPost corpus. In contrast, in the word embeddings trained on the Breitbart corpus, it is surrounded by such words as "horrid", "hater", etc. Such findings are rare, and their statistical significance should be proved on the exhaustive biased words lexicon and the word embeddings trained on larger datasets.

5 Conclusion and Future Work

We present experimental results of an approach for the automated detection of media bias using the implicit context of bias words, derived through fine-tuned word embeddings. Our key findings are:

1) Among the words with large distances after linear mapping, some can potentially induce bias. Their percentage is around 25% (if not to consider names, surnames, and short words that can be either abbreviations or noise, otherwise the ratio is about 15%).
2) In the small set of manually selected bias inducing words, median cosine similarity after the linear mapping is 0.6 which is even slightly higher than for the whole vocabulary. A direct relation to large distances also did not show on the words from the bias lexicon provided by Hube et al. [5].
3) There are no salient differences in the context of seed words apart from several words.

Obtained results are either point to the absence of a relation of bias and distant words or can be explained by the following flaws of the current project, which serve as future research directions. First, the data for training our word embeddings are relatively scarce. Intrinsic evaluation of the word embeddings trained on the Breitbart corpora shows low results. Our current evaluation methods do not reflect the actual suitability of the word embeddings for our specific task. Second, we did not test other word embedding models than Word2Vec, which might show a better overall performance, e.g., GloVe, BERT, Elmo, and Context2Vec [7]. We did also not integrate other features, such as lexical cues or sentiment. Third, bias inducing words are selected manually by a tiny group of non-native English speakers. Fourth, we based the comparison of context on the top 20 most similar words. But among these top twenty for one source, the similarity can be on average high and for another on average low.

We envision to integrate our method for bias word identification into a bias-aware news aggregator or visualizations [12].

Acknowledgments. This work was supported by the Hanns-Seidel-Foundation supported by the Federal Ministry of Education and Research of Germany.

References

1. Bruni, E. et al.: Multimodal distributional semantics. J. Artif. Intell. Res. **49**, 1–47 (2014)
2. Finkelstein, L. et al.: Placing search in context: the concept revisited. In: Proceedings of the 10th International Conference on World Wide Web, pp. 406–414 (2001)

3. Hamborg, F., Donnay, K., Gipp, B.: Automated identification of media bias in news articles: an interdisciplinary literature review. Int. J. Digit. Libr. **20**(4), 391–415 (2018). https://doi.org/10.1007/s00799-018-0261-y
4. Hamborg, F. et al.: News-please: a generic news crawler and extractor. In: Proceedings of the 15th International Symposium of Information Science (2017)
5. Hube, C., Fetahu, B.: Detecting biased statements in Wikipedia. In: Companion of the The Web Conference 2018 on The Web Conference 2018 - WWW 2018, Lyon, France, pp. 1779–1786. ACM Press (2018). https://doi.org/10.1145/3184558.3191640.
6. Kozlowski, A.C., et al.: The geometry of culture: analyzing the meanings of class through word embeddings. Am Sociol Rev. **84**(5), 905–949 (2019). https://doi.org/10.1177/0003122419877135
7. Melamud, O. et al.: Context2vec: learning generic context embedding with bidirectional LSTM. In: Proceedings of The 20th SIGNLL Conference on Computational Natural Language Learning, , Berlin, Germany, pp. 51–61. Association for Computational Linguistics (2016). https://doi.org/10.18653/v1/K16-1006
8. Mikolov, T. et al.: Efficient estimation of word representations in vector space. arXiv:1301.3781 [cs]. (2013)
9. Mikolov, T. et al.: Exploiting similarities among languages for machine translation. arXiv: 1309.4168 [cs]. (2013)
10. Recasens, M. et al.: Linguistic models for analyzing and detecting biased language, vol. 10 (2013)
11. Spinde, T. et al.: An integrated approach to detect media bias in German news articles. In: Proceedings of the ACM/IEEE Joint Conference on Digital Libraries in 2020. pp. 505–506. ACM, Virtual Event China (2020). https://doi.org/10.1145/3383583.3398585
12. Spinde, T. et al.: Enabling news consumers to view and understand biased news coverage: a study on the perception and visualization of media bias. In: Proceedings of the ACM/IEEE Joint Conference on Digital Libraries in 2020, pp. 389–392. ACM, Virtual Event China (2020). https://doi.org/10.1145/3383583.3398619
13. Spinde, T. et al.: MBIC – a media bias annotation dataset including annotator characteristics. In: Proceedings of the 16th International Conference (iConference 2021). Springer Nature, Virtual Event China (2021).
14. Spinde, T., Hamborg, F., Gipp, B.: Media bias in German news articles: a combined approach. In: Koprinska, I. (ed.) ECML PKDD 2020. CCIS, vol. 1323, pp. 581–590. Springer, Cham (2020). https://doi.org/10.1007/978-3-030-65965-3_41
15. Tan, L., et al.: Lexical comparison between Wikipedia and Twitter corpora by using word embeddings. In: Proceedings of the 53rd Annual Meeting of the Association for Computational Linguistics and the 7th International Joint Conference on Natural Language Processing,Beijing, China, (Volume 2: Short Papers), pp. 657–661. Association for Computational Linguistics (2015). https://doi.org/10.3115/v1/P15-2108.

The Model of Influence in Cybersecurity with Frames

Philip Romero-Masters(✉)

University of Wisconsin, Madison, WI 53706, USA
pmasters@wisc.edu

Abstract. The Model of Influence in Cybersecurity with Frames unifies the current literature around influence and media effects in cybersecurity messaging. Building on the Process Model of Framing Research by Scheufele, this new model applies directly to the cybersecurity area and provides a macro-level view to further researcher understand of cybersecurity influence and provide options for intervention by organizational security professionals. This analysis included 42 documents concerning the work of influencing users to engage in secure behavior covering topics in persuasion, user interface design, equivalency framing, managing, and understanding user perceptions, and exploring user mental models regarding cybersecurity. This review also investigates the use of framing in cybersecurity and the definitions needed to contextualize and understand research in cybersecurity that uses framing. This model is intended as a starting point with which to build a larger understanding of cybersecurity communication to address human factors in cybersecurity.

Keywords: Cybersecurity · Framing · Frames · Information security · Schema

1 Introduction

I surveyed the literature for documents about the framing of cybersecurity messages. This paper intends to synthesize the results of this investigation and organize the work of the field using a typology based upon Cacciatoire and Scheufele's work [1, 2]. This analysis centers upon the relationship between two distinct types of framing in the literature. Framing can refer to the mental representation of ideas and the shaping of media to convey ideas. A person possesses an individual frame of a topic, and this frame represents their understanding of that topic. The media possesses a media frame which shapes the opinion of the audience based upon choices in constructing the media artifact.

Studies commonly assert the human aspect of security presents the largest risk to organizations. Users tend to view secure practices as inconvenient which means organizations face the challenge of motivating users to perform secure behaviors to protect organizational interests and follow security policies. Addressing this challenge means researchers need to understand what factors influence policy, compliance, and behavior that might compromise security. Understanding the role of security communication preceding security-relevant-behavior represents a crucial area for supporting strong cybersecurity.

K. Toeppe et al. (Eds.): iConference 2021, LNCS 12646, pp. 225–234, 2021.
https://doi.org/10.1007/978-3-030-71305-8_18

This paper does three things. The paper reviews the ways researchers use frames in cybersecurity. The paper considers the other methods of influence being studied in cybersecurity and their relation to frames to propose a model of cybersecurity influence using frames. As a whole my paper provides a unique perspective centered on cybersecurity communication and how it can be changed rather than where it can be changed in (phishing, mobile app selection, passwords).

2 Theory

This paper focuses on the communication of cybersecurity information and uses the theoretical framework of framing as an analysis lens. I will explore the concept of framing and the typology crafted to disambiguate the use of the term [1, 2]. Following this, I will introduce the Process Model of Framing Research (PMFR) presented by Scheufele [1, 2]. I use the PMFR to structure the literature around the framing of cybersecurity messages. Scholars use framing in many ways which creates ambiguity around the term. This paper uses several definitions. The term framing developed from sociology and psychology. The literature on framing call for a refinement of the topic and the development of better models and has produced several ways to understand framing.

To address the ambiguity surrounding framing, I will describe the typology of framing presented by Scheufele [1, 2]. Scheufele and Tewksbury describe media frames as a macro-construct [3]. Media frames describe the way communicators present information to an audience. This conception of frames concerns the presentation and act of communication. Media frames further the colloquial idea of focusing not on what you say but rather how you say it. Media frames influence their audience.

Equivalence framing narrows the media framing definition to the effect of presenting the communication in logically equivalent forms. The term framing effect describes the phenomenon where describing equivalent options produces a cognitive bias to select the message option that avoids losses.

Framing with emphasis frames constitutes a selection of which information in a communication to emphasize. In contrast with equivalence frames, an emphasis frame does not present the same information as another emphasis frame [1, 2]. Emphasis framing selects certain information regarding a topic to bring to the forefront of the mind of a recipient. Emphasis frames guide a recipient's interpretation or sense of importance of a piece of information.

Media frames shape the reception of communication by the audience, and individual frames shape individuals understanding of the world. Individual frames are clusters of information that guide decision making in individuals [4]. These idea clusters, like schemas, exist within a person and shapes their understanding of the world. Scheufele presents a process model linking individual frames with media frames [1]. Media frames serve as the input for the frame building process that shapes individual frames. These individual frames shape an audience's values, behavior, and attributions of responsibility that ultimately end up shaping the media frames produced by journalists and other communicators.

The PMFR contextualizes and unifies the two broader types of framing: media frames and individual frames [1]. This model describes mass media and its impact on an audience and the impact the on journalists.

3 Methods

Using the literature on framing and cybersecurity, I generated a list of author keywords, controlled vocabulary, and Computing Classification System (CSS) terms for the ACM Digital Library to search with. These terms focused on different ways authors refer to cybersecurity or framing.

I conducted structured searches and included works on security communication, awareness training, or interventions aimed at organizational or information system end-users, studies on user behavior and characteristics and studies on social influence related to the topic of cybersecurity framing. These searches used the controlled vocabulary and author keyword fields and were not filtered by publication year. My searches aimed at gathering articles that used framing but also articles that may have used other ways to influence a user. This structured search aims to bring together framing research and other influence techniques to examine the way cybersecurity communication has been studied to build a conceptual model around framing [5] (Fig. 1).

Table 1. Structured search results

Database	Search results	Literature included
ACM Digital Library	238	28
Communication & Mass Media Complete	5	0
Business Source Complete	15	3
ProQuest's ABI/INFORM Collection	49	2
APA PsycInfo	68	9
Total	375	42

In total the results of my structured searches and the number of results (including book chapters, conference proceeding, articles, dissertations, thesis, and other literature) equals 375 documents and I selected 42 as relevant to this review. I reviewed the title and abstract of all search results. Not everything in the results was relevant. I kept studies regarding social influence, user behavior, and characteristics related to messaging.

I excluded results that focused on developers or development, focused on the experience of security professionals, focused on user reaction to malicious messages or phishing attempts that did not examine the role of the message, studies on training games, augmented reality, virtual reality, managerial focused studies, visualization studies, duplicate results, public communication surrounding cybersecurity, and organizational information sharing. I intended to collect as much work on message framing in my initial search as possible. Once I began reading the literature and developing an understanding of the ambiguity surrounding the term framing, I supplemented my search with additional search terms related to individual frames such as 'mental model,' 'script', and 'paradigms' to include studies related to individual frames but used different terminology.

My analysis includes 42 documents: 6 book chapters, 10 conference papers, 23 journal articles, 2 dissertations, and 1 thesis. Most of the articles included in this analysis come from the ACM Digital library. This topic is new and only developed over the past two decades but is growing quickly with many of the included articles being published over the past 5 years.

4 Results and Analysis

My survey of the research sought to find the work done on the framing of cybersecurity communication. Given the theoretical ambiguity of framing, many articles were included that made no explicit mention of framing but instead intersected conceptually. When authors use framing explicitly, they rely heavily on the early work on framing by Kahneman and Tversky [6–12]. Because conceptual ambiguity plagues the literature on framing thus, I introduced a typology of emphasis framing, equivalency framing and individual framing to clarify what authors mean when they are using a framing concept in their work.

I found that researchers use framing, but they also use other theories and constructs related to the idea of framing. I refer to this related literature as the non-framing literature. I will cover the use of framing in the literature I analyzed and what assumptions authors are using regarding framing, then I organize the non-framing literature included in this review around five themes. This analysis organized around framing and influence contributes a new perspective around communication and media effects rather than around cybersecurity context (i.e., phishing, mobile app selection, passwords).

4.1 Framing Literature

I analyzed each study for how framing appeared in the study. I specifically looked at the way the word framing was used in the document, the citation used by the study to introduce the term, and the operationalization of framing in the study design. I looked closely for the use of gains/loss terminology or the introduction of the framing effect to help me identify equivalent frames. To categorize a study as using emphasis frames, the study needed to change the message presented to a user in terms of content rather than presentation of losses versus gains. For individual frames, I looked for studies that examined the ideas within the mind of the user rather than in a piece of communication.

I found six documents that used equivalent framing [6–11]. Most studies using framing in cybersecurity use equivalent frames to test user action in response to framing message as losses or gains. I found only one study using emphasis frames [13]. I found one document that addressed individual frames and three 3 documents that employed mental models [14–17]. I group the study that used individual frames and studies that used mental models together because the concepts are similar.

4.2 Influence Literature

I found twenty-seven articles that did not explicitly use any framing approach but looked at aspects of cybersecurity communications related to influence. The other major concepts I found in the literature include perceptions, persuasion, and user-characteristics.

I found six documents that used the concept of perception, the process of perceiving or observing [18–23]. Perception refers to the beliefs of a user. Cybersecurity research incorporates perception as both an independent variable and a dependent variable [18, 22, 23]. Huang examined perception around security to inform security management practice [22]. Lee and Addae used perception as the independent variable to predict security behavior [18, 23]. Lee used the perception of autonomy, competence and relatedness, the essential needs of self-determination theory. Addae used perceived ease of use and usefulness, the user's belief that the system will be difficult to use and will help perform a task. [18, 23, 24]. Choong suggests that perception of a user-account's criticality and susceptibility influences password management tasks for users [20]. Hirshfield examines the role of suspicion in detecting cyber-attacks and defines suspicion as a "mismatch between what is being perceived and what one expects" [21].

I found five documents that used persuasive security features in their work [25–29]. Busch's work evaluated user reception of persuasiveness strategies and found sharing the statistics around organizational policy violations to be well liked [25]. Kankane's work on nudges found the salience nudge the only one to significantly increase the likelihood of keeping an auto-generated password [26]. Pope found the behavioral intent of users to violate policy decrease when more certainty, greater severity, and quicker celerity of sanction. Weirich and Sasse investigated methods to persuade users to adopt "proper password behavior" and recommend the use of fear appeals [28]. Successful fear appeals convince a person they can perform a recommend behavior to avoid a negative outcome [28]. In a similar line of inquiry Zhang found that negative consequence information, when salient and explicit, can deter unethical behavior in cybersecurity.

Three studies I found examine user characteristics and study the traits of users and their correlation to undesired cybersecurity behavior [30–32]. Kajzer explored seven personality traits to discover if personality plays a role in security message effectiveness and confirmed user characteristic do play a role. Jeske and Kajzer each explore one individual trait for a relationship with a security behavior [30, 31]. Jeske found that higher impulsivity correlated to lower likelihood to carefully deliberate before connecting to a wireless network on a mobile device. Li examined long-term-orientation and found long-term-orientation corresponds to less likelihood to engage in consequence delayed information security violations.

I identified two minor themes user interface design and security awareness training. I found three studies that addressed the role of user interface design and behavior in cybersecurity [33–35]. I found three studies on effective security training [36–38]. I found four documents that did not fit within the other themes identified [20, 39, 40]. I identified seven works as providing overviews of several topics and interpretations in the form of literature reviews and book chapters [41–47].

5 Discussion

I found six documents that use equivalent framing, and one example of emphasis framing. I found one document using individual framing and three documents using the concepts of mental models which are very similar to the idea of individual frames. This suggests that the equivalent framing approach is the most used in cybersecurity literature. I found twenty-seven articles that did not explicitly use any framing approach but looked at aspects of cybersecurity communications adjacent to framing. Other related major themes I found in the literature include perceptions, persuasion, and use characteristics.

I modified Scheufele's PMFR into the Model of Influence in Cybersecurity with Frames (MICF), see Fig. 1. I made this modification to better reflect the domain where cybersecurity frames appear, and incorporate themes identified in this review. Scheufele's original process model represents and describes research about mass media, whereas cybersecurity messages primarily exist within organizations. Cybersecurity messaging exist in a separate environment with different factors. These differences motivated my modification of the model. I use this model to describe cybersecurity framing and the related influence research.

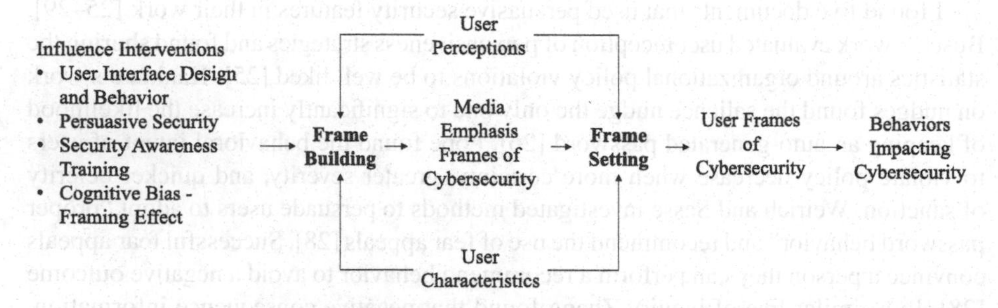

Fig. 1. The model of influence with cybersecurity frames

Using Scheufele's model as a base, I created a model of cybersecurity influence using frames. The model organizes the methods of influence present in the documents included in this analysis around the processes of frame building and frame setting. The sources of influence available to organizations feed into the building of emphasis media frames of cybersecurity both in general and in respect to specific issues. User characteristics and perceptions can also play into the construction of emphasis frame when influence is tailored around characteristics or existing perceptions; however, these factors do play into frame setting. The model illustrates influence effects being incorporated into a media frame presented by an organization that sets the user's individual frame on cybersecurity. It is this individual frame that guides user behavior impacting security. The model highlights multiple sources of influence using emphasis frames but includes equivalent frames as an influence intervention so as to be inclusive of specific effects such as the framing effect and to prevent misinterpretation of the emphasis media frame central to the model. Emphasis media frames at the center of the model represent the combined effect of design decisions, allowing for a macro-level view of the efforts to influence cybersecurity.

The typology based upon Cacciatoire and Scheufele's work allows the discussion of studies with different definitions of framing in an unambiguous way. The Barlow, Das Sauvik, Laaksonen documents each use a different interpretation around what framing is [6, 9, 15]. I cast aside the theoretical differences between the mental models and individual frames to better describe the research in the mental model-individual framing area. By utilizing the cybersecurity framing model, I can illustrate the relations between the themes in my analysis using both framing and non-framing.

Perception appears prominently in the literature. Researchers examine perception to strengthen security management activities [19]. Other research attempts to use perception to predict security behavior [18, 23]. Perception plays a role in specific security related tasks like password generation [20]. When not the focus of the study, perception manifests within other topics of inquiry in cybersecurity such as suspicion [21]. Media frames play an important role in shaping perception to influence users [48]. Perception plays multiple roles in the literature as both an independent and dependent variable. The cybersecurity framing model reflects this complex role by both being an outcome of individual frames and an input to the frame setting process. Using media frames to investigate and operationalize perception is a viable next step for the cybersecurity field.

Managing the behavior intention of users with persuasive security appears in many cybersecurity studies [25–29]. Persuasive security attempts to influence the recipient's intent and persuade them to perform some action rather than another. User personality traits and characteristics play a role in security message reception and behavior [30–32]. These investigations of the user intend to uncover intrinsic traits in users that effect their security behavior. The results indicate that personality affects security, which may mean security awareness interventions will need to be personalized [31]. Organizations can influence users via security awareness training, security user interface design and behavior, or persuasive security. User perception functions in a dual role as both an input for the frame setting process. User characteristics can play into the frame building process of media frames or the frame setting of individual frames.

The MICF model may aid future researchers in cybersecurity to contextualize studies with different uses of framing and framing relevant research in cybersecurity. I map the themes from the literature I analyzed onto the PMFR to reveal where the work is being done and its relation to framing. The MICF model provides structure for future empirical investigations around cybersecurity frames.

Most work in cybersecurity messaging fits within the regions of the model for inputs and outcomes, with little work in the process region. This reveals an opportunity to explore the process of frame building by security professionals, and the process of frame setting. These processes come before the framing outcomes. Studying the process region represent an opportunity to examine the relationship between security professionals' media frames and the individual frames of users. Exploring the processes of frame building and frame setting where information and influence flows from organizations to users may reveal interventions and means of creating positive cybersecurity outcomes with frames. A framing perspective presents an opportunity for understanding the process by which media frames are set into user frames that ultimately lead to negative security outcomes.

References

1. Scheufele, D.: Framing as a theory of media effects. J. Commun. **49**, 103–122 (1999). https://doi.org/10.1111/j.1460-2466.1999.tb02784.x

2. Cacciatore, M.A., Scheufele, D.A., Iyengar, S.: The end of framing as we know it… and the future of media effects. Mass Commun. Soc. **19**(1), 7–23 (2016). https://doi.org/10.1080/15205436.2015.1068811
3. Scheufele, D.A., Tewksbury, D.: Framing, agenda setting, and priming: the evolution of three media effects models: models of media effects. J. Commun. **57**(1), 9–20 (2007). https://doi.org/10.1111/j.0021-9916.2007.00326.x
4. Entman, R.M.: Framing: toward clarification of a fractured paradigm. J. Commun. **43**(4), 51–58 (1993). https://doi.org/10.1111/j.1460-2466.1993.tb01304.x
5. Webster, J., Watson, R.T.: Analyzing the past to prepare for the future: writing a literature review, MIS Q., **26**(2), xiii–xxiii (2002)
6. Barlow, J.B., Warkentin, M., Ormond, D., Dennis, A.R.: Don't make excuses! Discouraging neutralization to reduce IT policy violation. Comput. Secur. **39**, 145–159 (2013). https://doi.org/10.1016/j.cose.2013.05.006
7. Burns, A.J., Johnson, M.E., Caputo, D.D.: Spear phishing in a barrel: insights from a targeted phishing campaign. J. Organ. Comput. Electron. Commer. **29**(1), 24–39 (2019). https://doi.org/10.1080/10919392.2019.1552745
8. Chen, J., Gates, C.S., Li, N., Proctor, R.W.: Influence of risk/safety information framing on android app-installation decisions. J. Cogn. Eng. Decis. Mak. **9**(2), 149–168 (2015). https://doi.org/10.1177/1555343415570055
9. Das, S., Kramer, A.D.I., Dabbish, L.A., Hong, J.I.: Increasing security sensitivity with social proof: a large-scale experimental confirmation. In: Proceedings of the 2014 ACM SIGSAC Conference on Computer and Communications Security, Scottsdale, Arizona, USA, November 2014, pp. 739–749 (2014). http://doi.org/10.10/ggwmdd
10. Dennis, A.R., Minas, R.K.: Security on autopilot: why current security theories hijack our thinking and lead us astray. ACM SIGMIS Database DATABASE Adv. Inf. Syst. **49**, 15–38 (2018). http://doi.org/10.10/gdg2q3
11. Proctor, R.W., Chen, J.: The role of human factors/ergonomics in the science of security: decision making and action selection in cyberspace. Hum. Factors **57**(5), 721–727 (2015)
12. Tversky, A., Kahneman, D.: The framing of decisions and the psychology of choice. Science **211**(4481), 453–458 (1981). https://doi.org/10.1126/science.7455683
13. Johnston, A.C., et al.: Speak their language: designing effective messages to improve employees' information security decision making. Decis. Sci. Atlanta **50**(2), 245–284 (2019). https://doi.org/10.1111/deci.12328
14. Diesner, J., Kumaraguru, P., Carley, K.M.: Mental models of data privacy and security of indians extracted from texts. In: Conference Papers – International Communication Association, May 2005, pp. 1–13. http://search.ebscohost.com/login.aspx?direct=true&AuthType=ip,uid&db=ufh&AN=18655489&site=ehost-live&scope=site. Accessed 10 July 2020
15. Laaksonen, A.E., Niemimaa, M., Harnesk, D.: Influences of frame incongruence on information security policy outcomes: an interpretive case study. Int. J. Soc. Organ. Dyn. IT **3**(3), 33–50 (2014). https://doi.org/10.4018/ijsodit.2013070103
16. Qiu, C., Zhao, W., Jiang, J., Han, J.: A teaching model application in the course of information security. In: Proceedings of the 2011 Third International Workshop on Education Technology and Computer Science - volume 02, USA, March 2011, pp. 138–141. Accessed 10 July 2020
17. Raja, F., Hawkey, K., Hsu, S., Wang, K.-L.C., Beznosov, K.: A brick wall, a locked door, and a bandit: a physical security metaphor for firewall warnings. In: Proceedings of the Seventh Symposium on Usable Privacy and Security, Pittsburgh, Pennsylvania, July 2011, pp. 1–20. http://doi.org/10.10/fxp2bp
18. Addae, J.H., Sun, X., Towey, D., Radenkovic, M.: Exploring user behavioral data for adaptive cybersecurity. User Model. User-Adapt. Interact. **29**(3), 701–750 (2019). https://doi.org/10.1007/s11257-019-09236-5

19. Albrechtsen, E.: A qualitative study of users' view on information security. Comput. Secur. **26**(4), 276–289 (2007). https://doi.org/10.1016/j.cose.2006.11.004
20. Choong, Y.-Y.: A Cognitive-Behavioral Framework of User Password Management Lifecycle. In: Tryfonas, T., Askoxylakis, I. (eds.) Human Aspects of Information Security, Privacy, and Trust, vol. 8533, pp. 127–137. Springer, Cham (2014)
21. Hirshfield, L., et al.: The Role of Human Operators' Suspicion in the Detection of Cyber Attacks, pp. 1482–1499 (2019)
22. Huang, D.-L., Rau, P.-L., Salvendy, G.: A Survey of Factors Influencing People's Perception of Information Security. 2007, vol. 4553, pp. 906–915 (2007)
23. Lee, V.C.: Examining the Relationship between Autonomy, Competence, and Relatedness and Security Policy Compliant Behavior, Ph.D., Northcentral University, United States – Arizona (2015)
24. Davis, F.D.: Perceived usefulness, perceived ease of use, and user acceptance of information technology. MIS Q. **13**(3), 319–340 (1989). https://doi.org/10.2307/249008
25. Busch, M., Patil, S., Regal, G., Hochleitner, C., Tscheligi, M.: Persuasive information security: techniques to help employees protect organizational information security. In: Proceedings of the 11th International Conference on Persuasive Technology - Volume 9638, Salzburg, Austria, pp. 339–351 (2016). http://doi.org/10.10/ggwmfr
26. Kankane, S., DiRusso, C., Buckley, C.: Can we nudge users toward better password management? an initial study. In: Extended Abstracts of the 2018 CHI Conference on Human Factors in Computing Systems, Montreal QC, Canada, pp. 1–6 (2018). http://doi.org/10.10/ggwmcz
27. Pope, M.B.: Time orientation, rational choice and deterrence – an information systems perspective, ProQuest Information & Learning (2014)
28. Weirich, D., Sasse, M.A.: Persuasive password security. In: CHI 2001 Extended Abstracts on Human Factors in Computing Systems, Seattle, Washington, March 2001, pp. 139–140 (2001). http://doi.org/10.10/fkhtkx
29. Zhang, C., Simon, J.C.: "Ted" Lee, "An Empirical Investigation of Decision Making in IT-Related Dilemmas: Impact of Positive and Negative Consequence Information," J. Organ. End User Comput. Hershey, vol. 28, no. 4, p. 73 (2016). http://doi.org/10.10/f873bx
30. Jeske, D., Briggs, P., Coventry, L.: Exploring the relationship between impulsivity and decision-making on mobile devices. Pers. Ubiquitous Comput. **20**(4), 545–557 (2016). https://doi.org/10.1007/s00779-016-0938-4
31. Kajzer, M., D'Arcy, J., Crowell, C.R., Striegel, A., Van Bruggen, D.: An exploratory investigation of message-person congruence in information security awareness campaigns. Comput. Secur. **43**, 64–76 (2014). https://doi.org/10.1016/j.cose.2014.03.003
32. Li, Y., Zhang, N., Siponen, M.: Keeping secure to the end: a long-term perspective to understand employees' consequence-delayed information security violation. Behav. Inf. Technol. **38**(5), 435–453 (2019). https://doi.org/10.1080/0144929X.2018.1539519
33. Anderson, B.B., Jenkins, J.L., Vance, A., Kirwan, C.B., Eargle, D.: Your memory is working against you, Decis. Support Syst., **92**, 3–13 (2016). http://doi.org/10.10/ggjc9b
34. Jenkins, J.L., Anderson, B.B., Vance, A., Kirwan, C.B., Eargle, D.: More harm than good? how messages that interrupt can make us vulnerable. Inf. Syst. Res. **27**(4), 880–896 (2016). https://doi.org/10.1287/isre.2016.0644
35. Mathur, A.: A Human-Centered Approach to Improving The User Experience Of Software Updates, Thesis (2016). https://doi.org/10.13016/M2N220
36. Abawajy, J.: User preference of cyber security awareness delivery methods. Behav. Inf. Technol. **33**(3), 237–248 (2014)
37. Cuchta, T., et al.: Human risk factors in cybersecurity. In: Proceedings of the 20th Annual SIG Conference on Information Technology Education, Tacoma, WA, USA, September 2019, pp. 87–92 (2019). http://doi.org/10.10/ggwmch

38. Shaw, R.S., Chen, C.C., Harris, A.L., Huang, H.-J.: The impact of information richness on information security awareness training effectiveness. Comput. Educ. **52**(1), 92–100 (2009). https://doi.org/10.1016/j.compedu.2008.06.011
39. Papadaki, K., Polemi, D.: Collaboration and knowledge sharing platform for supporting a risk management network of practice. In: 2008 Third International Conference on Internet and Web Applications and Services, June 2008, pp. 239–244 (2008). http://doi.org/10.10/d2rvcz
40. Smith, S.W.: Security and cognitive bias: exploring the role of the mind. IEEE Secur. Priv. **10**(5), 75–78 (2012). https://doi.org/10.1109/MSP.2012.126
41. Briggs, P., Jeske, D., Coventry, L.: Behavior change interventions for cybersecurity. In: Little, L., Sillence, E., Joinson, A. (eds.) Behavior Change Research and Theory: Psychological and Technological Perspectives, San Diego, CA: Elsevier Academic Press, pp. 115–136 (2017)
42. de Bruijn, H., Janssen, M.: Building Cybersecurity Awareness: The need for evidence-based framing strategies. Gov. Inf. Q. **34**(1), 1–7 (2017). https://doi.org/10.1016/j.giq.2017.02.007
43. Houston, N.: The impact of human behavior on cyber security. In: Khosrow-Pour, M. (ed.) Multigenerational Online Behavior and Media Use: Concepts, Methodologies, Tools, and Applications, Hershey, PA: Information Science Reference/IGI Global, 2019, pp. 1245–1266 (2019)
44. Liu, X.M.: The cyber acumen: an integrative framework to understand average users' decision-making processes in cybersecurity. In: Yan, Z. (ed.) Analyzing Human Behavior in Cyberspace, Hershey, PA: Information Science Reference/IGI Global, 2019, pp. 192–208 (2019)
45. Pfleeger, S.L., Caputo, D.D.: Leveraging behavioral science to mitigate cyber security risk. Comput. Secur. **31**(4), 597–611 (2012). https://doi.org/10.1016/j.cose.2011.12.010
46. Tsohou, A., Karyda, M., Kokolakis, S.: Analyzing the role of cognitive and cultural biases in the internalization of information security policies. Comput. Secur., vol. 52, no. C, pp. 128–141, July 2015 (2015). http://doi.org/10.10/f82r6w
47. Williams, E.J., Beardmore, A., Joinson, A.N.: Individual differences in susceptibility to online influence: a theoretical review. Comput. Hum. Behav. **72**, 412–421 (2017). https://doi.org/10.1016/j.chb.2017.03.002
48. Nelson, T.E., Clawson, R.A., Oxley, Z.M.: Media framing of a civil liberties conflict and its effect on tolerance. Am. Polit. Sci. Rev. **91**(3), 567–583 (1997). https://doi.org/10.2307/2952075

Data and Privacy in a Quasi-Public Space: Disney World as a Smart City

Madelyn Rose Sanfilippo[1(✉)] and Yan Shvartzshnaider[2,3]

[1] University of Illinois, Champaign, IL 61820, USA
madelyns@illinois.edu
[2] NYU, New York, NY 10012, USA
[3] York University, Toronto, ON M3J1P3, Canada
yansh@yorku.ca

Abstract. Disney World has long been at the forefront of technological adoption. Walt Disney theme parks implement emerging technologies before other consumer or public spaces and innovates new uses for existing technologies. In contrast to public contexts with representative governance, Disney World is both a prototype and a functioning quasi-public smart city, wherein a private actor controls ICT adoption and data governance. As cities increasingly partner with private corporations in pursuit of smart systems, Disney provides a glimpse into a future of smart city practice. In this paper, we explore normative perceptions of data handling practices within Walt Disney World and discuss contextual differences from conventional cities. We consider what can be learned about privacy, surveillance, and innovation for other public applications, stressing the limitations of and potential social harms from Disney as a model for public services.

1 Introduction

Over the years, Walt Disney World (WDW) has innovated and employed emerging and futuristic technology to realize the "great big beautiful tomorrow, shining at the light of every day" [2]. Walt Disney envisioned the Experimental Prototype Community of Tomorrow (EPCOT) as the first smart city, though that label did not yet exist [16]. He envisioned that "[EPCOT] will take its cues from the new ideas and new technologies that are now emerging from the creative centers of American industry. It will be the community of tomorrow, that will never be completed" [5].

With the adoption of emerging technologies in WDW and other public spaces, critics have raised a number of privacy concerns.

Privacy is often rhetorically positioned in false trade-offs with efficiency and convenience, and thus sacrificed in favor of commercial over human interests. This results in potentially serious social repercussions and inequities [10], both within and between smart cities. Some smart cities' governance is purely public, drawing on community feedback and preferences through processes that incorporate many stages of revision, such as Seattle [1]. Other cities, such as Toronto [9],

K. Toeppe et al. (Eds.): iConference 2021, LNCS 12646, pp. 235–250, 2021.
https://doi.org/10.1007/978-3-030-71305-8_19

increasingly contract private partners to implement smart systems with varying degrees of transparency and consistency. It is difficult to predict outcomes for any specific smart city. Each path brings different governance models and decision-makers, in addition to local values, norms, harms, and benefits.

As cities pursue smart systems, in partnership with private companies, they evolve into quasi-public spaces. As an early adopter of numerous technologies, WDW provides useful perspective on the interplay among technology, people, and institutions in a context that narrowly represents the economic interests of private decision-makers and consumer preferences. In this context, individuals are pervasively quantified and surveilled, without consideration for broader social norms, rights, equity, or human autonomy.

In this paper, we examine WDW as a case study to explore both normative and institutional information governance challenges associated with: pervasive location monitoring, facial-recognition, data integration across contexts, and the seamlessness of smart experiences and interactions. Our analysis of WDW highlights a number of challenges, identified in terms of normative disagreement over particular smart systems between Disney and consumers, as well as with the general public. We discuss implications for technological adoption, including facial recognition, biometrics, and smart transportation systems, emphasizing that respecting privacy norms is important to retaining rights in smart urban environments. Further, we highlight the importance of feedback mechanisms on privacy norms and seamlessness in sociotechnical systems.

2 Background

An emerging body of scholarship employs institutional analysis to explore technology governance in public space, as way to complement existing legal and normative frameworks [8]. The governing knowledge commons (GKC) framework, built on institutional analysis approaches originating with Ostrom, facilitates descriptive and structural analysis of complex, layered, localized, and hybrid governing institutions around data, information, and technology. To address questions around privacy relative to public technology, recent studies integrate GCK with the contextual integrity (CI) framework [22], considering privacy as an appropriate flow of personal information [17], in order to address questions around privacy relative to public technology.

In this section, we provide an overview of: 1) governance of technological systems in public and quasi-public contexts, 2) privacy and security in smart cities, and 3) WDW as a smart city and a comparatively early adopter.

2.1 Technology Governance in Public

Privacy in public is essential to: individual autonomy, the relative invisibility of being one among many, safety in numbers, and the possibility of disinhibition within collective behaviors or experiences.

US law makes relatively clear distinctions between public spaces and the home, as a private space, dating back to *Bell v Maryland*, there is also a grey area. Quasi-public contexts, such as malls, airports, and amusement parks are privately owned and open to the public, as legally defined hybrid contexts [3]. Increasingly, these quasi-public contexts are privately policed and subject to significant surveillance, without public oversight or protections [3].

Quasi-public spaces raise numerous governance questions about the legitimacy of institutions, decision-making, and information flows in public-private partnerships in smart cities. As these technologies and systems are deployed, the implications of governance, relative to privacy, in public and quasi-public spaces, are becoming ever significant, particularly around questions of transparency. Important elements of governance in public spaces include: accessibility, understandability, and representativeness of the interests of the general public, both normatively and democratically [11,13]. The visibility and invisibility of these systems, which we discuss in the next section, is important relative to public expectations [6,17].

2.2 Privacy, Surveillance, and Smart Cities

Smart cities are arrangements between people, technology, and institutions. Smart cities are shaped by aspirational futurism and through socio-technical engineering, pursuing innovation, with little time dedicated to issues of privacy, socially normative expectations, or governance [7,15,26]. We define smart cities as public, semi-public, and quasi-public spaces in which information technologies provide feedback mechanisms or services enhanced beyond delivery.

Local trends in governance are just as important as trends in technological innovation, both to understand what is possible and what ought not to be replicated [20]. Privacy localism [20,21] manifests as governance or practices at the levels of states or regions, broadly, as well as in individual cities or neighborhoods. Notions of contested privacy in specific neighborhoods–as individuals negotiate how far the privacy of one's home extends and the extent to which neighborhoods are public or private [18] –illustrate the complexity and nuance of context.

The way we govern privacy, personal data flows, and data within social spaces illustrate the significance of learning from micro-level cases. This is particularly true around quasi-public spaces, where large populations of people interact, and in densely-populated urban areas that rush to transform into "smart-er" cities [26]. The notion of "Urban privacy," which integrates privacy in public with the normative frame and expectations of urban spaces [19] can help address the challenges of privacy governance in these spaces.

2.3 WDW as a Smart City

Immersive Disney spaces embrace techno-futurism, embedding innumerable data collection points throughout engaging experiences and pedestrian spaces. WDW has adopted many technologies and smart systems in a quasi-public space, prior to other applications in public. WDW was one of the earliest commercial applications of CCTV digital multiplexing to scale [4]; CCTV represents one of the first technical applications of pervasive surveillance in both Disney stores and parks. Before diving into the case study, we articulate the ways in which WDW is a smart city and the ways in which it compares to purely public smart cities or other conventional public spaces.

First, WDW is a quasi-public place, in which a private actor controls a large space open to consumers from the general public. Second, Disney is a quasi-public smart city that employs numerous digital technologies and multiple networks of sensors to enhance services and experiences, as well as to provide feedback. Under our definition, the relationships between people, technology, and institutions within Disney spaces constitute a quasi-public smart city, yet there are many contextual differences that clearly delineates appropriate conclusions from this study. In comparison to fully public contexts or public-private partnerships in other smart cities, WDW is distinct not only in private control and decision-making, but also due to normative distinctions, differences in objectives, and the unique history of Disney as a planned space.

Normative differences are rooted in context and interests; cities address social needs of local populations, while Disney is a commercial purveyor of entertainment. Intentional and planned spaces are also distinctly different from other public and urban spaces. Disney is similar to other quasi-public spaces and intentional communities, such as Irvine, CA which was thoroughly planned and engineering in pursuit of normative values that have shaped development and human interactions over time [12]. Governance implications for adoptions of parallel systems, in other, non-WDW, contexts and for distinct purposes, come with significant caveats. Despite status as an early adopter and characterization of WDW as a prototype smart city, outcomes and benefits may not compare between quasi-public spaces and conventional cities.

3 Methodology

The case study of WDW, as an early adopter of new technologies, examines implications of public-private partnerships in emerging smart cities. We empirically explore: cross-context data integration in practice at WDW, data collection and processing, social perceptions of privacy practices, and lessons about practice and governance from Disney.

We frame our privacy analysis of WDW information handling practices in terms of the contextual integrity (CI) framework. The CI framework captures information flows and norms using 5 essential parameters: senders, subjects, receivers of the transmitted information, type of information and transmission principle, which specifies the constraints imposed on the information flow.

We structured our analysis of norm formation, divergence in preferences among community members, and governance in effect, using the governing knowledge commons (GKC) framework [8]. GKC applies institutional analysis, including an underlying grammar to structure coding of strategies, norms, and rules, to the context of data, information, and technology as resources [22].

3.1 Empirical Assessment of Privacy Polices

Guided by the CI framework, we annotate words in each privacy statement matching CI parameters that prescribes information exchange and handling practices, as proposed in [24]. We then list all the prescribed information flow by a given statement. For example, the statement:

> We collect information using analytics tools, including when you visit our sites and applications or use our applications on third-party sites or platforms

prescribes a number of potential information flows. We can observe "Disney" as a recipient of the information, however, the statement omits several relevant parameters, namely, the type and subject of the information, and the sender of the information. The statement does include transmission principles, i.e. the two conditions under which the information is facilitated, which results in multiple potential information flows: when users visit the sites, when users using the apps, when users browse third party sites, and when users visit third party platforms. In our analysis, we compare each of the possible information flows to existing institutions to identify potential privacy violations or non-conformance to established practices.

3.2 Empirical Assessment of Surveillance and Perceptions

To empirically assess the prevalence and visibility of data collection within WDW, we systematically counted and categorized clusters of sensors that interact with apps and MagicBands, as well as cameras. We differentiate between visible sensors that visitors intentionally swipe their phones or MagicBands and those that are not transparently labeled or visibly identified. Sources for this analysis included official Disney blogs, coded and analyzed through content analysis, and the My Disney Experience app, through which we manually counted the sensors, identified by individual geo-located tags on embedded maps.

We evaluated stakeholder perceptions about technologies and information flows at WDW through sentiment analysis of text discussing privacy and data collection systems, as captured from blog posts. For this analysis, we differentiate between content: directly from Disney, endorsed by Disney, and Disney consumers and users of Disney systems. The first two categories were identified from Disney resources. To generate the third set of blogs, we identified the top 100 Google Page Rank results for "Disney blog", that: had at least 100 posts total, post at least once per month on average, and a disclaimer differentiating the blog from the Walt Disney Company. We examine these blogs as representations of Disney consumers and users of Disney systems, rather than as independent perspectives from the general public. A Python script collected a total of 12506 posts, from 112 blogs, that included keyword sets associated with systems of interest. This collection strategy included many false positives, which were then manually discarded by the investigators, as posts were further classified and tagged within NVivo.

Data were processed and modeled within R, using the packages: textclean, to normalize punctuation; tidytext, to format normalized posts in comparable units of analysis and assess word counts and frequency; SentimentAnalysis, to assess polarity, including in proximity to a custom dictionary that focuses on the technologies and information flows of interest; and sentimentr, to aggregate and compare sentiment measures. We measured the significance of differences in sentiment using Welch's t-tests, as the most accurate and effective measures given the characteristics of the blog post data set [23], and then noted confidence intervals for measures of polarity. For further reference, we included a table of reviewed blogs/posts considered in this study in this anonymous repository.

4 Analysis

This section: identifies key issues around information flows at WDW, presents an ontology of community members, analyzes their perceptions of specific systems, and examines WDW privacy policies and communications that inform consumer expectations.

4.1 Surveillance and Information Collection

User and behavioral data are critical information resources, from an institutional perspective. This is especially true at WDW, which employs a massive network of sensors and cameras across multiple systems to understand, predict, and influence consumers and throughout WDW. This includes documenting individuals': steps taken, time browsing shops, food and souvenirs purchased, lines waited in, and entertainment or attractions engaged with. Figures 1 and 2 illustrate the overall scale of this data collection across the four distinct theme parks within WDW, other public spaces at WDW, and WDW resorts as counts of the number of sensor clusters visitors actively connect with by categories of interaction.

Visible Collection. Highly visibile data collection at WDW includes ticket and security screenings at park gates. People knowingly wave their magic-bands over silver Mickey-shaped sensors when they enter the parks or their hotel rooms, which light up to indicate authentication. People actively scan their bands in the Fast-Pass system, have their picture taken, or make purchases. They download and engage with immersive apps to augment experiences throughout the parks. The aggregate scope of this visible, participatory data collection is immense and not always obvious. Many active interactions visitors have with sensors are unobtrusive, despite the transactional nature. Yet, many interactions are designed to seamlessly minimize visibility.

Seamless Collection. Seamless data collection includes step tracking and most security cameras, as well as complex cross-function and multi-use systems. The My Disney Experience and the former Shop Disney Parks apps make suggestions about things individuals looked at, but didn't buy, or where they could find souvenirs associated with favorite rides and characters. These recommendation systems connect digital platforms and the physical world, in real time.

MagicBands. MagicBands and step tracking represent newer streams of data collection about visitors, with many similarities to fitness trackers. MagicBands collect location data both through active (e.g. point-of-sale) and passive interaction (e.g. sensors, triangulation) [25]. The integrated MagicBand systems function based on the same RFID technology as luggage tracking, representing the first deployment of wearable RFID for the general public. Embedding RFID within Disney spaces allowed the MagicBands to integrate relatively seamlessly with many existing systems.

MagicBand sensors are pervasive all over the resort and theme park spaces, some unobtrusive, or even invisible. Invisible sensors are used both for tracking and engagement; for example, many attractions identify individuals' names via invisible sensors to greet them or personalize interactions. Figures 1 and 2 illustrate data collection via MagicBands, in which information flows are visible to visitors through their active engagement with these sensors, yet these represent the tip of the data iceberg. Sensors are relatively equally distributed across parks and categories, with most MagicBand interactions within customer service or through guest services staff (Fig. 1). However, there are significant differences between data collection within the amusement parks and other public contexts, including water parks or shopping areas, and hotels (Fig. 2), where guest services have greater prevalence and some categories of MagicBand interactions are non-existent. Customers view the parks and hotels as different contexts.

Apps. Apps provide another major means of data collection about park visitors, as well as the wider population of Disney customers. In addition to individual apps tailored to each Disney Park worldwide, the My Disney Experience app, and a Disney transportation app, there are various consumer directed apps like Disney+, Play Disney Parks, and a Shop Disney app. Data from these apps are integrated with other systems to assess traffic and interaction with various features in and across the parks.

Overall there are fewer location-based interactions via app than via MagicBands. Apps' location-based interactions are either enabled through smart phones' GPS capabilities or through proximity-based Bluetooth. The Play Disney Parks app directly integrates location data and behavioral data about children, as a protected population, with data from other sensors and systems within the park. Yet, this app does not have a unique privacy policy, despite its target population and leveraging permissions the following permissions: camera (take pictures and videos); approximate location (network-based); precise location (GPS and network-based); storage and photos/media/files (modify, delete, or read USB contents); view Wi-Fi connection information; download files without notification; receive data from Internet; view network connections; prevent device from sleeping; read Google service configuration and check Google Play license; run at startup; pair with Bluetooth devices and access Bluetooth

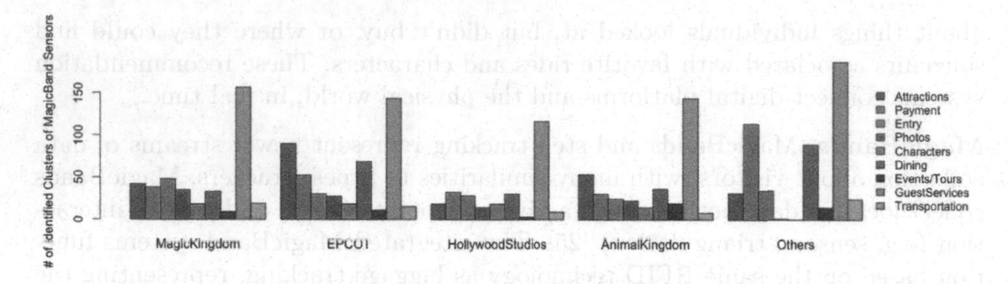

Fig. 1. Active engagement MagicBand sensors in 4 Disney theme parks and other quasi-public spaces at WDW

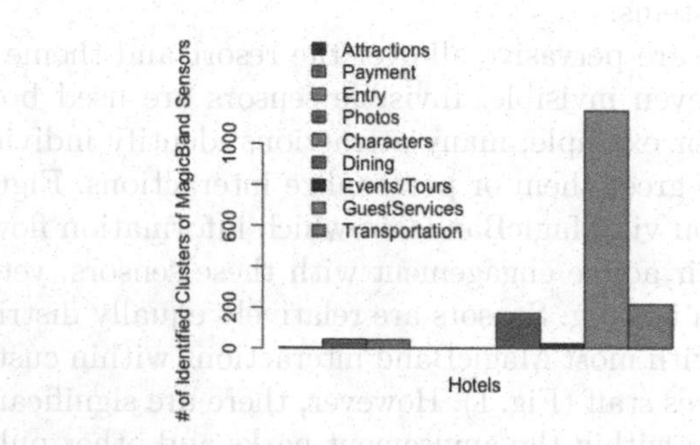

Fig. 2. Active engagement MagicBand sensors within Disney hotels

settings; full network access; toggle sync on and off; use accounts on the device; control vibration; and connect and disconnect from Wi-Fi.

Cameras. An extensive network of cameras facilitate a major means to collect data resources within WDW. Specifically, in addition to CCTV surveillance, there are a number of other unobtrusive cameras in quasi-public spaces, including those on rides or within attractions to capture the "action" and visitors at play. Some of these unobstrusive cameras are considered "automatic photographer machines." Automated photographers are distinct both from human photographers with cameras, who roam the parks, and surveillance cameras, for which footage and images are not distributed to visitors via any of their photography packages, such as PhotoPass, or accounts. The Disney Parks app does provide maps indicating how widely distributed these cameras are throughout the parks and where these systems are located, yet these cameras cannot be identified without searching the map.

Further, there are also visible cameras that document visitors during character interactions or photo opportunities that consumers knowingly choose to interact with. Sometimes, these cameras are accompanied by photographers or other humans in the loop, who link photos to individuals via their MagicBands. In other instances, however, photos are connected with individuals' accounts via facial recognition, to varying degrees of success.

4.2 Community Member Ontology

Relevant community members and stakeholders in the WDW case have distinct relationships to and with Disney, as well as interests associated with governance issues. Despite individuals' unique preferences, stakeholder groups' consensus on particular values and objectives are reflected in governance processes and outcomes.

Within the Disney organization diverse role-based groups–including approximately 70,000 union, salaried and non-union hourly employees at the Disney Parks without decision-making roles–span: interns, imagineers, cast members, musicians, a business office, and management roles, as well as many Disney subsidiary organizations. A major proportion of those hourly, non-unionized employees–who number approximately 43,000 employees–include veterans who fill security roles within the Disney parks.

Outside of the Disney organization, there are two distinct groups: (1) individual visitors and (2) business and organizational partners. First, among visitors, different stakeholder groups are represented, including: those associated with conferences and events, families with children, multi-generational families, local visitors versus tourists, adults without children, individuals with disabilities, techno-futurists, and military families. Second, businesses and organizations that partner with Disney, provide services or in supply chains, have very distinct interests and influence privacy and surveillance practices. Given the limited transparency about some of these relationships, their outsize influence on and role in data governance is likely to be surprising to many visitors. Third parties, completely distinct from the multifaceted Disney organization, include: Lyft, TSA, the Orlando International Airport (MCO), and the City of Orlando, who partner to provide smart transportation solutions and safety throughout transit; various hotel chains in and around Disney parks, specifically including joint properties with Marriott and the Four Seasons; and consumer products and retailers, including Ziplock, Target, AppleMusic, and ACE.

4.3 Stakeholders Perceptions

In order to understand the current state of governance around surveillance and personal data at Disney, it is important to have a sense of how goals and objectives diverge among stakeholders. We analyzed the perceptions of (a) visitors and Disney enthusiasts, (b) bloggers endorsed by Disney, and (c) the Disney organization on key governance issues around privacy and surveillance. Figure 3 illustrates perceptions of specific technological systems as polar sentiment expressed, noting significance via confidence interval error bars.

Perceptions of specific systems vary across these three stakeholder categories, with the greatest consensus demonstrated through comparably positive language employed to discuss apps, smart transportation systems, and voice recognition. Those individuals whose blogs are endorsed by Disney, unsurprisingly, generally frame their views more similarly to official Disney accounts, than to consumers. The most notable disagreements fall in framing of biometric technologies and facial recognition in WDW. Additionally, while endorsed blogs and official Disney blogs frame MagicBands and Smart Locks in similarly positive ways, mimicking the common divide between these groups and the broader population of users and visitors. Endorsed blogs describe recommendation systems and smart planning tools in similarly somewhat positive language to consumers, while Disney officially frames these technologies using more positive language.

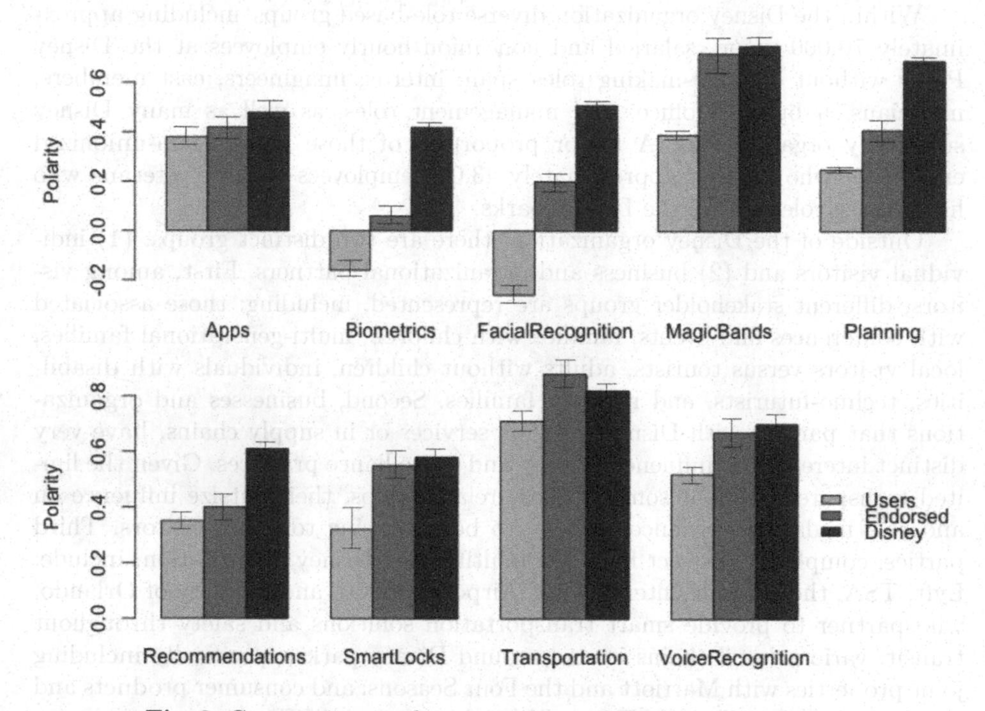

Fig. 3. Community member perceptions of WDW smart systems

The dataset did not include any posts expressing concerns about privacy or data governance associated with the application of facial and voice recognition technology in engaging and interactive attractions geared toward children, though there were blog posts analyzing and articulating the technological design. In contrast, some blogs about immersiveness, seamlessness, and recognition technologies outside of attractions express hesitation and concerns about whether people understand "concern over privacy," particularly with MagicBands outside of rides. This suggests a distinction between rides and more general experiences at WDW, within the minds of visitors.

Official Disney dialogue prioritizes security over privacy. Disney offers compelling and well packaged arguments about safety and security, while assuring visitors and readers, in vague terms, that biometric and surveillance data will be "secure." They do not discuss privacy or trade-offs made, leading to the very positive polar sentiments documented in Fig. 3. Language used around safety and security is also very assertive and positive in tone to communicate why visitors should trust them and need not worry.

Endorsed perspectives are similarly very positive overall; they communicate their trust in Disney and positive experiences, without interrogating trade-offs between privacy and security. Perspectives from parents and Disney youth and Disney marathon runners often continue to focus on their positive experiences, even when faced with questions about privacy or skepticism about public safety. For example, a question posed to the Disney Parks Moms Panel by a general visitor focused on biometric privacy: "Are fingerprints stored to your name or just to your MagicBand"? The response was framed in terms of safety and security; the only mention of privacy was in the link to the Walt Disney company's privacy policy.

The greatest disagreement between Disney and endorsed perspectives is about biometrics, when exploring these posts in aggregate. The general public frames biometrics and facial recognition as slightly negative, although their views of systems in aggregate are slightly positive, as depicted in Fig. 3; their views on safety and security as broad action arenas are very fragmented. Safety and security objectives are meaningful and important, given that there are so many children in these spaces; these objectives are well met. This gives people a sense of safety and many blogs discuss the same incidents in which security and surveillance identified individuals with handguns and prevented them from entering the parks.

Yet, there are, sometimes explicit, questions about whether there are more privacy preserving ways to manage these data flows. Many blog posts discuss or hint at privacy tensions, with most expressing uncertainty (e.g. "...though I don't know if..." or "I'm not really sure how..." data is stored or protected). While journalists have directly questioned data retention [14], bloggers suggest similar concerns in discussing long term tracking around MagicBands, including that they "don't expect to continue being watched... after [they] go home." Further, bloggers express greater trust in Disney than in privacy or security of technology in any context, including Disney technology. Similarly, consumers question the increasing presence of security personnel in these spaces, with third-party "back-up" from "more uniformed and plain-clothes police officers, security guards, and dogs patrolling the parks" at particularly busy times of year. In this sense, despite Disney's apparent trustworthiness, consumers don't necessarily trust Disney's third-party partners or understand the nature of those relationships around privacy and security.

4.4 Privacy Rules-on-the-Books

In this section, we present our analysis of information flows prescribed by the privacy policies and as explained to customers.

Given the large scope of the Walt Disney Company and diversity of subsidiaries, the privacy policy is highly generic and and lacks detail about data handling practices. For example, what types of information are collected is never disclosed, much less in what context or through what system. Disney's privacy policy describes practices in a general sense, only offering more details on what information is collected, processed and shared while engaging with Magic+ services (including MagicBands) in a separate, non-contractually binding My Disney Experience – Frequently Asked Questions (FAQ) page [27]. Many WDW information handling practices are performed without any meaningful user consent or rules guaranteeing consistent practice.

Although the privacy policy states that Disney collects information from visitors/users during any interaction with their devices or services, such as purchasing products or surveys, it omits what type of information is collected or about whom. It is ambiguous whether data collected is about the user, the account holder, or any other associated accounts or relatives. Not only are children included on accounts or able to access services through their parents' accounts, but accounts are also linked by design, for family reunions or school field trips. The FAQ fills in some of the details: "Your interactions provide us with information about the products and services you experience in the Parks; your wait time for rides, restaurants and other attractions; and similar types of information." From this, non-binding, statement, the implication is that data collection extends far beyond account holders and pertains to all interactions users and visitors have with Disney platforms or parks.

Similarly, Disney collects information "whether or not [users] are logged in or registered." Pervasive surveillance is implied by the privacy policy; information is acquired from other sources to "to update or supplement the information provided." An FAQ expands: "We also collect certain information from you while you are at select locations throughout the Resort through means other than the My Disney Experience website or mobile app. When you use your RF Device at touch points (e.g., for Disney Resort room entry, park admission, FastPass+, and purchases at select Resort locations), we are able to record your transaction and, when necessary, make the appropriate adjustment to your account."

Aggregation is as pervasive as surveillance and notably crosses contexts. The Disney privacy policy mentions using information for: personalization, optimization, and improvement of their services, as well as for targeted advertising. The policy lacks details on what type of information is shared, with whom, or under what conditions. Again, FAQs provide further details: "We will only share information about you that is collected automatically when your MagicBand is read by long-range readers with third parties for their marketing uses if you elect that we do so." This implies opportunities for users to opt-in to marketing uses, without making clear how or when.

The WDW case illustrates how informed consent, in its current form, through privacy policies can be reduced to a meaningless artifact from the general public point of view and a power manipulation tool on behalf of large corporations, like, Disney. For a privacy concerned member of the community it is extremely difficult, if not impossible, to understand precisely what type of information flows, to whom, about whom or what, and under what conditions. To piece this information together, the reader is expected to follow countless links, read length documents only to end up with an incomplete picture. The FAQ might seem to be a useful gesture on behalf of the company to help a concerned member, by providing more details and concrete scenarios, yet, it is not legally binding and is often as vague and incomplete as the privacy policy.

5 Discussion

In this section, we discuss potential implications of Disneyfication of smart cities. We identify lessons to be learned from WDW in comparison to adoption of new technologies, based on commercial interests and with private sector partners.

5.1 Lessons for Technological Adoption

As documented in Sect. 2, many of the technologies used at WDW, are ultimately integrated in smart cities and public spaces.

Facial Recognition. The use of facial recognition technologies is increasingly under public scrutiny. To address issues of concern to visitors, such as imprecision and bias, Disney maintains humans in the loop and triangulates against other data sets, such as to reunite children separated from their families. This illustrates a mechanism to address issues similar to those experienced in applications of facial recognition in quasi-public spaces, such as university campuses. However, humans in many contexts may provide other avenues for bias to creep in and claim justification through these systems. In this sense, an attempt to replicate this sociotechnical application of facial recognition would not translate to every context and would require critical evaluation of norms and expectations, prior to implementation.

Biometric Tracking. While Disney was first to employ biometric access for large public populations, the technology had long been perfected for access control to secure military bases and national laboratories. Yet the contexts have very different implications for stakeholder perceptions of appropriateness. Various blog posts analyzed were quick to explain that the data collected at the gates to Disney parks is not shared with law enforcement, unless legally obligated. These discussions reveal that Disney customers are more trusting of private corporations like Disney, with respect to their personal data, including fingerprints, than they are trusting of the government. There are significant issues of trust that must be overcome before employing fingerprint scanners in other public spaces and also appears analogous to trust issues associated with contact tracing.

Despite the parallel technological systems between urban public spaces and WDW as a quasi-public space, there are meaningful differences between private and public infrastructure with respect to: representation, inequality, values, and trust.

A major distinction between Disney and smart cities relates to community member perceptions of and trust in decision-makers by other stakeholders. Technologies do not operate in a vacuum, but rather are used and governed by people with distinct interests, needs, and expectations in particular contexts. Systems are not transferable with parallel outcomes across contexts.

5.2 Seamlessness in Sociotechnical Systems

Ubiquitous data collection quickly and seamlessly became integral defaults at WDW, challenging norms in absence of meaningful alternatives. As so often happens in techno-socially designed systems, analog alternatives diminish the quality of experience and are presented as a less attractive option, with time consuming procedures, little information, and reduced priority. For example, one may opt for an analogue ticket or not to use an app, but then is ineligible for particular experiences or must wait in stand by lines. Disney follows a worrisome trend, presenting false trade-offs between information collection and quality of service. Given the relationships between Disney and guests, as well as the success of Disney nudges in numerous contexts, opt-in would also likely work as well. This should serve as a warning to smart city advocates that face a much more challenging task relative to "opt-in" options.

Another increasingly common practice is the use of social nudging. Based on our analysis of blog posts, consumers seem to be more comfortable with nudges from Disney than other commercial partners, yet these consumers do not represent the general public. Nudges to encourage opting-in, when made by cheerful Disney characters are likely to be perceived as much less sinister than those from police officers. However, opting-in to data collection at WDW versus by law enforcement, within smart cities, are more similar in effect, than individuals realize, given the relationships between law enforcement and Disney. The implications of these information flows are thus, similarly problematic, particularly in an age where mistrust of law enforcement is increasingly pervasive.

Overall, WDW and smart cities represent different contexts with distinct system functions, perceptions of systems, values, and stakeholders' trust in decision-makers. Disney's data handling practices, generally align with consumer expectations–whether organically or due to extensive marketing–yet, many of their practices are inappropriate for publicly-accountable smart cities. While WDW functions as a smart city, it reflects very distinct norms, patterns of decision-making, and consumer preferences, rather than public needs. Even as cities partner with private sector firms, they should not assume that they can imitate Disney. Instead, they should likely question if those partnerships are appropriate and consider what types of governance are necessary to engender trust in decision-makers, data, and practices.

6 Conclusions

This paper examines normative perceptions of privacy, surveillance, and innovation in a case study of an emerging smart quasi-public space. Using WDW as our prototype smart city, we empirically compare two possible modes of governance: WDW's quasi-public model and conventional cities' publicly accountable model. We further identify the limitations and potential social harms of Disneyfication of conventional cities and the importance of contextual norms in cross-context data integration, data collection, and processing practices involving public-private partnerships. What is normatively appropriate for WDW is likely not appropriate for Oakland or New Orleans, just as what is appropriate for Atlanta is not necessarily right for Seattle.

Our institutional analysis shows the extent of data collection strategies in contrast with: normative customer perceptions and expectations; underlying trade-offs, marketing slogans, and corporate values; and privacy policies regarding data collection and sharing. While generalizability of this study is limited, by our focus on Disney and its consumers, as opposed to the general public, this highlights an important lesson: privacy is contextual. We also find that Disney illustrates two important procedural approaches to governance: the use of detailed social surveys to understand expectations, and a commitment to iterative reevaluation, including for negotiation of legitimate practices and information flows. Smart cities and other public contexts need to foster dialogue between all stakeholders even if they are not all involved in decision-making. Future studies of privacy governance in smart cities should explore feedback and evaluation mechanisms to better reflect local values and norms.

In her book *Privacy in Context* [17], Nissenbaum warns us of the "Tyranny of the Normal." Without careful analysis and consideration of harms and benefits of these practices, by the time we detect resulting raptures in social and cultural values, it might be too late, because "the new normal may be comfortably entrenched, but far from comfortably accepted". In this sense, other smart cities and public contexts should take caution in learning from WDW before citizens, too, become products and local governance is delegitimized.

References

1. AlAwadhi, S., Scholl, H.J.: Aspirations and realizations: the smart city of seattle. In: 2013 46th Hawaii International Conference on System Sciences, pp. 1695–1703. IEEE (2013)
2. Allen, R.: There's a Great Big Beautiful Tomorrow. Walt Disney (1964)
3. Button, M.: Private security and the policing of quasi-public space. Int. J. Sociol. L. **31**(3), 227–237 (2003)
4. Coleman, R., Sim, J.: From the dockyards to the Disney store: surveillance, risk and security in Liverpool city centre. Int. Rev. L. Comput. Technol. **12**(1), 27–45 (1998)
5. Disney, W.: EPCOT/florida film (1966). https://www.youtube.com/watch?v=sLCHg9mUBag
6. Ekbia, H., Nardi, B.: Heteromation and its (dis) contents: The invisible division of labor between humans and machines. First Monday (2014)
7. Elmaghraby, A.S., Losavio, M.M.: Cyber security challenges in smart cities: safety, security and privacy. J. Adv. Res. **5**(4), 491–497 (2014)
8. Frischmann, B.M., Madison, M.J., Strandburg, K.J.: Governing Knowledge Commons. Oxford University Press, Oxford, UK (2014)
9. Goodman, E.P., Powles, J.: Urbanism under google: Lessons from sidewalk Toronto. Fordham L. Rev. **88**, 457 (2019)
10. Heeks, R., Shekhar, S.: Datafication, development and marginalised urban communities: an applied data justice framework. Inf. Commun. Soc. **22**(7), 992–1011 (2019)

11. Johnston, M.: Good governance: Rule of Law, Transparency, and Accountability. United Nations Public Administration Network, New York (2006)
12. Kling, R., Lamb, R.: Bits of Cities: Utopian Visions and Social Power in Placed-Based and Electronic Communities. Rob Kling Center for Social Informatics (1996)
13. Kosack, S., Fung, A.: Does transparency improve governance? Ann. Rev. Political Sci. **17**, 65–87 (2014)
14. Mangu-Ward, K.: Mickey mouse is watching you. Slate (2013). https://slate.com/technology/2013/01/magicbands-disney-ceo-bob-iger-gives-smackdown-to-ed-markey-over-privacy-concerns.html
15. Martínez-Ballesté, A., Pérez-Martínez, P.A., Solanas, A.: The pursuit of citizens' privacy: a privacy-aware smart city is possible. IEEE Commun. Mag. **51**(6), 136–141 (2013)
16. Mosco, V.: City of technology: where the streets are paved with data. In: The Smart City in a Digital World (Society Now), pp. 59–95 (2019)
17. Nissenbaum, H.: Privacy in context: technology, policy, and the integrity of social life. Stanford University Press, Palo Alto, USA (2009)
18. Peel, M.A.: Between the houses: neighbouring and privacy. In: A History of European Housing in Australia, pp. 269–286. Cambridge University Press (2000)
19. Rubinstein, I.: Urban privacy. Research Privacy Group, NYU (2020)
20. Rubinstein, I., Petkova, B.: Governing privacy in the datafied city. Fordham Urb. L. J. **47**, 755 (2019)
21. Rubinstein, I.S.: Privacy localism. Wash. L. Rev. **93**, 1961 (2018)
22. Sanfilippo, M., Frischmann, B., Standburg, K.: Privacy as commons: case evaluation through the governing knowledge commons framework. J. Inf. Policy **8**, 116–166 (2018)
23. Sharma, R., Mondal, D., Bhattacharyya, P.: A comparison among significance tests and other feature building methods for sentiment analysis: a first study. In: International Conference on Computational Linguistics and Intelligent Text Processing, pp. 3–19 (2017)
24. Shvartzshnaider, Y., Apthorpe, N., Feamster, N., Nissenbaum, H.: Going against the (appropriate) flow: a contextual integrity approach to privacy policy analysis. In: Proceedings of the AAAI Conference on Human Computation and Crowdsourcing, vol. 7, pp. 162–170 (2019)
25. Stone, K.: Enter the world of yesterday, tomorrow and fantasy: Walt Disney world's creation and its implications on privacy rights under the magicband system. J. High Tech. L. **18**, 198 (2017)
26. Van Zoonen, L.: Privacy concerns in smart cities. Gov. Inf. Quart. **33**(3), 472–480 (2016)
27. Walt Disney Travel Company: My disney experience - frequently asked questions (faq). https://www.disneyholidays.com/walt-disney-world/faq/my-disney-experience/

Research on the Decision-Making Process of Civil Servant Resisting Open Data Based on Perceived Risks

Si Li[1] and Yi Chen[2](✉)

[1] Peking University, Beijing 100871, People's Republic of China
[2] Wuhan University, Wuhan 430072, Hubei, People's Republic of China
chenyi@whu.edu.cn

Abstract. Based on the cognition of stakeholders, this paper reveals the decision-making process of civil servants resisting open data based on perceived risks. Adopting grounded theory, this study interviewed 22 stakeholders to collect data and then identified four factors as the pillars of the theoretical framework: perception of risks, conservative organizational culture, insufficient external pressure and poor operability. After that, this paper constructed a model for the decision path, which explains the formation of perceived risks of civil servants, and it also explains how the perceived risks are transformed into resistance motivation and make a behavior decision. Based on the level of certainty, the decision environment can be divided into the resistance decision path under the determined environment and that under the uncertain environment. This study also summarizes that accountability and loss of interest are two types of risk that can be considered when civil servants decide not to open data.

Keywords: Open government data · Perception of risks · Grounded theory

1 Introduction

In the past decade, governments across the world have launched open government data (OGD) initiatives. However, the fourth edition of the Open Data Barometer Global Report indicated that "governments are slowing and stalling in their commitment to offer data; in some cases, progress has even been undone" [1]. In fact, many governments are reluctant to release data [2]. When discussing the reason for this phenomenon, current studies have indicated that the perceived risks of civil servants who are in charge of releasing data are the main barriers. Perceived risks can be defined as the disadvantage to themselves and the organization perceived by civil servants.

Current research papers have identified the risks that civil servants consider in OGD, discretely, including the risk of violating legislation by opening data, challenges around data ownership, and the potential for published data to be biased, misinterpreted or misused [3–6]. However, the classification and analysis of these risks are relatively complicated and lack conceptualization and summarization. More importantly, the decision of

K. Toeppe et al. (Eds.): iConference 2021, LNCS 12646, pp. 251–259, 2021.
https://doi.org/10.1007/978-3-030-71305-8_20

resistance is a complex process [6]. Few studies have provided a clear explanation as a whole for the formation of the decision-making process.

Sayogo and Pardo indicate that "the analysis and exploration of open data initiatives in different contexts and countries are expected to provide further insights for understanding this complex phenomenon" [7]. As a developing country, China is in the preliminary stage of OGD implementation, and the central government has issued government open data initiatives. As of April 2020, 130 local governments had launched an open data platform [8]. China is also in a situation where local government departments only respond nominally to higher level initiatives. Therefore, this study aims to reveal the decision-making process of civil servants resisting open data based on perceived risks in the Chinese context.

The research questions of this study are as follows:

(1) What risks can be considered when civil servants decide not to open data?
(2) How are the separate factors related and work together to lead to civil servants resisting open data?

2 Methodology

This study adopts grounded theory as the research method. Grounded theory was first proposed by Glaser and Strauss in 1967. When implemented, it rounded theory produces a theory that explains a pattern of behavior relevant to the participants or to the problems they are involved in [9].

2.1 Sampling and Interviews

The interview process was divided into two phases. In the first phase, this study adopts the method of purposeful sampling, which selects 9 samples related to the research topic. Researchers sought interviewees from data providers and OGD authorities. Data providers are government agencies responsible for making data available to the public as open data. OGD authorities' main responsibilities are to establish local OGD platforms, formulate local policies, coordinate various government agencies and assess the work of the data-providing departments.

Initially, the researchers only designed a broad outline to guide the interview containing some necessary questions but with the aim of allowing the respondents to express their opinions. The content of the interview mainly involved the current situation for open data, the performance of resistance to open data, and the risks they consider existing in the government's open data practice. The researchers also asked interviewees to express psychological mechanisms driving civil servants resisting open data. After each interview, the researchers collated the data promptly and coded them until the core concept "the decision-making process of civil servants resisting open data based on perceived risks" emerged.

In the second phase, based on the preliminary categories, theoretical sampling was conducted to collect more focused and detailed data, and more structured interviews were conducted with 13 new samples. The interviews focused on risks and the decision-making process of civil servants resisting open data. After these interviews, scholars' views as expressed in the news were added as a data supplement. When the material became repetitive and no new concepts appeared, we determined that theoretical saturation had been achieved. Then, the interviews were stopped.

From December 4, 2018, to February 12, 2019, the researchers interviewed 22 people engaged in OGD. These interviewees cover 6 cities and involve 17 different government departments. We conducted 17 interviews for a total of 21 h, and more than 200,000 words of text were obtained (in Chinese).

2.2 Basic Information of the Interviewees

Table 1 shows the basic information of the interviewees. Interviewees are participants or mentors in open data practice, which means they have a deep understanding of open data practice.

2.3 Coding

The coding process for this study includes open coding, selective coding and theoretical coding. The encoding software is NVivo12.

Through open coding, the core concept "the decision-making process of civil servants resisting open data based on perceived risks" emerged, which is the substance of what is going on in the data [9]. By selective coding, four core categories were obtained. Theoretical coding reveals the natural structure in substantive coding [9]. We review and extract cases from the research data. Based on the respondents' description of the logical relationship between concepts in each case, the researchers describe the relationship between the core categories and the concepts that they contain. The purpose is to clarify the timing relationship, the conditions of the association and the specific context for the correlation of all levels of coding. Subsequently, the characteristics and commonalities of each case were integrated, and the stages of path change were summarized by time integration induction [10].

2.4 Reliability and Validity

The reliability and validity of coding were ensured by the following methods. First, all the interview materials in this study were recorded and transcribed into written material to ensure that the data could be accurately reproduced. Second, the sample of this study was selected to be representative, which can ensure the comprehensiveness and relevance of the material. Third, after repeated comparison and analysis in the coding process, the concepts and the correlations among them have been repeatedly confirmed, and the coding has been found to be highly consistent. Finally, the concepts were tested by additional data (R21, R22, news), and no more new conceptual categories or relationships emerged. It can be considered that the existing categories are saturated.

Table 1. Basic information of interviewees

No.	Department	Data provider/OGD authority/Expert	Job title
R1	Network and Information Office	OGD authority	Civil servant
R2	Statistics Bureau	Data provider	Manager
R3	State-owned Assets Supervision and Administration commission	Data provider	Manager
R4	Civil Affairs Bureau	Data provider	Manager
R5	Civil Affairs Bureau	Data provider	Civil servant
R6	Agricultural Commission	Data provider	Manager
R7	Economic, Trade and Information Commission	OGD authority	Civil servant
R8	Economic, Trade and Information Commission	OGD authority	Civil servant
R9	Meteorological Bureau	Data provider	Manager
R10	Economic and Information Bureau	Data provider	Manager
R11	Economic and Information Bureau	Data provider	Civil servant
R12	Housing and Urban-Rural Development Bureau	Data provider	Civil servant
R13	Industry and Information Technology Commission	OGD authority	Civil servant
R14	Industry and Information Technology Commission	OGD authority	Civil servant
R15	Economic and Information Commission	OGD authority	Civil servant
R16	Health and Family Planning Commission	Data provider	Manager
R17	Health and Family Planning Commission	Data provider	Civil servant
R18	Economic and Information Commission	OGD authority	Civil servant
R19	Transportation Committee	Data provider	Civil servant
R20	University	Expert	Professor
R21	Transportation Committee	Data provider	Manager
R22	Tax Bureau	Data provider	Manager

3 Results

3.1 Core Categories and Their Dimensions

Perceived Risks. Civil servants believe that open data will bring risks to themselves and their institutions. Accountability risk and loss of interest risk are the two subcategories See Table 2.

Table 2. The elements of "perceived risks"

Subcategory	Dimensions of subcategories	Elements
Accountability risk	Accountability risk caused by data leakage	Involving personal privacy
		Involving state secrets
		Involving trade secrets
		Data with unclear ownership
		Not allowed to be opened by other laws
		The database may be attacked
	Accountability risk caused by data utilization	Users lose money by using low-quality data
		Users use data for illegal purposes
		Data are misinterpreted
		Combining different data sets leads to data leakage involving personal privacy
		Data are directly resold, not reused
		Data are illegally authorized by the user to others
		Data are used by foreign forces
	Risk of power being held accountable	Public and media accountability for data quality
		A wider range of government power is monitored
		Exposing flaws in work
		The conclusion drawn from the data is not consistent with that of the original government
Risk of loss of interest	–	Losing power and value
		Losing source of income in some institutions

Accountability Risk. Accountability risk means that civil servants are worried about the risk of being held accountable after the opening of data.

Data leakage. OGD emphasizes open raw data, which makes civil servants maintain easy misjudgment during data review, and some data restricted by laws may be unintentionally opened to the public. In these cases, the government agency, managers and staff could face legal liability and administrative punishment. "When government departments are unable to determine whether certain data can be released, they know that the responsibility is great, so they will not release these suspicious data easily" (R7).

Data utilization. Civil servants believe that it is difficult to control the data after it is opened, and if there are problems in the utilization, they will be held accountable. Respondents said, "if the results of users' analysis of the data are inconsistent with the government, who is responsible for this?" (R6) "Many people may use data to do illegal things" (R13).

Power being held accountable. As supervised parties, civil servants are worried that open data will expose problems in how their departments perform public duties. "As a law enforcement agency, if we release these raw data, will our law enforcement procedures or flaws in law enforcement be exposed?" (R22)

Risk of Loss of Interest. Some data-intensive departments have been reusing data for a long time, and they have become an income source for subordinate institutions and enterprises. Those institutions that are not fully funded by government agencies need to generate profits by reusing data to maintain their operations. As respondent 9 said, "many organizations of a central institution in China are not fully funded. They have to meet their performance targets every year, so these departments restrict the opening of data to the public". In addition, government departments see data as the embodiment of their power and value and fear that they will lose control over that data once they are opened (R15, R22).

Conservative Organizational Culture. Government departments have a culture of risk aversion. "Traditionally, civil servants are relatively conservative" (R1, R10, R11, R19). They have a tendency to avoid being blamed. As one interviewee said, "a hundred good things do not even count as credit, but if you have an accident, you should bear all the adverse consequences. In this way, they (civil servants) think that if they do more, they will make more mistakes; if they do less, they will make fewer mistakes; if they do not, it is good" (R20).

Insufficient External Pressure. Insufficient external pressure objectively means that there is a low cost for organizational resistance. The current environment—specifically, the laws, government management and the public—has not exerted sufficient external pressure on government departments and their staff (Table 3).

Poor Operability. The poor operability of open data means that government departments find the task of data review and release difficult. China has no clear regulations on what data can be opened and how to open it, including unclear open guidelines and vague data review standards (Table 4).

Table 3. The elements of "insufficient external pressure"

Category	Subcategories	Elements
Insufficient external pressure	Insufficient pressure from law and policy	Open data is not a mandatory obligation
	Insufficient pressure from government management	The power of the OGD authorities is weak
		Senior leadership does not pay enough attention
		The important work is not to open data but to share data between government departments
		Open data is weakly linked with performance
	Insufficient pressure from the public	There is little demand from the public
		The ecology of using data has yet to emerge

Table 4. The elements of "poor operability"

Category	Subcategories	Elements
Poor operability	Unclear guidelines	No clear definition of what can be published
		No specific rules on how to open data
	Unclear review criteria	Personal privacy is not defined clearly
		Trade secrets is not defined clearly
		The ownership of data is not specified
		Lack of desensitization rule
		Lack of classification standards

3.2 The Decision-Making Process of Civil Servants Resisting Open Data Based on Perceived Risks

After data collection and analysis, this study develops a framework that conceptualizes the decision-making process of civil servant resisting open data based on perceived risks. This model explains the formation of perceived risks of civil servants, and it also explains how the perceived risks are transformed into resistance motivation and make a behavior decision.

The resisting decision emerges in the data review stage. Based on the level of certainty, the decision environment can be divided into the resistance decision path under the determined environment and that under the uncertain environment, as shown in Fig. 1.

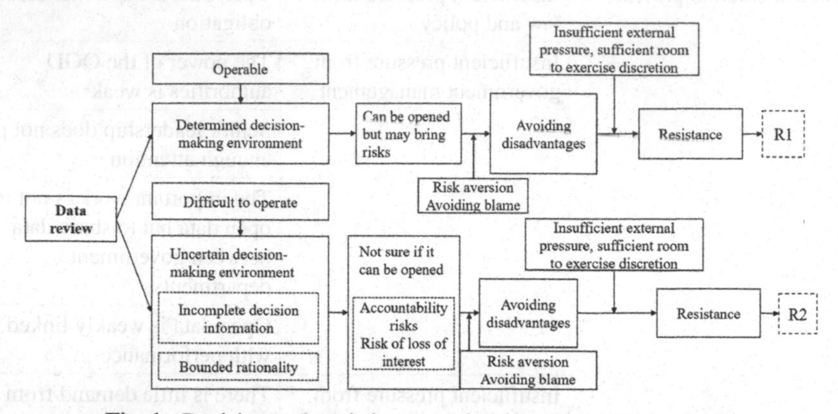

Fig. 1. Decision path resisting open data based on perceived risks

Path 1 shows that under the existing legal and policy frameworks, civil servants can determine that some data can be opened, but if they think that it will have adverse effects on them and their organization, then they will be motivated by avoiding disadvantages and will adopt the strategy of guarding the data. For example, high-value data may lead to the loss of interest. Under the objective environment of insufficient external pressure, institutions have greater discretion, and they will choose to not open data.

Path 2 traces the decision-making process of civil servants to resist open data in uncertain environments. Poor operability creates uncertainty for civil servants in opening data. In the review process, although the civil servants have sufficient professional knowledge, they still face complex problems. Within the foreseeable range, they may not know whether the data are within the scope of exemption, for example, due to a lack of ownership information for the data, which may lead to a situation in which data are made open that should not have been made open. In addition, under the principles for defining the data exemption standard, civil servants must judge whether the data involve personal privacy or business secrets and whether the data will affect social stability or economic security. In this case, whether the data can be opened depends on the professionalism of the civil servants, but people cannot always make fully rational decisions. These situations are challenging because decision information is poor, mistakes are costly. Traditionally, the culture of government departments has the tendency to avoid blame. The motivation to avoid disadvantages is naturally formed.

Under the objective conditions of insufficient external pressure, civil servants have greater discretion about whether to open data or not. "If I choose to not open, I am fine anyway. If I open up, I will take responsibility" (R15). In practice, to improve the legitimacy of the organization, they will respond by releasing data within the scope of their security.

4 Conclusion

Our key contribution to the OGD literature reveals the internal logic of civil servants' resistance to open data based on the perception of risks. The decision path has its development conditions, influencing factors, and evolutionary process. It is a new perspective for understanding the civil servant's attitude of resistance.

According to the literature review, previous studies lack conceptualization and summarization of the perceived risks. This study summarizes two categories of risks. One is the criticism and accountability that might stem from potential problems with data attributes in the process of opening data, using data and providing public supervision. The other is the loss of organizational vested interests due to open data; that is, data are the strategic resource of the organization, and the organization does not want to open it [11].

The perceived risk motive of civil servants reflects natural aspects of organizational behavior [11]. Although government departments should pursue public interest as agents of public power, as rational economic actors, they and their staff also have the characteristics of seeking advantages and avoiding disadvantages.

References

1. The fourth edition of the Open Data Barometer Global Report. https://opendatabarometer.org/4thedition/report/. Accessed 11 Oct 2020
2. Wirtz, B.W., Piehler, R., Thomas, M.J., Daiser, P.: Resistance of public personnel to open government: a cognitive theory view of implementation barriers towards open government data. Public Manage. Rev. **18**(9), 1335–1364 (2016)
3. Zuiderwijk, A., Janssen, M.: The negative effects of open government data-investigating the dark side of open data. In: Proceedings of the 15th Annual International Conference on Digital Government Research, pp. 147–152. Aguascalientes, Mexico (2014)
4. Zuiderwijk, A., Gasco, M., Parycek, P., Janssen, M.: Special issue on transparency and open data policies: guest editors' introduction. J. Theor. Appl. Electron. Commerce Res. **9**(3), 1–9 (2014)
5. Wang, H., Lo, J.: Adoption of open government data among government agencies. Government Inf. Quarterly **33**(1), 80–88 (2016)
6. Zuiderwijk, A., Janssen, M., Zhang, J., Puron-Cid, G., Gil-Garcia, J.R.: Towards decision support for disclosing data: closed or open data? Inf. Polity **20**(2–3), 103–117 (2015)
7. Sayogo, D.S., Pardo, T.A.: Understanding smart data disclosure policy success: the case of Green Button, http://citeseerx.ist.psu.edu/viewdoc/download;jsessionid=43378AF938C653F9C1D9D5A180D59ECF?doi=10.1.1.364.7555&rep=rep1&type=pdf. Accessed 11 Oct 2020
8. Lab for Digital and Mobile Governance. China local government data opening report (2020). https://www.sohu.com/a/410044259_468661. Accessed 11 Oct 2020
9. Glaser, B.G.: Advances in the Methodology of Grounded Theory: Theoretical Sensitivity. Sociology Press, Mill Valley (1978)
10. Langley, A.: Strategies for theorizing from process data. Acad. Manage. Rev. **24**(4), 691–710 (1999)
11. Ruijer, E., Detienne, F., Baker, M.J., Groff, J., Meijer, A.: The politics of open government data: understanding organizational responses to pressure for more transparency. The Am. Rev. Public Administration **50**(3), 260–274 (2020)

Open Government Data Licensing: An Analysis of the U.S. State Open Government Data Portals

Xiaohua Zhu(✉), Christy Thomas, Jenny C. Moore, and Summer Allen

University of Tennessee, Knoxville, TN 37996, USA
xzhu12@utk.edu

Abstract. Licenses are important for open government data (OGD) distribution for the purpose of reducing legal uncertainties and increasing data interoperability, especially in linked open data applications. Many standard licenses and open licenses have been created by national governments and OGD organizations to use with the publication and dissemination of OGD. Using a content analysis method, this study examines the U.S. state-level OGD licensing policies and practices. Results show that most state-level governments have OGD policies in the form of formal legislation, executive order, data policy, or terms of use/legal disclaimer, but only twenty-one states have licensing-related policies. Licenses, including a variety of standard open licenses and local custom licenses (most of which are not open licenses), are adopted by more than half of the state portals. The licensing practices are generally consistent with the licensing policies, but inconsistencies can also be identified. The study calls for the state governments to prioritize their OGD licensing policies and practices.

Keywords: Open government data · Licensing of Open Data · Public domain · Creative Commons License · Linked open data

1 Introduction

In the past decade, as the open government data (OGD) movement focusing on government transparency and data reuse expanded globally, governments worldwide have been establishing web portals for government data publication and dissemination for benefits including civic engagement, business innovation, and scientific discovery [1]. However, the very concept of "open" is not always clear to all OGD stakeholders. A definition widely used by OGD advocates is the Open Knowledge Foundation (OKF)'s "Open Definition," which states that knowledge/content/data "is open if anyone is free to access, use, modify, and share it—subject, at most, to measures that preserve provenance and openness" [2]. In other words, OGD should be free of almost any restrictions. However, some scholars warn us against this overly "liberal" view [3] because there are legal and technical complications around the access, use, redistribution, and reuse of these digital assets. In fact, it is a common practice for OGD providers to use licenses with the datasets they distribute on OGD portals. The types of licenses being used and their purposes vary (see Sect. 2), but the use of standard public licenses is often supported and

K. Toeppe et al. (Eds.): iConference 2021, LNCS 12646, pp. 260–273, 2021.
https://doi.org/10.1007/978-3-030-71305-8_21

even recommended by high-level OGD policies such as the E.U.'s *Directive on Open Data and the Reuse of Public Sector Information* (often called the PSI Directive) [4].

In the U.S., building on former U.S. President Obama's 2013 Open Data Policy, the U.S. government enacted the *Open Government Data Act* in January 2019 [5]. The law mandates that data assets published by federal government agencies should be machine-readable, in an open format, and available under an open license that does not impose "restrictions on copying, publishing, distributing, transmitting, citing, or adapting such asset" [6]. This act governs federal agencies only, consistent with the U.S. copyright law that does not allow the federal government to claim copyright for works it produces. However, this prohibition of government works' copyrightability does not apply to U.S. state or municipal governments; and the legal status of state government information has traditionally been ambiguous and various among states [7], which may extend to the intellectual property ownership of government data.

As license practices are prevalent in the OGD domain, in this paper, we examine the U.S. state-level OGD licensing policies and practices with these questions:

RQ 1: What form of OGD licensing policy does each state-level government (including the District of Columbia) have, if any?

RQ 2: What kind of licenses are used with datasets on state-level OGD portals, if any? To what degree do these licenses conform to the Open Definition?

RQ 3: Are the state licensing practices consistent with their licensing policies?

Answers to these questions are important to the development of subnational and local government OGD policies and the reuse of OGD in real-world applications. A meaningful way to use the OGD, especially in the domain of Linked Open Data, is to create mashups—merging different datasets to create new knowledge or services [8]. Since legal restrictions need to be analyzed before mashup, standard licenses are necessary to enable automated processing [9].

2 Literature Review

2.1 Legal Status of Government Data

In Tim Berners-Lee's 5-star rating system for Linked Open Data, the first star is given to open license, that is, open data should be first and foremost "available on the web (whatever format) but with an open licence" [10]. However, the use of intellectual property licenses in disseminating OGD has been considered by some as essential [3, 10] and others as a barrier [11, 12]. The underlying reason for the divergent view is likely the complex and ambiguous ownership of government data.

In countries like the U.S., the government (e.g., the U.S. federal government) cannot copyright the works generated by its employees; therefore, much government information is in the public domain for free public access, with certain exceptions, such as copyright transferred to the government [13]. This is often a basic legal argument that OGD advocates use to support the OGD movement [14]. Some scholars even argue that government data is not copyrightable even if other government works are, because as a universal rule, a work must be original to be copyrightable. Since most government data represents facts about the real world, it lacks the required originality [14–16].

Moreover, some scholars argue that governments should not claim copyright on their OGD databases because the arrangement of a database must be original to be copyrightable, while government databases usually are not [16, 17]. Therefore, "licenses may not constitute effective copyright protections in many OGD scenarios" [16].

So, why are licenses often used with OGD? At the outset, OGD composition is quite complicated (not just numbers and facts), and no one should assume that all government data is in the public domain [18]. In some countries, including some European countries, the government can legally own the works it produces. Moreover, even though numbers and facts are not copyrightable, particular compilations of data or facts may be [14]. Furthermore, in countries where the database right exists to recognize and protect the investment in compiling a database, governments can have *sui generis* property right for their OGD databases [16]. Therefore, in the current "copyright default" culture, without an explicit license, "it is safer to assume that data are legally locked-up preventing any kind of reuse (or copy)" [18]. Another important argument for license use is that, as mentioned, from a technical perspective, explicitly stated, machine-readable licenses can enable automated processing, which is critical to the reuse and mashup of datasets in the big data and linked data community [9].

The ownership issue is especially tricky regarding U.S. state-level government information. As mentioned, the copyright status of state government works is complicated and varies widely in different states. Some states have the clear legal authority to claim copyright to their works, in which case state government agencies can decide whether particular works are released into the public domain; in contrast, more states do not have clear legal guidance or policy regarding their information [7, 19]. It is worth noting that most states in the U.S. have a public records law that, similar to the federal Freedom of Information Act (FOIA), allows the public to access the state's records; however, these laws also vary significantly and most often are ambiguous on the relationship between access and copyright [19]. This means, in most states, the public can access and use the public records, but to reuse government information, users "do need to seek permission from [the] creator (agency)" [20]. However, it is difficult and even impossible to "locate agency policy- or decision-maker who can or will make a decision regarding rights and access" because even within the state government, there is often no comprehensive awareness or understanding of these issues [20]. State copyright experts suggest using licenses as one of the solutions [20].

A close examination of both sides of the argument shows the real debate is not whether to use licenses, but what kind of licenses should be used for what purpose—to impose restrictions on OGD or grant permissions. Mockus and Palmirani [21] noticed a "license culture" in the OGD community, arguing that OGD organizations usually adopt a license as an *admission* (giving out rights) rather than a *contract* (restricting rights/interests), especially in the E.U. A common belief is the middle ground of entirely open data and strict licensing to protect both the users and creators [3]. Scholars, practitioners, and policymakers suggest using standardized licenses to facilitate optimal access and achieve efficiency and interoperability [3, 4, 13, 18, 22], but the lack of clear OGD policies often creates challenges [20, 23].

2.2 Open Licenses

According to the Open Knowledge Foundation, open data "must be in the public domain or provided under an open license" and "[a]ny additional terms accompanying the work (such as a terms of use, or patents held by the licensor) must not contradict the work's public domain status or terms of the license" [2]. Essential to the "open license" concept are nine specific rights associated with openness, namely: use, redistribution, modification, separation, compilation, non-discrimination, propagation, application to any purpose, and no charge [2]. Similarly, the U.S. *OGD Act* defines open license as "a legal guarantee that a data asset is made available— "(A) at no cost to the public; and (B) with no restrictions on copying, publishing, distributing, transmitting, citing, or adapting such asset" [6]. The E.U.'s *Directive on Open Data and the Reuse of Public Sector Information* encourages E.U. member states to use standard, open, public licenses to ensure that OGD is "freely accessed, used, modified and shared by anyone for any purpose" [4].

Different versions of Creative Commons (CC) Licenses have been widely used in OGD distribution. Created by Lawrence Lessig to promote the free-culture movement, CC copyright licenses have developed into a suite of different, customizable licenses for general purposes [24]. CC licenses' four basic elements—Attribution (BY), Non-Commercial (NC), No Derivative Works (ND), and Share Alike (SA)—are also the requirements often used by the OGD community. Share Alike means requiring derivative works to have the same or similar license as the original work does. Even an open license can allow the requirement to attribute and share alike [2]; therefore, CC BY and CC BY-SA can be considered open licenses [18]. In contrast, using OGD for commercial purposes should be allowed because one of the significant benefits of OGD is to create new businesses and innovative services [25]. Therefore, licenses with NC are not open. Probably more desirable to OGD advocates is the CC0, a statement rather than a license *per se*, allowing creators to surrender their rights associated with their works to the public domain [24]. Some OGD advocates believe CC0 to be the best tool to release OGD [11, 14].

Another set of licenses often used with OGD was created specifically for open data by the Open Data Commons (ODC), a project under the Open Knowledge Foundation [26]. Different from CC licenses, they cover both copyright and database rights, and therefore are suitable for countries with database rights [18]. ODC Attribution (ODC-BY) license is similar to CC BY; ODC Open Database License (ODbL) is similar to CC BY-SA but more restrictive; Public Domain Dedication License (PDDL) is similar to CC0.

In addition to the licenses or dedications mentioned above, many governments and non-governmental organizations around the world have created their own licenses that govern OGD use and reuse, including U.K.'s *Open Government License* and *Non-Open Government License*, Canada's *Open Government License*, France's *License Ouverte*, Germany's *Data license Germany – attribution – version 1.0* and *2.0*, Italy's *Italian Open Data License*, Norway's *Norwegian Licence for Open Government Data* (NLOD), and Uruguay's *Uruguay Open Data License*, to name a few. Many national licenses were developed with the considerations of their unique copyright and database protection laws [16]. The U.K.'s *Open Government License* is often considered to be a good example among standard local licenses [10, 16].

In a global survey of OGD licensing on national data portals, Mockus and Palmirani [21] found that 56% of all 435,683 datasets investigated are covered by licenses (most of which are standardized local licenses), 27% of them are covered by legal notices in lieu of licenses, and only 17% have no licenses or legal notices/conditions at all. According to this study, the top six most-used licenses or legal notices are Canada's *Open Government License 2.0*, the U.S.'s Data Policy Statements, France's *License Ouverte*, the U.K.'s *Open Government License v3.0*, the E.U.'s legal notice, and CC-BY 4.0 [21]. This study also shows that U.S. federal OGD datasets are all covered by legal notices.

There exists a good number of studies that evaluate various aspects of government OGD portals. One of the common criteria concerns data licenses [8, 27, 28]. Some scholars argue OGD should be "license-free," that is, the data should not be subject to intellectual property regulations. But as stated, the OGD movement has a licensing culture, and licenses are often used as an admission of users' rights [21]. A study evaluating U.S. municipal OGD portals found 25 out of 34 portals provided CC licenses, PDDL, or other licenses that grant rights of reusing data [28].

Compatibility or interoperability between different licenses is particularly critical to Linked Open Data applications; therefore, many standard OGD licenses provide a compatibility provision [16]. For example, the U.K.'s standard licenses were designed to be interoperable with CC licenses [29]. In Mockus and Palmirani's study [21], the top six most-used licenses or legal notices are all compatible. However, many other licenses, including some versions of CC licenses, are not compatible, which creates a significant barrier for Linked Open Data mashups and OGD reuse [18, 21].

3 Method

We used a content analysis method in this study to collect and analyze state-level OGD policies and licenses used on state OGD portals. The analysis of OGD policies treats the state as the unit of analysis, while the portal analysis uses OGD portals as the unit of analysis (note that some states have more than one portal). Due to the dynamic nature of website content, data should be captured within a short timeframe in a web-based content analysis [30]. To ensure reliability, we completed dataset collection and license sampling within two weeks, from March 24th to April 6th, 2020. The policy collection was also conducted during a short period.

3.1 Collection of OGD Policies

Data policy is a broad concept that includes various forms of data-related laws, rules, and decisions. Within the OGD context, data policy adopted by state or local governments may cover some or all aspects of OGD production, gathering, processing, (re)distribution, (re)use, and preservation. In this study, we collect information about licensing policy that governs the (re)use and (re)distribution. Licensing policy is often part of the OGD policy, and therefore, for each of the 50 U.S. states and the District of Columbia (DC), we searched for OGD policies following the sequence of

(1) state statute,
(2) executive order,
(3) other forms of formal policy issued by the state government, including data policy, data plan, OGD handbook, etc., and
(4) other data-specific policies within the OGD portal, including copyright or other legal disclaimers, terms and conditions of use, and the portal's About section.

For (1)–(3), we mainly used Google in-site search on each state government website and the state's legal materials summarized by the Legal Information Institute [31]. For (4), we only collected policies regarding data on the portal, excluding terms of use that cover the entire state government website, policies that did not mention the data on the portal, and terms of use created by the software platform, such as Socrata and OpenGov, instead of the government entities.

3.2 Sampling and Analysis of OGD Portals, Datasets, and Licenses

We first conducted an exhaustive search of state-level OGD portals (including District of Columbia) in the U.S. and identified 77 websites that claimed to publish government data for their states. During this survey process, basic information of each site was recorded, including its name, agency, URL, number of datasets, categories of datasets, platform, functionalities, and more. A closer examination of all sites found many of them did not provide open datasets (roughly defined, in this study, as data that are machine-readable and in downloadable open formats), which excluded aggregated data reports, descriptive statistics as embedded tables, or summaries of census data. After eliminating those and some additional sites that were redundant, non-functional, or only providing a handful of downloadable datasets (which were usually buried deeply in the sites), 54 OGD portals were included in the study. Four states, Alabama, Georgia, Ohio, and South Dakota, did not have any OGD portals; seven states, including Alaska, Hawaii, etc., had two portals each. The 54 portals varied widely in terms of platform, number of datasets, data formats, and functionalities.

For each of the OGD portals in the study, a number of datasets were sampled with some exceptions, following the procedures below:

1. For portals that allow sorting by "Most recent", the 30 most recently updated datasets as reflective of the data portal's current practice regarding data license and the 20 oldest datasets as reflective of the early practice of the portal were chosen.
2. For portals on which recently updated information was not available and for portals with under 100 datasets, all datasets were sampled.
3. For portals on which recently updated information was not available and portals with over 100 datasets, datasets were selected by categories and a number of datasets were collected from each category to reach a total number of 50.

For each dataset collected, a license was searched for, first in the metadata license field, and then in any field in the metadata. If a license could not be identified, a search was made for legal notices or any other disclaimer(s) attached to the dataset.

After recording all the licenses, we analyzed the terms of licenses, giving particular attention to the non-standardized local licenses (also called bespoken licenses [22]), hereafter called custom licenses. Each custom license was compared to the open license definition elements cited in Sect. 2.2, including

- use—allowing free use,
- redistribution—allowing redistribution of the licensed work including sale,
- modification—allowing the creation of derivatives,
- separation—allowing any part of the work to be freely used, distributed, or modified separately from any other part,
- compilation—allowing the licensed work to be distributed along with other distinct works,
- non-discrimination—*not* discriminating against any person or group,
- propagation—rights apply to all whom the work is redistributed to,
- application to any purpose—allowing use, redistribution, modification, and compilation for any purpose, and
- no charge—*not* imposing any monetary remuneration [2].

We also checked the presence of some requirements allowed by the Open Definition, including

- attribution—*may* require distributions of the work to include attribution of contributors,
- share alike—*may* require distributions of the work to remain under the same or a similar license, and
- integrity—*may* require modified versions of a licensed work carry a different name or version number [2].

In addition, we examined the presence of some elements that should not, but often appear, in custom licenses, including

- applicable law,
- liability statement, and
- privacy policies [18].

It should be noted that the analysis focused on descriptive identification of absences/presence of features rather than a more comprehensive audit of licensing policies.

4 Results

4.1 Licenses and State OGD Policies

We specifically looked at legislations that can serve as the foundation of establishing state OGD portals and different types of policies governing the portals or data on the portal. We found many terms of use are about liability, but some do include data use and disseminate policies. Table 1 shows the overall results of the policy OGD analysis. Note that the term *license* is used in a broad sense, referring to guidelines or rules regarding the use of license with datasets, or the permission or restrictions on using, reusing, and distributing data.

Table 1. OGD Policies of 51 US State-level Governments

Type of policies	Number of States and Names of States	License Conditions
Law	23: AZ, AR, CT, DC, HI, IL, IN, KS, MD, MS, NE, NH, NJ, NM, NC, ND, OK, OR, TX, UT, WA, WI, WY	6 mention licenses: CT: no restrictions government use HI: can use license to allow copy, distribute, display, or create derivative with conditions IL: no license requirement or restrictions on use NH: OGD is license-free NJ: license-free, may reuse, and not subject to copyright restrictions OR: no license requirements
Executive Order	6: CT, DE, LA, NY, PA, RI	1 mentions license—PA, asking to develop policies on ownership and licensing of open data
Data policy, plan, or handbook	14: AK, CA, CT, DE, DC, FL, IL, IA, ME, MD, MO, NH, NY, PA	8 mention licenses: CA: Open licenses CT: Public asset for the public good DE: Freely available to everyone; use and republish DC: Unencumbered by license restrictions ME: Free public use and no commercial resale; require attribution MO: Requires disclaimers for attribution & liability NY: No restrictions on reuse PA: Free & no restrictions

(*continued*)

Table 1. (*continued*)

Type of policies	Number of States and Names of States	License Conditions
Terms of Use	16: A.K., AZ, CT, DE, DC, GA, HI, ID, IN, NE, ND, OK, OR, RI, WV, WI	10 mention licenses: AK: Claim copyright; usage beyond fair us needs permission DE: Public data DC: CCO 1.0 Universal GA: Open record by law HI: Each dataset may have copyright ID: Reserve full intellectual property rights IN: Limited, non-exclusive non-commercial use NE: Very limited non-commercial use; usage beyond copying & printing prohibited OK: Reformat, repurpose, and reuse in different ways RI: Public information; require attribution; may have copyright or other restrictions WV: Individual use only; no resale, no redistribution for charge

Among the 51 state-level governments, 14 states (AL, CO, KY, MA, MI, MN, MT, NV, OH, SC, SD, TN, VT, and VA) do not have any OGD policies at all. Thirty-seven states have OGD policies, including 18 states that have multiple policies. Sixteen states have some kind of OGD policy, but none mention licensing issues (AZ, AR, IA, FL, KS, LA, MD, MS, ND, NM, NC, TX, UT, WA, WI, and WY). That leaves 21 states with 25 license-related OGD policies (CT, DE, DC, and PA all have 2 policies). We then used these policies to compare with the licenses on their state OGD portals.

4.2 Availability of License

Table 2 shows the overall results of the availability of licenses on the OGD portals. In this table, we separated the portals whose datasets all have licenses from those with licenses for only some datasets.

The most common practice for using a standard license is one legal notice/disclaimer: “License: public domain,” which is similar to the practice of data.gov, the U.S. federal government’s OGD portal. The standard licenses used on these portals include different

Table 2. License Availabilities on 54 State-Level OGD portals

License availability	Number of States	Type of license	Names of States
None	22	n/a	AK1, AR, CO, ID2, IL2, KS1&2, KY, LA, ME, MN, MS, MT, NE, NV, NH, NM, NC, RI, SC1&2, WV
Some datasets	20	One standard license (6) Multiple standard licenses (7) Multiple custom licenses (5) Standard + custom (2)	IN, MD, MI1, NJ, NY, UT CT, IL, IO, MO, OR, TX, VT AZ, FL, HI1, ID1, TN MI2, WA
All datasets	12	One standard (7) Multiple custom (5)	CA, DE, DC, MA, ND, OK, PA AK2, HI2, VA, WI, WY

versions of CC0, CC BY, ODC-ODbL, PDDL, and occasionally ODC-BY. Some portals use both the "public domain" legal notice and a CC0 license or a PDDL. There is one appearance of GNU Free Documentation License, which is usually used for software documentation.

A close examination of nearly 100 custom licenses identified from 12 portals shows a diverse pool of statements with varying lengths, formats, styles, and elements. Most of them are silent on most or all Open Definition elements like free use (reuse) and redistribution. Many of them contain restrictions on the purpose of use, including a number of non-commercial requirements and statements like "only for informational purposes." We also found some licenses claim the dataset to be public information, so standard license could and should have been used. Generally speaking, very few local custom licenses conform to the Open Definition; these licenses are mainly used to restrict the purposes of use and serve as liability statements.

4.3 Licenses and OGD Policies

Considering the license practices in the context of OGD policies, we found, among the 22 portals that do not use licenses at all, 7 of them (CO, KY, MN, MT, NV, and two SC portals) belong to states with no OGD policies and 7 of them (AR, two KS portals, LA, MS, NM, and NC) belong to states where OGD policies are silent on the licensing issues. In these cases, the lack of policy may explain the absence of a license. There are 8 portals that belong to states that have established licensing policies, and in many of these cases, licenses should have been used. For example, IL and NH have OGD legislations stating OGD is open for public use. ME's OGD policy says data is for free public use, but no commercial resale is allowed; and it requires users to cite the source (attribution). RI has terms of use that claim data is public information but requires attribution; and some materials may have copyright or other restrictions. The other four states, AK, ID,

NE, and WV, have terms of use or legal disclaimers that restrict data use; in these cases, the terms of use or disclaimers may arguably serve as a license but in practice are very inconvenient to users.

Among the 32 portals that use licenses, 22 of them use existing standard licenses, and 12 use custom licenses (with two overlaps). Among portals using standard licenses, 12 have open or no-restriction licensing policies, 4 have no OGD policies, and 6 are silent on licensing issues in their OGD policies. This shows some portals do try to follow the best licensing practices despite the lack of specific policies. We also noticed some inconsistencies between licensing practices and OGD policies. For example, DC's terms of use claim to use CC0, but CC BY is used instead. IN's terms of use claim data are for non-commercial use, and yet CC BY license is adopted. MO's data policy requires users to add a disclaimer for attribution and liability when using the data, but the portal uses a "public domain" disclaimer and ODbL license.

The 12 states that use custom licenses to varying degrees show a different pattern. Only 4 of them have license policies, which claim full intellectual property rights to their data or indicate copyright protection may exist. The policies can serve as their rationale for using custom licenses. Three of these states do not have OGD policies and 4 are silent on licensing issues, which may explain why they do not use standard licenses.

Overall, there are 21 states with 25 license-related OGD policies. In practice, these policies are followed but inconsistencies do exist. For those states with no OGD policies or with policies silent on licensing issues, licenses, sometimes standard licenses, are still used.

4.4 Summary of Results

Results of the study answer the research questions as follows: Most (37 out of 51) state-level governments do have some kind of OGD policy, in the form of formal legislation, executive order, data policy, or terms of use/legal disclaimer; but only 21 states have licensing-related policies. Licenses, including different standard open licenses and some local custom licenses (most of which are not open licenses), are adopted by more than half (32 out of 54) of the state portals. The licensing practices are generally consistent with the licensing policies, but inconsistencies can also be identified.

A direct relationship between licensing policy availability and the licensing practice was not identified, but a rough analysis shows portals with specific data policies or guidelines are probably more likely to use standard open licenses. Lack of policies did not stop some portals from publishing OGD under standard open licenses, but without the relevant policies, more states choose not to use any licenses with their data.

5 Discussions

This analysis shows state portals' licensing practices have not been consistent. A somewhat surprising finding is that only seven portals use one public license or one legal disclaimer for all their datasets consistently. There may be some truth in Lee's argument that "it's possible that governments do not conduct due diligence regarding the legal status of the subject data" [16]. Data ownership and restrictions have been ambiguous and need more attention from policymakers. Without an explicit licensing policy devised by experts, inconsistency will likely persist.

And there is another kind of inconsistency. Many portals use license(s) for some of the datasets but not all, and there may be several reasons for that. Usually, a state-level portal relies on individual government agencies to provide licenses with their datasets, but agencies may neglect this responsibility or lack the legal knowledge to select licenses. In some cases, only older datasets on the portals do not have licenses, probably due to the lack of licensing policies at the time. If so, this type of inconsistency reveals some problems in reinforcing OGD policies. That is, even when a licensing policy is in place, portal managers may need to help agencies with licensing issues and manage their dataset licenses retrospectively.

Previous studies show all U.S. federal datasets and most U.S. municipal OGD portals utilized open licenses [21, 28]. Compared with U.S. federal and municipal-level portals, state portals are overall less adequate in their licensing practices. Are state portals inferior to city portals in licensing practice because they are more cautious of their intellectual property ownership and government information misuse? Or do state governments have fewer incentives to distribute government data than big cities do? Another possible reason is that some state governments rely on cities to achieve OGD goals instead of focusing on their own portals. It would be interesting to compare these OGD portals further to reveal other potential distinctions and find explanations for the differences.

Practical OGD licenses should not only be standardized but also be explicit and machine-readable to ensure interoperability for dataset mashup [9]. Although intellectual property rights and data ownership can be complicated in U.S. states, many countries have provided good examples of developing standard licenses that address their copyright law and database protection law [16, 22, 29]. Even if state governments cannot devise localized standard, machine-readable licenses, they probably should consider using CC licenses or ODC licenses rather than create custom licenses. This analysis shows that local custom licenses can be a concern, with various, inconsistent formats and content that should not be placed in a license, such as liability and privacy statements [18]. Issues of using standard licenses also exist; for example, some portals use standard open license(s) for some datasets and a legal disclaimer of public domain for others, which is unnecessary and will only add confusion and uncertainty. In addition, using a legal notice in place of an open license is arguably problematic for Linked Open Data applications because notices are not standardized and machine-readable, even though it is the practice of the U.S.'s federal OGD portal. Some of the problems may be the result of lacking details and specifications in licensing policies, as discussed in the literature [20, 23]. In general, the portals would greatly benefit from following the existing international standard and best practices on license use [29].

6 Conclusions

Establishing and maintaining OGD portals is a complex and challenging project, in which licensing is only a small component. Nevertheless, the importance of this component cannot be overlooked. Standard, open licenses are critical for reducing legal uncertainties and increasing data interoperability, especially in the Linked Open Data applications.

One of the major issues is the lack of state-wide, specific licensing policies. And this may be true for other aspects of OGD portals—without a formal and specific policy in place, interoperability and compatibility issues are likely to appear and hinder the further development of the portal and any practical, significant utilization of OGD. Therefore, we strongly suggest that U.S. state and local governments develop OGD policies or clarify/specify the existing OGD policies, as an essential step to realize the benefits of OGD fully.

References

1. U.S. Executive Office of the President: Memorandum for the heads of executive departments and agencies (2013). https://obamawhitehouse.archives.gov/sites/default/files/omb/memoranda/2013/m-13-13.pdf. Accessed 1 Oct 2020
2. Open Knowledge Foundation: Open Definition. https://opendefinition.org/od/2.1/en/. Accessed 1 Oct 2020
3. Khayyat, M., Bannister, F.: Open data licensing: more than meets the eye. Inf. Polity **20**(4), 231–252 (2015). https://doi.org/10.3233/IP-150357
4. Union Directive (E.U.) 2019/1024 of the European Parliament and of the Council of 20 June 2019 on open data and the reuse of public sector information. http://data.europa.eu/eli/dir/2019/1024/oj/eng. Accessed 1 Oct 2020
5. SPARC: Passed into Law: OPEN Government Data Act (S. 760/ H.R. 1770). https://sparcopen.org/our-work/open-government-data-act/. Accessed 1 Oct 2020
6. Foundations for Evidence-Based Policymaking Act of 2018, P.L. 115–435 (2018). https://www.congress.gov/115/bills/hr4174/BILLS-115hr4174enr.pdf. Accessed 1 Oct 2020
7. Courtney, K.K.: The state copyright conundrum: what's your state government's rule on copyright? College and Res. Libraries News **79**(10), 571–574 (2018). https://doi.org/10.5860/crln.79.10.571
8. Attard, J., Orlandi, F., Scerri, S., Auer, S.: A systematic review of open government data initiatives. Government Inf. Quarterly **32**(4), 399–418 (2015). https://doi.org/10.1016/j.giq.2015.07.006
9. Krötzsch, M., Speiser, S.: ShareAlike your data: self-referential usage policies for the semantic web. In: Aroyo, L., et al (eds.) The Semantic Web – ISWC 2011, pp. 354–369. Springer (2011). https://doi.org/10.1007/978-3-642-25073-6_23
10. Berners-Lee, T.: Linked data. https://www.w3.org/DesignIssues/LinkedData.html. Accessed 1 Oct 2020
11. Janssen, K.: The influence of the PSI directive on open government data: an overview of recent developments. Government Inf. Quarterly **28**(4), 446–456 (2011). https://doi.org/10.1016/j.giq.2011.01.004
12. Dulong de Rosnay, M., Janssen, K.: Legal and institutional challenges for opening data across public sectors: towards common policy solutions. J. Theor. Appl. Electron. Commerce Res. **9**(3) (2014). https://doi.org/10.4067/S0718-18762014000300002
13. Okediji, R.L.: Government as owner of intellectual property? Considerations for public welfare in the era of big data. Vanderbilt J. Entertainment Technol. Law **18**(2), 331–349 (2016)
14. Tauberer, J.: Open government data: The book (2nd ed) (2014). https://opengovdata.io/. Accessed 1 Oct 2020
15. Ford, B.: Open wide the gates of legal access. Oregon Law Rev. **93**, 534–570 (2014)
16. Lee, J.-A.: Licensing open government data. Hastings Business Law J. **13**(2), 207–240 (2017)

17. Scassa, T.: Public transit data through an intellectual property lens: lessons about open data. Fordham Urban Law J, **41**(5), 1759 (2016)
18. Morando, F.: Legal interoperability: Making open (government) data compatible with businesses and communities. Italian J. Library, Archives Inf. Sci. **4**(1), 441–452 (2013)
19. Messenger, A., Pitman, D.: Can states use copyright to restrict the use of public records? Commun. Lawyer **29**(3), 4–8 (2013)
20. Bartlett, B., Bonfiglio, J., Kasianovitz, K.: State government information and the copyright conundrum. LRL Webinar, National Conference of State Legislatures (2014). http://www.ncsl.org/documents/lrl/Copyright-Conundrum-2014.pdf. Accessed 1 Oct 2020
21. Mockus, M., Palmirani, M.: Open government data licensing framework. In: Kő, A., Francesconi, E. (eds.) Electronic Government and the Information Systems Perspective, pp 287–301. Springer (2015). https://doi.org/10.1007/978-3-319-22389-6_21
22. Korn, N., Oppenheim, P.C.: Licensing open data: a practical guide (2011). http://www.discovery.ac.uk/files/pdf/Licensing_Open_Data_A_Practical_Guide.pdf. Accessed 1 Oct 2020
23. Okamoto, K.: Introducing open government data. The Reference Librarian **58**(2), 1–13 (2016). https://doi.org/10.1080/02763877.2016.1199005
24. Creative Commons: About CC licenses (2019). https://creativecommons.org/about/cclicenses/. Accessed 1 Oct 2020
25. Borgesius, F.Z., Gray, J., van Eechoud, M.: Open data, privacy, and fair information principles: towards a balancing framework. Berkeley Technol. Law J. **30**(3), 2073 (2016)
26. Open Knowledge Foundation: Open data commons: Legal tools for open data. https://opendatacommons.org. Accessed 1 Oct 2020
27. Thorsby, J., Stowers, G.N.L., Wolslegel, K., Tumbuan, E.: Understanding the content and features of open data portals in American cities. Government Inf. Quarterly **34**(1), 53–6a (2017). https://doi.org/10.1016/j.giq.2016.07.001
28. Zhu, X., Freeman, M.A.: An evaluation of U.S. municipal open data portals: a user interaction framework. J. Assoc. Inf. Sci. Technol. **70**(1), 27–37 (2018). https://doi.org/10.1002/asi.24081
29. Mewhort, K.: Creative Commons Licenses: Options for Canadian open data providers (2012). https://cippic.ca/en/cc-for-open-data. Accessed 1 Oct 2020
30. McMillan, S.J.: The microscope and the moving target: the challenge of applying content analysis to the World Wide Web. J. Mass Commun. Quarterly **77**(1), 80–98 (2000)
31. Legal Information Institute: Listing by Jurisdiction. https://www.law.cornell.edu/states/listing. Accessed 1 Oct 2020

Information Systems as Mediators of Freedom of Information Requests

Daniel Carter[1(✉)] and Caroline Stratton[2]

[1] Texas State University, 601 University Dr, Old Main 102, San Marcos, TX, USA
dcarter@txstate.edu

[2] Florida State University, Tallahassee, FL, USA
cstratton2@fsu.edu

Abstract. While Freedom of Information requests play an important role in government oversight, the process remains largely untheorized, especially in relation to the role of information systems. To address this gap, we conducted an exploratory study using a random, stratified sample of 96 municipalities in one state. Our findings suggest that information systems play multiple mediating roles in shaping and affording access to government records, and that this mediation influences the outcomes of the FOI process. Our work has practical implications for transparency advocates, IS designers, and other information professionals.

Keywords: Freedom of information · Open government data · Information systems

1 Introduction

Freedom of Information (FOI) laws are crucial for democratic oversight. While the increasing digitization of government records suggests important advantages for FOI process—potentially decreasing the time taken to submit and process requests and increasing the value of the information obtained—the existing, scant literature on FOI largely ignores the role played by information systems (IS). Given that the information requested through the FOI process may be created, stored, and delivered digitally, ignoring IS is a significant oversight in theorizing FOI.

To address this gap, we conducted an exploratory, inductive analysis of the FOI process in municipalities in the state of Texas. We submitted FOI requests to a population-based sample of municipalities and report on mediations of IS that were made visible through the request process, such as the electronic submission of FOI requests and software configurations shaping records provision. We contribute to existing literature by providing an empirical view of current processes and indicating how IS figure into theorizing the FOI process.

Our primary finding is that IS play multiple mediating roles, such as in requestor-government interactions and recordkeeping as it relates to FOI requests, thereby influencing the outcomes of the FOI process. IS may enable and/or constrain provision of requested information in a particular format, depending on organizational limitations,

K. Toeppe et al. (Eds.): iConference 2021, LNCS 12646, pp. 274–282, 2021.
https://doi.org/10.1007/978-3-030-71305-8_22

software availability and configuration, and user agency, among other factors. We bring IS to the fore, setting the stage for future theorizing about the FOI process that better reflects widespread computerization in government, in conversation with the organizational structures and operational decisions surrounding this topic. These areas of concern have the potential to impact professional practice and legal and political action.

In the following section, we review literature related to both FOI and Open Government Data (OGD). We include literature on OGD because of its potential to guide the focus of our exploratory work and to allow for contrasts to be drawn with distinct features of FOI processes. We then review our methods and findings, highlighting indications of IS use, before discussing the mediating roles played by IS in FOI processes and how this finding might complicate current theories.

2 Background

The current state of public access to government data in the U.S. can be traced to the Freedom of Information Act (FOIA), signed in 1966 by Lyndon Johnson. FOIA established the judicially enforceable right for citizens to access information produced by government agencies. The act was intended to promote an open and democratic society, in which citizens are able to hold the government accountable and make informed political decisions [11]. In 1996, Bill Clinton signed the Electronic Freedom of Information Act Amendments (E-FOIA), which acknowledged the growing role of information technology in governmental organizations and extended FOIA access to include electronic records.

FOIA and E-FOIA established an ideological and legal precedent for the concept of Open Government Data (OGD), which additionally calls for government information to be made preemptively available. An influential conference organized by the Open Knowledge Foundation [17] put forward a set of principles describing open data, including specifications that data be complete, primary, timely, accessible, machine processable, non-discriminatory, non-proprietary and license-free. Many of these principles were implemented in Barack Obama's 2013 executive order, with one consequence being a shift from the access of government information for purposes of accountability to the provision of government data for "more speculative uses" [20]. Yu and Robinson [23] argue that the definition and implementation of OGD have shifted focus away from public accountability and toward "politically neutral public sector disclosures that are easy to reuse."

Thus, while both FOI and OGD are strongly associated with the concept of open government and the promotion of democratic participation and oversight, they differ in the value claims associated with them. Where FOI is generally discussed in relation to supporting democratic processes and making government actions visible to citizens, OGD is also discussed in relation to values such as encouraging innovation and generating economic value [4, 12]. The U.S. Office of Management and Budget, for example, argues that making government data "accessible, discoverable, and usable by the public can help fuel entrepreneurship, innovation, and scientific discovery" [5]. This value claim has consequences for the development of IS for delivering OGD, for example highlighting the need for release of "raw, unprocessed information that allows individuals to reach their own conclusions" [23]. This emphasis, in part, results in technical guidelines such as the storage and provisioning of data in machine-readable formats such as CSV and XML [1].

Another difference between OGD and FOI is the attention paid in the literature to the systems that create, organize and provide access to data as well as the organizational structures around these systems. Because OGD directives assume the goal of economic value and call for the creation of new technical systems such as the U.S. government's data.gov portal, studies have paid particular attention to the design and evaluation of such systems. Studies in fields such as information management, accordingly, have produced taxonomies of the barriers to OGD that, broadly, differentiate between those related to providers and users. Crusoe and Melin [7] review the extensive literature on OGD and group barriers to providers into social (e.g., a risk averse culture that fears negative repercussions from data release [13]), technical (e.g., poor data quality [8]) or lack of skill [3]) and legal (e.g., conflicts between release of data and individuals' privacy interests [9]).

In contrast to the numerous sociotechnical analyses of barriers to OGD, the existing literature on FOI tends to be sparse and highly pragmatic. Scholars have discussed FOI as a potential method for data collection [10, 14, 16, 19, 22] and considered its consequences from a legal perspective [6, 18]. However, as Luscombe and Walby [15] point out, there are no existing theoretical frameworks that explain the workings of FOI and how these interact with state power and information. In response to this gap, Luscombe and Walby put forward three frameworks: the live archive (which would see FOI in relation to memory and the pursuit of social justice), obfuscation (which would see FOI as masking political power by holding up a false sense of transparency) and actor-network theory (which would focus on the microprocesses that comprise the FOI process). Luscombe and Walby's work hints at several roles for IS in FOI, such as in electronic records management, disclosing information in "unworkable formats," making digital records unsearchable, and in cooperation and contestation with people and objects.

With these understandings of the FOI process, further informed by sociotechnical analyses of OGD, we pose the research question:

In what ways is IS made visible in the FOI process, and in what ways does it mediate the provision of information?

3 Method

To explore the role of IS in FOI processes, we submitted the same public information request (PIR) to a random sample of 96 of the 1217 municipalities in the state of Texas. Because states tend to have a small number of large cities and a large number of small cities, we utilized stratified sampling in order to capture key differences related to the distribution of municipalities' populations. To construct the strata, we followed Angel and Blei's [2] technique, partitioning cities into six groups, such that each group contained roughly twice the number of cities in the previous group.

We designed the submitted PIR by considering phenomena that occur in all municipalities, regardless of location or size. We also sought to avoid requesting information about controversial, confidential, or proprietary phenomena that might cause the municipality to appeal, deny or limit the request. As such, we requested six months of records for city code enforcement for the period of January 1, 2019 through June 30, 2019. Our request letter sought a list of code compliance cases, including location, description of the case, date the case was initiated, date the case was resolved, and the case's disposition. Following standard journalistic practice, we requested that responsive information be returned in a spreadsheet, database, or delimited text file rather than as PDFs or paper records.

We submitted PIRs in January and February 2020, with all submissions occurring on Sundays in order to avoid potential discrepancies in processing time. When submitting requests, we recorded the date of the request, the method used to make the request (e.g., web form, email), and the software platform used to make the request, if applicable. As municipalities responded to our requests, we recorded the first date they contacted us, the date that we received data or the request was otherwise completed, as well as any notes about our communication.

4 Findings

4.1 IS Mediating Interaction Between FOI Requestors and Government

Our first contact with IS mediating the FOI process emerged at the point of making a request, shaping the interaction between FOI requestors and government officials. Table 1 summarizes how we submitted requests through specific technologies as well as the responses we received, according to the population groups of municipalities in the study.

Table 1. Summary of FOI requests and responses

Population group	<2000	2k–7k	7k–19k	19k–65k	65k–149k	>149k	Total
Requests submitted	**16**	**16**	**16**	**16**	**16**	**16**	**96**
By online portal	0	0	0	5	5	10	20
By web form	0	3	3	3	4	1	14
By email	16	13	13	8	7	5	62
Responses received	**10**	**10**	**12**	**16**	**15**	**15**	**78**
Records received	2	5	10	14	14	12	57
Request denied	0	0	0	1	0	0	1
No responsive records	8	3	0	0	0	0	11
Request withdrawn due to cost	0	2	2	1	1	3	9
No response received	**6**	**6**	**4**	**0**	**1**	**1**	**18**

Twenty of our 96 requests were routed through an online portal accessed via municipal website. All municipalities utilizing a portal to manage FOI requests belonged to the three largest population groups in our sample. Notably, every portal we encountered relied on the same commercial software, GovQA. We submitted fourteen of the 96 requests via web forms on municipal websites, which we expected to route requests to an appropriate party.

The remaining 62 FOI requests were sent by email. Municipal websites indicated an appropriate email recipient for FOI requests in most of the 62 cases. We encountered seven municipalities in our sample that did not have a web site, all in the smallest population group. In these cases, we referred to directory contact information to call these municipalities and ask how to best submit a FOI request. Most of the email addresses that we sent FOI requests to were associated with a municipal web domain (e.g., of the format citysecretary@city.gov); however, municipal employees also directed us to submit requests via email to addresses using email providers such as Gmail and Hotmail as well as addresses associated with internet service providers such as Verizon and AT&T. The 13 municipalities directing our requests to email services lacking local hosting were all in the group of smallest municipalities in our study.

In response to requests, we received digital records from 57 municipalities via email, online portal delivery, or referral to an open data portal. One municipality denied our request citing a legal exemption based on undue burden to staff, eleven reported having no responsive records, and we withdrew nine requests because of the reported cost of scanning paper records. Eighteen municipalities did not respond to our requests or deliver records as indicated.

4.2 IS Mediating the Disclosure of Information in Response to FOI Requests

The 57 responses that we received demonstrated additional mediation of IS in disclosing information in response to FOI requests. Table 2 summarizes features of the records we received, according to the population groups of municipalities in the study.

Table 2. Summary of records received responsive to FOI requests

Population group	<2000	2k–7k	7k–19k	19k–65k	65k–149k	>149k	Total
Records received	**2**	**5**	**10**	**14**	**14**	**12**	**57**
PDF – digital output	0	1	2	5	2	5	15
PDF – scanned	1	1	2	6	1	0	11
Excel or CSV file	1	3	6	3	11	6	30
Text file	0	0	0	0	0	1	1
Containing all requested data	0	4	7	10	9	11	41
Containing extraneous data	0	2	4	8	8	6	28

Our request specified that we sought responsive records in the form of a spreadsheet, database, or delimited text file rather than as a PDF; however, a large portion of the responsive records we received (26 of 57) were formatted as PDFs. We further distinguish between those PDFs written as digital output from software used for recordkeeping (15 in total) as opposed to those produced by scanning physical records (11 in total). 15 of the 26 responses not adhering to our requested format indicated that the agency could not or would not provide records in the requested format because of incompatibility with IS and/or recordkeeping practices.

IS mediation in disclosing information also emerged in the content of records: 41 of 57 responses contained all of the data we had requested, while about half (28 of 57) contained data we had not requested. These features of the responsive records point to recordkeeping practices with IS that may or may not easily lend themselves to finding, editing, and formatting records as requested. First, recordkeeping practices with IS capture particular attributes of phenomena and exclude others. Second, the configuration of software and user agency to find, edit, and format records shapes what is ultimately disclosed.

Figure 1 illustrates the flow of requests and their outcomes. Figure 2 depicts the FOI process from request to disclosure with indications of IS mediations as they emerged through our findings. We discuss the mediations of IS in the following section.

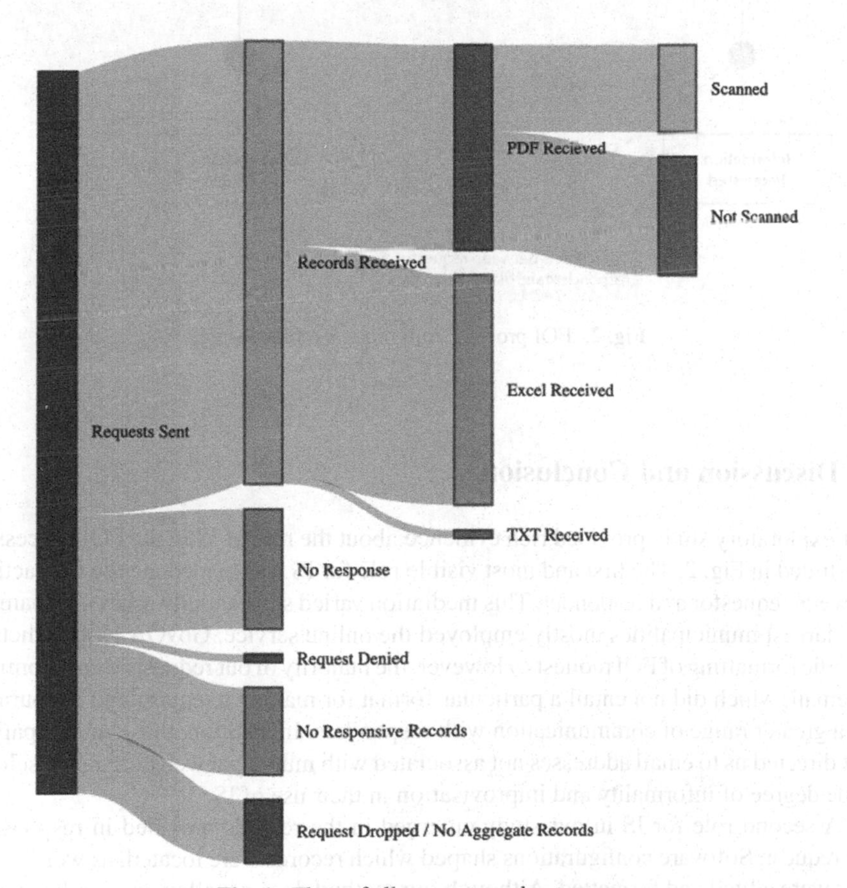

Fig. 1. Flow of all requests and outcomes

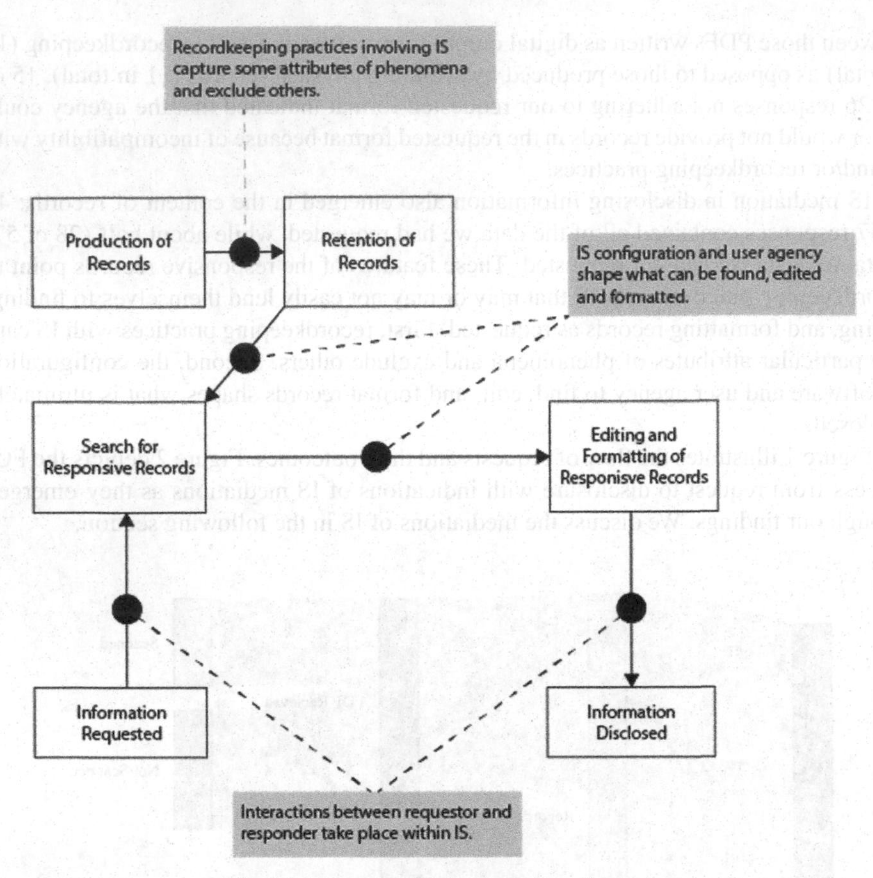

Fig. 2. FOI process from request to disclosure

5 Discussion and Conclusion

Our exploratory study provided rich evidence about the role of IS in the FOI process, as illustrated in Fig. 2. The first and most visible role for IS was to mediate the interactions between requestor and responder. This mediation varied substantially across our sample. The largest municipalities mostly employed the online service, GovQA, which dictates specific formatting of FOI requests. However, the majority of our requests were submitted by email, which did not entail a particular format for making a request and encouraged for a greater range of communication with responders. In addition, those municipalities that directed us to email addresses not associated with municipal web domains disclosed some degree of informality and improvisation in their use of IS.

A second role for IS in our study emerged in the records provided in response to our request. Software configurations shaped which records were located, as well as how they were edited and formatted. Although our method does not allow us to exhaustively characterize records omitted because they could not be found by the responding official, our communication to clarify and negotiate requests with officials indicated that finding what we sought could be limited by software configuration and user agency. This second mediating role was also evident in the quantity of records we received that were not in the format we had requested, as well as in the portion of records containing extraneous data. The portion of records we received in formats we had not requested suggest multiple contributing factors: recordkeeping practices that were not fully digital, officials uninterested in or even inhibiting information reuse and repackaging, and users lacking agency to edit and format information as requested.

A third role for IS in recordkeeping surfaced in the responsive information we received. The extent to which municipalities keep digital records of phenomena, as well as the attributes of those phenomena they choose to record digitally, influence what information can be disclosed and how. The responsive records we received that did not contain all of the data we had requested illustrate this mediation. Further work might pursue theorizing record creation and IS, particularly as it has implications for FOI and OGD.

In closing, our exploratory study indicates that IS play multiple mediating roles in the FOI process that influence how and what information is ultimately disclosed. Studying and theorizing the FOI process in relation to IS can expand on limited scholarly work in this area and point to practical lessons for transparency advocates, IS designers, and government officials. First, we suggest that transparency advocates prioritize disclosure of machine-readable information in FOI legislation. Utilizing the rationale from OGD of the benefits of reuse and repackaging information may be a compelling rationale. Second, we recommend that IS designers explicitly address the agency and skill of public sector employees. Third, we propose that government officials incorporate training about IS within FOI training, as it influences the output of the process and relates to compliance with the law. This study offers a starting point for scholarship and action to better understand and harness IS towards government transparency and public benefit.

References

1. Afful-Dadzie, E., Afful-Dadzie, A.: Liberation of public data: exploring central themes in open government data and freedom of information research. Int. J. Inf. Manag. **37**(6), 664–672 (2017). https://doi.org/10.1016/j.ijinfomgt.2017.05.009
2. Angel, S., Blei, A.M.: The spatial structure of American cities: the great majority of workplaces are no longer in CBDs, employment sub-centers, or live-work communities. Cities **51**, 21–35 (2016)
3. Barry, E., Bannister, F.: Barriers to open data release: a view from the top. Inf. Polity **19**(1,2), 129–152 (2014). https://doi.org/10.3233/IP-140327
4. Bertot, J., Gorham, U., Jaeger, P., Sarin, L., Choi, H.: Big data, open government and e-government: issues, policies and recommendations. Inf. Polity **19**(1,2), 5–16 (2014). https://doi.org/10.3233/IP-140328
5. Burwell, S.M., VanRoekel, S., Park, T., Mancini, D.J.: M-13–13 memorandum for the heads of executive departments and agencies; subject: open data policy-managing information as an asset. Office of Management and Budget (2013)
6. Cordis, A.S., Warren, P.L.: Sunshine as disinfectant: the effect of state freedom of information act laws on public corruption. J. Public Econ. **115**, 18–36 (2014)

7. Crusoe, J., Melin, U.: Investigating open government data barriers. In: Parycek, P., et al. (eds.) EGOV 2018. LNCS, vol. 11020, pp. 169–183. Springer, Cham (2018). https://doi.org/10.1007/978-3-319-98690-6_15
8. Dawes, S., Helbig, N.: Information strategies for open government: challenges and prospects for deriving public value from government transparency. In: Wimmer, M.A., Chappelet, J.-L., Janssen, M., Scholl, H.J. (eds.) EGOV 2010. LNCS, vol. 6228, pp. 50–60. Springer, Heidelberg (2010). https://doi.org/10.1007/978-3-642-14799-9_5
9. Dulong de Rosnay, M., Janssen, K.: Legal and institutional challenges for opening data across public sectors: towards common policy solutions. J. Theor. Appl. Electron. Commer. Res. **9**(3), 1–4 (2014)
10. Greenberg, P.: strengthening sociological research through public records requests. Soc. Curr. **3**(2), 110–117 (2016). https://doi.org/10.1177/2329496515620646
11. Halstuk, M.E., Chamberlin, B.F.: The freedom of information act 1966–2006: a retrospective on the rise of privacy protection over the public interest in knowing what the government's up to. Commun. Law Policy **11**(4), 511–564 (2006). https://doi.org/10.1207/s15326926clp1104_3
12. Janssen, K.: Open government data and the right to information: opportunities and obstacles. J. Commun. Inf. **8**(2) (2012). https://www.ci-journal.net/index.php/ciej/article/view/952
13. Janssen, M., Charalabidis, Y., Zuiderwijk, A.: Benefits, adoption barriers and myths of open data and open government. Inf. Syst. Manag. **29**(4), 258–268 (2012)
14. Keen, M.F.: The freedom of information act and sociological research. Am. Sociol. **23**(2), 43–51 (1992)
15. Luscombe, A., Walby, K.: Theorizing freedom of information: the live archive, obfuscation, and actor-network theory. Gov. Inf. Q. **34**(3), 379–387 (2017). https://doi.org/10.1016/j.giq.2017.09.003
16. Oltmann, S.M., Knox, E.J., Peterson, C., Musgrave, S.: Using open records laws for research purposes. Libr. Inf. Sci. Res. **37**(4), 323–328 (2015)
17. Open Government Data. https://opengovernmentdata.org/. Accessed 7 Apr 2020
18. Pozen, D.E.: Freedom of information beyond the freedom of information act. U. Pa. L. Rev. **165**, 1097 (2016)
19. Savage, A., Hyde, R.: Using freedom of information requests to facilitate research. Int. J. Soc. Res. Methodol. **17**(3), 303–317 (2014). https://doi.org/10.1080/13645579.2012.742280
20. Schrock, A.R.: Civic hacking as data activism and advocacy: a history from publicity to open government data. New Media & Society (2016). https://doi.org/10.1177/1461444816629469
21. Shepherd, E., Stevenson, A., Flinn, A.: Freedom of information and records management in local government: help or hindrance? Inf. Polity **16**(2), 111–121 (2011). https://doi.org/10.3233/IP-2011-0229
22. Walby, K., Larsen, M.: Access to information and freedom of information requests: neglected means of data production in the social sciences. Qual. Inquiry **18**(1), 31–42 (2012). https://doi.org/10.1177/1077800411427844
23. Yu, H., Robinson, D.G.: The new ambiguity of open government. UCLA L. Rev. Discourse **59**, 178 (2011)

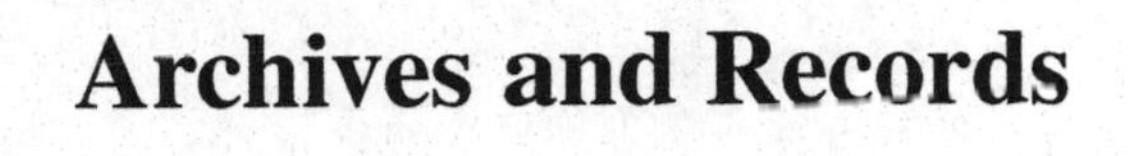
Archives and Records

Towards a Human Right in Recordkeeping and Archives

Kathy Carbone[1], Anne J. Gilliland[1], Antonina Lewis[2], Sue McKemmish[2], and Gregory Rolan[2](✉)

[1] UCLA, Los Angeles, CA 90095, USA
kcarbone@g.ucla.edu, gilliland@gseis.ucla.edu
[2] Monash University, Melbourne, Australia
draylewis@gmail.com, {sue.mckemmish,greg.rolan}@monash.edu

Abstract. The global *Rights in Records for Refugees (R3)* and the *Rights in Childhood Recordkeeping in Out-of-Home Care* in Australia research projects have both surfaced the role that rights in recordkeeping and archives might play in actualising the human rights of refugees of all demographics, and Care-experienced children and adults, including Australian Indigenous children and the Stolen Generation. Each has centred and privileged the experiences of those who are disempowered and unable to exercise their human rights in major part due to governmental and institutional recordkeeping policies, practices and technologies. Each has also taken a participatory and critical approach, applying testimonial and instrumental warrant analysis. In this paper, we first demonstrate how rights in records are critical to actualizing human rights and self-determination. We then map and discuss the convergences and divergences of key findings of the two projects, with reference to their contextual differences, similarities and overlaps. Based on this comparison, we propose a new global human rights framework encompassing three high-level sets of rights in records, recordkeeping and archives: Recordkeeping and Archival Sovereignty and Participation; Disclosure and Access; and Privacy and Safe Recordkeeping Places.

Keywords: Rights · Records · Recordkeeping · Archives

1 Introduction

Contemporary discourses of the postmodern information society often focus on issues of commercial data access and privacy [1], or the biases or unfairness of algorithmic uses of data [2–4]. However, this treatment often sidesteps the deeper, societal issues concerning the ways in which 'authoritative' records and recordkeeping[1] pervade and determine people's lives.

[1] The term Recordkeeping refers to the entirety of conceiving, creating, managing, and deriving utility from records in a continuum of use. It subsumes records management and archival administration; embracing the design of sociomaterial systems that deal with records (see [5]).

K. Toeppe et al. (Eds.): iConference 2021, LNCS 12646, pp. 285–300, 2021.
https://doi.org/10.1007/978-3-030-71305-8_23

Records, as persistent representations of the events that shape human activity [6], are instruments that are integral to identity, memory, and cultural heritage, as well as safety, security, wellbeing and accountability [7]. They are always in a process of becoming [8] across space and through time, continuously influencing and controlling the trajectories of people's lives. Fundamentally, recordkeeping and archives are instruments for defining and exercising power [9]. Records created and managed by the state, institutions, and other organisations, become 'authoritative' resources in a societal structuration [10, 11] that can see the displacement, marginalisation, disempowerment, and disenfranchisement of individuals, families, communities, and entire peoples [12].

An individual or collective inability to control or access records by or about oneself renders it impossible to actualize or assert inalienable human rights that are acknowledged and guaranteed under the UN Declaration on Human Rights [13] and a host of other instruments. Recordkeeping and archives have always involved a complex interplay between governance, politics, law, ethics, societal expectations and information management. Today, however, rapidly evolving technical capabilities and information system design have significantly and strategically enhanced the effectiveness of official recordkeeping, often circumscribing the ability of the least powerful to act or self-actualize in unprecedented ways. It is therefore necessary to ask whether specific rights relating to records and recordkeeping are needed at this critical juncture in societal and technological development, and if so, whether those rights rise to the level of fundamental human rights.

To unpack this question, we report on a comparative study on the synergies and differences between two research projects that are investigating recordkeeping and archives within different, although sometimes overlapping, domains of egregious human rights violations. The *Rights in Records for Refugees* (R3) initiative is concerned with the roles that recordkeeping and archives play in the lives of refugees and others who have been forcibly displaced at different times and from different places around the globe.[2] The Australian *Lifelong Rights in Childhood Recordkeeping in Out-of-Home Care* project is concerned with how recordkeeping and archives impact Care experienced children and adults, including Australian Indigenous children and Stolen Generations.[3] Both projects employed instrumental and testimonial warrant [9] as the basis for their findings.

The study compared the starting points for each project: in the global, transnational context of Rights in Records for Refugees, and the liberal democratic national context of the Australian research, as well as the specific recordkeeping and archives needs of the two communities linked to their respective contexts and circumstances. It explored

[2] All are referred to for brevity in this paper as 'refugees', but in full recognition of the several status categories into which an individual who has experienced forced displacement or fled due to actual or credible fear of persecution might fall or be placed.

[3] The term Out-of-Home Care encompasses a variety of alternative accommodation arrangements currently including foster care, kinship care, residential and group homes largely run by the private and not-for-profit sector, independent living arrangements, and other forms of placement. Historically the term covers institutional Care including orphanages and children's homes run by states, churches and other charitable bodies. We acknowledge that this term is not the preferred terminology of all persons with lived Care experience. We use the capitalized term Care 'to denote the ironic connotations of manifestly uncaring treatment, without continually enclosing the word in quotation marks' [14].

possible synergies in the projects' animating arguments, operating assumptions, and conceptual frames of reference. Analysis and mappings of the two suites of rights in recordkeeping and archives produced by the projects revealed areas of overlap and significant differences.

Given that both projects take a rights-based approach, this paper begins by posing the question: Why rights? Discussion of their diverse contexts and circumstances, and an overview of each project follow. This leads into an exploration of the comparative study's findings on the projects' synergies and differences. Analysis and mapping of the major outcomes of the projects, the *Refugee Rights Framework* and the Australian *Charter of Lifelong Rights in Childhood Recordkeeping in Out-of-Home Care,* identify how and why they converge or diverge. Based on these findings, we conclude that there is an imperative for further research in additional contexts, with the goal of identifying, supporting and gaining international acknowledgment of an individual and collective Human Right in Recordkeeping and Archives.

2 Why Rights?

A diachronic (through-time) perspective makes clear the ways in which recordkeeping technologies establish and perpetuate bureaucratic structures that and amplify societal inequities [9]. Such structures are predicated on information and recordkeeping systems – from manual mechanisms of recording, surveillance, and data aggregation, to those that are digital, automated, biometric or remotely sensed or operated. Many are first commercially developed and deployed under the guise of helping the vulnerable or marginalised, before being applied to ever-increasing segments of the population [12].

This rapid intensification of surveillance and control has led to considerable reflection in the recordkeeping field. Under scrutiny is the implication or even culpability of recordkeeping professionals and system designers, as well as best professional practices, and socio-technical systems. The aim is to understand how they systematise and reinforce societal inequities and how these might be rectified [2, 9–12]. These realisations have emerged through historical analysis; advocacy and activism from impacted communities calling for change to recordkeeping regimes; and contemporary recordkeeping research in partnership with a wide range of individuals, families, and communities caught up in various bureaucratic systems over generational timescales [15]. It is against this background of critical reflection that the need for rights in recordkeeping and archives has emerged, potentially complemented by ethics of care approaches.

A 'common standard' for the "inherent dignity and of the equal and inalienable rights of all members of the human family" was codified in the United Nations Declaration of Human Rights [13]. Since then, a host of other international and national declarations, covenants, and other instruments have been developed that articulate specific rights for different communities – for example, for Children, Indigenous Peoples, and Refugees [16–18]. In recent years, there have also been many calls for information rights [19–23]. Some of these have been partially actualised through jurisdictional juridical instruments. For example, the 2016 European General Data Protection Regulation (GDPR) provides controls for the management of personal data held by others, but falls short as an instrument for general information rights, including the right to explanations of decision-making [24].

However, for some, rights-based approaches for pursuing individual and collective wellbeing may be problematic, culturally antithetical, or even alienating. Examples include Indigenous peoples that may view the granting of rights as being constitutive of an imposed colonial regime and therefore reject the very idea of western legal 'rights' [25, 26], and LGTB advocates that may reject externally imposed structures of identity [27]. Additionally, jurisdictions may measure equity differently, or privilege collective interests over individual autonomy. Human rights frameworks also have their own limitations and unintended consequences [28–30].

Moreover, critiques of rights-based approaches surface when exploring the philosophical basis of societal structuration. This discourse is often reduced to the juxtaposition of the normative *ethics of justice* that pervades institutions in western liberal democracies with the contextual *ethics of care* [31]. The former assumes that rules are applied fairly and impartially to people with standing and means in an objective manner, the latter that decision-making should nurture and support our relational interdependencies in a subjective way. Both approaches have deontological foundations but may serve different societal needs.

In community archives contexts, the ethics of care has been posited as a more appropriate model to address social justice issues [32]. However, the justice/care binary is arguably a false dichotomy when it is broadly applied beyond communities that have the autonomy to direct or document their own affairs and memories. In governmental and institutional settings, an exclusive ethics of care approach fails to address pressing issues relating to agency and accountability, and cannot enforce better behaviour by bad actors or deal with those who simply do not care [33]. As was noted when the UN Declaration on Human Rights was implemented,

> *"… it is essential, if man is not to be compelled to have recourse, as a last resort […] that human rights should be protected by the rule of law"* [13]

Thus, a rights-based approach to recordkeeping and archives, complemented by ethics of care, is "essential" where there are demonstrably uncompliant power hierarchies that give rise to a legalistic and often adversarial structuration of modern society that disenfranchises individuals and alienates them from their rights. Significantly, investigation of human rights violations usually exposes recordkeeping issues at their core [34–36]. Given that records – and, particularly official, bureaucratic records – both represent and circumscribe people and their lived experience during and beyond their lives, it is vital that individuals whose lives are most affected have mandated avenues to become active participatory agents in recordkeeping and archives [37, 38].

While, not necessarily failsafe or unproblematic, rights-based frameworks are designed to be incorporated in legislation and policy development, and offer an infrastructure to offset imbalances of power. Transnational and national rights frameworks can play a significant role in influencing and rallying discourse and action across political and power spectrums. They provide a language that is familiar to law and policy makers, and which is capable of crossing jurisdictional boundaries by facilitating "a political space" rather than presenting "static standards" [39]. Indeed, the success of diverse social movements has been due to the strong support of rights frameworks as an effective, albeit imperfect, means of translating between community concerns and political discourse (for example, [37, 40–42]).

3 Rights in Refugee Records

3.1 The Context

Forcible displacement is an age-old phenomenon. The intersection of displacement, documentation and recordkeeping, technology, border regimes, and legal actions in the modern world can be traced back at least to the early twentieth century when passports requiring personal photographs were first introduced during World War I by European countries for security reasons. By the end of the twentieth century, machine-readable passports limited the ability of desperate displaced people to flee across borders with passports of countries that no longer existed, or with borrowed or altered travel documents.

With worldwide numbers of refugees at a level today that has not been seen since World War II, governments and aid agencies have increasingly turned to emerging technologies and their corporate developers for assistance. Implementation of biometrics in travel documents and other personal records is widespread. For example, DNA is being used by UNHCR in refugee camps to establish a base identity record for every person, regardless of age, who seeks to enter; smartcard technology is employed to distribute money, food and other aid instead of cash, as well as for geolocating and restricting movement of cardholders. Algorithmic decision-making regarding asylum is being deployed at borders; and widespread and undisclosed digital sharing of personal data about refugees is occurring between nations, aid agencies, and international police and intelligence agencies [43, 44]. In addition, digital records that refugees may have created themselves via social media or mobile phones are subject to scrutiny and may be used against them, while digital copies they might have taken of documents that they could not carry physically or safely, and stored on their phones or Cloud-based accounts, may be dismissed as inadmissible [45].

Nations seeking to limit the numbers of asylum seekers they admit or to prevent forcibly displaced persons from returning have also increased and complexified their requirements regarding the presentation of records to authorities and/or prevented access to records held by official agencies and archives. Records production and family reunification are further complicated by traditional and cultural recordkeeping practices as well as interdependencies between documents. This is particularly the case in countries of displacement and refuge in the Middle East where women and children may be precluded from accessing or contributing to the creation of personal or family records, especially birth and other records used to determine citizenship [46–49], thereby also contributing to a growing population of stateless persons. In other cases, recordkeeping and archival infrastructure in conflict-torn or failed states is unable to function effectively; has been destroyed; or simply lacks the technological capacity to be of much assistance to citizens who cannot physically travel to the location of the record. Refugees' literacy, language and records expertise may present additional barriers.

3.2 The Project

Refugees, therefore, are often impossibly situated in terms of being able to carry or access the documents they need in order to identify and support themselves, reunify their families and communities, rebuild their lives, return to their homelands, or seek reparations. The *Refugee Rights in Records (R3)* Initiative (https://informationasevidence.org/refugee-rights-in-records) was begun in 2016 by researchers at the University of California, Los Angeles and the University of Liverpool to surface and try to ameliorate the imbrication of these concerns in historical and contemporary refugee and migrant crises worldwide.

The project has employed multiple, disparate sources of warrant:

1. Government and aid agency reports and statements as well as articles and refugee stories published in the Arabic and English language press, primarily since 2014:

   Close readings have identified the specific documents that refugees and other displaced peoples are often lacking in displacement spaces and countries that have taken them in; the inter-dependencies between these documents; and, most importantly, how the absence of these documents has affected or continues to affect the immediate lives of refugees and other displaced persons and their well-being even after their eventual settlement or repatriation.
   Stories told by and about refugees that relate to topics such as identification, identity, memory, evidence, recognition, family reunification, the carrying or accessing of personal documents, and other issues that have a bearing on recordkeeping, whether official or personal, have also been extracted and analysed.

2. Forums organized or co-organized in Africa, Europe, Australia and the United States in locations that have ongoing and/or historical experiences with displacement, refugee flows or refugee resettlement:

   Attendees have included current and former refugees, staff from local aid agencies as well as international NGOs and UN bodies, human rights activists, lawyers, literacy experts, archivists and other information professionals from national as well as human rights archives, historians and migration scholars and artists. Presentations and discussions at each forum have been captured and analysed.

3. Prominent internationally-agreed upon instruments and other authoritative statements of human, civil and information/data rights:

   These have been analysed to identify how and to what extent they address or implicate records and recordkeeping issues that affect refugees.

The findings of these analyses, not surprisingly, focus on issues of personal agency; privacy; security; access; and records and records' archives creation, as well as education and expertise regarding records and records production. Based on these findings, a *Refugee Rights in Records Framework* has been developed to address the situations of refugees, their families and descendants over time, generations, geography and jurisdictions, and their enduring needs to be able to actualize their human and civil rights [45].

4 Rights in Childhood Recordkeeping in Out-Of-Home-Care in Australia

4.1 The Context

In Australia, almost 50,000 children, nearly one in every 100, are in Care [50]. Aboriginal and Torres Strait Islander children are ten times more likely than their non-Indigenous counterparts to be in Care relative to their numbers in the general population, forming a new generation of stolen children [51, 52].

Care Leavers, their allies, and community organisations have advocated for transformational changes in the Out-of-Home Care sector, including childhood recordkeeping, for over 30 years. Their rights-based activism, supported by the acknowledgement of the child as a human being with rights and agency in the United Nations Convention on the Rights of the Child [16], highlighted the failure of the child welfare system to care for Australia's most vulnerable children, and the complicity of critical failures in recordkeeping and archival systems (for example [53–55]).

The 1997 Australian Human Rights and Equal Opportunity Commission's inquiry on the Stolen Generation was a response to advocacy by Aboriginal and Torres Strait Islander peoples and their allies. Its *Bringing Them Home* report on the Stolen Generation [53] documents the embedded colonialism and racism in policies on the removal of 50,000 Indigenous children from their families. Activism relating to Indigenous Australian children caught up in the Care system today is part of a larger movement for acknowledgement of the unceded sovereignty of the "First Nations of the Australian Continent" [56].

This advocacy in the Out-of-Home Care sector and the First Nations Sovereignty movement has highlight the weaponisation of data, information and recordkeeping systems, and the limited inclusion of rights in records in most jurisdictions in Australia. They demand recognition of lifelong rights in childhood recordkeeping and Indigenous Data Sovereignty [41].

4.2 The Charter

The transdisciplinary *Rights in Childhood Recordkeeping in Out-of-Home-Care* project began in 2017 and incorporates historical and contemporary analysis of the child welfare system and Out-of-Home Care, exploration of the emergence of child rights; and the role of recordkeeping and systems design in their actualisation. The *Charter of Lifelong Rights in Childhood Recordkeeping in Out-of-Home Care* is a key outcome of the project.

The Charter's development was endorsed by the 2017 National Summit on Setting the Record Straight for the Rights of the Child [57], convened by Monash University in partnership with Care Leavers Australasia Network (CLAN), an advocacy and support group for older Care leavers; the Child Migrants Trust; Connecting Home, a service for the Stolen Generation; the CREATE Foundation which represents children and young people in Care; and researchers from Monash University, Federation University Australia, and the University of Melbourne.

The summit partners and participants imagined transformed, participatory recordkeeping for children and young people in Care that would document their lives; develop a sense of identity and belonging; keep them connected with family and community; and address their questions about who they are, where they come from, and why they are in Care. They envisaged a shift away from organisation-centric records of control and surveillance towards child-centred recordkeeping. Indigenous Australian participants emphasised the role recordkeeping should play in truth telling and connecting to their rich heritage and country.[4]

Testimonial warrant was sourced from inquiry testimony and advocacy of Care leavers and members of the Stolen Generations; the voices of children in Care represented in reports of CREATE, State Child Commissioners and Guardians, Indigenous service and advocacy organisations, and research findings; and works authored or performed by Care leavers and Stole Generation, including histories, memoirs, truth telling and artwork. Sources of instrumental warrant included UN instruments, Australian federal and state legislation, standards, charters, policies and guidelines. CLAN's Charter of Rights in recordkeeping provided another key source of warrant [58].

The rights identified in the Charter support actualisation of the rights in the United Nations Convention on the Rights of the Child [16] and the UN Declaration on the Rights of Indigenous People [17]. The Charter, detailed in Fig. 2 below, references broad framing rights associated with human rights and self-determination:

- Participatory rights in developing frameworks, legislation, policies and processes that impact them an in related decision making;
- Memory rights, including the right to be forgotten;
- Identity rights to cultural, family and self-identity; to know who you are, where you belong and to practice your culture; and
- Accountability rights to hold society, governments and service providers to account for their actions.

Its core principles are child safety and wellbeing,[5] including the cultural safety of Aboriginal and Torres Strait Islander children, and self-determination linked to archival agency and autonomy. In articulating an extensive warrant-based suite of rights, the Charter aims to support community advocacy for adoption of a more extensive suite of rights in records across all jurisdictions through legislative and policy change.

[4] The summary of emerging themes is drawn from two videos produced at the Setting the Record Straight for the Rights of the Child National Summit 8–9 May 2017 (Setting the Record Straight and An Aboriginal Perspective.

[5] Child safety and wellbeing as defined in the Australian Human Rights Commission's Children's Rights report (2017) includes '(i) A right to be heard; (ii) Freedom from violence, abuse and neglect; (iii) The opportunity to thrive; (iv) Engaged citizenship; and (v) Action and accountability for these commitments'.

5 Comparison of the Projects and Outcomes

5.1 Animating Argument

The comparative study found that the animating argument of both projects is that records and recordkeeping have become weaponized in ways that make it impossible for those who are disenfranchised and disempowered to actualize and attain their inalienable human rights. It concluded that the ability of refugees and Care experienced children and adults, including Indigenous peoples, to exercise such rights is severely impacted by governmental and institutional recordkeeping and archiving policies, practices and technologies.

This situation directly affects their families and relatives and their descendants regardless of location, time, or generation. The continuing removal of Indigenous Australian children from their families and communities and the separation and detention of Indigenous children from Latin American countries, together with lack of adequate translation of non-written Indigenous languages at the US border, is also related to the ongoing impact of racism, colonisation, dispossession and denial of sovereignty in contravention of the UN Declaration on the Rights of Indigenous Peoples [53, 59, 60].

5.2 Operating Assumption

The operating assumption of both projects is that fundamental inequities underlie the implementation of recordkeeping regimes and technologies. These inequities are part of a wicked worldwide problem, in which an inability to exercise sovereignty in data, information and records affects us all, but impacts disproportionately those who have been disempowered, marginalised and disenfranchised. Consequently, in addressing these inequities, both projects privilege living experience and testimony over instrumental warrants that derive from prevailing power structures.

Emanating from the Refugee Rights in Records Initiative is a new interdisciplinary approach dubbed Human Security Informatics (HSI). HSI argues that current digital infrastructures and priorities fail to grapple with complexity and contingency or to meet the identity, memory, cultural, information, evidence and accountability needs of those at the margins. This argument aligns with records continuum theory and the rationale of the Lifelong Rights in Childhood Recordkeeping project [61]. Fundamental to the notion of human security as a paradigm is that it places people who are marginalized and vulnerable at the centre, and:

- Seeks to develop practices and technologies that confront and redress systemic inequities, biases and inaccessibilities (e.g., issues of power, privilege, control, granularity, trust, sustainability, security, archival autonomy);
- Works to prevent these vulnerable populations being experimented on by technology developers and to increase the capacity for personal privacy and security;
- Tackles problems of lack of institutional will, coordination and capacity;
- Calls out and addresses ethical and policy concerns; and
- Identifies and addresses the long-term effects and affects of recordkeeping, archives and information practices and infrastructures [62].

5.3 Conceptual Framework

The conceptual frame of reference for both projects is Records continuum theory. It grapples with the complexity, diversity and contingency of recordkeeping and archives. From a continuum perspective, the Archive begins with the motivation and capacity to create content and continues through all possible uses and reformulations of that content over time and space. It offers a transformative definition of records and recordkeeping that encompasses the multiple forms records take. It integrates recordkeeping and archiving processes throughout a record's lifespan, and underpins the concept of archival autonomy – the ability for individuals and communities to participate in societal memory, to find their own voice, and to become participatory agents in recordkeeping and archiving for identity, memory and accountability purposes [38].

5.4 Convergences and Divergences: Mapping the Rights

Bidirectional mapping of the suites of rights produced by the two projects were undertaken to identify convergences and divergences, and as a form of cross-validation to locate any important gaps in either suite of rights.

As detailed in Figs. 1 and 2, the *Refugee Rights Framework* (The Framework) identifies 9 clusters of recordkeeping and archives rights, while the *Australian Charter of Life-long Rights in Childhood Recordkeeping in Out-of-Home Care* (The Charter) identifies 3 broad clusters. Colour coding is used to indicate the degree of alignment.

There are interdependencies between the recordkeeping rights in each Figure. Some of these rights facilitate other specified rights. For example, consider asylum seekers who fled without travel and identity documents, or whose documents were lost or stolen during their journey to safety. The Framework's *Rights Regarding Records Expertise* would ensure a 'records advocate' to locate, acquire, explain, and challenge their records—enabling them to utilize their right to seek asylum. Similarly, the *Disclosure* right in the Charter provides Care leavers with information about where records about them are located and enables them to exercise their access right.

In spite of the divergent starting points and contexts of the two projects, and the circumstances of the two cohorts, the mapping identifies large areas of full and partial overlap. The Framework does not separate out framing rights like the Charter, but the *Cultural, Self-Identity and Family* Rights, and the *Right to Have a Record Created* map against the Charter's framing rights of *Participation, Identity and Memory*, as well as its *Participatory Recordkeeping* Rights.

The significant areas of divergence are directly related to aspects of the different contexts and circumstances of the two cohorts in the research projects. There is a gap in the Charter relating to the Framework's *Rights to have a Record Created*, *Rights Regarding Records Expertise*, and *the right to access one's record according to one's own literacy, modality, writing or signing system.* The Charter's *Participatory Recordkeeping Rights* are not all included in the Refugee Rights Framework but there is partial mapping with the Framework's *Consultation Rights* and *Rights to Know*. There is a gap in the Framework relating to *Privacy Rights.*

A further point of difference between the Framework and the Charter is the inclusion in the latter of both individual and collective recordkeeping and archives rights. The UN

Rights to have a record created:

- The right to be provided with a universally recognized identity document upon request
- The right to have a birth certificate, and to have both parents' names listed on that birth certificate if the father is deceased or otherwise unable to be present at his child's birth, if the mother requests it
- The right for family members and other dependents to a process for issuing a death certificate when there is no body after a certain amount of time

Rights to know:

- Prior to a record about oneself being created, the right to be fully informed about why it is being created, what it will contain, what it may be used for now and in the future, and how it will be secured
- The right to know that a record about oneself exists, where, why, and who can see it and under what circumstances
- The right to know if there is a classified record or data impeding an action one is trying to complete

Rights regarding records expertise:

- The right to be provided, and at no cost, with the index terms or other metadata necessary for locating and retrieving records about oneself
- The right to request and be provided with a records advocate or other expert in locating, introducing and challenging records
- The right to have a records expert testify regarding the historical and bureaucratic circumstances surrounding the creation, management, reproduction, translation and reliability of records about oneself that are introduced in asylum and immigration adjudications, return, restitution and other actions

Cultural, self-identity and family rights in records:

- The right to have one's cultural or community recordkeeping practices recognized in legal, bureaucratic and other processes that depend upon the introduction of records
- The right to have one's self-identity acknowledged in records about oneself, including, but not limited to name, gender, and ethnicity

Right to respond and to annotate (right to rectification):

- The right to respond to and include a permanent annotation on records about oneself

Refusal and deletion rights:

- The right to refuse to participate in the creation of a record about oneself or to resist being recorded if there is a credible fear that doing so will compromise one's human rights or those of others
- The right to request deletion of a record or deletion of data or metadata about oneself from a record if that record, data or metadata would compromise one's human rights

Accessibility, reproduction and dissemination rights:

- The right to access records about oneself, including those that are still otherwise subject to legal or other closure periods
- The right to access one's record according to one's own literacy, modality, writing or signing system
- The right to guaranteed safe, secure, timely and low or no-cost access to relevant records about oneself upon request
- The right to receive copies of records about oneself, and to specify the form and format of those records, or else to be given a clear explanation as to why one may not
- The right to transmit or share records about oneself

Consultation rights:

- The right to be consulted regarding how, where and when records about oneself are preserved or archived, made available for archival research, or disposed of
- The right to be consulted when and why another party, including family members, requests access to a record about oneself

Personal recordkeeping rights:

- The right to a secure personal recordkeeping/archival space
- The right to a safe, secure, and trusted infrastructure for managing, preserving, certifying, and transmitting one's documents

Legend: Full Overlap – Partial Overlap – No Overlap

Fig. 1. Refugee rights in records framework

Declaration on the Rights of Indigenous Peoples 2007 specifies both. This and consideration of the possible relevance of collective rights to communities of non-Indigenous children and young people in Care, and Care leavers resulted in their inclusion in the Charter. As a result of these findings, both projects are now considering how to address

Participatory Rights in Recordkeeping including:
• Setting recordkeeping and archival frameworks (metadata, classification, categorisation, description), making policies (appraisal, access, disclosure, keeping places), and participation in decision making about legal and administrative processes (note: a collective and individual right) • Creating records about you in organisational settings; and your own personal records • Deciding or consenting to what is recorded in organisational and archival systems about you, how your records are used and who has access to your records • Intervening in the record (right of reply/setting the record straight/truth telling) • Deciding how long to keep records, and in what form
Disclosure and Access Rights to:
• Lifelong access to your records (including rights to receive copies, timely and low-cost access; and special accelerated access where circumstances required this) • Have a say in intergenerational access • Know and be informed of where your records are held • Understand the type(s) of records held about you • Be informed of when and why others are given access to your records • Consent to use of your records by others • Know when and why records about you are destroyed
Privacy and Safe Recordkeeping Rights to:
• Individual and collective privacy as understood in your culture and worldviews • Not to have your records used for other than their original agreed purpose without consent • Safe and secure recordkeeping infrastructure, processes and systems • Safe and secure keeping places for archival records

Legend: Full Overlap – Partial Overlap – No Overlap

Fig. 2. Charter of lifelong rights

the gaps. For example in consultation with partners and participants, we will consider inclusion of rights regarding records expertise, and a right to the creation of specified vital records in the Charter.

Given the degree of disempowerment experienced by refugees, and frequent denial of their human rights, the recordkeeping *Consultation Rights*, *Rights to Know*, and *Rights regarding Records Expertise* are essential foundation rights. These meet their immediate needs for sufficient knowledge to satisfy records production requirements; to have input into who can access their personal records, including their DNA, through time and across jurisdictions; to be able to view and obtain copies of those records themselves; and to be able to keep and transport legally admissible copies of their own documentation across jurisdictions.

By contrast, in Australia, following years of activism and advocacy and a series of inquiries, the human rights of children and young people in Care are now referenced in national standards, legislation, charters of rights, and other local instruments in all Australian states and territories, and child-centred approaches and policies are being introduced in all jurisdictions. Indeed, there is some exemplary Care services provision. However, despite these advancements, the transformational changes needed to achieve a child-centred sector are a work-in-progress. Truth telling about the past, transformation of childhood recordkeeping and archives, structural reform, and cultural change driven by ethics of care principles are complementary in this endeavour. While the Australian experience may not be universal, a recordkeeping human right could provide support for Care activists around the globe.

Campaigning continues relating to the recognition of recordkeeping and archives rights, with a focus on participatory rights rather than consultation rights and rights to know. Recordkeeping rights for the Stolen Generation and Indigenous Australian children caught up in the Indigenous child welfare sector today are part of a larger movement for recognition of data sovereignty [63], broadly defined and thus paralleling the expansive records continuum definition of records.

6 Conclusion: Towards Human Rights in Records, Recordkeeping and Archives

In conclusion, we return to the animating argument of both projects – that records and recordkeeping have become weaponized in ways that make it impossible for those who are disenfranchised and disempowered to actualize and attain their inalienable human rights and often, also, other civil rights and privileges. This comparative research study reinforces the need to address fundamental inequities underlying the implementation of recordkeeping regimes and technologies. Comparison of the two projects highlights the critical need for recordkeeping rights-based approaches to address issues of recordkeeping and archival agency and accountability, as well as the dehumanising effects and affects of not being able to exercise one's human rights, such as those relating to self-determination, identity, culture, memory and accountability.

We believe this offers a compelling argument for the need to enforce better recordkeeping and archival behaviour and embed an ethics of care in bureaucratic cultures and legal structures that can behave in uncaring and dehumanising ways; and to prioritize recordkeeping and archival research relating to these situations that affect so many millions of people worldwide. The wicked nature of the challenge calls for international collaboration across multiple and diverse sectors and the use of multiple methods to research rights-based instruments, laws and policies. Such ethical and human-centred initiatives will involve the coming together of professional and lived expertise and experience. In relation to our own projects, we see the need for further research in relation to an, as yet, unexplored overlapping area, the deliberate separation of refugee and migrant children from their families at borders which often leads to them being caught up in the Out-of-Home Care sector.

Finally, the research conducted in the two projects, complemented by the strong synergies revealed by the comparative research study, provide strong indicators of the imperative for an individual and collective warrant-based Right in Records, Recordkeeping and Archives as a Human Right that would address archival autonomy, participatory recordkeeping; disclosure, access and expertise; and privacy and safe recordkeeping places for communities adversely impacted by recordkeeping and archival governance frameworks, laws, policies, technologies and systems.

Acknowledgements. The *Rights in Records by Design Project* was funded through an Australian Research Council (ARC) Discovery Grant DP170100198.

We would like to acknowledge all of the participants in our research, and the emotional, intellectual, professional, and artistic generosity of many organisations and individuals around the globe who shared their time and knowledge, on or off the record, who have made this work possible.

References

1. Nissenbaum, H.: Respecting context to protect privacy: why meaning matters. Sci. Eng. Ethics. **24**, 831–852 (2018)
2. Boyd, D., Crawford, K.: Critical questions for big data: provocations for a cultural, technological, and scholarly phenomenon. Inf. Commun. Soc. **15**, 662–679 (2012)
3. Noble, S.U.: Algorithms of Oppression: How Search Engines Reinforce Racism. NYU Press, New York (2018)
4. Hagendorff, T.: The ethics of ai ethics: an evaluation of guidelines. Mind. Mach. **30**(1), 99–120 (2020). https://doi.org/10.1007/s11023-020-09517-8
5. McKemmish, S., Upward, F., Reed, B.: Records continuum model. In: Bates, M., Maack, M. (eds.) Encyclopedia of Library and Information Sciences, pp. 4447–4459. CRC Press, Boca Raton (2010)
6. Yeo, G.: Records, Information and Data: Exploring the Role of Record-Keeping in an Information Culture. Facet Publishing, London (2018)
7. Evans, J., McKemmish, S., Rolan, G.: Critical approaches to archiving and recordkeeping in the continuum. J. Crit. Libr. Inf. Stud. **1** (2017).
8. McKemmish, S.: Evidence of me. Arch. Manuscr. **24**, 28–45 (1996)
9. Ketelaar, E.: Recordkeeping and social power. In: McKemmish, S., Piggott, M., Reed, B., Upward, F. (eds.) Archives: Recordkeeping in Society, pp. 277–298. Centre for Information Studies, Charles Sturt University, Wagga Wagga (2005)
10. Giddens, A.: The constitution of Society: Outline of the Theory of Structuration. Polity Press, Cambridge (1984)
11. Upward, F.: Structuring the records continuum part two: structuration theory and recordkeeping. Arch. Manuscr. **25**, 10–35 (1997)
12. Gilliland, A., McKemmish, S., Rolan, G., Reed, B.: Digital equity for marginalised and displaced peoples. In: Proceedings of the Association for Information Science and Technology, pp. 572–574. Wiley Online Library, Melbourne (2019)
13. United Nations: Universal Declaration of Human Rights. New York, USA (1948)
14. Wilson, J.Z., Golding, F.: Latent scrutiny: personal archives as perpetual mementos of the official gaze. Arch. Sci. **16**(1), 93–109 (2015). https://doi.org/10.1007/s10502-015-9255-3
15. Gilliland, A., McKemmish, S., Lau, A. (eds.): Research in the Archival Multiverse. Monash University Publishing, Clayton (2017)
16. United Nations: Convention on the Rights of the Child (1990)
17. United Nations Commission on Human Rights: United Nations declaration on the rights of indigenous peoples. United Nations, New York (2007)
18. United Nations: Conventions and protocol relating to the status of refugees. Geneva, Switzerland (1951)
19. Kelmor, K.M.: Legal Formulations of a Human Right to Information: Defining a Global Consensus (2016)
20. O'Neal, J.R.: "The Right to Know": Decolonizing Native American Archives (2015)
21. McDonagh, M.: The right to information in international human rights law. Hum. Rights Law Rev. **13**, 25–55 (2013)
22. Britz, J., Lor, P.: The right to be information literate: the core foundation of the knowledge society. Innov. J. Appropr. Librariansh. Inf. Work South. Afr. 8–24 (2010)
23. Rodotà, S.: Data protection as a fundamental right. In: Gutwirth, S., Poullet, Y., De Hert, P., de Terwangne, C., Nouwt, S. (eds.) Reinventing Data Protection?, pp. 77–82. Springer, Dordrecht (2009). https://doi.org/10.1007/978-1-4020-9498-9_3
24. Wachter, S., Mittelstadt, B., Floridi, L.: Why a right to explanation of automated decision-making does not exist in the general data protection regulation. Int. Data Priv. Law. **7**, 76–99 (2017)

25. Indigenous peoples as subjects of international law/edited by Irene Watson. Routledge, Abingdon, Oxon; New York (2018)
26. Carbone, K., Gilliland, A.J., Montenegro, M.: Rights in and to records and recordkeeping: fighting bureaucratic violence through a human rights-centered approach to the creation, management and dissemination of documentation. Educ. Inf. 1–24 (2020)
27. Mertus, J.: The rejection of human rights framings: the case of LGBT advocacy in the US. Hum. Rights Q. 1036–1064 (2007)
28. Alen, A., Vande Lanotte, J., Verhellen, E., Ang, F., Berghmans, E., Verheyde, M.: A commentary on the United Nations Convention on the Rights of the Child, article 7: the right to birth registration, name and nationality, and the right to know and be cared for by parents/Ineta Ziemele. Martinus Nijhoff, Leiden, The Netherlands (2007)
29. Parton, N.: Child protection and safeguarding in England: changing and competing conceptions of risk and their implications for social work. Br. J. Soc. Work. **41**, 854–875 (2011)
30. Todres, J.: Emerging limitations on the rights of the child: the UN convention on the rights of the child and its early case law. Colum. Hum. Rts. Rev. **30**, 159 (1998)
31. Botes, A.: A comparison between the ethics of justice and the ethics of care. J. Adv. Nurs. **32**, 1071–1075 (2000)
32. Caswell, M., Cifor, M.: From human rights to feminist ethics: radical empathy in the archives. Archivaria **81**, 23–43 (2016)
33. Held, V.: Care and justice, still. In: Engster, D., Hamington, M. (eds.) Care Ethics and Political Theory, pp. 19–36. Oxford University Press, Oxford (2015)
34. Harris, V.: Seeing (in) blindness: South Africa, archives and passion for justice. Archifacts 1–13 (2001)
35. Ketelaar, E.: Truths, memories and histories in the archives of the international criminal tribunal for the former Yugoslavia. In: van der Wilt, H., Vervliet, J., Sluiter, G., Houwink ten Cate, J. (eds.) The Genocide Convention. The Legacy of 60 Years, pp. 201–221. Martinus Nijhoff Publishers, Leiden; Boston (2012)
36. Swain, S.: History of Australian inquiries reviewing institutions providing care for children (2014)
37. Gooda, M.: The practical power of human rights: how international human rights standards can inform archival and recordkeeping practices. Arch. Sci. **12**, 141–150 (2012)
38. Evans, J., McKemmish, S., Daniels, E., McCarthy, G.: Self-determination and archival autonomy: advocating activism. Arch. Sci. **15**(4), 337–368 (2015). https://doi.org/10.1007/s10502-015-9244-6
39. Libesman, T.: Indigenous child welfare post bringing them home: from aspirations for self-determination to neoliberal assimilation. Aust. Indig. Law Rev. **19**, 46–61 (2016). https://doi.org/10.2307/26423302
40. Bamblett, M., Lewis, P.: Detoxifying the child and family welfare system for Australian Indigenous peoples: self-determination, rights and culture as the critical tools. First Peoples Child Fam. Rev. J. Innov. Best Pract. Aborig. Child Welf. Adm. Res. Policy Pract. **3**, 43–56 (2007). https://doi.org/10.7202/1069396ar
41. Golding, F.: "Problems with records and recordkeeping practices are not confined to the past": a challenge from the royal commission into institutional responses to child sexual abuse. Arch. Sci. **20**(1), 1–19 (2019). https://doi.org/10.1007/s10502-019-09304-0
42. O'Rourke, M.: Ireland's experience of memorialisation in the context of serious violations of human rights and humanitarian law: a submission to the United Nations Special Rapporteur on the promotion of truth, justice, reparations and guarantees of non-recurrence. Justice for Magdalenes Research, Galway (2020)
43. Latonero, M., Kift, P.: On digital passages and borders: refugees and the new infrastructure for movement and control. Soc. Media Soc. **4**, 2056305118764432 (2018)

44. Cummings, E.: Digital Technology Development in Support of Refugee Needs: A Literature Review. Center for Information as Evidence (2018)
45. Gilliland, A.J., Carbone, K.: An analysis of warrant for rights in records for refugees. Int. J. Hum. Rights. **4**, 483–508 (2020)
46. Alalawi, S.: 'Struggles with Documents': A Review of Documentation Issues Affecting Refugees and Displaced Persons from Arab Middle East Countries. Center for Information as Evidence (2018)
47. Alalawi, S.: A Review of the Role of Records and Recordkeeping in Dilemmas Faced by Refugee Wives in Polygamous Marriages. Center for Information as Evidence (2019)
48. Alalawi, S.: The Impact of Child Marriage on Female Refugees and Other Forcibly Displaced Persons and the Roles Played by Records and Recordkeeping: A Review. Center for Information as Evidence (2019)
49. Jiménez, K.: Documentation and Recordkeeping Issues Affecting Refugees in Turkey: A Review. Center for Information as Evidence (2018)
50. Australian Institute Of Health And Welfare: Child protection Australia 2017–18, Children in out-of-home care. Australian Institute Of Health And Welfare, Canberra, Australia (2019)
51. Tilbury, C.: The over-representation of indigenous children in the Australian child welfare system. Int. J. Soc. Welf. **18**, 57–64 (2009)
52. Wahlquist, C.: Indigenous children in care doubled since stolen generations apology | Australia news | The Guardian. https://www.theguardian.com/australia-news/2018/jan/25/indigenous-children-in-care-doubled-since-stolen-generations-apology. Accessed 19 Sept 2018
53. Australian Human Rights Commission: Bringing them home: National Inquiry into the Seperation of Aboriginal and Torres Strait Islander Children from their Families. Australian Human Rights Commission, Sydney, Australia (1997)
54. Munro, E.: The Munro review of child protection: Final report, a child-centred system. Department for Education, London, UK (2011)
55. Royal Commission: Final report. Royal Commission into Institutional Responses to Child Sexual Abuse. Commonwealth of Australia, Canberra, Australia (2017)
56. National Constitutional Convention: Uluru Statement from the Heart (2017)
57. Setting the Record Straight: Setting the Record Straight: For the Rights of the Child. https://rights-records.it.monash.edu/. Accessed 19 Feb 2019
58. Care Leavers Australasia Network: A Charter of Rights to Childhood Records (2020)
59. Moreton-Robinson, A.: The white possessive: property, power, and indigenous sovereignty. University of Minnesota Press, Minneapolis (2015)
60. McKemmish, S., et al.: Decolonizing recordkeeping and archival praxis in childhood out-of-home care and indigenous archival collections. Arch. Sci. **20**(1), 21–49 (2019). https://doi.org/10.1007/s10502-019-09321-z
61. Rolan, G., Phan, H.D., Evans, J.: Recordkeeping and relationships: designing for lifelong information rights. Presented at the Proceedings of the 2020 ACM Designing Interactive Systems Conference, Eindhoven, Netherlands (2020). https://doi.org/10.1145/3357236.3395519
62. Gilliland, A.J., Lowry, J.: Human security informatics, global grand challenges and digital curation, no. 14 (2019). https://doi.org/10.2218/ijdc.v14i1.636.
63. Maiam Nayri Wingara: Maiam Nayri Wingara. https://www.maiamnayriwingara.org. Accessed 20 Oct 2020

Is This Too Personal? An Autoethnographic Approach to Researching Intimate Archives Online

Jennifer Douglas(✉)

School of Information, University of British Columbia, Vancouver, BC V6T 1Z1, Canada
jen.douglas@ubc.ca

Abstract. This short research paper uses a personal research story to explore how autoethnographic methods can provide a guide to navigating the complicated ethics of researching online. Drawing on the author's personal experience of bereavement and her subsequent research in online grief communities, the paper demonstrates how autoethnography can provide a lens for identifying points of tension, conflict and vulnerability in online research. The paper concludes by advocating for compassionate research [1] and shows how autoethnographic inquiry supports its development.

Keywords: Autoethnography · Personal research · Research ethics · Intimate online archives

1 Introduction: A Research Story

This paper tells a research story in order to explore the benefits of an autoethnographic approach to researching intimate records online. As researchers in several disciplines have shown, scholarly work that centres the personal experiences and feelings of the researcher often faces significant resistance. Denise Turner [2] writes about "research you cannot talk about:" the kind of research one's colleagues do not want to discuss in the hallways or that causes one to be avoided at conferences. Arthur Bochner [3], a communications scholar whose research agenda shifted drastically following the death of his father, discusses how the apparatuses of theory and academia provide researchers with a veneer of objectivity, of distance, of authority, and therefore of *safety,* that is quickly threatened once we begin to let the personal into our research. Sociologist Kari Lerum [4] uses the term "academic armor" to refer to the "physical and psychological means through which professional academics protect their expert positions or jurisdictions" (p. 470) or, to put it another way, the way we use our position and privilege to hold ourselves separate and protected from those whom we study, from whose personal stories and traumas we might profit professionally.

My own research has been the kind of research people do not want to talk about [5]. It has involved a breaking down of my academic armour and has caused me to feel vulnerable and at times exposed. It has frequently required me to strip the veneer of

K. Toeppe et al. (Eds.): iConference 2021, LNCS 12646, pp. 301–307, 2021.
https://doi.org/10.1007/978-3-030-71305-8_24

objectivity and distance that Bochner refers to, and I have experienced both anxiety and opportunity as a result. As an archival scholar, my personal experience has profoundly affected my professional life, my research agenda, my ideas about archives, what they are for and what they can do, and it has become a defining feature of my methodological approach to the questions I ask about recordkeeping and griefwork, the focus of my most recent research project. In this paper, I want to explore how personal experience – and the intertwining of my personal life with my research life – has impacted my understanding of the ethics of research, and my efforts to find a methodological approach that aligns with that understanding. By telling the story of how a research project evolved from personal experience, this paper advocates for a personal approach to data collection, analysis and understanding.

2 A Personal Story

If I want to start at the beginning, I have to tell my own story. I am never sure I want to. I worry it will seem self-indulgent, or that I will appear to be exploiting personal tragedy for professional gain, or that others will find me not serious, not rigorous, not sufficiently *scholarly*. But try as I might to extricate my story from my research, I cannot. This is part of the dilemma I am trying to communicate through the telling of this research story. Talking about my research on recordkeeping and griefwork does not feel right without telling my story, but telling my story never feels right in *this* kind of context: an academic context. As you read this, I would like to invite to you keep in mind ideas about academic armour, about research that cannot be talked about and about the safety net of keeping the personal out of the professional.

My story starts in 2012 when I discovered the types of online grief communities that sparked my ongoing interest in the relationships between recordkeeping and grief work. I discovered these communities not in my capacity as an archival scholar, but as a grieving mother, whose baby had recently been stillborn. The grief experienced by the parents of babies who were stillborn is often described as disenfranchised grief [6] because it is not well understood by those who have not experienced it; bereaved parents report feeling that their grief is minimized and that they themselves are often avoided and silenced, even by their friends and family [5]. Feelings of disenfranchisement may push the bereaved to seek community and support outside their usual networks; in so doing, they may turn to online communities to find people with the same or similar experiences and a corresponding need to connect [7]. This was the case for me. In deep grief, I sought out a community of parents online who understood the pain and bewilderment I was feeling and supported me through it as I in turn learned to support them.

2.1 Transitioning from Research Questions to Research Action

After some time, my personal participation in online communities sparked a professional interest, and I grew increasingly interested in the types of memory work being performed by bereaved parents online. Because of how I started down this research path, however, I found moving on from this point to be extremely difficult. Once I needed to shift from research questions to research actions I had to confront serious issues that directly related to my personal involvement in the communities I was studying and my connections to them. I found myself asking: How do I talk about this community ethically? How do I approach this topic with the sensitivity and care it deserves and requires? How can I think about the things I want to think about without exposing this community to unwanted scrutiny? Participants in these communities would sometimes refer to the non-bereaved as 'muggles;' they talked about being misunderstood by the 'muggles,' and about being judged and isolated. How would my research feed into this type of judgment and misunderstanding? Being a participant *in* the community, and not just an observer of it, has clearly affected the way I understand the role of the researcher and the impact of research *on* the community. It has strongly affected my understanding of the complicated ethics of undertaking such research.

When I began to try to plan a study of the online communities I was interested in, I consulted my university's research ethics website. I discovered that, under Canada's Tri-Council Policy on behavioural research ethics, I was not required to seek ethical approval to study the community websites. Because the websites, and all the information shared on them, were publicly available, i.e., they were not password protected and were open to all visitors, no "reasonable expectation of privacy" existed.

3 Rethinking Research Ethics Online

Recently – and especially in relation to social media as it is used by activists and historically marginalized communities [8] – the need for a more nuanced understanding of online space as both public *and* private has been acknowledged. Malin Svenginsson Elm [9] describes how "even if a certain internet medium admittedly is public" (p. 77) it might not *feel* public to its users. Dorothy Kim [10] draws an analogy between Twitter and Times Square, suggesting that even if we understand Twitter to be a kind of Times Square, it should still be possible for people to have a private conversation there. Private experiences occur in public spaces, and social media and other online researchers are increasingly realizing that privacy needs to be understood as something that is contextual and negotiated. Danah boyd [11] argues, for example that "privacy isn't a state of a particular set of data. It's a practice and a process" that must be "actively negotiated" with "agency" in mind (np).

An example drawn from the online grief communities I have focused on can illustrate this point: it is common practice for bereaved parents of stillborn babies to post photographs of their babies to social media, including their own personal blogs and platforms like Instagram and Facebook. How private or public these photographs are understood to be, though, depends not on the nature of the record (i.e. that it is a photograph of a stillborn baby) but how it is used and situated by the parents. For example, the public/private nature of a photograph of a mother holding her stillborn child used on a website to raise awareness of stillbirth might be understood very differently than a photograph posted by a bereaved parent on their personal blog on the anniversary of their baby's passing.

In an article that discusses notions of publicity and privacy in a community of 'mommy bloggers,' Aimee Morrison [12] introduces the concept of 'intimate publics,' which she adapts from the work of Lauren Berlant [13]. Morrison outlines the steps that particular communities take to create spaces that are both public *and* intimate. For example, mommy blogging relies on the openness of the web to build community, but it seeks to build a particular kind of community, where emotional reciprocity between writers and readers, and between readers and other readers, is key. To create this kind of space, Morrison explains, mommy bloggers rely on network versus broadcast modes of transmission to create spaces that have "a sense of boundedness and small scale relative to the Internet's generally wide open…culture" (p. 44). Mommy bloggers use deliberate tactics to attract, cultivate and protect a particular type of community, managing blog settings (in particular searchability) and using pseudonyms strategically not to hide online but conversely to permit themselves to be more open. The result of this strategic approach to network building is the creation of "a space apart" (p. 49), an intimate public where community members are both *in* public and *not*.

A method Morrison advocates for working within these communities is known as feminist cyberethnography, which Usha Zacharias and Jane Arthurs [14] describe as "a methodology in which the researcher participate[s] fully in the online environment" (p. 204). In a feminist cyberethnology, the researcher "becomes both the object and subject of her own study through self-reflective engagement with the context and inhabitants" of online communities [15] (p. 211). She is not a participant-observer, but a participant-experiencer. Mary Walstrom [16], who studies online support groups, suggests that such an engaged participatory approach is the most ethical approach to studying vulnerable online communities and is a primary means of earning trust and credibility from the community. Morrison's work, and the work on feminist cyberethnography that it led me to, offered me a way back into my research, and showed me a path I thought I could follow, carefully, and treading lightly, to do the work I wanted to do in the most respectful way possible. In keeping with the feminist cyberethnographic approach to studying online communities, I began to conduct an autoethnographic study of my own personal experiences and archiving practices within the communities I joined and/or helped to create. This study helped me determine how to move outward; it helped me to think not only about how and why recordkeeping and grief work are interrelated, but also where potential ethical tension might exist in the research process.

4 The Potential of Autoethnographic Approaches

Autoethnography can be defined as "ethnographic research, writing, story, and method that connect the autobiographical and personal to the cultural, social, and political" [18 np]. Wall [18] defines autoethnography as "a qualitative research method that allows the author to write in a highly personalized style, drawing on his or her experience to extend understanding about a social phenomenon." Autoethnography, she further explains, is "linked to growing debate about reflexivity and voice in social research" and allows a researcher to "acknowledge the inextricable link between the personal and the cultural and to make room for nontraditional forms of inquiry and expression" (abstract).

Taking an autoethnographic approach to understanding the ethics of researching online grief communities has involved asking myself a series of questions, including: 1) How is recordkeeping part of my grief work? 2) What kinds of records have I made and kept? 3) How are these records embedded in online grief communities? 4) What are the implications of that embeddedness? As I observe and reflect on my own recordkeeping habits and participation in online grief communities, I am attuned to times where I feel discomfort, where points of tension arise, and to where and how I draw the boundaries of public and private as well as when and why I feel vulnerable.

To illustrate this process, I can describe how I have shared parts of my personal blog in various research presentations. My blog is an integral part of my personal archive. It is the primary means through which I worked through the grief of losing my daughter. I viewed it from the start as a record, not only of my grief, but also of her short life, and of everything I had wanted for her had she lived. Eventually, I also viewed is as a record of the friendship and support that grew in the community that formed around it. In a presentation I gave at the Personal Digital Archives (PDA) conference in 2017, I included a slide with a screenshot of my first blog post. The blog is written under a pseudonym and at PDA, I blurred out the pseudonym. I explained then that it was not necessarily to protect myself from scrutiny, but because from *my* blog, a reader may be led to the blogs of others like me, none of whom have consented to have their words re-broadcast out of context. Several months later, at another conference, I included the same slide but this time also blurred out most of the text. I had realized that someone might google a string of words from the screenshot and find their way to the live blog in a second. The question of visibility and how much to reveal was a point of tension, a point where I identified vulnerability, and where the way that I defined vulnerability changed over time.

I had another slide at PDA that included a photograph of my daughter taken post-mortem. It is common in the online communities I have participated in to share these types of photos; it is *not* common, however, to show postmortem photographs of our loved ones at conferences, with colleagues and strangers. I agonized over whether I wanted to share the photograph. I thought: if I want to research these communities, if I want to look at, study, share and talk about the record of someone else's grief work, should I not be prepared to share mine, too? At the last minute, I cut most of the photograph out of the slide. I told the audience something I thought was true then, and it was, at least partly. I said: "I thought I could show it, but in the end, it was too much: not for me, but for *you*." I talked about a point that Denise Turner and Rebecca Webb [19] make about the ethics of bereavement research: that sometimes what does not feel vulnerable for the *subject* of research, can make the researcher or the consumer of that research feel vulnerable themselves. At PDA, I told myself I was protecting the audience from feeling scared or traumatized by a photograph of a dead baby. However, after the presentation, in my field notes, I wrote: "Who is vulnerable? The audience or me? I go back and forth on this. Vulnerability is in *how the records are looked at*, not that they're looked at. My deepest concern is will people think my baby is creepy? I can't control how they see her. But I also felt proud, a little defiant and resistant showing those screenshots. I need to think more about how vulnerability *works*."

5 Conclusion: Compassionate Research

My autoethnographic study began with a concern to "do right by" [20] those whom I wished to study. Uncomfortable with the idea that the communities I knew and experienced had "no reasonable expectation of privacy," I was more inclined to agree with Elm

[9], that "even if users are aware of [the possibility of] being observed by others, they do not consider the possibility that their actions and interactions may be documented and analyzed in detail at a later occasion by a researcher." I understood – viscerally – that "if the content was created for one certain audience and context, the transmission of this content to other contexts may upset the creator." Amani Hamdan [21] argues that autoethnographers possess a kind of "privileged knowledge" that permits them to "provide an insider account and analysis of weaved power structures that an outsider cannot dismantle" (p. 587). As Douglas and Mills [22] point out, "the ability of the ethnographic researcher to convey a particularly personal knowledge is key to the method's success;" drawing on Wall's [17, 23, 24] work, they argue that autoethnographic methods are "especially appropriate" (p. 263) when a researcher needs to access or acquire emotional and intimate knowledge.

In the end, I decided not to pursue my study of online communities. I simply could not find a way to justify my research interests that would also honour parents' expectations of their participation in the community [25]. Since then, I have endeavoured to adopt a "compassionate research" [26] approach; Ellis's [1] notion of compassionate research is grounded in a "relational ethics of care" and aims "to honour, care for and support" research participants (p. 57). When I apply this to my understanding of the vulnerabilities of online communities, influenced by my autoethnographic approach, I realize there is no way to adequately honour, care for and support the range of voices and experiences shared in the online communities I know. Influenced, too, by Tillman-Healy's [26] "friendship as method," I require myself to think of research participants (or would-be research participants in this case) as an audience and to "struggle to write both honestly and empathically for them" (p. 735). There is, I have decided, no way to do so with integrity, without ignoring the intense discomfort I have felt when I have subjected my own participation in online grief communities to scrutiny.

Perhaps this research story appears to end in failure. I spent years trying to write about the novel and inspiring memory work that transpires in the communities I know and ultimately, I have little to show, professionally, for that time. I do not think of it as failure. Through this research I learned to drop my "academic armor," to research compassionately, and to honour – and maybe also preserve – a much-loved and much-needed intimate public.

References

1. Ellis, C.: Manifesting compassionate autoethnographic research. Int. J. Qual. Res. **10**(1), 54–61 (2017)
2. Turner, D.: 'Research you cannot talk about': A personal account of researching sudden, unexpected child death. Illness Crisis Loss **24**(2), 73–87 (2015)
3. Bochner, A.P.: It's about time: narrative and the divided self. Qual. Inquiry **3**(4), 418–438 (1997)
4. Lerum, K.: Subjects of desire: academic armour, intimate ethnography, and the production of critical knowledge. Qual. Inq. **7**(4), 466–483 (2001)
5. Author, N.: [citation removed for purposes of anonymization]
6. Lang, A., Fleiszer, A.R., Duhamel, F., Sword, W., Gilbert, K.R., Corsini-Munt, S.: Perinatal loss and parental grief: the challenge of ambiguity and disenfranchised grief. Omega **63**(2), 183–196 (2011)

7. Swartwood, R.M., McCarthy Veach, P., Kuhne, J., Lee, H.K., Ji, K.: Surviving grief: an analysis of the exchange of hope in online grief communities. Omega **63**(2), 161–181 (2011)
8. Velte, A.: Ethical challenges and current practices in activist social media archives. Am. Archivist **81**(1), 112–134 (2018)
9. Elm, M.S.: How do various notions of privacy influence decisions in qualitative internet research? In: Markham, A., Baym, N.K. (eds.) Internet Inquiry: Conversations about Method, pp. 69–93. Sage Publications (2009)
10. Kim, D.: Social media and academic surveillance: the ethics of digital bodies. Model View Culture. https://modelviewculture.com/pieces/social-media-and-academic-surveillance-the-ethics-of-digital-bodies. Accessed 10 Oct 2020
11. Boyd, D.: What is privacy? It's also not so simple. The Medium. Medium https://medium.com/message/what-is-privacy-5ed72c66aa86. Accessed 10 Oct 2020
12. Morrison, A.: 'Suffused by feeling and affect': the intimate public of personal mommy blogging. Biography **34**(1), 37–55 (2011)
13. Berlant, L.: The Queen of America Goes to Washington City: Essays on Sex and Citizenship. Durham, NC, Duke UP (1993)
14. Zacharias, U., Arthurs, J.: Introduction: feminist ethnographers in digital ecologies. Fem. Stud. **7**(2), 203–204 (2007)
15. Gajjala, R., Rybas, N., Altman, M.: Epistemologies of doing: E-merging selves on line. Feminist Media Stud. **7**(2), 208–213 (2007)
16. Waltsrom, M.K.: Ethics and engagement in communication scholarship: analyzing public, oninc support groups as researcher/participant-experiencer. In: Buchanan, E.A. (ed) Readings in Virtual Ethics: Issues and Controversies, pp. 174–202. Information Science Publishing, London (2004)
17. Ellis, C.: Autoethnography. In: Given, L.M. (ed) The Sage Encyclopedia of Qualitative Research Methods. Sage, Thousand Oaks (2008)
18. Wall, S.: An autoethnography on learning about autoethnography. Int. J. Qual. Methods **5**(2), 146–160 (2006)
19. Turner, D., Webb, R.: ethics and/or Ethics in qualitative social research: negotiating a path around and between the two. Ethics Soc. Welfare **8**(4), 383–396 (2012)
20. Hobbs, C.: Personal ethics: being an archivist of writers. In: Morra, L.M., Schagerl, J. (eds.) Basements and Attics, Closets and Cyberspace: Exploration in Canadian Women's Archives, 181–192. Waterloo, ON, Wilfrid Laurier Press (2012).
21. Hamdan, A.: Autoethnography as a genre of qualitative research: A journey inside out. Int. J. Qual. Methods **11**(5), 585–606 (2012)
22. Douglas, J., Mills, A.: From the sidelines to the center: reconsidering the potential of the personal in archives. Arch. Sci. **18**(3), 257–277 (2018). https://doi.org/10.1007/s10502-018-9295-6
23. Wall, S.: Easier said than done: Writing an autoethnography. International Journal of Qualitative Methods **7**(1), 38–53 (2008)
24. Wall, S.: Toward a moderate autoethnography. Int. J. Qual. Methods **15**(1), 1–9 (2016)
25. Franzke, A.S., Bechmann, A., Zimmer, M., Ess, C., and Association of Internet Researchers: Internet Research Ethical Guidelines 3.0 (2020). https://aoir.org/reports/ethics3.pdf. Accessed 10 Oct 2020
26. Tillman-Healy, L.M.: Friendship as method. Qual. Inq. **9**(5), 729–749 (2003)

Assessing Legacy Collections for Scientific Data Rescue

Hilary Szu Yin Shiue, Cooper T. Clarke, Miranda Shaw, Kelly M. Hoffman, and Katrina Fenlon(✉)

College of Information Studies, University of Maryland, College Park, USA
kfenlon@umd.edu

Abstract. Widespread investments in facilitating reuse and reproducibility of scientific research have spurred an increasing recognition of the potential value of data biding in unpublished records and legacy research materials, such as scientists' papers, historical publications, and working files. Recovering usable scientific data from legacy collections constitutes one kind of *data rescue*: the usually urgent application of selected data curation processes to data at imminent risk of loss. Given growing interest in data-intensive research, and a concomitant movement toward computationally amenable collections in memory institutions, scientific data repositories and collecting institutions would benefit from systematic approaches to assessing and processing legacy collections with the specific goal of retrieving reusable or historically valuable scientific data. This paper suggests a preliminary framework for assessing legacy collections of research materials for the purpose of data rescue. Developed through three case studies of agricultural research collections held by the United States Department of Agriculture's National Agricultural Library, this framework aims to guide data rescue initiatives in agricultural research centers, and to provide conceptual framing for emerging conversations around data rescue across disciplines.

Keywords: Data curation · Data rescue · Data reuse

1 Introduction

Publicly accessible data have proliferated in the past decade due to distributed investments in the reproducibility and reuse of scientific data—through research repositories, deposit requirements, data curation services, data journals, data citation systems, etc. These investments in research and data curation infrastructure have largely focused on active or recent data collections. However, individual researchers amass significant collections of data and research materials over the course of their careers. For currently active scientists with long careers behind them, most of these materials have not been subject to curation. In parallel, research centers and collecting institutions have accumulated massive, unprocessed collections of data, publications, documentation, and other records of prior work. These legacy collections contain data and documentation of indefinite—but potentially significant—value: both as evidence of historical research and as potentially reusable data to support new scientific inquiries.

K. Toeppe et al. (Eds.): iConference 2021, LNCS 12646, pp. 308–318, 2021.
https://doi.org/10.1007/978-3-030-71305-8_25

Stakeholders across the research landscape share an increasing recognition of the potential value of data that bide in unpublished records and collections of legacy research materials, such as scientists' papers, historical publications, and working files [1, 2]. Beyond constituting the scientific record, historical data may support new longitudinal analysis, meta-analyses, computational modeling, and cross-disciplinary research [3–6]. While data curation research and practice have advanced considerably in the past decade, there is a need for systematic approaches to retrospective data curation in the context of recovering reusable data from historical collections.

Like many collecting institutions, data repositories, and scientific research centers, the United States Department of Agriculture's National Agricultural Library (NAL) has accumulated a mass of legacy collections in various stages of processing, acquired through donations from scientists or in the wake of lab closures. This research is motivated by the accumulation of agricultural legacy collections, the increase in data-intensive research across all disciplines, and the attendant movement toward computationally amenable collections in memory institutions [7].

Data rescue is a term for the urgent application of data curation processes to salvage usable data from at-risk collections of materials. In information sciences, this term most commonly denotes crowdsourced efforts to capture and preserve data disappearing from the Web; the term has gained uptake in agriculture, climatology, and other scientific domains as a handle for a range of strategies for efficiently processing data from at-risk collections in various contexts.

This paper offers a preliminary framework for assessing legacy collections of research materials for the purpose of data rescue in agriculture. Developed through three case studies of research collections held by NAL, this framework aims to guide data rescue initiatives at NAL and other agricultural research centers. This framework synthesizes existing guidance for collection processing, with the goal of distilling key properties for data rescue in agriculture. In addition, the framework aims to elaborate foundational concepts to support emerging conversations around data rescue across disciplines.

2 Background

Data rescue "refers to efforts that enable the sustained use of data that otherwise might go unused" [1], often entailing a combination of data curation processes to (a) identify and extract data as such from within collections of documents; and (b) perform systematic conversions from at-risk formats into more sustainable—and often computationally amenable—formats [8, 9]. Data rescue efforts may engage various curation processes, including digitization, data recovery (the retrieval of deleted or damaged digital data, e.g., as from a corrupted hard drive), metadata creation, and efforts to enhance data quality or add value to data [10]. The term *data rescue* gained prominence during distributed efforts to salvage data related to climate change during administrative turnover after the 2016 presidential election, through the crowd-sourced "Data Refuge" initiative [11]. While climatology is at the heart of distributed data rescue, data rescue initiatives have been deployed in diverse fields including astronomy, geology, and pharmacology.

Since much legacy data that has survived has done so "more by circumstance than by design" [12], data rescue tends to be resource intensive [13, 14]. Curating data can be more complex than curating traditional archival materials, as context tends to need significant documentation [15]. Archival science has provided data rescuers with relevant guidance, including the application of the "More Product, Less Process" [16] principles to "minimal data curation" [15]. Tiered systems, like those created by Emory University [17] and Cornell University Library [18] define different levels of processing. Our assessment framework synthesizes these sources to distill key properties in the context of agricultural research collections.

Agriculture is characterized by distinctive data practices, and agricultural researchers confront major barriers to accessing and reusing data [19]. Across disciplines, data rescue initiatives face common obstacles, including the volume of potentially reusable data and the variability of data quality. Beyond these common obstacles, agriculture and other fields dominated by "small science" face distinctive challenges [14, 20]. Agricultural research encompasses work across an unusually broad disciplinary spectrum, highly diverse data types, and a high degree of collaboration across institutions and sectors [19]. These factors complicate any systematic approach to curation. In addition, agricultural research includes work done by extension services, so many of the outcomes of agricultural research are published as grey literature, in the form of reports, blog and social media posts, educational materials, videos, etc. The field's reliance on grey literature may pose challenges for the completeness of legacy collections, as data and important documentary materials may be difficult to locate, access, verify, or reconcile outside of conventional systems of publication and preservation. Finally, field work and observational research tend to produce data that are difficult or impossible to recreate or replicate, which lends them high potential value and priority in data rescue efforts. However, work in some agricultural domains has been shaped by the necessity of confirmatory research for scientific progress and agricultural practice. Because confirmatory research is not always accorded the same value as research resulting in fundamentally novel scientific outcomes, these data are not always prioritized for curation.

3 Assessing Legacy Collections

This project conducted three case studies of historical collections of scientists' papers and data, to identify strategies for efficiently assessing data-rich collections for data-rescue processing. Two are historical collections of paper materials, and one is a more recent collection of born-digital materials. The data being 'rescued' in the course of case studies is intended for inclusion in the USDA's open access data repository, Ag Data Commons. The three cases are as follows:

- Frederick Vernon Coville Blueberry Records (1907–1938): This collection represents the USDA blueberry records of Frederick Vernon Coville, documenting the earliest crosses of commercial blueberries. The collection is 6 linear feet or 24 boxes of material, including hand-written fieldnotes, typed notes, and tabular observations. Figure 1 shows an example of a loose-leaf page from this collection (box 2, 1909). Figure 2 shows a photographed notebook page.
- Wilbur Olin Atwater Papers (1891–1906): Stemming from Atwater's research in the chemical composition of foods, dietary studies, and the respiration calorimeter. The

collection includes 900 handwritten sheets of tabular data. Figure 3 shows an example datasheet from this collection.

- The Rufus Chaney collection (1989–2014): Donated to NAL in 2019 by retired USDA agronomist Rufus Chaney, this is a born-digital collection of Chaney's impactful soil science research, which includes raw data sets, related publications, and analysis files. The collection includes 262 files, including raw data sets and system files. Figure 4 shows raw data in a plain text file from this collection.

In each case, we conducted an exploratory survey of the collection's scope and scale. Building on existing frameworks for data appraisal (described below), we created a draft assessment framework and applied the framework to each collection, in an effort to identify and document data rescue priorities and challenges for each collection. Those collection properties that emerged as essential for data rescue, in conversation with NAL staff and domain experts, were distilled as the essential criteria for quickly assessing potentially data-rich collections. The resulting framework is summarized below and fully detailed in [24].

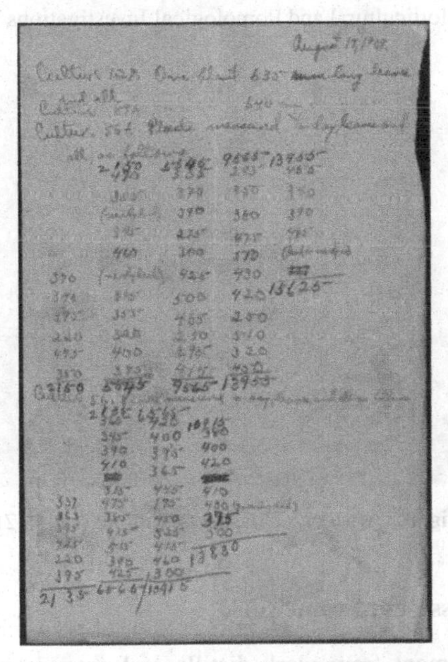

Fig. 1. An example loose-leaf page of Frederick Vernon Coville blueberry records (1909)

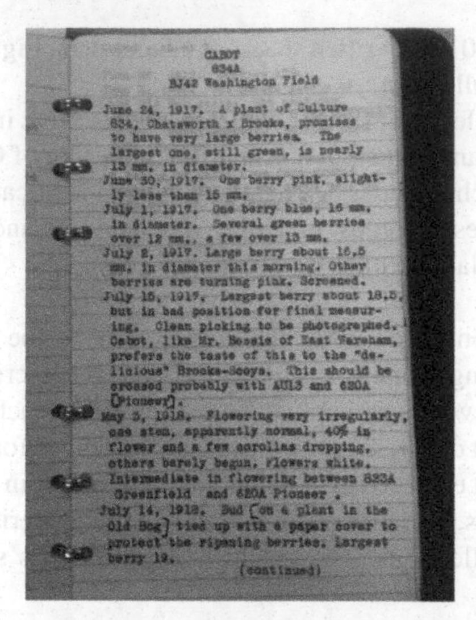

CABOT
634A
BJ42 Washington Field

June 24, 1917. A plant of Culture 634, Chatsworth x Brooks, promises to have very large berries. The largest one, still green, is nearly 13 mm. in diameter.
June 30, 1917. One berry pink, slightly less than 15 mm.
July 1, 1917. One berry blue, 16 mm. in diameter. Several green berries over 12 mm., a few over 13 mm.
July 2, 1917. Large berry about 16.5 mm. in diameter this morning. Other berries are turning pink. Screened.
July 15, 1917. Largest berry about 18.5, but in bad position for final measuring. Clean picking to be photographed. Cabot, like Mr. Bessie of East Wareham, prefers the taste of this to the "delicious" Brooks-Sooys. This should be crossed probably with AU13 and 620A [Pioneer].
May 3, 1918. Flowering very irregularly, one stem, apparently normal, 40% in flower and a few corollas dropping, others barely begun. Flowers white. Intermediate in flowering between 823A Greenfield and 620A Pioneer.
July 14, 1918. Bud [on a plant in the Old Bog] tied up with a paper cover to protect the ripening berries. Largest berry 19.

(continued)

Fig. 2. Photographed notebook page from "Frederick Vernon Coville Blueberry Notes | USDA Bureau of Plant Industry Horticultural and Pomological Investigations Records | Number 1"

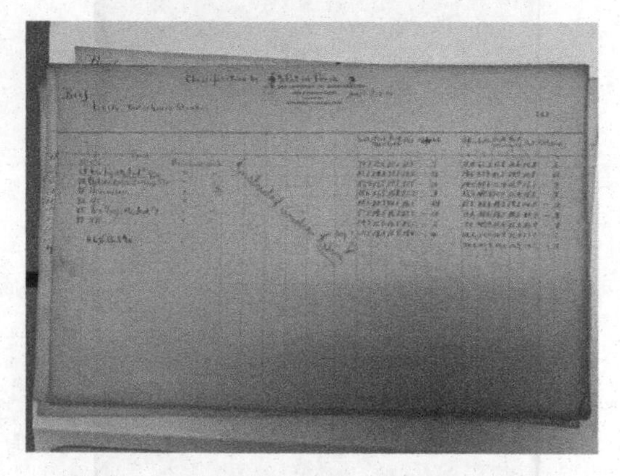

Fig. 3. Atwater's datasheet example (no. 742)

3.1 Preliminary Assessment Framework

The preliminary assessment framework distills and extends properties from several sources in data curation research and practice, including the Data Curation Profiles Toolkit [21], the Cornell Digital Processing Framework [18], and a United States Geological Survey data appraisal process [22]. In particular, the framework builds upon one provided in [23], for understanding the potential reuse value of scientific data as *analytic potential*. Analytic potential frames reuse value as stemming from two factors: (a) data's preservation readiness—whether, in line with the Open Archival Information System reference model, the data include representation, context, reference, provenance, and

```
1   27.0000   55.800  32.6000  5.66  0.250  17  78 LA097001X  271.0  1.80774  21500   3520 227.00  5490  37700   7320  258.0  0.915
2   28.4000   58.100  32.2000  6.45  0.210  21  78 LA097002X  273.0  2.19000  19300   4590 244.00  5380  36700   7410  228.0  0.894
3   24.2000   41.300  17.9000  5.83  0.170  15  78 LA097003X  219.0  1.46879  17300   2580 193.00  4490  32800   6830  245.0  0.919
4   12.7000   43.600  16.3000  4.86  0.150  12  78 LA097004X  126.0  0.76751  13000   1470 170.00  3050  21400   4730  179.0  0.916
5   27.2000   69.500  29.0000  6.30  0.250  21  78 LA097005X  340.0  2.14670  27500   4910 353.00  7000  49900   9250  336.0  0.910
6   17.7000   49.500  15.5000  6.10  0.240  17  78 LA097006X  180.0  1.33243  15400   2680 220.00  3700  28300   6040  196.0  0.910
7   21.2000   54.800  20.0000  5.96  0.190  21  78 LA097007X  305.0  1.51165  21900   3580 338.00  5960  40100   7850  292.0  0.924
8   27.4000   78.700  26.7000  5.96  0.420  24  78 LA097008X  389.0  2.01423  25100   4160 496.00  6700  45000   8850  317.0  0.904
9   27.3000   55.800  20.0000  5.50  0.340  18  78 LA097009X  309.0  2.22462  31600   3420 136.00  8020      .  11000      .  0.909
10  12.1000   24.700   7.6000  6.51  0.160   9  78 LA097010X   91.0  0.60388  12100   1540  61.50  2700      .   5330      .  0.909
11  20.2000   61.400  17.9000  4.48  0.070  18  78 LA097011X   19.8  0.83000  19700   2410   7.61  5140      .   4130      .  0.915
13  13.7000   26.900  10.5000  4.96  0.160   8  78 LA097013X   81.0  0.56492  12600   1100  25.80  2860      .   5520      .  0.918
14  28.2000   75.800  28.1000  4.68  0.150  26  78 LA097014X  300.0  1.47269  38300   2170 425.00  7380      .  11200      .  0.908
15  26.1000   56.400  23.3000  6.26  0.450  14  78 LA065001X  178.0  0.85322  21300   3170 197.00  5300      .   8820      .  0.904
16  36.5000   93.400  28.8000  6.29  0.680  20  78 LA065002X  269.0  0.96231  31700   3790 448.00  7710      .  10900      .  0.931
17  25.8000   66.800  19.9000  5.83  0.720  15  78 LA065003X  179.0  0.79089  22500   3040 186.00  5580      .   8300      .      .
18  34.2000  101.000  25.5000  6.09  1.000  18  78 LA065004X  308.0  1.54282  28600   4300 339.00  7150      .  10800      .      .
19  35.2000   86.300  27.3000  6.88  0.500  20  78 LA065005X  263.0  1.11036  29400   4200 362.00  7350      .  10900      .  0.923
20  29.2000   46.300  17.8000  5.86  0.380  10  78 LA065006X  137.0  0.61946  12100   2640 314.00  3000  20900   4920  153.0  0.916
21  34.5000   66.800  21.7000  6.13  0.440  14  78 LA065007X  163.0  0.88050  14200   2930 524.00  3600  24800   6110  164.0      .
23  48.4000  133.000  40.0000  6.02  0.640  24  78 LA065009X  419.0  2.06878  32100   5850 542.00  8270  65400  11700  292.0  0.920
24  50.0000  144.000  44.3000  6.22  0.600  24  78 LA065010X  439.0  1.90514  34300   6050 480.00  8840  70500  12200  321.0  0.916
25  53.9000  132.000  43.6000  5.74  0.720  24  78 LA065011X  477.0  2.31033  33300   6520 492.00  8750  52900  11200  298.0  0.899
26  53.0000  138.000  43.8000  6.91  0.600  24  78 LA065012X  473.0  1.44542  35100   6850 928.00  9520  67800  11100  263.0  0.922
27  53.3000  129.000  41.0000  6.37  0.710  29  78 LA065013X  481.0  2.12332  32800   6550 512.00  8050  68500  11000  262.0  0.893
28  48.3000  122.000  42.5000  6.28  0.790  25  78 LA065014X  434.0  2.91031  30000   6860 450.00  7340  61600  10600  253.0  0.915
29  56.8000  121.000  42.8000  6.11  0.490  27  78 LA065015X  453.0  1.58957  32800   5670 509.00  8550  66500  10900  265.0      .
31  60.2000  142.000  56.4000  7.54  1.000  32  78 LA065017X  461.0  2.24799  36600  11300 528.00  9870  79000  11500  358.0  0.922
32  49.9000  182.000  57.4000  7.56  0.950  31  78 LA065018X  447.0  2.08046  44400  10900 375.00 11300      .  15100      .      .
33  52.0000  173.000  55.3000  7.52  0.980  26  78 LA065019X  456.0  2.06878  44400  10100 480.00 11100      .  14300      .  0.910
34  47.6000  151.000  50.3000  7.37  0.780  19  78 LA065020X  466.0  2.09215  44400   8570 348.00 11100      .  14700      .  0.904
35  46.6000  152.000  50.3000  7.18  0.780  16  78 LA065021X  444.0  2.16228  44400   8540 335.00 11200      .  14300      .  0.898
36   4.8000   12.600   3.8200  5.19  0.170   7  78 LA053001X   64.0  0.71297   7340    562 135.00   787      .   1960      .  0.903
37   4.8200    9.340  -0.6000  4.70  0.240   9  78 LA053002X   69.0  0.74414   5380    533  37.70   826      .   2480      .  0.913
38   5.8000   13.600   1.3000  4.52  0.180  10  78 LA053003X   63.0  0.74803  11600    506 290.00   858      .   2080      .  0.911
```

Fig. 4. A raw data in plaintext file format in Chaney's born-digital collection

fixity information to support long-term use; and (b) their fit for purpose, or the alignment of a dataset with methods and tools for a given application [23]. Both of these factors are understood in the context of a range of possible user communities.

The framework is intended for use in the initial, exploratory phase of data rescue, to determine priorities for data rescue, explain the potential value of data rescue processes, anticipate potential obstacles to processing, and begin to assess the labor and resources required for different levels of processing. This framework defines a set of 18 factors that together determine the costs and value of processing a collection to recover research data for reuse. Because ultimate data processing decisions are necessarily contextualized by the resources and priorities of the institutions undertaking data rescue, this framework does not offer any prescriptive guidance or formula for weighing factors against one another. Here, we dwell only on the factors identified in bold in Table 1, which constitute the main conceptual contributions of our framework. The highlighted factors are defined in Table 2.

Table 1. List of factors in data rescue assessment

Reuse value	Fit for purpose	Extent
Historical value	**Obstacles to recovery**	Completeness
User communities	Data objects	Sensitivity
Stakeholders	Associated publications	Access and use constraints
Reuse objects	Relevant collections	Rarity or uniqueness
Historical objects	Reproducibility	Priorities

Table 2. Definitions of selected factors

Factor	Definition
Reuse value	What are the intended, demonstrated, anticipated, or plausible reuse opportunities for the collection? Note that this question is *not* limited to the data alone, but also encompasses potential reuse of other facets of the collection, including methodological or contextual documentation, tools, or protocols
Historical value	What is the potential historical value of the collection? What important or noteworthy scientific approaches, results, or advances are documented or evidenced by the data? Note that this question is *not* limited to the data alone, but also encompasses other facets of the collection, including methodological or contextual documentation, tools, or protocols
User communities	Who are the potential data users? If data rescue is being conducted for open-ended future reuse, user communities should be evaluated based on the originating community of the data, any explicitly indicated audiences for the research, and a meta-analysis of related fields conducted by a curator in consultation with domain experts [23]
Stakeholders	Other than direct users, what groups, institutions, or communities may have an ongoing interest in the data? Who has invested in the data? Who would be impacted by use or reuse of the data?
Reuse objects	Are there specific components of the collection that carry reuse opportunities? Are there specific components that are amenable to reuse? Components may be material or abstract; they may correspond to a subset of *data objects*
Historical objects	Are there specific components of the collection that carry historical value? Components may be material or abstract; they may correspond to a subset of *data objects*. While historical value and reuse potential are certainly interwoven, this factor distinguishes data as potential evidence for *science* from data as potential evidence for the *history of science*, as such data may have different processing entailments. For example, it may be sufficient for data intended as *historical objects* (and not *reuse objects*) to be digitally readable without being computationally amenable
Fit for purpose	To what extent are the data ready or suitable for actual or potential uses identified in *reuse value* and *historical value*? How much additional documentation, interpretation, and processing are required to prepare data? What level of scientific, technical, or research expertise would additional processing entail? This factor takes into account other factors including *data objects*, *completeness*, and *access and use constraints*
Obstacles to recovery	What are the anticipated or observed obstacles to recovering data from the collection? This question builds on *fit for purpose* and other factors to invite data rescuers to inventory potential obstacles. Obstacles to data recovery may result from a very wide range of properties of the collection, including its physical condition; the quality, completeness, and forms of data in the collection; digital file formats; extant documentation; and the approachability or understandability of the collection to unfamiliar or non-expert users or curators

3.2 Framework Applied to Collections

When applied to our case study collections, the framework reveals significant differences in how the collections manifest reuse and historical value, as well as significant differences among obstacles to data recovery. By applying the factors described above to each collection, we paint contrasting portraits of the pros and cons confronting data rescue for the collection. Table 3 provides a structured, comparative view of two portraits; for the full comparison of all three cases, see [24].

Table 3. Framework applied to two collections

Factor	Coville collection	Chaney collection
Reuse value	Documents cultivars still in contemporary cultivation; genetics research using longitudinal pedigree information; confirmatory research of blueberry cultivation practices	Confirmatory research, e.g. analytical steps; longitudinal study in soil science; genetics research using crop cultivars information; interoperate with other crop data
Historical value	Significant contributions to blueberry domestication: early fertilizers, use of acidic soil and cold treatment for blueberry cultivation	During Chaney's 48-year career as an agronomist at USDA-ARS, his research made significant contributions to the study of heavy metals present in soil and their uptake in crops, the application of biosolids to cropland (collaboration with FDA and EPA), and phytoextraction of contaminated soil
User communities	Horticultural scholars; genetic scientists; general public	Soil scientists; plant scientists; crop scientists; environmental scientists; biosolids scientists
Stakeholders	USDA Agricultural Research Service (ARS), commercial blueberry growers	USDA ARS, the Food and Drug Administration (FDA), and the Environmental Protection Agency (EPA)
Reuse objects	Detailed pedigree information for both released and unreleased cultivars, including parent cultivar names of well-known cultivars, years of release, and plant characteristics and inheritance	Raw data sets, Analytics system files (.sas)
Historical objects	Complete collection (see *fit for purpose*)	Complete collection (see *fit for purpose*)
Fit for purpose	Collection is incompletely digitized and documented, so is not fit for purpose as historical evidence (*historical value*). Further processing, including transcription of handwritten notes, would support *reuse value*	To support *historical value* interpretation and documentation would be necessary to connect data sets to documentation and publications. To support *reuse value* verification of variable names would also be necessary

(*continued*)

Table 3. (*continued*)

Factor	Coville collection	Chaney collection
Obstacles to recovery	• Mix of analog and digitized materials • Fragility of analog materials • Loose leaf pages vulnerable to loss of original order • Determining number and completeness of datasets within documents • Mix of handwritten and typed data • Mixed data types • Structured data embedded in unstructured text • Inconsistent or incomplete metadata (missing column headers, empty fields) • Determining priorities requires expert consultation • No documentation • Linking data to publications	• Linking data to relevant publications • Missing context and metadata require expert consultation • Determining completeness within files, overlap among files • Access to outmoded software originally used to create the files • File format verification and migration

4 Discussion and Future Work

This framework synthesizes existing frameworks for processing collections, with the goal of distilling key properties for data rescue: to support efficient assessment of the challenges and priorities for recovering usable data from at-risk collections of agricultural materials. Our framework extends that of *analytic potential* [23] in the following ways, specifically to support data rescue from legacy collections:

- We distinguish Potential *Reuse Value* from Potential *Historical Value.* This is to acknowledge dual aspects: (a) the potential to support novel scientific inquiry through new methods of analysis, combination with new data, comparison, replication or reproduction, etc.; and (b) the potential to support historical studies of science, by evidencing processes, origins, and outcomes of historically significant research.
- We distinguish *User Communities* from *Stakeholders.* When data may be considered to have both reuse and historical value, then stakeholders must be considered beyond potential user communities.
- We distinguish *Reuse Objects* and *Historical Objects.* These factors acknowledge that reuse and historical value are not necessarily evenly distributed across a collection of research materials, and that certain data objects within a collection may carry greater value, and therefore represent greater priorities for processing.
- We expand on *Fit for Purpose* with *Obstacles to Data Rescue.* In the case of legacy collections, the data residing in these collections are rarely fit for open-ended reuse

(because they were created to serve specific uses of the data producers). To support reuse, data may undergo various levels of processing, including documentation, digitization, transcription, cleaning, and enhancement. By identifying as many obstacles to data rescue as possible, data curators may choose how "fit" data need to be for the open-ended possibilities of reuse, given the context of their anticipated user communities and stakeholders.

This framework is intended as a starting point for ongoing development. Further case studies and interviews with agricultural researchers and curation professionals will evaluate the assessment framework in more institutional and research contexts. Future work will aim to contribute guidance on the specific roles of curators and domain experts in data rescue workflows, and to investigate the outcomes impact of data rescue initiatives on scientific reproducibility, data reuse, and public access to science.

Acknowledgements. This research was conducted by Fellows in the Digital Curation Fellowship program, supported by Non-Assistance Cooperative Agreement #58-8260-6-003 between the University of Maryland and the United States Department of Agriculture (USDA), Agricultural Research Service (ARS), National Agricultural Library with funding provided by the USDA, ARS, Office of National Programs.

References

1. Downs, R.R., Chen, R.S.: Curation of scientific data at risk of loss: data rescue and dissemination. In: Curating Research Data Volume One: Practical Strategies for Your Digital Repository, pp. 263–277. Association of College and Research Libraries (2017)
2. Wippich, C.: Preserving science for the ages—USGS data rescue. In: Preserving science for the ages–USGS data rescue (USGS Numbered Series No. 2012–3078; Fact Sheet, Vols. 2012–3078). U.S. Geological Survey (2012). https://doi.org/10.3133/fs20123078
3. Rountree, R.A., Perkins, P.J., Kenney, R.D., Hinga, K.R.: Sounds of western north Atlantic fishes—data rescue. Bioacoustics **12**(2–3), 242–244 (2002). https://doi.org/10.1080/09524622.2002.9753710
4. Fisher, M.: Making the case for evidence-based agriculture. CSA News **59**(5), 4–11 (2014). https://doi.org/10.2134/csa2014-59-5-1
5. Oden, J.T., et al.: National Science Foundation Advisory Committee for Cyberinfrastructure Task Force on Grand Challenges (2011). https://www.nsf.gov/cise/oac/taskforces/TaskForceReport_GrandChallenges.pdf
6. Cragin, M.H., Palmer, C.L., Carlson, J.R., Witt, M.: Data sharing, small science and institutional repositories. Philos. Trans. Royal Soc. A Math. Phys. Eng. Sci. **368**(1926), 4023–4038 (2010). https://doi.org/10.1098/rsta.2010.0165
7. Padilla, T., Allen, L., Frost, H., Potvin, S., Elizabeth, R.R., Varner, S.: Final Report – Always Already Computational: Collections as Data (Version 1), Zenodo, 22 May 2019. https://doi.org/10.5281/zenodo.3152935
8. Brunet, M., Jones, P.: Data rescue initiatives: bringing historical climate data into the 21st century. Climate Res. **47**(1), 29–40 (2011). https://doi.org/10.3354/cr00960
9. Wyborn, L., Hsu, L., Lehnert, K., Parsons, M.A.: Guest editorial: special issue: rescuing legacy data for future science. GeoResJ **6**, 106–107 (2015). https://doi.org/10.1016/j.grj.2015.02.017

10. Vearncombe, J., Riganti, A., Isles, D., Bright, S.: Data upcycling. Ore Geol. Rev. **89**, 887–893 (2017). https://doi.org/10.1016/j.oregeorev.2017.07.009
11. Janz, M.M.: Maintaining access to public data: Lessons from Data Refuge [Preprint]. LIS Scholarship Archive (2018). https://doi.org/10.31229/osf.io/yavzh
12. Griffin, E.R.: When are old data new data? GeoResJ **6**, 92–97 (2015). https://doi.org/10.1016/j.grj.2015.02.004
13. Fallas, K.M., MacNaughton, R.B., Sommers, M.J.: Maximizing the value of historical bedrock field observations: an example from northwest Canada. GeoResJ **6**, 30–43 (2015). https://doi.org/10.1016/j.grj.2015.01.004
14. Specht, A., Bolton, M., Kingsford, B., Specht, R., Belbin, L.: A story of data won, data lost and data re-found: the realities of ecological data preservation. Biodivers. Data J. **6**, e28073 (2018). https://doi.org/10.3897/BDJ.6.e28073
15. Lafferty-Hess, S., Christian, T.-M.: More data, less process? The applicability of MPLP to research data. IASSIST Q. **40**(4), 6 (2017). https://doi.org/10.29173/iq907
16. Greene, M., Meissner, D.: More product, less process: revamping traditional archival processing. Am. Archivist, **68**(2), 208–263 (2005). https://doi.org/10.17723/aarc.68.2.c741823776k65863
17. Waugh, D., Russey Roke, E., Farr, E.: Flexible processing and diverse collections: a tiered approach to delivering born digital archives. Arch. Rec. **37**(1), 3–19 (2016). https://doi.org/10.1080/23257962.2016.1139493
18. Faulder, E., et al.: Digital Processing Framework [Report] (2018). https://ecommons.cornell.edu/handle/1813/57659
19. Cooper, D., et al.: Supporting the Changing Research Practices of Agriculture Scholars. Ithaka S+R (2017). https://doi.org/10.18665/sr.303663
20. Brouder, S., et al.: Enabling open-source data networks in public agricultural research (CAST Commentary No. QTA2019–1). Search Results Web Result with Site Links Council for Agricultural Science and Technology (2019). https://www.cast-science.org/wp-content/uploads/2019/05/QTA2019-1-Data-Sharing.pdf
21. Carlson, J.: The Data Curation Profile Toolkit: The Profile Template. Purdue University Libraries/Distributed Data Curation Center (2010).https://doi.org/10.5703/1288284315653
22. Faundeen, J.L., Oleson, L.R.: Scientific Data appraisals: the value driver for preservation efforts, vol. 6 (2007). https://www.pv2007.dlr.de/Papers/Faundeen_AppraisalsValue_for_Preservation.pdf
23. Palmer, C.L., Weber, N.M., Cragin, M.H.: The analytic potential of scientific data: Understanding re-use value. Proc. Am. Soc. Inf. Sci. Technol. **48**(1), 1 (2011). https://doi.org/10.1002/meet.2011.14504801174
24. Hoffman, K.M., Clarke, C.T., Shiue, H.S.Y., Nicholas, P., Shaw, M., Fenlon, K.: Data rescue: an assessment framework for legacy research collections (2020). https://doi.org/10.13016/1zmx-ghhq

Multi-generational Stories of Urban Renewal: Preliminary Interviews for Map-Based Storytelling

Myeong Lee[1(✉)], Mark Edwin Peterson[2], Tammy Dam[3], Bezawit Challa[1], and Priscilla Robinson[4]

[1] Department of Information Sciences and Technology, George Mason University, Fairfax, USA
mlee89@gmu.edu
[2] Cultural Studies, George Mason University, Fairfax, USA
[3] College of Information Studies, University of Maryland, College Park, USA
[4] Independent Researcher, Asheville, USA

Abstract. Urban renewal was a project of the American government that aimed to reconstruct poorly-managed neighborhoods. Because community-level data that shows underlying mechanisms of urban renewal has not been curated systematically, due to the complexity and volume of relevant archival collections, we aim to digitally curate property acquisition documents from the urban renewal projects that affected the Southside neighborhood of Asheville, North Carolina, in the form of a map-based, interactive web application. This paper presents part of the interview analysis to understand how Asheville citizens of different generations remember their neighborhood before and after the gentrifications that they have experienced. The result of this analysis provides design implications for the archival system we are developing by revealing generation-driven value structures of potential users.

Keywords: Digital curation · Urban renewal · Computational archival science · Value-sensitive design

1 Background

During the 1960s and 1970s, the federal government of the United States began a nationwide initiative of urban renewal that aimed to reconstruct so-called "blighted" neighborhoods across the country. Billions of dollars were spent to purchase real estate and move people to make way for new development. Working with local housing authorities, the government determined cases in which cities and towns could benefit from the complete redevelopment of neighborhoods considered unsightly and poorly-managed. Some officials used condemnation to seize property when they encountered resistance, and these urban renewal projects in general offered people low prices for their homes, with tremendous pressure to sell [1]. The process displaced local communities and destroyed many of the businesses that served them. The public housing projects and

K. Toeppe et al. (Eds.): iConference 2021, LNCS 12646, pp. 319–326, 2021.
https://doi.org/10.1007/978-3-030-71305-8_26

more expensive developments that replaced these neighborhoods often did not include many of the community elements of what had been lost in the urban renewal process.

This process occurred after several decades of what was known as the "Great Migration," when millions of African Americans had moved from the rural South to towns across the nation, including major cities such as New York, Houston, and Chicago [2]. Many historians and social scientists have described how the urban renewal projects at both the local and federal levels often ended up focusing their attention on breaking up African-American neighborhoods specifically [3, 4]. Urban renewal repeatedly resulted in moving African Americans further away from downtown business districts, displacing vibrant communities in the process. Government projects built to provide new housing often suffered from neglect or mismanagement, when they went up at all, so that any initial enthusiasm for urban renewal among African American communities quickly diminished [5, 6]. Many of these regions are now attractive city districts after going through multiple gentrifications. However, few families who lived in these areas before urban renewal remain, and little evidence survives of the former communities or their history.

One example of urban renewal having dramatic, complicated effects on the lives of African Americans occurred in the town of Asheville, in the mountains of western North Carolina. Though many African-American families had moved to Asheville in the twentieth century to take jobs in the regions' developing industry, town leaders began to encourage tourism as early as the 1920s. By the 1960s, there were great concerns about maintaining an attractive downtown for visitors, as well as ongoing racial tensions among different groups of residents [7]. These factors contributed to enthusiasm for several urban renewal projects over the next two decades, the largest in the East Riverside neighborhood and Southside business district. Urban renewal in Southside would ultimately destroy over 1,100 homes, 8 apartment complexes, a hospital, and many businesses. Affecting more than four hundred acres, this urban renewal project in a relatively small town was the largest of its kind in the Southeastern United States.

The interplay of systematic racism and urban renewal has attracted a lot of attention recently [8, 9]. Scholars, high school teachers, activists, and journalists are just some of the groups interested in understanding this history. Comparisons with real estate redlining and racial property covenants show urban renewal to be part of the effects that systematic racism had on people across the country in the late twentieth century [10]. Many records of urban renewal property sales also contain information for genealogists and those seeking to uncover cases of personal injustice. However, while some individual stories are available, community-level data that shows the underlying mechanisms of urban renewal projects has not been curated in a systematic way, because of the complexity and large volume of documents in the relevant archival collections. This is certainly the case with the Southside project. The work of planning, appraisal, and property sales created over one hundred boxes of documents. Almost eighty linear feet of archival material survive in the copies held by the special collections of the University of North Carolina-Asheville, but the researchers who make their way to the library have limited access and little guidance on how to understand the collection.

Our project aims at building an archival system that captures people's stories during urban renewal using the documents of the acquisition process from over nine hundred properties in the Southside neighborhood. People have different interpretations of the meaning of some of the documents, and they also have very individual needs in their use of a wide range of material that could be considered personal. Any digital interface that contains such sensitive data connected to different populations and examined by various types of users should be treated with a high level of care and precision.

In order to be cognizant of the ways that all the stakeholders would approach the final archival system, we have adopted the value-sensitive design (VSD) methodology with a focus on understanding their values and potential conflicts surrounding the future system. As part of such efforts, we conducted interviews with potential users from different generations and ethnic groups to illustrate their value structures on topics relevant to the issues of urban renewal. We analyzed the interview data using both the ground-theory and top-down approaches. The analysis results will guide the development of tools available to users of the final digital archive. Among the diverse aspects of the value structures found from the interviews, we report our preliminary findings on values and perspectives that stem from the different generations in Asheville. The results of the analysis, along with other aspects of value structures, will inform the design of the archival system.

2 Related Work

With historians and cultural critics paying greater attention to the subtle mechanisms of racism that plague American society, there has been growing recognition of the importance of federal real estate and lending policies in directing who lives where and what that means for prosperity. Some early work has illustrated this history in Asheville [11]. In recent years, research on race and real estate in America has been supported by a growing number of digital humanities tools to track changes in housing [12, 13]. Little scholarship exists, however, to understand potential stakeholders' values and knowledge or the ways researchers might put the resources to use for this kind of long-standing collection. The urban renewal archive used by our project [14], in particular, focuses on legal processes of how each property was acquired by housing authority during the period as well as the human stories that are embedded in the legal documents. By understanding Ashville citizen's different understanding of the urban renewal project as well as the future data system, this paper sheds light on the building blocks that inform the design of historically-sensitive data systems.

General discussions of the need for honest assessments of the values of stakeholders involved in digital projects have become a common concern in a variety of fields. As scholars develop historical collections and exhibits [15, 16], musical archives [17], and even natural resource tools [18], they have been working to see what people might think of the resources and how they might be used to meet user needs. Our project, dealing with government actions and records, has many similarities with the ways that institutions interact with local communities. This kind of project has been informed by research in value sensitive design [19], community building [20], community engagement [21], and enacting the political [22] in the development of helpful digital tools in the realm of community-based research [23]. Based on the methodologies in these topics, we systematically work out the ways that digital tools provide effective and impactful access to complicated resources from the archives [24]. By working through the different uses of the archival tool and researching the existing views that diverse stakeholders have about the complicated issues of urban renewal, along with ways they might be able to deepen their understanding, we will make this collection of documents the basis for a valuable digital resource.

3 Approach

To properly understand potential responses to a historically-sensitive data platform such as this, as well as ongoing tensions between stakeholders, interviews followed a set protocol design based on previous work on identifying user values [25]. Starting with The Digital Curation and the Local Community Workshop in April 2017, we identified key protocols that could form tensions surrounding such a system based on a literature review. These dimensions have been included in the list of topics to be covered in each interview, though the sessions have been kept semi-structured, giving subjects opportunities to include their own thoughts on the experiences of urban renewal in their neighborhoods and what it meant for them.

Throughout the inception of the project, the recruitment of participants has relied on a snowball sampling approach where our primary contacts directed us to the target, hard-to-reach participant population. Through this process, we were able to recruit participants from very different participant pools to conduct in-depth semi-structured interviews. This paper reports two analyses of these interviews: one participant (P1), a white male in early 40s, is a current resident, born and raised in Asheville; the second participant (P2), an African-American male in his 70s, was a former resident of Asheville who experienced urban renewal when it happened.

To help compare the interviewees' different perspectives that are based on generational gaps, we mapped quotes regarding urban renewal about specific Southside locations from their memories on the actual map. Statements in the interviews were also coded when they addressed themes that connected to the identified concepts of VSD or general familiarity with the issues at hand, in order to illustrate the different value structures at play among the stakeholders in different age groups.

4 Multi-generational Stories

After completing the analysis of two interviews, we were able to begin identifying how people from different backgrounds could see the issues differently. Because "urban renewal" that P1 remembered was only the one that happened in the 1990s and early 2000s, for example, with regard to the Hilcrest area (south of downtown Asheville), P1 noted that in Hilcrest of the 2000s, "*There's only one way in and one way out and like these neighborhoods that are just like geographically isolated so that there's no intermingling.*" P2 who remembered multiple urban renewal projects in Asheville illustrated Hilcrest as a place of getting together: "*Even though it was segregated, all the white and black, everybody intermingled. [...] everybody knew everybody. [...] Urban renewal program came in and started dividing, not only did it divide the community of Blacks and destroy that, it destroyed quite a bit of the white community too.*" Younger generations see the results of these changes while older people who experienced these changes first-hand have a clearer idea of how the processes of urban renewal shaped the community and pushed development in certain directions.

General comments from these two participants surfaced their *sense of place* about various locations in Asheville, which shows us different conceptions of the lived environment and the ways that it affected the people of the city. P1 often commented on neighborhoods in terms of the failures of local government. He says that the downtown festival "Bele Chere" did not bring the desired sense of refinement, nor did it diminish the visible homelessness and drug use downtown until the gentrification of the late 1990s. According to him, the bus station in Pritchard Park was not considered safe. Many Black families had to leave their homes because of rising taxes. P2, on the other hand, focuses much more on the changes and losses that people experienced, especially in terms of schooling and community cohesion. According to him, before urban renewal in Asheville, children of both races used to play together on French Broad Avenue, though African Americans were separated into their own churches and schools. Stephens Lee High School was the only all-Black high school in the city, but the students were later dispersed to other schools. Even in neighborhoods with a strong sense of segregation, urban renewal had the effect of breaking groups further apart, so that all the people of the city no longer even knew each other.

The two participants expressed different views on the meaning of the future archival system that we are designing. P2, having experienced urban renewal, valued the *educational possibilities* of access to the housing authority records for Asheville. "*I don't know if it does anything for the debt, but it would give them a great timeline to see what our community, a thriving community, the black community, looked like and what it became, and what, of course, it is now, non-existent.*" For him, it was especially important to let the Asheville community see their own history. For P1, neutral handling of the information was deemed more important than any interpretation. "*I don't think it matters to me whether they sold that house for $5,000, $10,000, or $3,000 [...] I think if you bring an agenda, I feel like it dilutes the value a little bit. I think the idea is to let the history speak for itself.*" Though he does note that, "*Local history is something that people are really into here.*"

5 Value Structures of Users

Interview analysis to find shared themes of concern provides us with a framework of community values in understanding the effects of urban renewal and the capabilities that need to be considered in the system design. Previous work listed some user concerns that could be found in historically-sensitive data platforms, and these dimensions were implemented as protocols in all interviews about design of a digital archive created from this collection. These dimensions are: *accessibility control, individual position, authenticity in digitization, neutrality, boundary, guardianship, collection development, benefits of digitization, design needs, perceptions of urban renewal, and transparency in digitization* [26]. By analyzing the interview data using this framework, it was possible to identify clear boundaries between the shared values of different generations and ethnic groups today who have been affected by Asheville's past of urban renewal. For instance, in the discussions with P1 and P2, it was clear that P2 wanted any digital archive to allow open access, even though he had greater concerns about privacy issues than the younger interviewee. His responses also made it clear that he had a greater concern for education opportunities in the neighborhood than the general living conditions. P1 felt that a digital archive would be valuable for distinct stakeholders among users but had little to say on how to serve their needs. He also considered his own views less representative of any unified group. Both participants understood that social and cultural segregation were issues in contemporary Asheville with links to thc past, but there were clear generational splits on what this meant for a person's sense of place in the community.

This shows that the values of the potential stakeholders reflected on the future archival system are multidimensional and often conflicting. Citizens from different generations indicated the community's symbolic boundaries, which were formed based on the priority in values (i.e., educational vs. living conditions), temporal focus (i.e., future opportunity vs. current status), and spatial focus (i.e., educational/cultural institutions vs. business/residential areas). Their different experiences of urban renewal partly shaped their sense of places, which might have shaped their focus on particular values. Despite the different values, both of them emphasized the importance of social and cultural segregation as an issue that the future system needs to take into account. This indicates that the digital archival platform for the urban renewal collection can be a useful boundary object that can be used by diverse stakeholders [27].

6 Discussion and Future Work

A clear limitation of the analysis so far is that the generational differences are based on only two interviews. However, this work shows clear directions for further research. The analysis of the interviews provides meaningful data about values and concerns that need to be taken into account when designing the features of the final archival tool. As we continue to work with more people from the community and see further patterns, it will be clearer how user groups differ in certain ways, and it will be possible to discern how to address the needs of various populations as we add more depth to the understanding of which values touching on the issues of historic urban renewal are shared and which ones divide people. The differences, tensions, and commonalities between stakeholders' values found from the interviews will be a basis to create a useful boundary object for the communities.

Acknowledgments. This project was approved by the Institutional Review Board (IRB) from the University of Maryland at College Park and George Mason University. This research was supported by the Research Improvement Grant (RIG-II) from the University of Maryland's iSchool. Also, we are grateful to faculty and students, Dr. Richard Marciano, Edel Spencer, Yuheng Zhang, Shiyun Chen, Mahitha Kalyani, Richard Bool, and Lauren Schirle, for their support for this project.

References

1. Pfau, A., Sewell, S.: Newburgh's "last chance": the elusive promise of urban renewal in a small and divided city. J. Plan. Hist. **9**(3), 146 (2020)

2. Reed, C.: Knock at the Door of Opportunity: Black Migration to Chicago, 1900–1919. Southern Illinois University Press, Carbondale (2014)
3. Retzlaff, R.: Connecting public school segregation with urban renewal and interstate highway planning: the case of Birmingham Alabama. J. Plan. Hist. **19**(14), 259 (2020)
4. Doucet, B.: Deconstructing dominant narratives of urban failure and gentrification in a racially unjust city: the case of Detroit. Tijdschrift voor Economische en Sociale Geografie **111**(4), 638 (2020)
5. Bouknight, A.: Casualty of progress: the ward one community and urban renewal, Columbia, South Carolina, 1964–1974. Master's thesis, University of South Carolina, p. 20 (2010)
6. Nickoloff, S.: Urban renewal in Asheville: a history of racial segregation and black activism. Master's thesis, Western Carolina University, p. 6 (2015)
7. Nickoloff, S.: Urban renewal in Asheville: a history of racial segregation and black activism. Master's thesis, Western Carolina University, pp. 13–16 (2015)
8. Connolly, N.: The Southern Side of Chicago: Arnold R. Hirsch and the Renewal of Southern Urban History. J. Urban Hist. **46**(3), 505–510 (2020)
9. Logan, C.: Historic Capital: Preservation, Race, and Real Estate in Washington. University of Minnesota, D.C. Minneapolis (2017)
10. Taylor, K.: Race for Profit: How Banks and The Real Estate Industry Undermined Black Homeownership. University of North Carolina, Chapel Hill (2019)
11. Twilight of a Neighborhood: Asheville's East End. https://www.nchumanities.org/galleries/twilight-neighborhood-ashevilles-east-end. Accessed 17 Apr 2020
12. Renewing Inequality. https://dsl.richmond.edu/panorama/renewal/. Accessed 17 Apr 2020. See also Mapping Inequality, https://dsl.richmond.edu/panorama/redlining/. Accessed 17 Apr 2020
13. Mapping Decline. https://mappingdecline.lib.uiowa.edu/map/. Accessed 09 Jan 2021
14. Housing Authority of the City of Asheville. https://toto.lib.unca.edu/findingaids/mss/housing_authority_city_asheville/HACAFindingAidPart7.html. Accessed 17 Apr 2020
15. Lin, J., Milligan, I., Oard, D., Ruest, N., Shilton, K.: We could, but should we? Ethical considerations for providing access to GeoCities and other historical digital collections. In: Proceedings of the 2020 Conference on Human Information Interaction and Retrieval, pp. 135–144 (2020)
16. Schofield, T., Smith, F., Bozoglu, G., Whitehead, C.: Design and plural heritages: composing critical futures. In: Proceedings of the 2019 CHI Conference on Human Factors in Computing Systems, pp. 1–15 (2019)
17. Su, N., Stolterman, E.: A design approach for authenticity and Technology. In: Proceedings of the 2016 ACM Conference on Designing Interactive Systems, pp. 643–655 (2016)
18. Su, N., Cheon, E.: Reconsidering nature: the dialectics of fair chase in the practices of American Midwest hunters. In: Proceedings of the 2017 CHI Conference on Human Factors in Computing Systems, pp. 6089–6100 (2017)
19. Fox, S., Le Dantec, C.: Community historians: scaffolding community engagement through culture and heritage. In: Proceedings of the 2014 Conference on Designing Interactive Systems, pp. 785–794 (2014)
20. Crivellaro, C., Taylor, A., Vlachokyriakos, V., Comber, R., Nissen, B., Wright, P.: Re-making places: HCI, 'community building' and change. In: Proceedings of the 2016 CHI Conference on Human Factors in Computing Systems, pp. 2958–2969 (2016)
21. Corbett, E., Le Dantec, C.: The problem of community engagement: disentangling the practices of municipal government. In: Proceedings of the 2018 CHI Conference on Human Factors in Computing Systems, pp. 1–13 (2018)

22. Crivellaro, C., Comber, R., Dade-Roberston, M., Bowen, S., Wright, P., Olivier, P.: Contesting the city: enacting the political through digitally supported urban walks. In: Proceedings of the 33rd Annual ACM Conference on Human Factors in Computing Systems, pp. 2853–2862 (2015)
23. Corbett, E., Le Dantec, C.: Exploring trust in digital civics. In: Proceedings of the 2018 Designing Interactive Systems Conference, pp. 9–20 (2018)
24. Underwood, W., Marciano, R.: Computational thinking in archival science research and education. In: 2019 IEEE International Conference on Big Data, pp. 3146–3152 (2019)
25. Lee, M., et al.: Heuristics for assessing computational archival science (CAS) research: the case of the human face of Big Data project. In: 2017 IEEE International Conference on Big Data, pp. 2262–2270 (2017)
26. Lee, M., Chen, S., Zhang, Y., Spencer, E., Marciano, R.: Toward identifying values and tensions in designing a historically-sensitive data platform: a case-study on urban renewal. In: Chowdhury, G., McLeod, J., Gillet, V., Willett, P. (eds.) iConference 2018. LNCS, vol. 10766, pp. 632–637. Springer, Cham (2018). https://doi.org/10.1007/978-3-319-78105-1_72
27. Leigh Star, S.: This is not a boundary object: reflections on the origin of a concept. Sci. Technol. Human Values **35**(5), 601–617 (2010)

The Politics of Digitizing Art and Culture in Vietnam: A Case Study on *Matca Space of Photography* in Hanoi

Emma Duester(✉)

Royal Melbourne Institute of Technology (RMIT), Hanoi, Vietnam
emma.duester@rmit.edu.vn

Abstract. The nature of work in the art and cultural sector in Hanoi, Vietnam, is changing. The new generation of cultural professionals is harnessing digital technology to display art and cultural collections in innovative and creative ways. Digitization today is not only about creating 'hidden' digital archives but, instead, about curating digital art and culture experiences that are publicly accessible. This allows a way to preserve culture, which can be digitally displayed in a contemporary format. The paper presents findings from a case study on *Matca Space for Photography* (Matca), including semi-structured interviews, secondary data analysis, and a digital ethnography of Matca's digital platforms. The current study highlights the challenges and opportunities associated with digitization in Vietnam. While there are challenges with digitization due to a lack of technical resources and human resources, using digital platforms can allow cultural professionals an agency to present Vietnamese art and culture to local and international audiences. This has the potential to redress the imbalance in representation and redefine digital orientalism.

Keywords: Digitization · Art and cultural sector · Vietnam

1 Introduction

This paper investigates the current nature of digitization in the art and cultural sector in Hanoi, Vietnam. It will use a case study on one particular non-profit, independent art organization in Hanoi, *Matca Space for Photography* (Matca), in order to show how cultural professionals are digitizing their art collections and how they are utilizing digital platforms to publicly display these collections. The last five years have seen the uptake of digital technologies for the digitization of collections in Hanoi. However, the art and culture sector is only just begining to utilize digital platforms for the public display of art collections. Even though digitization itself is not a new phenomenon, the digital technologies that are being used to publicly display these digitized collections today are new. Moreover, the transformation this is having on the work of Vietnamese cultural professionals is also new.

The current state of the creative industries across Asia, including China, Japan and South Korea is well-documented in the literature [12, 14, 15, 20] and, more specifically,

K. Toeppe et al. (Eds.): iConference 2021, LNCS 12646, pp. 327–338, 2021.
https://doi.org/10.1007/978-3-030-71305-8_27

in relation to other countries in South East Asia, such as Thailand, Singapore and Malaysia [4, 6, 28]. However, Vietnam seems to have been left out of this discourse. Importantly, the Vietnamese government has now positioned the creative industries as a major sector for national development and for international cooperation [18, 26]. However, there is a need to assess the role of digital technology in the creative industries in Vietnam, as this transition and development is helping to transform the nation's image from 'Made in Vietnam' to 'designed, innovated and created in Vietnam'. This generation of cultural professionals is showcasing this new national image by harnessing digital technologies, including digital tools, apps and platforms.

It is important to investigate the opportunities and challenges of digitization that cultural professionals are facing in Vietnam. In particular, research is required on the take-up of digital technologies for preserving and displaying art and cultural collections in Hanoi. On one hand, digitization provides a means of sustainable preservation and can provide a way to overcome issues to do with archiving and preserving cultural heritage. It can also allow cultural professionals an agency to present Vietnamese art and culture to local and international audiences. This has the potential to redress the imbalance in representation and to redefine digital orientalism, which exists in the differing amounts of content available online, the quality of content due to level of access to high-quality technologies and algorithms that perpetuate bias towards content from the West. However, on the other hand, there are challenges in terms of human resources and technical resources, such as the need for skilled professionals dedicated to this role and equipment that can be used to properly digitize artworks.

The research uses a case study on Matca in order to investigate how they use digitized content to curate their digital platforms (website and Facebook page), how digitization can be a sustainable solution for the precarities they face in their work (lack of funding, technical resources and human resources), and how digital display can provide a way to regain control of the narrative on Vietnamese photography. This case study is significant in demonstrating the broader transition towards digitization of art and culture in Hanoi, as Matca is at the forefront in utilizing digital technologies to display their photography collections, in using digital platforms for displaying Vietnamese photography and in building a community of dialogue and critique both locally and internationally.

The two hypotheses of the current study are as follows:

1) There have been many developments in digitization over the last five years in Hanoi, but there are country-specific challenges due to lack of funding, human resources and technical resources that are impeding developments in digitization.
2) Digitization and digital display allow Vietnamese cultural professionals to regain control over the narrative on Vietnamese art and culture.

The two research questions of the current study are as follows:

1) What are the opportunities and challenges in digitizing art and cultural collections for Matca Space for Photography?
2) How are Vietnamese cultural professionals using digitization and digital display in order to present *their* narrative on contemporary Vietnamese art and culture?

2 Methods

2.1 Methodology

The current study used an 'explanatory' case study by exploring "cause-effect relationships and how events happen[ed]" [27]. The case study was carried out between January and September 2020. Due to the length of time and specificity on one organization, the case study approach provided an in-depth and detailed understanding of Matca's current state of digitization and how they are digitally displaying this work. The case study approach provided a way of understanding the developments and challenges Matca is currently facing in the process of digitization. This highlights a politics of digitization that is specific to Vietnam.

A range of methods were used in the case study in order to triangulate the results and to provide more breadth. "While it offers depth and specificity, case study research also offers breadth and diversity in terms of methods of data collection and analytical techniques" [2]. This case study included semi-structured interviews, secondary data analysis, and a digital ethnography of Matca's digital platforms. The data was then analyzed from each method of data collection. After data analysis, the findings were combined in order to deduce key themes. "This offer[ed] the possibility of several different layers of analysis which [could] reveal several different perspectives, with the added benefit of triangulation of the results" [2].

By using interviews, digital ethnography and secondary data analysis over a nine month period (January–September 2020), the case study identified how Matca is digitizing its collections and the challenges in undertaking digitization projects, such as utilizing search engine optimization for their digital platforms. More broadly, the case study identified the effects of digitization on representation, curation and preservation of art and culture in Hanoi. It also highlighted the precarious nature of work in the art and cultural sector today, especially for an independent, non-for-profit art organisation. This is a timely case study, as it also revealed how art organisations have been effected by the COVID-19 pandemic and how this has changed the pace of and the way of digitizing collections.

2.2 Data Collection

The case study included three modes of data collection:

(1) Semi-structured interviews with Matca and other art organizations in Hanoi
(2) Digital ethnography of Matca's website and Facebook page
(3) Secondary data analysis on Matca

Three semi-structured interviews were carried out with the co-founders of Matca. These took place in January, June and July 2020. Pseudonyms are used in this paper for the interviewees. The three interviews will be referred to as 'Interview 1', 'Interview 2' and 'Interview 3' in the Results and Discussion section. Twenty semi-structured interviews were also conducted with other art organisations across Hanoi, carried out between May and September 2020, in order to provide context for what Matca is experiencing in terms of the opportunities and challenges in digitization as well as how they use their digital platforms. The digital ethnography involved collecting data from Matca's website and Facebook page from March until April 2020. Screenshots were taken of a selection of posts during this time period. Algorithms that are applied to Facebook posts meant that it was not possible to collect all posts during this time period. Posts were selected in relation to three themes: (1) emotion/connection, (2) increased digitization and (3) international connection. This limitation was mitigated by the use of triangulation with other methods. These Facebook posts were then analysed in terms of how they digitized and displayed art as well as how they used their Facebook page to communicate to local and international audiences. Secondary data and publications on Matca were collected and analysed from June until August 2020, including newspaper articles, art and culture magazine articles, past interviews published online with the co-founders and content from their website.

3 Relevant Literature

3.1 The Politics of Digitization in the Art and Cultural Sector

This section addresses the impact of digitization on the curation, preservation and representation of art and culture. A lot of the literature discusses the changes to audience engagement and participation, the changes to art itself and the impact of digitization on the art market [3, 9, 19]. For instance, Abassi, Vassilopoulou and Stergioulas argue that digital technologies have become commonplace in the creative industries, with "new materials, processes and tools for creative practices" [1]. Digital tools allow art spaces to be able to show their collections to audiences globally and to show artworks from past and present exhibitions. However, research is required on the nature of work of cultural professionals, in terms of how their work practices are changing, how they are digitizing their collections and what challenges they are facing in this process.

It is important to look at this in relation to what is happening in a developing country in South East Asia, as a lot is written about in relation to Western countries [5, 17, 21, 25]. The current study will redress this imbalance. Furthermore, it will also include discussion on the challenges of digitization, in response to much of the literature that perceives the shift towards digitization as largely positive [1, 24]. This paper identifies the challenges of digitization, including the lack of funding and resources, issues with the digitization divide and politics of representation over the narrative of Vietnamese art and culture.

Systematic protocols and standard practices need to be put in place in order for digitization to be a sustainable solution for preservation and representation of art and culture in Vietnam. As Fanea-Ivanovici argues, "sustainable digitization" is required and certain requirements are necessary for effective digitization [11]. This is critical in order to avoid a further digitization divide globally. Fanea-Ivanovici considers the need for open-access archives and museum collections so that audiences can have equal access to digital art and cultural content [11]. However, this does not address cultural professionals' need for equal access to digital technologies. Fanea-Ivanovici argues it is important to discuss this issue in order to ensure a "sustainable and inclusive growth" in the art and cultural sector [11] and to ensure more equal representation of art and culture globally.

It is important to consider the role of cultural professionals, who become agents in the digital curation of art collections. They can decide which pieces from their collections to publicly display and the themes or narrative they wish to convey. This links to Derrida and Prenowitz who argue that institutions have "the power to interpret the archive" [10]. Digital platforms allow cultural professionals to become mediators of content and to have an agency with which to convey their own narrative on art and culture.

The issue has to do with power over access and representation on a global scale. For instance, the *Louvre Museum* or the *Victoria & Albert Museum* can digitally display and make their collections accessible by using the latest digital technologies and employ skilled staff dedicated to this role, due to availability of human resources and funding. This is in contrast to the issues with budget, technical resources and human resources experienced in developing countries such as Vietnam. This can hold back or slow down the digitization process and, hence, further increase the digitization divide. This creates an imbalance in representation in art and culture globally. As Chaumont argues, "preserving becomes the privilege of the hegemony; where technological-advanced countries get to define, choose and provide cultural material for the rest of the world" [9]. This is why it is important to identify what work is being done to digitize and publicly display art and culture in Vietnam and to understand how digitization can provide a way to overcome this imbalance in representation.

3.2 The Current State of Digitization in the Art and Cultural Sector in Vietnam

Across Asia, discussion is taking place on policy around preserving cultural heritage [7, 18, 26]. However, there is very little discussion on digitization processes in Hanoi and still little policy discussion on this area with regards to agreement on professional standards and practices. This is the case even though a lot of art and culture is transitioning to online platforms, being transferred to digital formats and an increasing amount of exhibitions are being held online. Furthermore, while there have been studies on the digital disruption of the design, book publishing, and print industry sectors [5, 21, 25], relatively less has been researched on the art and culture sector. As Miles and Green argue, there are few studies of innovation in the creative industries, especially with regards to technological innovation and the digitization of content that is driving major changes in the creative industries [17].

Digitization projects have been taking place in the art and cultural sector in Hanoi, including scanning images, objects and photographs for preservation or in order to create digital archives. However, these digital archives and collections have not been used for public engagement but, rather, solely for preservation and archiving. For this reason, art and cultural organizations are only just beginning to work on the public display of and access to digitized collections. The collections of many state-run art and cultural institutions in Hanoi are not publicly displayed on a digital platform because the means for digitization onto apps or digital platforms has yet to be utilized effectively. Whereas, independent art organisations have had to innovate in order to survive and so are harnessing digital technologies more readily. There is a need for more provision of access to artworks and artifacts as well as in making it more accessible to the public.

Matca has developed its online platforms for the public digital display of photography, for publishing a photography journal and in order to create a community around photography that connects photographers, curators, researchers and critics in Hanoi, Vietnam and internationally.

This demonstrates that there has been an evolution in the use of digital technologies for work in the creative industries, especially with regards to non-profit art organizations in Hanoi. This case provides evidence of the opening up of multiple channels for international connections as well as for the display, archiving, and distribution of collections. This is allowing a more seamless transition between the initial digitization of content and the public display of content via digital means. This has been accelerated due to the pandemic, as cultural professionals have had to utilize digital platforms for work and for communicating with their audience.

4 Results and Discussion

Matca is an independent, not-for-profit art organization in Hanoi. It was co-founded by two professional photographers. Matca started in 2016 as an online photography journal. The reason for starting online was that, as Phuong says, "starting online is without question for something lacking in Vietnam" (Interview 3) as this is "the most feasible and cost-effective option" (Interview 2). This means that their work online has always been a priority. They use their website to publish an online photography journal and they also connect to their audience through Facebook, Instagram, Twitter and Youtube. Facebook is their main social media platform. However, their website is their main focus. Phuong reiterates this point by saying that "everything we do is linked back to our website. Besides disseminating information about activities in the real world, our social media are used to share published articles [from their website]" (Interview 2). As their online following grew between 2016 and 2018, they started organizing offline events in various spaces, such as coffee shops. They decided they needed a permanent physical space, which provided the impetus to open a permanent exhibition and event space. At this point, "48 Ngoc Ha came into being in 2019" (Interview 1).

They face many challenges with surviving as an art organization. This is largely because, as Phuong says, "everything we do [at Matca] is non-profit and free to access" (Interview 3). They receive funding in order to make it sustainable. "Matca does serious creative work but it doesn't generate money" (Interview 3). They must apply for grants in order to fund projects. For instance, Matca received a grant from the Danish embassy in 2018. When talking about funding from embassies, Phuong says "they [embassies] used to invest a lot in contemporary art. As Vietnam moves forwards and there is economic growth, so most of the funding has stopped" (Interview 3). This means there are fewer funding opportunities. Due to this, they have had to find other sources of funding and collaborate internationally on projects.

They also publish books and catalogues in a physical and digital format in order to sell on their premises and online. Matca published a bilingual photography journal, written in Vietnamese and English, entitled *Makét*. They have also published a bilingual photobook, entitled Hà Nôi, which was published in 2020. The first edition, *Makét 01: A Vietnamese Photography Village,* was published in 2019. It was produced by Matca in partnership with Lao Dong Publishing House. As outlined on Matca's website, "[t]he periodical publication Makét aims to explore and document the transforming photography scene in Vietnam" [16]. This is a physical and digital record - and an archive in itself - of how the photography scene has evolved over time. The production of this publication included scanning photographs, objects and artifacts, digitally archiving and storing materials as well as then organizing and arranging materials for the book. Alongside their photo-articles on their online journal and publicly displaying content on Facebook, these two publications provide another example of how they digitize and publicly display Vietnamese photography.

Matca publishes articles regularly for their online journal on their website. Through this work and publishing an online journal, Matca aims to nurture the local photography community, give recognition to photographers and bring contemporary Vietnamese photography to a global audience. They call their online journal an "open archive" (Interview 2 and Interview 3) - an archive of the story and record of the development of photography in Vietnam and the South East Asia region. The content is digitized and made publicly accessible, which means it is open and accessible to all. They feel it is important for the archive to be open, as many collections in Hanoi are currently hidden from the public. As Phuong says, "for a couple of years I have been receiving emails and interviews - our website is the only place to find photography in Vietnam. So we have it as open so everyone can access it" (Interview 3). With this initiative, they want to be at the forefront of the transition towards making more cultural content publicly accessible.

The mission and purpose of Matca is to raise awareness on Vietnamese photography, to develop the conversation on Vietnamese photography as well as to give photographers a digital and physical space to show their artwork publicly. They do this with physical events but, more often, through digitization and public display of artists' work on their website. This content is then re-posted and shared on their social media accounts. Through this work, Phuong and Duc wish to show that "[t]here is more to Vietnam and Vietnamese photography than ao dai and Ha Long bay. We don't have any specific agenda other than to showcase the diversity [in Vietnamese photography] that has always existed" (Interview 2). The commercial and easily-sharable images of traditional clothes and tourist destinations are commonly promoted in relation to Vietnamese photography internationally. Digitization can enable Vietnamese cultural professionals, like Phuong and Duc, to recreate and control the narrative on Vietnamese photography, as it involves their own choice of artworks and their own curation of digital platforms with which to disseminate to a global audience.

A key part of their mission is to redress the imbalance of representation on photography globally. "A lot of information online regarding photography is Western-centric, while local insights are often shared in casual conversations over coffee. It's hard for an emerging photographer to break into that clique and learn from those around them. There's also a certain division among people who pursue different genres of photography, say social documentary versus fashion. Professional opportunities remain rather limited. You're either a photojournalist or a commercial photographer" [16]. Matca chooses to represent those artists who are not represented due to an imbalance of representation on Vietnamese photography globally. Matca wants to change this through their work and they are trying to redefine the narrative from their own perspective. "At Matca we acknowledge and feature local photographers as well as visual artists out there whose compelling works are often overlooked in today's context." [16].

The reason for establishing Matca was to provide a digital platform and a physical space for independent artists in order to show professional Vietnamese photographers' work publicly. This type of 'outlet' is lacking in Vietnam: "in the local context where resources for independent artists remain scarce, Matca reflects our vision to offer emerging photographers the opportunity to showcase their works as well as broaden public understanding of this visual art form" [23].

They know this is required from first-hand experience as photographers themselves: "As working Vietnamese photographers ourselves, we empathise with how hard gaining visibility and recognition can be" (Interview 1). In an interview with Vincetera in 2017, Duc said "it's difficult to be an independent photojournalist in Vietnam today because it isn't really a profession yet" [23]. There is a concern around photography as it is not regarded as a skilled profession, which reflects the broader issue of the nature and precarity of work in the creative industries in Vietnam. As Phuong says "they [the photographers Matca works with] are not professional photographers in the sense that they don't earn a living from their photography practice – most have to juggle their passion project, day job, and family life. But we're continually surprised by their level of commitment without any financial reward and feel prompted to amplify their visions" (Interview 2). This precarity in the nature of work in the creative industries is experienced globally; however, this case study has identified additional factors that add to the precarity of work in the art and cultural sector in Hanoi.

Matca wants to ensure that photography in Vietnam is seen as a profession. This means nurturing critique and discussion online as well as organizing meaningful events. They understand digital platforms (when utilized in an effective way) can be the most effective way to achieve this aim. This relates to when they state that their website is used to "build a collective platform giving recognition to a growing number of artists" [16]. This platform for photography is both digital and collective, allowing people to contribute and participate through comment, critique and dialogue. The website is created and used in a way that gives recognition to these artists, in order to ensure photography is perceived as a profession and that it is presented in a professional manner online.

Their Facebook page actively encourages comment, dialogue and critique. They say they curate their Facebook page in order to create a conducive space for this kind of dialogue on photography (Interview 1). They can control the narrative in *their* way on Vietnamese photography but, also, they can create an environment for intellectual discussion on photography in order to obtain its professional status.

Furthermore, there was a change in layout and style of posts on Matca's Facebook page during lockdown. Overall, the Facebook page turned into a virtual exhibition space. This helped to diversify the function of their Facebook page, as it became a resource for art content and work opportunities for artists as well as becoming a community centre for information about the COVID-19 virus. During this time, Matca was able to develop a communication strategy in order to maintain connection with their audience and allowed them time to develop the digital curation of images, text and videos on Facebook.

Their use of digital platforms means that they can raise awareness of Vietnamese photography globally. "Over the years, we have established a strong online presence with over 18,000 organic followers on Facebook and Instagram. Even though statistics are not everything, by utilizing the readily available digital tools, we can see the emergence of a very niche audience with a sustained interest in photography" (Interview 2). During lockdown, Matca's Facebook page included posts about international events, jobs and resources; the posts and work became more transnational during the lockdown, showing how Matca was trying to invite Vietnamese artists to participate in international events and collaborative projects but also how Matca wanted to show their audience of professional photographers in Hanoi and Vietnam about exhibitions and events happening internationally.

With their work online, they want to attract and connect both a local and an international audience. Overall, the local audience (of professional photographers, curators, journalists and critics) is their priority but they also want to attract and connect with an international audience. "Digital platforms are used to engage with existing audiences and attract new audiences, through the use of hashtags, search engine optimization, interactions such as likes, comments and shares, and so forth. Based on the statistics, most are aged 18–34 from Vietnam with a minority from Southeast Asian countries and English speaking countries such as the US, the UK, and Australia" (Interview 2).

The time during the pandemic has had an impact on their work with digitization, as this time shifted their focus more to online and helped them to accelerate the digitization of their collections. In relation to the lockdown in Hanoi during March and April 2020, Phuong says "the lockdown happened when we were rebuilding the website. Having to stop most other activities including the physical space and our own work, we were able to finalize and launch the new site in early April. Like most other creative spaces during this time, we shifted our focus to the online sphere, in particular producing content on the website and social media (Interview 2).

The time during the pandemic has led to an increase in digital work. As a result, more aspects of their work and events now happen online. They have adapted their practice and have responded to the situation with, for example, allowing a way for photographers in Hanoi to still work. "We reached out to photographers to interview how they are utilizing their time at home" (Interview 2). During lockdown, they began to digitize more aspects of their work, such as uploading photographs of working from home, uploading images of events across the world, documenting (visually through photographs) the installation and de-installation of exhibitions. They provided their audience with a journey through their everyday work practices. This contrasts to beforehand, whereby their Facebook page was used solely as a communication tool to share updates on events.

However, there are challenges in digitization and using digital platforms, which Phuong raises when discussing the obstacles they faced during lockdown (during March and April 2020). "As content creators, we feel that both photographers and us should not feel pressured to immediately react to the current situation, given that we essentially only have two people on the team, and there is already a lot to be done" (Interview 2). Even though they did a lot of work during lockdown in terms of digitization, there is still more they would like to develop. However, they mention the lack of staff and resources. Phuong says they "haven't had resources to find someone with the skills" and "we don't know how to use algorithms and make use of these tools. This is something we will look into. There are just two people here. So we haven't had time to look into it"

(Interview 3). As mentioned earlier, the main challenges are in association with funding and resources – there are challenges around both a lack of technical resources and human resources. Moreover, Phuong says that it is "not as simple as – put this online and it's fine" (Interview 3). This shows that digital and social media strategy requires skills, expertise and time. Even the new generation of cultural professionals are struggling to utilize digital platforms due to challenges of lack of human resources, technical resources and funding. This will hamper developments in digitization, digital display and, hence, regaining control in representation of the narrative on Vietnamese art and culture.

5 Conclusion

The findings from the case study show that Matca has used digital platforms in order to establish and maintain a professional community around photography. Digitization provides a sustainable solution for them in order to do this, due to a lack of funding and resources. Digital platforms allow Vietnamese photography to reach international audiences and provide a way for Vietnamese cultural professionals to define the narrative on Vietnamese photography.

The time during the pandemic has given Matca the chance to digitize more of their collections, to draw closer relationships with their audience (due to the increased posts and an increased amount of digitized collections during lockdown) and to widen their scope of activities to include international events and connections. Together, this has resulted in the professionalization of these digital platforms for work. These developments in digital work can be taken forward into work practices in the future; this moment has created more efficient and meaningful development of these technologies in the art and culture sector in Hanoi. Furthermore, these digital spaces are no longer just an add-on to the physical space. As a result of these changes, there is now more synergy between the physical and digital space. These developments have also provided a more widespread acceptance and validation of Facebook as a legitimate tool for work.

This increased digitization makes these online spaces feel more like 'real' spaces. For instance, Matca posted photographs of whole-room exhibition views during the lockdown so the audience could see all the artworks in one room. This made the audience feel as though they were going around the exhibition and made the audience feel as though they were there in person. They also digitized other content as well as their art collections, in order to provide a narrative to their communication. The time during lockdown made them think, work and operate at a more international level. Posts about international workshops and online events were in abundance at this time of lockdown. This responds to a broader development in art organizations across Hanoi, which have been trying to become more internationally focused rather than only focusing on an audience in Vietnam.

It is important to discuss the challenges that cultural professionals in Hanoi are facing in the digitization process, as this is distinct from other Asian countries as well as from Western countries. These challenges include the lack of human resources in terms of skilled personell, lack of funding, and lack of technical resources such as 3D scanners in order to properly digitize cultural objects. There is a need for more advanced digitization technologies, such as 3D scanners and 3D cameras. Broadly speaking, this is required in order to avoid the most powerful museums becoming even more influential globally whilst leaving behind those that cannot afford to make their collections publicly accessible online. The digitization divide will increase if sustainable solutions are not found.

However, the opportunities of digitizing art and cultural collections for Hanoi, and Vietnam more broadly, are twofold. Firstly, digitization of content means that more people can access and view art and cultural collections, both relating to traditional culture and contemporary culture, at the local and international level. Secondly, Vietnamese cultural professionals can shape their own narrative on what is contemporary or traditional Vietnamese culture. This connects with the broader shift in Vietnam – moving away from the narrative and image associated with the phrase 'made in Vietnam' to a new narrative and image associated with the phrase 'designed, innovated and created in Vietnam'. This shift is important for the preservation and representation of Vietnamese cultural heritage and contemporary culture.

References

1. Abassi, M., Vasssilopoulou, P., Stergioulas, L.: Technology roadmap for the creative industries. Creative Ind. J. **10**(1), 40–58 (2017)
2. Adolphus, M.: How to undertake case study research (2018). https://www.emeraldgrouppublishing.com/archived/research/guides/methods/case_study.htm. Accessed 20 Sept 2020
3. Arora, P., Vermeylen, F.: Art markets. In: Towse, R., Hanke, C. (eds.) Handbook of the Digital Creative Economy Cultural Economics. Edward Elgar Publishing, Cheltenham (2013)
4. Barker, T., Beng, L.: Making creative industries policy: the Malaysian case. Kajian Malaysia **35**(2), 21–37 (2018)
5. Benghozi, P.-J., Salvador, E., Simon, J.-P.: The race for innovation in the media and content industries: Legacy players and newcomers. In: Bouquillion, P., Moreau, F. (eds.) Digital Platforms and Cultural Industries, pp. 21–40. Switzerland Peter Lang, Bern (2018)
6. Bhatiasevi, V., Dutot, V.: Creative industries and their role in the creative value chain – a comparative study of SMEs in Canada and Thailand. Int. J. Entrep. Innov. Manag. **18**(5–6), 466–480 (2014)
7. British Council, Cultural and Creative Hubs in Vietnam 2018–2021. https://www.britishcouncil.vn/en/programmes/arts/cultural-creative-hubs-vietnam. Accessed 23 Mar 2020
8. Caves, R.: Contracts between Art and Commerce. J. Econ. Perspect. **17**(2), 73–84 (2003)
9. Chaumont, P.: Reshaping the experience in art: digitization and 3D archives. https://digicult.it/news/reshaping-the-experience-of-art-digitization-and-3d-archives/. Accessed 05 Sept 2020
10. Derrida, J., Prenowitz, E.: Archive Fever: A Freudian Impression. Diacritics **25**(2), 9–63 (1995)
11. Fanea-Ivanovici, M.: Culture as a prerequisite for sustainable development: an investigation into the process of cultural content digitisation in Romania. Sustainability **10**(1), 18–59 (2018)
12. Gu, X., O'Connor, J.: A new modernity? The arrival of creative industries in China. Int. J. Cult. Stud. **9**(3), 271–283 (2006)
13. Henry, C.: Entrepreneurship in the Creative Industries: An International Perspective. Edward Elgar Publishing, Cheltenham (2007)
14. Keane, M.: Creative Industries in China: Art, Design and Media, 1st edn. Polity Press, Cambridge (2013)
15. Kim, T.: Creative economy of the developmental state: a case study of South Korea's creative economy initiatives. J. Arts Manag. Law Soc. **47**(5), 322–332 (2017)

16. Matca Homepage. https://matca.vn. Accessed 02 Oct 2020
17. Miles I., Green L.: Hidden innovation in the creative industries, NESTA report, pp. 1–81, July 2008
18. Nguyen, D. et al.: Vietnam's Future Digital Economy: Towards 2030 and 2045. Commonwealth Scientific and Industrial Research Organisation, Brisbane, Australia (2019)
19. Oudea, E.: Digital Technology is Transforming the Art World (2017). https://impakter.com/digital-technology-transforming-art-world/. Accessed 05 Sept 2020
20. Oyama, S.: In the Closet: Japanese Creative Industries and their Reluctance to Forge Global and Transnational Linkages in ASEAN and East Asia. ERIA Discussion Paper Series (2019).
21. Porter, M., Heppelman, J.: How Smart, Connected Products Are Transforming Competition. Harvard Business Review, November 2014 Issue. https://hbr.org/2014/11/how-smart-connected-products-are-transforming-competition. Accessed 20 Mar 2020
22. Pratt, A.: Creative cities: the cultural industries and the creative class. Geografiska Annaler: Series B - Human Geography **90**(2), 107–117 (2008)
23. Pugh, L.: Linh Pham and Matca: Beyond Wedding Photography and Selfies in Vietnam. Vincetera (2017). https://vietcetera.com/en/linh-pham-matca-beyond-wedding-photography-and-selfies-in-vietnam. Accessed 22 Dec 2020
24. Sherratt, T.: Hacking Heritage: Understanding the Limits of Online Access. The Routledge International Handbook of New Digital Practices in Galleries, Libraries, Archives, Museums and Heritage Sites (2018)
25. Thorén, K.: Towards an instrument for measuring strategic motives for corporate ventures. Int. J. Entrepreneurship Innov. **15**(1), 265–278 (2014)
26. Vietnam Government, Viet Nam Sustainable Development Strategy for 2011–2020 (2019). https://www.chinhphu.vn/portal/page/portal/English/strategies/strategiesdetails?categoryId=30&articleId=10050825. Accessed 18 Dec 2019
27. Yin, R.: Case Study Research: Design and Methods. Sage, Thousand Oaks (2003)
28. Yue, A.: Cultural governance and creative industries in Singapore. Int. J. Cult. Policy **12**(1), 17–33 (2006)

Creating Farmer Worker Records for Facilitating the Provision of Government Services: A Case from Sichuan Province, China

Linqing Ma[1] and Ruohua Han[2](✉)

[1] School of Information Resource Management, Renmin University of China, Beijing 100872, China
malinqing2010@126.com

[2] School of Information Sciences, University of Illinois Urbana-Champaign, Champaign, IL 61820, USA
rhan11@illinois.edu

Abstract. Farmer workers constitute a unique social group in China. Due to the limitations of their rural *hukou* (*hukou* designates the household registration status of Chinese citizens), they often cannot take full advantage of government services compared to urban *hukou* holders, and the lack of authoritative documentation on their basic information, skills, and employment histories creates more barriers to their access to public services. Government institutions have thus proposed to create records for farmer workers to better facilitate their service utilization, but only few such projects exist and more empirical research is necessary for understanding them. This paper presents an empirical case study of the Sichuan Province's government digital repository for farmer worker records. Based on a qualitative analysis of public information and data collected on-site in Sichuan, the paper presents a detailed account of the background, development trajectory, and challenges of the repository project. The analysis of the case yields three insights. First, building the repository as part of a larger business system to support business needs enables the improvement of services and the enhancement of records for farmer workers to stimulate each other. Second, the case's approach to responsibility allocation in farmer worker records management illustrates the advantages of assigning the human resources and social security department as the leading department that handles main tasks. However, its circumvention of the complexities of region-based allocation also created other challenges. Third, farmer worker records projects can be immensely labor-intensive, and securing robust, consistent government support is crucial.

Keywords: Farmer workers · Migrant workers · Farmer worker records · Government services · Digital repository

1 Introduction

In China, farmer workers (also called migrant workers; in this paper, we use the translation "farmer workers" following [1]) constitute a unique social group. Developing

K. Toeppe et al. (Eds.): iConference 2021, LNCS 12646, pp. 339–347, 2021.
https://doi.org/10.1007/978-3-030-71305-8_28

alongside the continuous transfer of labor from rural areas to cities driven by the implementation of economic and social reforms since the late 1970s, the number of farmer workers reached 290.7 million in 2019, making up 20.8% of the Chinese mainland population [2]. Farmer workers are individuals whose *hukou* is rural but have been doing non-farming-related jobs for six months or longer within or outside of where their *hukou* was issued [2]. *Hukou* designates the household registration status of a Chinese citizen. It can be categorized as rural or urban, which differentiates one's eligibility for state-provided welfare and is mostly allocated by place of birth with stringent controls on rural/urban conversion [3]. Thus, although farmer workers may live and work in cities, their rural *hukou* prevents them from enjoying the full benefits that urban *hukou* holders have in using public services.

As the struggle with the *hukou* system continues for farmer workers, China's central government and local governments have put a series of policies and regulations in affect to improve the quality and equity of public services for the group (e.g., [4, 5]). One proposed area of work is for the government to create records for farmer workers. Applying for public services such as employment aid and healthcare often requires providing authoritative, sufficient records that can authenticate one's basic information, work experiences/skills, and other relevant information [6, 7]. Unfortunately, farmer workers often lack such records due to the inadequate recordkeeping practices of their employers (mostly private organizations or individuals) [1, 8], which creates barriers to accessing government services. Therefore, the government is in the position to take responsibility for creating records for farmer workers, but only a few local governments have made progress so far. Among these initiatives, there are notable differences in what areas of information are recorded (e.g., social security status, employment, personal credit) as well as how responsibilities of creating and maintaining records are allocated across different local governments/departments [9]. Overall, the establishment of farmer worker records by government institutions is still in an emergent and experimental stage in China with very little empirical research on how such activities are carried out.

This paper reports preliminary findings of an empirical case study of the government digital repository for farmer worker records established by Sichuan Province ("the Sichuan Repository"). The study is part of an ongoing research project that broadly explores documentation practices centered on farmer workers. Previously completed work examined the community archives of farmer workers [10] and the role of records in the identity transformation of farmer workers [11]. This study follows [11] and joins global conversations on official recordkeeping practices that affect temporary workers (e.g. [12]). The purpose of this paper is to present the development, outcomes, and challenges of the Sichuan Repository project and the insights that can be gained from Sichuan's experiences.

2 Related Work

In this paper, we define farmer worker records (based on the definition of "records" in [13]) as recorded information about farmer workers (e.g., general background, employment history) that is created, received, and maintained by government institutions and/or other organizations in pursuance of legal obligations or in transactions of business relevant to farmer workers. As the management (including creation) of farmer worker records in China is an emerging phenomenon, the existing body of research is relatively small. Most studies are conceptual papers focused on outlining management challenges and proposing potential solutions.

Scholars have identified three major challenges in managing farmer worker records. One is farmer worker records are often created by and distributed across multiple government departments. Since departments may follow different procedures/standards when they generate and collect information about farmer workers most relevant to their operations, there can be overlap or gaps in their information collection, and the lack of interoperability across departmental information systems can inhibit sharing and integration [14–16]. The other two challenges are related to how farmer workers frequently move from one short-term job to another across regions. First, a high degree of mobility means keeping farmer worker records up-to-date is crucial for ensuring services can be matched to current needs based on the latest information, but it also makes updating information in time quite difficult [9, 14, 17]. Second, because most regions have not yet established digital systems for farmer worker records, it is hard to transfer digital records when farmer workers relocate between locations with and without such systems [14, 18].

Discussions on how to address these challenges are largely centered on how records management tasks and responsibilities should be coordinated. Some emphasized allocating responsibilities by region: across the management lifecycle of farmer worker records, the region where a worker's rural *hukou* was issued should be responsible for the creation, organization, retention, and digitization of records, while the region where the worker is employed should be responsible for updating their records during their length of stay [15, 17, 19, 20]. There is no consensus on whether "region" should be defined on the provincial, prefecture, county, or township level (see [15, 19, 21]). Others suggested assigning duties by department (based on its functions) without specifying region-related considerations. For instance, departments governing corresponding occupations/areas of work of farmer workers should be responsible for the creation of records, while integrating and updating information should be the duties of the human resources and social security department [14] or the archives administration [22].

Additionally, most studies share two ideas about building systems for farmer worker records: an independent, dedicated system should be built specifically for the management of farmer worker records [14, 15, 17, 19, 21, 23, 24], and the ideal goal is to establish a unified national system for managing farmer worker records in China [14, 15, 17, 19, 21, 25, 26].

While existing conceptual papers have provided a useful starting point for understanding farmer worker records management, the topic calls for more empirical research to understand completed/in-progress projects, examine key challenges and explore potential solutions, and identify and share best practices. This study on the Sichuan Repository is a step in this direction.

3 Methods

The Sichuan Repository was selected as an extreme case that highlights "the most unusual variation in the phenomena under investigation" [27]: it differs in nature from most existing projects and diverges from some main ideas in the literature. First, it is the only provincial-level repository for farmer worker records in China to date, which gives it the unique advantage of utilizing provincial resources but also the exacerbated challenge of large-scale coordination. Second, Sichuan's approach to allocating management responsibilities follows neither a region-based nor a department-based "split" as the literature usually suggests. Therefore, this unique case can potentially build on and challenge existing knowledge on the topic.

The study is a qualitative analysis of publicly-available information on the repository as well as data gathered via participant observation and a focus group interview at the Department of Human Resources and Social Security of Sichuan Province ("the HRSS"), which manages the repository. As external researchers, we started with a close reading of relevant government documents and news reports to obtain background knowledge of the case. Next, a site visit to the HRSS was completed in August 2020. Data was collected by observing demonstrations of the Sichuan Repository and conducting a semi-structured interview with a focus group of respondents from the HRSS. Interview questions focused on the process of building the repository, how records were created/collected and updated, the integration of information across departments and regions, and future development. The site visit data was analyzed using a general inductive approach [28], which is suited for qualitative analysis with relatively focused questions.

4 The Case of the Sichuan Repository

4.1 Background

Over 25 million farmer workers are Sichuan *hukou* holders (with 14 million working within the province and 11 million working outside), accounting for around 10% of all farmer workers in China [29]. Policy documents and press reports indicate Sichuan has long valued farmer workers as a core part of the workforce and is highly committed to improving public services for farmer workers by incorporating such efforts into its provincial development strategy [5, 29], which supports prioritizing and pooling resources for relevant projects. This includes building the provincial Sichuan Repository to create and manage farmer worker records that are foundational for providing comprehensive, targeted, reliable, and sustainable government public services for farmer workers [29].

4.2 Scope and Responsibilities

The project planned to create and manage records for all Sichuan-*hukou*-holding farmer workers regardless of where they reside/work and handle the full management lifecycle of these records. In this way, Sichuan circumvented region-based allocation issues of assigning different management tasks across the lifecycle to different regions because it simply covered everything on its own. Additionally, the HRSS led the project instead of assigning different responsibilities to various departments.

4.3 Development Trajectory and Information Collection and Maintenance

The Sichuan Repository was not built as a fully independent project. Instead, it has always been a part of a larger system called the System for Employment and Entrepreneurship Services ("EES System"). The EES System is the business system of the Sichuan Employment Service Administration Bureau ("the ESAB," a bureau under the HRSS) that enables the provision of employment- and entrepreneurship-related services for the public.

EES System Version 1 was completed in 2014 and updated in 2018 into Version 2. A component of ESS System Version 1 was a back-end database called the Rural Labor Force Real-name Database ("the Database"). From 2014 to 2018, somewhat of a precursor of the Sichuan Repository first existed in the form of a large subset of data of the Database and its supporting infrastructure and resources. After the release of EES System Version 2 in 2018, the Sichuan Repository developed into its current form and was officially designated as the governmental digital repository for the farmer worker records of Sichuan Province. Thus, a full account of the Sichuan Repository's story should begin in 2014.

In 2014, EES System Version 1 was implemented across all levels (prefecture, county, and township) of government employment service units in Sichuan Province, which included the infrastructure for the then-unpopulated Database. Afterwards, a highly labor-intensive process to collect information for the Database was initiated. The aim was to register all members of the Sichuan rural labor force—Sichuan rural *hukou* holders aged 16–60 and able to work [30]. Sichuan-*hukou*-holding farmer workers is a large subset of this group and therefore fully captured in the scope of collection.

Information collectors, who had been hired full-time or part-time specifically for the job, went door to door to gather information within each village-level (one level under the township level) region. Information was registered on paper forms for qualifying individuals (family members may provide information about those who are not present during the process), which included basic information (e.g., name, gender, birthdate, phone number), current employment, training and skills, and employment needs and preferences. Forms from the village-level regions were collected and verified by upper township-level department employees, who manually inputted the information into EES System Version 1. Official guidance requires conducting this process on a seasonal basis to keep information updated [30], but this proved to be difficult in practice and some areas often lagged behind.

As the process went on, EES System Version 2 was implemented in 2018. The major update was providing a new module that is connected to all databases storing registered information of individuals in the system, including the Database. The module enabled the integration and retention of information from various sources and supported uploading and storing additional digital documents for each registered individual. For farmer workers, HRSS integrated and added existing social security and training certification information from several internal bureaus to their registered information.

Apart from the door-to-door method used for Version 1, EES System Version 2 supports three new ways of information collection/updating. First, under the coordination of the HRSS, employees can use the new module to pull a predefined scope of information from six external departments (e.g., civil affairs, poverty alleviation and development, education) and store it in the EES System as necessary when providing services for farmer workers. Second, employees can use the module to upload and store digital scans of credential documents (e.g., identification cards, licenses, certificates) and other paper files submitted by farmer workers. Third, farmer workers can use the ESAB's official mobile app to submit information, which is uploaded to the EES System after verification. Unique to the Sichuan Repository, these methods can reduce the burden of collecting/updating information, with the third method also facilitating a certain degree of remote updating that may help address the mobility issue of farmer workers. This enhanced set of information, infrastructure, services, and resources is the latest iteration of the Sichuan Repository.

4.4 Outcomes and Next Steps

By July 2019, Sichuan Province has created records for over 35 million members of its rural labor force in the Sichuan Repository, which includes the records of almost 25 million farmer workers. These records have supplied valuable information for providing tailored aid in job searching, personalized training services, entrepreneurship support, and assistance in protecting labor rights. The continued challenges are the information collection process remains difficult and costly (with maintaining seasonal updates being especially demanding), and information integration from departments other than the HRSS is still limited at this stage.

As a component of the EES System, the repository currently only facilitates HRSS-related services. The long-term plan for Sichuan Province is to connect the repository to additional provincial business systems to support a wider range of public services for farmer workers. At present, a project to integrate the Sichuan Repository with the Platform for Farmer Worker Services built by the Sichuan Provincial Big Data Center is already underway.

5 Discussion and Conclusion

Several insights on farmer worker records work can be gained from this case study.

First, the way the Sichuan Repository was built raises interesting questions about the prevalent idea in the literature that creating and managing farmer worker records should require its own independent, dedicated system. Two notable points from the case are:

- The repository benefited from being constructed as a component of the larger EES System because it can take advantage of the infrastructure and resources that support the evolvement of the EES System as a whole. For example, since the module added to EES System Version 2 is shared across all component databases, it enhances the records of and the records creation/updating processes for all clients, including farmer workers. Therefore, this linked structure can ensure that the growth and maintenance of farmer worker records is supported by the same conditions as all groups served by the system.
- Since the repository was developed specifically for supporting the needs of a corresponding business system (at least in its initial stage), the alignment between providing

services via the system and collecting/maintaining relevant information is very strong. This may help to form a virtuous cycle in which services and records for farmer workers can stimulate each other's improvement driven by need. Information can be easily collected/updated directly in the course of providing services with EES System Version 2, which strengthens the farmer worker records in the repository; in turn, the strengthened records can also enhance the system's ability to provide faster and more personalized services for farmer workers.

Overall, the benefits illustrated by these points may invite deeper thinking about what creating an "independent, dedicated system" for farmer worker records really means and how it may compare to other construction pathways.

Second, the case's approach to responsibility allocation in farmer worker records management enriches our understandings of the issue. An assumption of region-based allocation is it should be easier for a region in which farmer workers are currently working to update their records, hence the need to identify and assign updating duties to such regions. Sichuan bypassed the issue at this stage by handling the entire records management lifecycle of all Sichuan-*hukou*-holding farmer workers (regardless of working location) by itself, but the downside is it is indeed challenging to update records for out-of-province farmer workers, even with the mobile app. Thus, how to better approach region-based allocation requires more research. From a department-based allocation perspective, Sichuan's way of centralizing main tasks (especially records creation) to the human resources and social security department may be more practical than the suggestions in the literature. Since a significant part of public services commonly used by farmer workers is provided by this department, information collected to support its operations can cover the main scope of farmer worker records, with a small part of additional information needed from other sources. In sum, the case deepens our understandings of responsibility allocation but also underscores the necessity of further exploration.

Finally, the case highlights the importance of having robust and consistent government support for farmer worker records projects. It is clear that the Sichuan Repository takes extraordinary, continuous effort to build and maintain, which is especially apparent in the essential but astonishingly labor-intensive door-to-door information collection process that continues to this day. The project was able to secure resources because Sichuan Province views enhancing public services for farmer workers as part of the long-term strategic development of the region [29]. Thus, the positioning of service improvement for farmer workers in government development planning may play a vital role in how sophisticated and sustainable a farmer worker records projects can eventually be. Outstanding projects will necessarily be rooted in a rightful acknowledgment of the outstanding importance of farmer workers.

In conclusion, the case of the Sichuan Repository broke new ground in the management of farmer worker records with its creative, incremental, and dedicated pathway of building an impressive farmer worker records repository. We hope Sichuan's story can provide inspiration for future farmer worker records projects and spark new ideas for further scholarly research on the topic.

Acknowledgements. This work is supported by the National Social Science Foundation of China (Grant No.16CTQ036). Thank you to our HRSS respondents for sharing their time and knowledge.

References

1. Xie, L., Feng, H., Ma, L.: Records and farmer workers-a unique Chinese case. Arch. Rec. **40**(3), 259–280 (2019). https://doi.org/10.1080/23257962.2017.1392936
2. The 2019 farmer workers survey report 2019 年农民工监测调查报告. https://www.stats.gov.cn/tjsj/zxfb/202004/t20200430_1742724.html. Accessed 04 Oct 2020
3. Chan, K.W.: The Chinese hukou system at 50. Eurasian. Geo. Econ. **50**(2), 197–221 (2009). https://doi.org/10.2747/1539-7216.50.2.197
4. Opinions of the State Council on further improving services for farmer workers 国务院关于进一步做好为农民工服务工作的意见, https://www.gov.cn/zhengce/content/2014-09/30/content_9105.htm. Accessed 04 Oct 2020
5. Sixteen measures to strengthen services for farmer workers 加强农民工服务保障十六条措施. https://www.sc.gov.cn/10462/10464/10797/2018/12/7/10464400.shtml. Accessed 04 Oct 2020
6. Liao, J.: An exploration on creating comprehensive records for farmer workers 浅析建立健全农民工档案. China Arch. **11**, 42–43 (2007)
7. Zhang, Y.: Some thoughts on the creation of records for farmer workers 对建立农民工档案的几点思考. Arch. Manag. (02), 85–87 (2011). https://doi.org/10.15950/j.cnki.1005-9458.2011.02.036
8. Feng, H.: The value of archives in modern identity 当代身份认同中的档案价值. J. Renmin. Univ. China **29**(01), 96–103 (2015)
9. Li, Z.: Discussion on the construction and improvement of farmer workers records 浅议农民工档案的建立与完善. Shandong. Arch. **01**, 27–28 (2011)
10. Ma, L., Ma Y., Zhang W.: Practice and reflection on the construction of community archives under the background of social transformation—Based on a case study of the Pi Cun Working Culture and Art Museum 社会转型背景下社群档案馆建设的实践与思考——基于皮村打工文化艺术博物馆的个案研究. Arch. Sci. Bull. (03), 67–71 (2018). https://doi.org/10.16113/j.cnki.daxtx.2018.03.018
11. Xie, L., Feng, H., Ma, L.: The function of records in the process of identity transformation—a case study of farmer workers in China 转型身份认同过程中档案的功用——以中国农民工群体为例. Arch. Sci. Bull. (01), 4–8 (2019). https://doi.org/10.16113/j.cnki.daxtx.2019.01.001
12. Garcia, P.: Documenting and classifying labor: the effect of legal discourse on the treatment of H-2A workers. Arch. Sci. **14**(3–4), 345–363 (2014). https://doi.org/10.1007/s10502-014-9230-4
13. Association of Records Managers and Administrators (ARMA) International: Glossary of records and information management terms. 3rd edn. ARMA International, Lenexa (2007)
14. Wang, P.: Research on the construction of farmer worker records based on the equalization of public service 公共服务均等化视角下的农民工档案建构策略研究. Arch. Constr. (08), 34–37, 42 (2018)
15. Xie, L.: Research on the management mechanism of farmer worker records under the background of the new urbanization strategy 新型城镇化背景下的农民工档案管理模式研究. China Mark. **12**, 85–87 (2013)
16. Zhang, J.: Analysis on the information management of farmer worker records 农民工档案信息化管理探析. Hum. Resour. Dev. (13), 19–20 (2019). https://doi.org/10.19424/j.cnki.41-1372/d.2019.13.013
17. Wang, L.: Analysis on the management mechanism of farmer workers records from the perspective of urbanization in Guangxi 广西城镇化视角下农民工档案管理模式探析. Lantai World. (17), 75–76 (2014). https://doi.org/10.16565/j.cnki.1006-7744.2014.17.026

18. Cao, A.: On issues in farmer worker records management 农民工档案管理问题初探. J. Luohe Vocat. Technol. Coll. **11**(04), 185–187 (2012)
19. Cheng, X.: An exploration of the records of a new type of migrants—farmer workers 中国新型流动人员——农民工档案初探. Master's thesis, Shandong University (2010)
20. Li, H.: Analysis on the management mechanism of farmer workers records 农民工档案管理模式探析. J Changsha Railw. Univ. (Soc. Sci.) **12**(04), 279–280 (2011)
21. Wang,Y., Huang,Y.: Research on farmer worker records under the background of the national strategy of urbanization 城镇化国家战略背景下农民工档案问题探析. Lantai World (07), 17–20 (2016). https://doi.org/10.16565/j.cnki.1006-7744.2016.07.05
22. Zhang, G.: Research on the current situation and innovative mechanism of farmer worker records management under the background of the integration of urban and rural areas 城镇一体化背景下农民工的档案管理现状与创新模式研究. Abil. Wisdom **28**, 242–244 (2016)
23. Zhang, D.: Feasibility analysis of establishing farmer worker records 建立农民工档案的可行性分析. Leg. Syst. Soc. **02**, 503–504 (2007)
24. Yuan, L.: On the practical function of farmer worker records based on the 12th Five Year Plan 基于十二五规划谈农民工档案的现实作用. Lantai World. (18), 52–53 (2011). https://doi.org/10.16565/j.cnki.1006-7744.2011.18.051
25. Jiang, H.: Research on the management model of farmer worker records by township governments, supplemented by employer organizations 以乡政府为主用工单位为辅的农民工档案管理模式探讨. Ningxia J. Agric. For. Sci. Technol. **53**(09), 102–103 (2012)
26. Gu, K.: The management of farmer worker records and safeguarding rights 农民工档案管理与依法维权问题探析. Hum. Resour. Dev. (05), 49–51 (2019). https://doi.org/10.19424/j.cnki.41-1372/d.2019.05.023
27. Mills, A.J., Durepos, G., Wiebe, E.: Encyclopedia of case study research. Sage, Thousand Oaks (2010). https://doi.org/10.4135/9781412957397
28. Thomas, D.R.: A general inductive approach for analyzing qualitative evaluation data. Am. J. Eval. **27**(2), 237–246 (2006). https://doi.org/10.1177/1098214005283748
29. Why providing services for farmer workers is vital in Sichuan 服务保障农民工为什么是四川的战略性工程, https://www.sohu.com/a/325380649_207224, Accessed 07 July 2019
30. Interim measures for the dynamic management of information on the real-name system of rural labor in Sichuan Province 四川省农村劳动力实名制信息动态管理暂行办法. http://rsj.panzhihua.gov.cn/jyfwglj/zcwj/1081774.shtml Accessed 04 Oct 2020

Case Study on COVID-19 and Archivists' Information Work

Deborah A. Garwood(✉) and Alex H. Poole

College of Computing and Informatics, Drexel University, Philadelphia, PA 19104, USA
dgarwood@drexel.edu

Abstract. This paper presents preliminary findings from an exploratory, qualitative case study bounded by the city of Philadelphia. The case study brings the literature on information work (IW) to bear for the first time on archives and special collections repositories. Empirical interview data on archivists' information work at five medical history collections, pre- and post- pandemic onset, suggests that institutional and personal conditions surrounding COVID-19 prompted archivists to change their information work tasks in phases, first shifting office tasks to remote work under quarantine, then to hybrid work contexts. We explore an information work model including work purposes, work tasks, and work roles. The model shows how tasks of collection management, reference services, and outreach constitute the context and purpose for archivists' information work. The paper details how hybrid work tasks and hybrid work contexts emerged.

Keywords: Archives · Information work · Pandemic

1 Introduction

Declared a public health emergency of international concern (PHEIC) on January 30, 2020 by the World Health Organization (WHO), the COVID-19 virus outbreak continues to disrupt interconnected global and local knowledge systems at the end of 2020.[1] In parallel to local and global health information networks, the work of information professionals at cultural heritage institutions is a vital component of non-pharmaceutical interventions that support human and animal life and sustain the environment [1, 2].

Websites archivists develop for people to record and share their stories about the pandemic contribute to a reservoir of knowledge useful now and for posterity. In the US, for example, archivists at the National Library of Medicine (NLM) are collecting first-hand accounts of COVID-19 as part of NLM's web archiving repository for Global Health Events [3, p. 496].[2] Similarly, the Society of American Archivists (SAA) is actively collecting first-hand accounts and reporting on similar national initiatives.[3] Archivists remain engaged in information work even in the face of ubiquitous institutional closures and hybrid, often tentative and contingent re-openings.

[1] https://www.who.int/director-general/speeches/detail/who-director-general-s-statement-on-ihr-emergency-committee-on-novel-coronavirus-(2019-ncov).

[2] https://archive-it.org/collections/4887.

[3] https://www2.archivists.org/news/2020/archivists-rally-to-document-covid-19.

K. Toeppe et al. (Eds.): iConference 2021, LNCS 12646, pp. 348–357, 2021.
https://doi.org/10.1007/978-3-030-71305-8_29

1.1 Case Study Synopsis

This paper considers how the pandemic's onset impacted the information work of nine archivists at special collections repositories in one urban center. It is part of an exploratory qualitative case study that brings the literature on information work (IW) to bear for the first time on archives and special collections work [4–7].[4] The case study contributes both theoretically in testing and applying the IW framework and empirically through two sets of semi-structured interviews spanning pre- and post-COVID-19 onset in Philadelphia. The first round included all 11 participants in the sample and took place from 10 November 2019 to 12 March 2020 at work settings. The second round with nine participants occurred from 7 August to 24 September 2020 via videoconferencing.

We first briefly review the germane literature. Next, we outline the study's methods. Third, we present our paper's results. Fourth, we discuss the theoretical and empirical implications. Finally, we note limitations and suggest areas for future research.

2 Literature Review

At a time when information that transcends disciplinary borders is useful to pandemic research, archivists' information work is enabling regular professional life to continue [1, 2, 8, 9]. More than three decades of research on information work has enlarged the concept of what information is, who uses it, how it is organized and accessed, and how tasks, lines of work, biography, emotion, and resources operationalize shared responsibilities [4–6, 8–19]. Information work (IW) literature suggests that work, a broad concept, becomes specified through contextual description and definition which, in turn, renders tasks as visible or invisible [4, 16, 17, 20]. Work tasks entail the construction, performance, and evaluation of information needed, sought, and used [16, 19]. In the information work setting, more specifically, tasks are theorized as the performance of an objective task description according to one's internalized comprehension of it [16]. IW concepts regarding work purpose, work roles, work tasks and their performance imply that qualitative methods are suitable for information work research on archival work settings bounded by Philadelphia [4, 8, 9].

3 Methods

The case study relies on qualitative research methods in the naturalistic paradigm [21, 22]. Purposive sampling and analysis influenced by Constructivist Grounded Theory inform interview techniques and coding [23–25]. Textual data in this paper was drawn from semi-structured interviews with archivists in late summer 2020. The author recorded interviews using the Zoom videoconferencing platform and fully transcribed recordings with iTunes and Word's "dictate" tools. Trustworthiness for the study accrues through repeated contact with participants and peer debriefings; a research journal and analytic memos are hedges against, and acknowledgment of, researcher bias

[4] The special collections repositories in the case study include archives, manuscripts, rare books, medical materiel, and records in multiple format types. I refer to case study participants as archivists because archival science applies to information work in these settings.

[21]. Findings are bounded rather than generalizable: trustworthiness forms a basis for transferability to other settings at the discretion of the reader.

4 Findings

Findings center on second round interview data with nine of 11 participants. Four phases demarcate transitions from a pre-closure work context (the period of first interviews) to a hybrid work context by the second interviews. Some archivists anticipated institutional closures due to pandemic onset while others did not. In March 2020, as P06 put it, "We didn't know at that time how long we would be closed". Uncertainty motivated archivists to reconfigure work tasks in overlapping phases (Table 1).

Table 1. Phases of pre-closure, preparation, transition to quarantine, and hybrid work tasks

Phase	Pre-closure, preparation, transition to quarantine, hybrid work tasks		
Phase 1 Interview 1	Pre-closure 11–2019 to 3–2020		
Phase 2		Preparation 1–2020 to 3–2020	
Phase 3		Transition to work under quarantine 3- to 6–2020	
Phase 4 Interview 2			Hybrid work tasks and work contexts 6- to 9–2020

4.1 Phase 1 and Phase 2: Pre-closure and Preparation

Participants comment on the timing of institutional closure and their preparation for closure (Table 2).

Preparation for Closure. Intern P03 anticipated closure by six weeks. She consolidated two projects, one external to the workplace, and prioritized completing it one week before the institution's three-day notification of closure. At the same workplace, temporary archivist P11 moved project files to two external drives one week ahead and transported them home; most of her tasks involved digital systems for storage and website work. Archivist P08, a fulltime staff of one, received notification three days before departure but had anticipated closure; she made plans to handle reference questions remotely. Archivist P01's work tasks primarily involved tours and reference services. Expecting a two-week hiatus, she bundled physical files and "bottom of the pile" tasks for creating databases on three days' notice of closure. Repository director P05 and archivist P06 received three hours' notice to depart; they took what physical files they could carry – and cleaned out the employee refrigerator.

Table 2. Pre-closure work tasks and preparation for closure

Pre-closure work tasks	Preparation for closure	Participant code
Received three-day notification but anticipated quarantine by six weeks	Consolidated two projects at the workplace and prioritized completing one before departure	P03
Received three-day notification but anticipated quarantine by one week	Readied equipment and electronic files to work from home	P11
Received three-day notification but anticipated quarantine by about three weeks	Made plans to communicate with researchers about reference services	P08
Received three-day notification to depart and quarantine	Bundled physical files and made plans to create databases	P01
Received three hours' notification to depart and quarantine	Quickly scooped up a few physical files	[P05, P06]

Pre-closure work and preparation channeled into performing work tasks under quarantine.

4.2 Phase 2 and Phase 3: Preparation and Transition to Work Under Quarantine

Under quarantine, the context for work tasks shifted dramatically from the office to the home (Table 3). Three participants were already connected to their institutions' remote servers (P02, P08, P09). Four reported a moderately easy transition to remote set up within one week (P01, P03, P10, P11), while two participants at another institution had email access but suffered a three month wait to get remote server access (P05, P06).

Table 3. Work context under quarantine

Quarantine work context	Tasks	Participant code
Work context at home	Already set up for remote work	P02, P08, P09, P11
	Moderately easy transition to remote access in one week	P01, P03, P10
	Challenging transition, remote access took three months	P05, P06

Quarantine Work Context. P02, head archivist of 20 years, easily adapted to performing work tasks at home; the institution's remote access to electronic collections dated from "years earlier". P08 likewise reported that her transition to performing work tasks from home on a work-issued laptop was "fairly easy" after her institution improved existing infrastructure for snow days to boost remote access. P09's well-organized home workspace attested to an easy transition as well. P11, finally, adjusted within a week. By contrast, P01 experienced frustration, communicating by phone with IT staff over a whole week merely to connect to her work server. Conversely, though they had remote access, P03 felt discomfort and P10 temporary disorientation while adjusting to working from home. While waiting three months for remote access, P05 and P06 resourcefully used email for reference services, created bridge projects for visitor services staff, curated a digital exhibition, and collaborated with colleagues in other departments.

4.3 Phase 4: Hybrid Work Tasks and Hybrid Work Contexts

As Sects. 4.1 and 4.2 demonstrate, phasing work tasks to a quarantine context entailed technological, psychological, and social adjustment. Enabled to access repository collections, email, and virtual conferencing platforms to perform work tasks under quarantine between April and June, participants phased in hybrid home-remote and in-office work tasks from June to August 2020 while institutions remained closed. Hybrid tasks involved four lines of work: communicating, coordinating, and collaborating with colleagues, creating remote work for staff whose work was exclusively on-site, adapting reference services for the hybrid context, and making on-site visits for reference work and to check on collections (Table 4).

Table 4. Hybrid work tasks and hybrid work contexts

Hybrid work context	Tasks	Participant code
Performed work tasks remotely and on-site	Communicated and collaborated with colleagues	All
	Created remote work for visitor services staff and transcribed catalog cards	P05, P06
	Adapted reference services to remote and hybrid delivery	All
	Brief on-site visits to fulfill reference requests and check on physical collections	All
	Planned for reopening and performing on-site information work tasks	All

Communicated, Coordinated, and Collaborated with Colleagues. P01 communicated weekly with her volunteers, several of whom are elderly and/or live alone. In similar spirit, P02 regularly checked in with his colleague who was on maternity leave. P08 attended remote departmental meetings concerning remote work policies and employee well-being. P05 and P06 met virtually and often, finding a productive work rhythm by

April 2020 while awaiting remote connection, which came only in June 2020. The four interviewees from the same work setting, P03, P09, P10, and P11, held frequent wellness check-ins during the first weeks of quarantine.

Created Remote Work for Staff Normally On-Site. P05 created information work for visitor services staff; they transcribed 19th century medical student notebooks showing future research potential. Both P05 and P06 collaborated with colleagues in other departments on transcriptions of 19th and early 20th century museum catalog cards.

Adapted Reference Services to Remote and Hybrid Delivery. P01 "canceled the calendar" for tours of the building, VIP events, and on-site research appointments. Like all participants, she addressed research questions via email and took notes on documents to consult on returning to the archives. P02 meanwhile discovered under quarantine welcome opportunities to pursue reference questions using digitized collections—and to do so in more depth than was possible when immersed in her hectic physical work context. Similarly, P08 notified researchers that reference services were open even if the archives were closed; like P02, she relied heavily on previously digitized resources for detailed requests. For frequently asked questions, finally, P08 directed users to a webpage she had previously set up for the purpose.

Conversely, P03 and P10 were unaccustomed to performing work tasks in the home. The separation between home and work contexts felt meaningful and required effort—and time—to overcome. Once reoriented, however, both were highly productive: P03 completed a substantial metadata project for 19th century medical student doctoral theses, and P10 resumed reference services for questions that called on her extensive knowledge of collections. Like P02 and P08, P10 and her colleagues relied on previously digitized resources for reference requests.

Made Brief On-Site Visits. By May or June 2020, participants had leveraged state and institutional safety protocols to make brief on-site visits to scan or retrieve materials necessary for reference questions and to check on the physical status of collections.

Planned for Reopening and On-Site Work Tasks. By July 2020, all participants had planned for on-site work tasks according to state and institutional reopening protocols.

4.4 Hybrid Work Tasks Through December 2020

As of their second interviews conducted in August and September 2020, participants had maintained hybrid work contexts and hybrid work tasks while adapting to staff reductions (Table 5). P05 and P07, both directors, left their positions in July 2020 (P07 declined the second interview). P09 anticipated the loss of temporary staff and the intern. Only P01 and P06 performed on-site reference services and tours, though at reduced capacity.

Hybrid Re-openings. P01 and P06 were glad to reconnect with colleagues and regain proximity to their repositories' physical collections. P01 remarked poignantly, "There's an emotional stake in everything that's here". P01 scheduled a minimum of on-site reference appointments and tours of the physical building, whereas P06 followed elaborate

Table 5. Hybrid work contexts and tasks anticipated through December 2020

Hybrid work context	Task	Participant code
Conduct work tasks through end of 2020	Glad to be on-site, missed physical collections and colleagues	P01, P06
	Scheduling a minimum of on-site visits and tours	P01
	Following elaborate protocol (state & institution) for on-site reference services	P06
	Catching up on reference interview work that could not be done remotely, physical care of collections, administrative responsibilities due to staff reductions	P01, P06, P03, P09, P10, P11
	No on-site reference interviews yet	P02, P08, P09, P10, P11
	Continuing remote information work through end of 2020	All
	Might be terminated	P11
	Planning to virtualize as many services as possible	P09
	Staff reductions, related and not related to COVID-19, increase responsibilities for which staff are not trained	P06, P03, P09, P10, P11

state and institutional protocol to ensure safe on-site reference services. P06 also caught up on her backlog of reference interview tasks, such as scanning and shipping materials, and returned to projects put on hold during quarantine. In addition, P06 tended physically to collections and to a limited amount of processing given staff reductions in July that conferred on her temporary administrative responsibilities. At the same repository, P03, P09, P10, and P11 also found their work tasks much altered by loss of two staff.

Plans to Continue Working Remotely from Home. Given institutional policy to work from home if possible, P02 plans to continue her remote work tasks. P08's remote work tasks persist through the end of 2020; her department is developing a policy for remote work tasks. "It's been interesting to see what we can do without the physical documents," she notes. Temporary staff member P11 works exclusively with electronic documents and digitized heritage resources. She plans to continue remote tasks.

Adapting Work Tasks to Staff Reductions. Grant funding that subsidizes P11 runs out at the end of 2020. Given that challenge and the unlikely prospect of replacing two staff who left over the summer, P09 is planning to virtualize as many services as possible and pare back tasks she shares with P10, the reference specialist and remaining full time staff member.

Hybrid Work Tasks and Hybrid Work Contexts in 2020. Participants' phasing of work tasks due to COVID-19 quarantine beginning in March 2020 enabled hybrid work tasks to emerge in hybrid work contexts post-quarantine. P06 remarked, "it was a good way to get us to think about how much we can actually do remotely, but it was very challenging". In starker terms, P09 observed, "the pandemic has accentuated what our needs are and accentuated our limited resources".

5 Discussion

This paper surfaces the complex, emotional, time-consuming work archivists perform professionally in the midst of the pandemic [4, 8, 9]. Interview data explores how archival work tasks traverse the boundaries between processing collections, invisible to users, and reference services and outreach, the visible output of their work [17, 20]. Building on Huvila's [4] premise that work purpose is a useful context for tasks, Table 6 introduces an information work model for the case study.

Table 6. Information work model for the case study

Work (purpose)	Work tasks (outcomes)	Work roles (viewpoints)
Reference services (visible work)	Reference services	Director, Librarian, Archivist, Administrator, Head archivist, Staff-of-one archivist, Reference specialists
Outreach (visible work)	Donor relations	
	Social media	
	Tours	
	Student orientation day	
Processing collections (invisible work)	Physical arrangement of collections	Archivist, Curator, Managing archivist, Processing archivist, Project archivist, Student intern
	Writing finding aids and description	
	Digitizing resources	
	Metadata	
	Curating website reference content	
	Curating online exhibitions	
	Creating transcription tasks for intra-institutional students and staff	

6 Conclusion

The exploratory qualitative case study applies information work concepts to archivists' information work tasks at five medical history collections in Philadelphia. This paper analyzes second round interview data to explore how nine archivists phased work tasks while transitioning from office to home under quarantine from March through September 2020. These interviewees managed information work within their institution and external to it. Indeed, their information work shifted between visibility and invisibility as hybrid work contexts for hybrid work tasks emerged. Future research may explore archivists' information work and dynamics of visible and invisible work at archives and special collections repositories.

References

1. National academies of sciences, engineering, and medicine, exploring lessons learned from a century of outbreaks: readiness for 2030. In: Proceedings of a Workshop, p. 25391. National Academies Press, Washington, D.C. (2019)
2. Miller, M.A., Viboud, C., Balinska, M., Simonsen, L.: The signature features of influenza pandemics—implications for policy. N Engl. J. Med. **360**(25), 2595–2598 (2009). https://doi.org/10.1056/NEJMp0903906
3. Greenberg, S.J.: Resilience, relevance, remembering: history in the time of coronavirus. JMLA **108**(3) (2020). https://doi.org/10.5195/jmla.2020.986
4. Huvila, I.: Work and work roles: a context of tasks. J. Documentation **64**(6), 797–815 (2008). https://doi.org/10.1108/00220410810912406
5. Cox, M.: An exploration of the practice approach and its place in information science. J. Inf. Sci. **38**(2), 176–188 (2012). https://doi.org/10.1177/0165551511435881
6. Savolainen, R.: Information behavior and information practice: reviewing the "umbrella concepts" of information-seeking studies. Libr. Quart. **77**(2), 109–132 (2007). https://doi.org/10.1086/517840
7. Huvila, I.: How a museum knows? structures, work roles, and infrastructures of information work. J. Am. Soc. Inf. Sci. Tech. **64**(7), 1375–1387 (2013). https://doi.org/10.1002/asi.22852
8. Hogan, T.P., Palmer, C.L.: "Information work" and chronic illness: Interpreting results from a nationwide survey of people living with HIV/AIDS. Proc. Am. Soc. Info. Sci. Tech. 42(1), n/a-n/a (2006). https://doi.org/10.1002/meet.14504201150
9. Strauss, A., Corbin, J.: Managing chronic illness at home: three lines of work. Qual. Sociol. **8**(3), 224–247 (1985). https://doi.org/10.1007/BF00989485
10. Fulton, C., Henefer, J.: Information practice. In: Bates, M.J., Maack, M.N. (eds.) Encyclopedia of Library and Information Sciences, 3rd edn., pp. 2162–2171. CRC Press (2018)
11. Lloyd, A.: Learning to put out the red stuff: becoming information literate through discursive practice. Libr. Quart. **77**(2), 181–198 (2007). https://doi.org/10.1086/517844
12. Lloyd, A.: Informing practice: information experiences of ambulance officers in training and on-road practice. J. Documentation **65**(3), 396–419 (2009). https://doi.org/10.1108/00220410910952401
13. Lloyd, A.: Framing information literacy as information practice: site ontology and practice theory. J. Documentation **66**(2), 245–258 (2010). https://doi.org/10.1108/00220411011023643
14. Wenger, E.: Communities of Practice: Learning, Meaning, and Identity. Cambridge University Press, Cambridge (1998)

15. Wenger, E., McDermott, R.A., Snyder, W.: Cultivating Communities of Practice: A Guide to Managing Knowledge. Harvard Business School Press, Boston (2002)
16. Byström, K., Hansen, P.: Conceptual framework for tasks in information studies. J. Am. Soc. Inf. Sci. **56**(10), 1050–1061 (2005). https://doi.org/10.1002/asi.20197
17. Veinot, T.C.: "The eyes of the power company": workplace information practices of a vault inspector. Libr. Quart. **77**(2), 157–179 (2007). https://doi.org/10.1086/517842
18. McKenzie, P.J.: A model of information practices in accounts of everyday-life information seeking. J. Documentation **59**(1), 19–40 (2003). https://doi.org/10.1108/00220410310457993
19. Byström, K., Lloyd, A.: Practice theory and work task performance: how are they related and how can they contribute to a study of information practices. Proc. Am. Soc. Info. Sci. Tech. **49**(1), 1–5 (2012). https://doi.org/10.1002/meet.14504901252
20. Star, S.L., Strauss, A.: Layers of silence, arenas of voice: the ecology of visible and invisible work. Comput. Support. Coop. Work **8**(1–2), 9–30 (1999). https://doi.org/10.1023/a:1008651105359
21. Lincoln, Y.S., Guba, E.G.: Naturalistic Inquiry. SAGE Publications, Beverly Hills (1985)
22. Mellon, A.: Naturalistic Inquiry for Library Science: Methods and Applications for Research, Evaluation, and Teaching. Greenwood Press, New York (1990)
23. Miles, M.B., Huberman, A.M., Saldaña, J.: Qualitative Data Analysis: A Methods Sourcebook, 4th edn. SAGE Publications, Los Angeles (2020)
24. Charmaz,K.: Constructing Grounded Theory, 2nd edn. SAGE Publications, London, Thousand Oaks (2014)
25. Rubin, H., Rubin, I.: Qualitative Interviewing: The Art of Hearing Data, 2nd edn. SAGE Publications, Thousand Oaks (2005)

Research Methods

A Meta-review of Gamification Research

Ping Zhang[1], Jian Tang[2](✉), and Eunmi (Ellie) Jeong[1]

[1] Syracuse University, Syracuse, NY 13244, USA
{pzhang,ejeong01}@syr.edu
[2] Central University of Finance and Economics, Beijing 100081, China
jiantang@cufe.edu.cn

Abstract. Gamification has gained significant attention from academia and industry in the recent decade. Correspondingly, there has been a prolific publication record on gamification research. This research aims to explore a landscape view of gamification research through a meta-review, a systematical assessment of a collection of gamification literature reviews. The meta-review addresses four research questions on (1) literature review scope, (2) application domains, (3) review types, and (4) review foci in 48 reviews published from January 2018 to June 2020. This research contributes to a high-level overview of the state of development in the gamification field. It also demonstrates a process of conducting meta-reviews.

Keywords: Gamification · Gamification review · Meta-review

1 Introduction

Gamification is characterized by using game elements to serve serious purposes [1]. Gamification also refers to designs that afford similar experiences and motivations as games do [2]. The first documented use of gamification as a term appeared in 2008 or earlier, but the term was officially defined by academia researchers in 2011 [1, 3]. Over the past 10 years, a large and fast-growing body of literature has demonstrated the wide applications of gamification in many domains [4]. In the broad information field, gamification has been applied to the design and evaluation of information artifacts [5], to change users' attitudes [6], motivation [7], and behaviors [8]. With the prolific publications of primary studies on gamification, such as empirical investigations, design cases, and prototype evaluations, a growing number of literature reviews have been conducted to develop conceptual understandings, synthesize empirical evidence, and identify topics and gaps for future investigations. Many of these literature reviews show particular aspects, streams, or domains of the literature (e.g., [9, 10]), yet limited effort was exerted to leverage the findings of multiple literature reviews to form a landscape view of gamification research. A review of existing literature reviews can be insightful at a high level to depict the state of the art of the gamification research.

Umbrella review and meta-review are two frequently used methods to conduct reviews of literature reviews. An umbrella review integrates relevant evidence from multiple systematic reviews to address a narrow research question [11]. A meta review is the systematic evaluations of reviews [12] and can have multiple objectives and address a wide range of research questions.

K. Toeppe et al. (Eds.): iConference 2021, LNCS 12646, pp. 361–373, 2021.
https://doi.org/10.1007/978-3-030-71305-8_30

To our best knowledge, there are few published reviews of literature reviews on gamification research with the exception of two umbrella reviews [13, 14] aiming to address specific research questions. Considering the developing state of the gamification field, an evaluation of existing literature reviews will be beneficial for researchers and practitioners to gain a landscape view of this field. Reviews analyzed in a meta-review may use qualitative and quantitative primary studies as sources, and they could be of any review types and address any research questions. We believe that such a meta-review has not been published yet. Thus, we expect that this meta-review could describe gamification research and present novel findings in this field by addressing the following research questions.

RQ1: What is the scope of reviews? Such as year spans, and numbers of primary studies reviewed?
RQ2: What application domains are covered in the reviews?'
RQ3: What review types have been conducted?
RQ4: What has been reviewed? Such as field of study, factors, and components of nomological net?

2 Research Method

2.1 Paper Selection Procedure and Criteria

A literature review (or "review" for the rest of this paper) is a scholarly publication that surveys published primary studies. Literature reviews help readers understand the existing body of knowledge regarding a specific topic or a selected field. In this study, published literature reviews of gamification research are selected and analyzed to answer the four research questions.

Multiple reviews confirmed that the first academic publications on gamification research were published around 2010 [14–16] (but see one exception in Fig. 1 below). Although we found that the earliest reviews on gamification research were published in 2014 [17], we believe that a more recently published review is more likely to include available primary studies spanning from early years to close to the time the review is published. For example, a review published in 2020 likely to be able to include all primary studies published from 2010 to when the review was conducted. Given that our purpose is to explore the current developing state of gamification research, this meta-review considered only those reviews published since 2018 to ensure coverage of many recent primary studies.

Searches were conducted using a variety of search terms that include "gamif*" ("gamification" or "gamified" if wild card is not supported) and "review" in the topic field of Web of Science and in the title, abstract and keywords fields of Science Direct. The following screening criteria were applied to the search results: (1) written in English, (2) peer-reviewed, (3) has a gamification focus even if gamification is not the only theme, and (4) a review paper that synthesized the results of multiple primary studies.

The resulting pool of reviews, after removing duplicates, was further examined in full text with semantic filtering criteria to ensure that the included reviews are of high quality. We reviewed the best practices in the medical field because this field has well-accepted systematical review protocols, specific review procedures, or highly cited methodology cases [12], then developed the following filtering criteria: (1) explicitly shows the intention of the authors to review or summarize the literature (e.g., has review, overview, or meta-analysis in the title, abstract or section headings), (2) demonstrates a clear set of objectives or research questions, (3) presents a methodology that allows us to judge the review types (see Sect. 3.3 below for review types), (4) has more than one primary study on gamification, and (5) presents the synthesis of findings beyond those provided by single studies.

All types of reviews were included in this study due to the relatively young age yet fast-growing nature of gamification research. After screening and filtering, a total of 48 reviews (see the list at the end) were included for data extraction and analyses.

2.2 Data Extraction and Quality Control

Data were extracted from various facets to answer the research questions. Two researchers independently examined the reviews, extracted data from each review, and then compared the results. Where there was a disagreement between the two researchers, discussions led to the final agreement.

3 Findings

The 48 reviews were published in a total of 42 different outlets. Five journals had more than one review: International Journal of Serious Games (3), Computers & Education (2), Computers in Human Behavior (2), JMIR Mental Health (2), and Technology, Knowledge, and Learning (2). This suggests that there are no concentrated outlets to publish gamification reviews.

3.1 RQ1. What is the Scope of the Literature Reviews?

The scope considers (a) methods of finding primary studies, (b) sources (search engines or databases) used to identify primary studies, (c) minimum, maximum, median, mean, and standard deviation of primary studies per review, and (d) year spans of published primary studies. Collectively, these results can provide an overview of the extensiveness of the coverage of gamification reviews.

Except for two narrative reviews that did not explicitly specify the search procedure and method, the remaining reviews all used keyword searches with search engines, e-libraries, and scientific databases. Additional ways of finding primary studies, though only used sparsely, are reported in Table 1 (a).

Three reviews did not specify any engines or databases. Many reviews used more than one search engine or database. The top 10 most used search engines and databases are reported in Table 1 (b). Google Scholar and Scopus are the most popular sources for data gathering in the reviews. This may have to do with their extensive coverage and relatively wide accessibility. Overall, scientific databases are still the major outlets that researchers use for primary study collection.

For the number of primary studies per review, the minimum, maximum, median, mean, and standard deviation are 2, 1164, 27, 70, and 175, respectively. The drastic differences in the number of primary studies covered in a review may be due to several factors, such as available sources, available time and effort, review type, application domain, the year span of publications covered, and inclusion criteria.

Table 1. Summary of review scope

(a) Methods of collection		#Review
(Some reviews used multiple methods)	Keyword search	46
	Convenient sampling	1
	Forward, backward	3
	Journal hand-search	2
	Snowballing	4
(b) Top 10 sources of primary studies		#Review
	Google Scholar	20
	Scopus	20
	ACM digital library	16
	ScienceDirect	16
	Web of Science	16
	IEEE	13
	EBSCO	11
	PsycINFO	9
	Springer	9
	PubMed	8

Figure 1 shows the reported year spans of the primary studies in the 48 reviews. Note that Review #30 claimed to have covered 203 works on the implementation of gamification from 2001 to 2018, including journal publications, books and book chapters, thesis, conference papers, and company reports. It did not provide identification of each work for us to validate. For this reason, we reported the year span authors claimed. For the ending years of primary studies, one review reported the latest primary studies were published in 2013, and one review has one latest published primary study in 2020. The majority of the 48 reviews included primary studies published the earliest from 2011 to 2014 and the latest from 2015 to 2019.

3.2 RQ2. What Application Domains Are Reviewed?

The analysis of the reviews confirmed that gamification has been applied in many domains. Table 2 summarizes the application domains covered by the reviews. The domain classification is the result of the review compilation, and each review has one primary domain. When two potential domains are involved in a review, such as gamification in healthcare professional education, the review is coded with the central theme (which is education in this case).

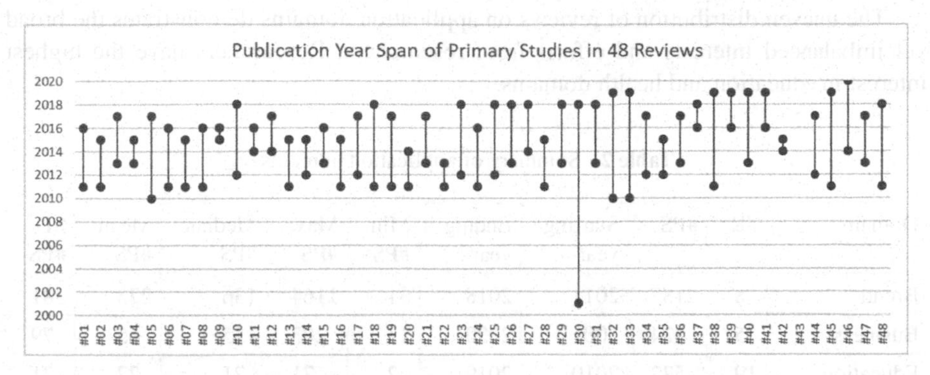

Fig. 1. Publication year span of primary studies in the 48 reviews

A broad domain means that reviews did not specify a particular domain, did not intend to make distinctions, or chose to be general rather than domain specific. Reviews for broad domains tend to cover more primary studies than reviews for a specific domain. Among the eight reviews in broad domains, the number of primary studies in each review ranged from 31 to 1164, with a median, mean, and standard deviation of 136, 273, and 381, respectively. The total number of primary studies covered for the broad domain is 2187 (note: a summarized number of primary studies may include duplicates of which we have no way of identifying), spanning from 2011 to 2018.

The business domain has five reviews on the following topics: production and logistics operation, organization and enterprise, online consumption, and broad businesses, such as finance, marketing, and human resource management.

The education domain has various foci, such as learning in K-12, pedagogical instruction, learning in different majors in higher education, professional education, or corporate training. Among all the specific domains (business, education, health, sustainability, and other), education encompassed the largest number of reviews (19) and largest number of primary studies (522). Specifically, four reviews examined the application of gamification in the general education field based on 155 primary studies, and the remaining 15 reviews considered particular areas of education, such as e-learning, information systems (IS) education, corporate training, higher education, medical education, and computer-assisted learning by utilizing 367 primary studies.

The health domain includes healthcare- or medical-related studies. A total of 12 reviews handled 11 specific issues, such as biopsychosocial, dietary habits of children, depression care, physical exercise, healthcare, mental health, and psychiatric disorder, with 221 primary studies published from 2010 to 2019. Compared to other specific domains, reviews in the health domain tend to cover smaller number of primary studies.

The sustainability domain has only two reviews based on 14 primary studies ranging from 2010 to 2017. One review focused on water governance with five primary studies, and the other focused on water and energy with nine primary studies.

Finally, two reviews in the "other" domain used 97 primary studies. One review covered e-governing (e-participation, citizen engagement in policy making) and used 56 primary studies ranging from 2012 to 2018. The other review investigated information systems (IS) research based on 41 primary studies published between 2011 and 2018.

The uneven distribution of reviews on application domains demonstrates the broad yet imbalanced interests and efforts from researchers. Researchers have the highest interest in education and health domains.

Table 2. Summary of application domains

Domain	#R	#PS	Starting year	Ending year	Min #PS	Max #PS	Median #PS	Mean #PS	SD #PS
Broad	8	2187	2011	2018	31	1164	136	273	381
Business	5	324	2001	2020	13	203	35	65	79
Education	19	522	2010	2019	2	71	21	27	21
Health	12	221	2010	2019	2	70	8	17	20
Sustainability	2	14	2010	2017	5	9	7	7	3
Other	2	97	2011	2018	41	56	49	49	11
All	48	3365	2001	2020	2	1164	27	70	175

Note: R = Reviews, PS = Primary Studies in a review

3.3 RQ3. What Review Types Have Been Conducted?

Paré et al. [11] introduced a typology of nine literature review types to clarify the existing confusion regarding the term "review." The nine types are categorized by four overarching goals, including summarization of prior knowledge, data aggregation or integration, explanation building, and critical assessments. Table 3 lists the overarching goals and names of review types under each goal. Detailed definitions and explanations of review types are omitted due to space limit and can be found in [11].

We found the type information claimed by the authors of two reviews were less accurate, so we recoded them based on our understanding. For each of the remaining 46 reviews, we confirmed the authors' claims if review types were specified or we decided the review type of a review after examining its characteristic. Except for critical review, all nine other types were found in the 48 reviews, where one review utilized two review types.

The most utilized overarching goal is summarizing prior knowledge, which appeared in 41 (84%) reviews. Under this goal, descriptive review is the most used review type that appeared in 22 (45%) reviews, followed by narrative reviews (12, 24%) and scoping reviews (7, 14%).

Six reviews addressed the data aggregation or integration goal. Among the six, meta-analysis appeared in three reviews: two in education domain and one in broad domain. Umbrella review appeared in two reviews, one in education, and one in the business domain. Qualitative systematic review appeared in one review in the business domain. This review also used descriptive review type.

Explanation building is a deeper level of analysis than summarization or aggregation. Two reviews are with this goal, one for each of the two review types. The theoretical review covered 41 primary studies from 2011 to 2018 to build theoretical foundations for gamification research. The realistic review included six primary studies from 2013 to 2018 to examine the methodological approaches to gamification implementation.

The screwed pattern of relatively shallow reviews is an indication that gamification research remains on the uptrend trajectory. Greater efforts should be invested to analyze the existing investigations, integrate current efforts, find explanations, and develop theories.

Table 3. Summary of review types

Overarching goal	Type	#Review	%Review
Summarization of prior knowledge	Narrative review	12	24%
	Descriptive review	22	45%
	Scoping review	7	14%
Data aggregation or integration	Meta-analysis	3	6%
	Qualitative systematic review	1	2%
	Umbrella review	2	4%
Explanation building	Theoretical review	1	2%
	Realist review	1	2%
Critical assessment of extant literature	Critical review	0	0%
	Total	49	100%

Note: One review used two review types

3.4 RQ4. What Has Been Reviewed?

This question indicates the specific aspects focused by the reviews. A scheme with three categories of review foci at different abstract levels was iteratively developed after examining all the 48 reviews.

The first level is on a field of study. A review with this focus would provide an overview of the entire research field that may include but not limited to publication trends and outlets, prolific authors and outlets, application domains, methodological approaches, platforms or apps, and/or specific research topics. The second level focuses on research topics as reflected by various variables, components, and factors in research investigations. This level of reviews covers multiple factors instead of only one and often examines the relationships among these factors. The third level focuses on one specific variable, component, or factor of research investigations. Such reviews concentrate on only one aspect of research investigations. Table 4 presents some typical research questions that would be addressed at each of the three levels.

Table 4 shows that almost a quarter of the reviews (12, 25%) attempted to depict fields of studies. More than half (27, 56%) of the reviews concentrated on multiple

Table 4. Summary of review foci

Level	Typical RQs	#Review	%Review
I. Field of Study: Provide an overview of the entire field such as: publication trends and outlets, prolific authors, application domains, methodological approaches, platforms or apps, participant characteristics, and research topics	[18]'s: 1. Methodological 2. Underlying theoretical models 3. Platforms or apps 4. Participant's levels of education and the most common game mechanics 5. Potential effects of implementation in various fields of education 6. Unexplored future research avenues	12	25%
II. Clusters of factors: Cover multiple factors or components or variables in a study, such as components in a nomological net	[19]'s: 1. Most used gamification elements 2. Effects on learners' behavior 3. Factors that must be considered for designing effective gamification	27	56%
III. Specific Factors Design elements	Major considerations undergirding the design of digital badges [9]	2	4%
Effects/Outcomes	Effects on motivation and engagement [20]	4	8%
Effectiveness	Overall effectiveness for cognitive bias modification [21]	2	4%
Methodologies	Qualitative approach to gamification implementation (practice) within an information ecology [22]	1	2%

factors. However, these reviews did not focus on the same factors or same number of factors. Examples of such factors included gamification design and their effectiveness, gamification strategies, design elements and outcomes, and game elements in relation to self-determination theory. Nine reviews focused on single factors in their reviews, including design elements, effects and impact, effectiveness, and methodology issues. Some factors cannot be studied alone; thus, the number of reviews at this level is relatively small, and not all potential factors are found in single factor reviews.

4 Discussions

The 48 reviews covered primary studies that appeared in a wide publication timespan and indicate that gamification studies have rapidly grown within the past 10 years.

The 42 publication outlets where the 48 reviews were found suggest that there are no concentrated dissemination channels yet for gamification reviews. The various application domains of gamification suggests that gamification is regarded more like an add-on research topic in traditional disciplines to improve user experiences, attitudes, or behaviors.

Despite the fast development and broad applications, gamification research seems still in a developing stage. One evidence is the dominant number of relatively shallow reviews: summarization of prior knowledge is the first step to building some understanding of the research progress. Aggregation and integration can inform continued efforts. Furthermore, explanation building and critical assessment can provide in-depth understanding of the gamification research. Conducting deep reviews beyond summarization and aggregation would require researchers' explicit intentions, as well as more established empirical evidence, ideally grounded by theories.

5 Limitations and Future Efforts

This paper contributes to the understanding of the gamification field by providing a high-level overview of the gamification research. The findings of this study can support researchers to identify and pursue future research opportunities in the field.

Our findings are limited by the data collection method. We only searched in two databases and limited to reviews published since 2018. Adding more databases or increasing the year span can generate a larger collection of reviews to be covered in a meta-review, which may lead to more comprehensive investigations. Another limitation is the aggregation of numbers of primary studies. Since not all reviews provided identifications of their primary studies, we do not have a way to remove the duplicate primary studies among different reviews.

There are several future directions. Gamification field is still in its early stage of development and requires more substantial efforts to enhance its conceptual clarity, theoretical development, and design strategies. Thus, it may be too early to discuss the potential of gamification field on its own, but it can be studied in the future once some issues are addressed. Literature reviews with overarching goals of building explanation and critical assessments can help identify issues need to be addressed, establish consensus, and pave a foundation for gamification research to move forward.

Future research can also expand the current descriptive meta-review study to conduct other deeper meta-reviews to examine in-depth the many factors and domains related to gamification research. For example, it is well realized that there are no agreements on conceptual definitions and classifications of gamification design elements. An in-depth meta-review may shed light and build potential resolutions.

Acknowledgments. This work was partially supported by the National Natural Science Foundation of China (71904215) and the Ministry of Education, Humanities and Social Sciences Council in China (18YJCZH160).

References

1. Deterding, S., Dixon, D., Khaled, R., Nacke, L.: From game design elements to gamefulness: defining gamification. In: Proceedings of the 15th International Academic MindTrek Conference on Envisioning Future Media Environments - MindTrek '11, Tampere, Finland, pp. 9–15 (2011). https://doi.org/10.1145/2181037.2181040
2. Hamari, J.: Transforming homo economicus into homo ludens: a field experiment on gamification in a utilitarian peer-to-peer trading service. Electron. Commer. Res. Appl. **12**(4), 236–245 (2013). https://doi.org/10.1016/j.elerap.2013.01.004
3. Deterding, S., Sicart, M., Nacke, L., O'Hara, K., Dixon, D.: Gamification: using game-design elements in non-gaming contexts, Vancouver, BC, Canada, pp. 2425–2428 (2011). https://doi.org/10.1145/1979482.1979575
4. Kasurinen, J., Knutas, A.: Publication trends in gamification: a systematic mapping study. Comput. Sci. Rev. **27**, 33–44 (2018). https://doi.org/10.1016/j.cosrev.2017.10.003
5. Morschheuser, B., Hassan, L., Werder, K., Hamari, J.: How to design gamification? A method for engineering gamified software. Inf. Softw. Technol. **95**, 219–237 (2018)
6. Hamari, J., Koivisto, J.: 'Working out for likes': an empirical study on social influence in exercise gamification. Comput. Hum. Behav. **50**, 333–347 (2015). https://doi.org/10.1016/j.chb.2015.04.018
7. Suh, A., Cheung, C.M.K., Ahuja, M., Wagner, C.: Gamification in the workplace: the central role of the aesthetic experience. J. Manag. Inf. Syst. **34**(1), 268–305 (2017). https://doi.org/10.1080/07421222.2017.1297642
8. Buckley, P., Doyle, E.: Individualising gamification: an investigation of the impact of learning styles and personality traits on the efficacy of gamification using a prediction market. Comput. Educ. **106**, 43–55 (2017). https://doi.org/10.1016/j.compedu.2016.11.009
9. Facey-Shaw, L., Specht, M., van Rosmalen, P., Borner, D., Bartley-Bryan, J.: Educational functions and design of badge systems: a conceptual literature review. IEEE Trans. Learn. Technol. **11**(4), 536–544 (2018). https://doi.org/10.1109/TLT.2017.2773508
10. Morschheuser, B., Hamari, J., Koivisto, J., Maedche, A.: Gamified crowdsourcing: conceptualization, literature review, and future agenda. Int. J. Hum. Comput. Stud. **106**, 26–43 (2017). https://doi.org/10.1016/j.ijhcs.2017.04.005
11. Paré, G., Trudel, M.-C., Jaana, M., Kitsiou, S.: Synthesizing information systems knowledge: a typology of literature reviews. Inf. Manag. **52**(2), 183–199 (2015). https://doi.org/10.1016/j.im.2014.08.008
12. Closs, S.J., et al.: Towards improved decision support in the assessment and management of pain for people with dementia in hospital: a systematic meta-review and observational study. Health Serv. Deliv. Res. **4**(30), 1–62 (2016). https://doi.org/10.3310/hsdr04300
13. Khan, A., Boroomand, F., Webster, J., Minocher, X.: From elements to structures: an agenda for organisational gamification. Eur. J. Inf. Syst. 1–20 (2020). https://doi.org/10.1080/0960085X.2020.1780963
14. Rozman, T., Donath, L.: The current state of the gemification in e-learning: a literature review of literature reviews. J. Innov. Bus. Manag. **11**, 5–19 (2019)
15. Albertarelli, S., et al.: A survey on the design of gamified systems for energy and water sustainability. Games **9**(3), 38 (2018). https://doi.org/10.3390/g9030038
16. Chan, G., Arya, A., Orji, R., Zhao, Z.: Motivational strategies and approaches for single and multi-player exergames: a social perspective. PeerJ Comput. Sci. **5**, e230 (2019). https://doi.org/10.7717/peerj-cs.230
17. Hamari, J., Koivisto, J., Sarsa, H.: Does gamification work? – a literature review of empirical studies on gamification. In: 2014 47th Hawaii International Conference on System Sciences, Waikoloa, HI, January 2014, pp. 3025–3034 (2014). https://doi.org/10.1109/HICSS.2014.377

18. Zainuddin, Z., Chu, S.K.W., Shujahat, M., Perera, C.J.: The impact of gamification on learning and instruction: a systematic review of empirical evidence. Educ. Res. Rev. **30**, 100326 (2020). https://doi.org/10.1016/j.edurev.2020.100326
19. Antonaci, A., Klemke, R., Specht, M.: The effects of gamification in online learning environments: a systematic literature review. Informatics **6**(3), 32 (2019). https://doi.org/10.3390/informatics6030032
20. Alsawaier, R.S.: The effect of gamification on motivation and engagement. Int. J. Inf. Learn. Tech. **35**(1), 56–79 (2018). https://doi.org/10.1108/IJILT-02-2017-0009
21. Zhang, M., Ying, J., Song, G., Fung, D.S., Smith, H.: Attention and cognitive bias modification apps: review of the literature and of commercially available apps. JMIR Mhealth Uhealth **6**(5), e10034 (2018). https://doi.org/10.2196/10034
22. van der Poll, A.E., van Zyl, I.J., Kroeze, J.H.: A systematic literature review of qualitative gamification studies in higher education. In: Rønningsbakk, L., Wu, T.-T., Sandnes, F.E., Huang, Y.-M. (eds.) ICITL 2019. LNCS, vol. 11937, pp. 486–497. Springer, Cham (2019). https://doi.org/10.1007/978-3-030-35343-8_52

Appendix: 48 Reviews Included in the Study

1. Osatuyi, B., Osatuyi, T., de la Rosa, R.: Systematic review of gamification research in IS education: a multi-method approach. Commun. Assoc. Inform. Syst. **42**, 95–124 (2018)
2. Kasurinen, J., Knutas, A.: Publication trends in gamification: a systematic mapping study. Comput. Sci. Rev. **27**, 33–44 (2018). https://doi.org/10.1016/j.cosrev.2017.10.003
3. Subhash, S., Cudney, E.A.: Gamified learning in higher education: a systematic review of the literature. Comput. Hum. Behav. **87**, 192–206 (2018). https://doi.org/10.1016/j.chb.2018.05.028
4. de Freitas, S.: Are games effective learning tools? A review of educational games. Educ. Technol. Soc. **21**(2), 74–84 (2018)
5. Albertarelli, S., et al.: A survey on the design of gamified systems for energy and water sustainability. Games **9**(3), 38 (2018). https://doi.org/10.3390/g9030038
6. Facey-Shaw, L., Specht, M., van Rosmalen, P., Borner, D., Bartley-Bryan, J.: Educational functions and design of badge systems: a conceptual literature review. IEEE Trans. Learn. Technol. **11**(4), 536–544 (2018). https://doi.org/10.1109/TLT.2017.2773508
7. de A Souza, M.R., Veado, L., Moreira, R.T., Figueiredo, E., Costa, H.: A systematic mapping study on game-related methods for software engineering education. Inf. Softw. Technol. **95**, 201–218 (2018). https://doi.org/10.1016/j.infsof.2017.09.014
8. Bozkurt, A., Durak, G.: A systematic review of gamification research. In Pursuit of homo ludens. Int. J. Game Based Learn. **8**(3), 15–33 (2018). https://doi.org/10.4018/IJGBL.2018070102
9. González-González, C., Río, N.G., Navarro-Adelantado, V.: Exploring the benefits of using gamification and videogames for physical exercise: a review of state of art. Int. J. Interact. Multimedia Artif. Intell. **5**(2), 46–52 (2018). https://doi.org/10.9781/ijimai.2018.03.005
10. Magista, M., Dorra, B.L., Pean, T.Y.: A review of the applicability of gamification and game-based learning to improve household-level waste management practices among schoolchildren. IJTech **9**(7), 1439–1449 (2018). https://doi.org/10.14716/ijtech.v9i7.2644
11. Zhang, M., Ying, J., Song, G., Fung, D.S., Smith, H.: Gamified cognitive bias modification interventions for psychiatric disorders: review. JMIR Ment Health **5**(4), e11640 (2018). https://doi.org/10.2196/11640
12. Zhang, M., Ying, J., Song, G., Fung, D.S., Smith, H.: Attention and cognitive bias modification apps: review of the literature and of commercially available apps. JMIR Mhealth Uhealth **6**(5), e10034 (2018). https://doi.org/10.2196/10034

13. Gorbanev, I., et al.: A systematic review of serious games in medical education: quality of evidence and pedagogical strategy. Med. Educ. Online **23**(1), 1438718 (2018). https://doi.org/10.1080/10872981.2018.1438718
14. Garett, R., Young, S.D.: Health care gamification: a study of game mechanics and elements. Technol. Knowl. Learn. **24**(3), 341–353 (2018). https://doi.org/10.1007/s10758-018-9353-4
15. Dias, L.P.S., Barbosa, J.L.V., Vianna, H.D.: Gamification and serious games in depression care: a systematic mapping study. Telematics Inform. **35**(1), 213–224 (2018). https://doi.org/10.1016/j.tele.2017.11.002
16. Alsawaier, R.S.: The effect of gamification on motivation and engagement. Int. J. Inf. Learn. Tech. **35**(1), 56–79 (2018). https://doi.org/10.1108/IJILT-02-2017-0009
17. Alhammad, M.M., Moreno, A.M.: Gamification in software engineering education: a systematic mapping. J. Syst. Softw. **141**, 131–150 (2018). https://doi.org/10.1016/j.jss.2018.03.065
18. Treiblmaier, H., Putz, L.-M.: University of Applied Sciences, Upper Austria, P. B. Lowry, and Virginia Tech, "Setting a Definition, Context, and Theory-Based Research Agenda for the Gamification of Non-Gaming Applications. AIS Transactions on Human-Computer Interaction, pp. 129–163 (2018). https://doi.org/10.17705/1thci.00107
19. Tondello, G., Premsukh, H., Nacke, L.: A theory of gamification principles through goal-setting theory. In: The 51st Hawaii International Conference on System Sciences (2018). https://doi.org/10.24251/HICSS.2018.140
20. Tamayo-Serrano, P., Garbaya, S., Blazevic, P.: Gamified in-home rehabilitation for stroke survivors: analytical review. Int. J. Serious Games **5**(1), 1–26 (2018). https://doi.org/10.17083/ijsg.v5i1.224
21. Stepanovic, S., Mettler, T.: Gamification applied for health promotion: Does it really foster long-term engagement? A scoping review. In: Twenty-Sixth European Conference on Information Systems (ECIS2018), Portsmouth, UK (2018)
22. Hinton, S., Wood, L.C., Singh, H., Reiners, T.: Enterprise gamification systems and employment legislation: a systematic literature review. Australas. J. Inf. Syst. **23**, 1–24 (2019). https://doi.org/10.3127/ajis.v23i0.2037
23. Baptista, G., Oliveira, T.: Gamification and serious games: a literature meta-analysis and integrative model. Comput. Hum. Behav. **92**, 306–315 (2019). https://doi.org/10.1016/j.chb.2018.11.030
24. Rodrigues, L.F., Oliveira, A., Rodrigues, H.: Main gamification concepts: a systematic mapping study. Heliyon **5**(7), e01993 (2019). https://doi.org/10.1016/j.heliyon.2019.e01993
25. Hassan, L., Hamari, J.: Gamification of e-participation: a literature review. In: 52nd Hawaii International Conference on System Sciences, pp. 3077–3086 (2019)
26. van der Poll, A.E., van Zyl, I.J., Kroeze, J.H.: A systematic literature review of qualitative gamification studies in higher education. In: Rønningsbakk, L., Wu, T.-T., Sandnes, F.E., Huang, Y.-M. (eds.) ICITL 2019. LNCS, vol. 11937, pp. 486–497. Springer, Cham (2019). https://doi.org/10.1007/978-3-030-35343-8_52
27. Antonaci, A., Klemke, R., Specht, M.: The effects of gamification in online learning environments: a systematic literature review. Informatics **6**(3), 32 (2019). https://doi.org/10.3390/informatics6030032
28. Koivisto, J., Hamari, J.: The rise of motivational information systems: a review of gamification research. Int. J. Inf. Manage. **45**, 191–210 (2019). https://doi.org/10.1016/j.ijinfomgt.2018.10.013
29. Helmefalk, M.: An interdisciplinary perspective on gamification: mechanics, psychological mediators and outcomes. IJSG **6**(1), 3–26 (2019). https://doi.org/10.17083/ijsg.v6i1.262
30. Wanick, V., Bui, H.: Gamification in management: a systematic review and research directions. Int. J. Serious Games **6**(2), 57–74 (2019). https://doi.org/10.17083/ijsg.v6i2.282

31. Cheng, V.W.S., Davenport, T., Johnson, D., Vella, K., Hickie, I.B.: Gamification in apps and technologies for improving mental health and well-being: systematic review. JMIR Ment Health **6**(6), e13717 (2019). https://doi.org/10.2196/13717
32. Rozman, T., Donath, L.: The current state of the gemification in e-learning: a literature review of literature reviews. J. Innov. Bus. Manag. **11**, 5–19 (2019)
33. Chan, G., Arya, A., Orji, R., Zhao, Z.: Motivational strategies and approaches for single and multi-player exergames: a social perspective. PeerJ Comput. Sci. **5**, e230 (2019). https://doi.org/10.7717/peerj-cs.230
34. Albertazzi, D., Ferreira, M.G.G., Forcellini, F.A.: A wide view on gamification. Technol. Knowl. Learn. **24**(2), 191–202 (2018). https://doi.org/10.1007/s10758-018-9374-z
35. Aubert, A.H., Medema, W., Wals, A.E.J.: Towards a framework for designing and assessing game-based approaches for sustainable water governance. Water **11**(4), 869 (2019). https://doi.org/10.3390/w11040869
36. Muangsrinoon, S., Boonbrahm, P.: Game elements from literature review of gamification in healthcare context. J. Technol. Sci. Educ. **9**(1), 20–31 (2019). https://doi.org/10.3926/jotse.556
37. Alomari, I., Al-Samarraie, H., Yousef, R.: The role of gamification techniques in promoting student learning: a review and synthesis. J. Inf. Technol. Educ. Res. **18**, 395–417 (2019). https://doi.org/10.28945/4417
38. Martinho, D., Carneiro, J., Corchado, J.M., Marreiros, G.: A systematic review of gamification techniques applied to elderly care. Artif. Intell. Rev. **53**(7), 4863–4901 (2020). https://doi.org/10.1007/s10462-020-09809-6
39. Wang, A.I., Tahir, R.: The effect of using Kahoot! for learning – a literature review. Comput. Educ. **149**, 103818 (2020)
40. Tobon, S., Ruiz-Alba, J.L., García-Madariaga, J.: Gamification and online consumer decisions: is the game over? Decis. Support Syst. **128**, 113167 (2020). https://doi.org/10.1016/j.dss.2019.113167
41. Zainuddin, Z., Chu, S.K.W., Shujahat, M., Perera, C.J.: The impact of gamification on learning and instruction: a systematic review of empirical evidence. Educ. Res. Rev. **30**, 100326 (2020). https://doi.org/10.1016/j.edurev.2020.100326
42. Alabdulakareem, E., Jamjoom, M.: Computer-assisted learning for improving ADHD individuals' executive functions through gamified interventions: a review. Entertain. Comput. **33**, 100341 (2020). https://doi.org/10.1016/j.entcom.2020.100341
43. Chow, C.Y., Riantiningtyas, R.R., Kanstrup, M.B., Papavasileiou, M., Liem, G.D., Olsen, A.: Can games change children's eating behaviour? A review of gamification and serious games. Food Qual. Prefer. **80**, 103823 (2020). https://doi.org/10.1016/j.foodqual.2019.103823
44. Warmelink, H., Koivisto, J., Mayer, I., Vesa, M., Hamari, J.: Gamification of production and logistics operations: status quo and future directions. J. Bus. Res. **106**, 331–340 (2020). https://doi.org/10.1016/j.jbusres.2018.09.011
45. Larson, K.: Serious games and gamification in the corporate training environment: a literature review. TechTrends **64**(2), 319–328 (2019). https://doi.org/10.1007/s11528-019-00446-7
46. Khan, A., Boroomand, F., Webster, J., Minocher, X.: From elements to structures: an agenda for organisational gamification. Eur. J. Inf. Syst. 1–20 (2020). https://doi.org/10.1080/0960085X.2020.1780963
47. Sailer, M., Homner, L.: The gamification of learning: a meta-analysis. Educ. Psychol. Rev. **32**(1), 77–112 (2019). https://doi.org/10.1007/s10648-019-09498-w
48. Howard, M.C., Gutworth, M.B.: A meta-analysis of virtual reality training programs for social skill development. Comput. Educ. **144**, 103707 (2020). https://doi.org/10.1016/j.compedu.2019.103707

Research Agenda-Setting in Medicine: Shifting from a Research-Centric to a Patient-Centric Approach

Ania Korsunska(✉)

Syracuse University, Syracuse, NY, USA
akorsuns@syr.edu

Abstract. Traditional approaches to research agenda-setting focus on researchers and their ability to review and synthesize literature, identify gaps, prioritize their ideas, and find the resources to make them a reality. Recent initiatives in medical research have shifted the focus away from the researcher to other stakeholders. Through a series of semi-structured interviews with medical researchers, we illustrate both the traditional researcher-centric as well as the novel patient-centric approaches. The patient-centric approach allows patients to contribute their diverse perspectives and pose unique questions, which can direct more impactful research agenda-setting. This paper provides insights into how medical research agendas are established, what factors impact decision-making and how an innovative use of crowdsourcing can refocus attention on the patient and their needs.

Keywords: Research agenda-setting · Patient-centric approach · Crowdsourcing

1 Introduction

Traditionally, the scientific method dictates that new research should build on prior work, filling in gaps, synthesizing or testing prior findings [13, 24, 29]. Yet there are many incentives and constraints that affect researchers' decisions, shape their individual careers and, in aggregate, the direction of scientific discovery more broadly [9]. This means that both individual research agendas and broader disciplinary focuses do not necessarily align with what, in theory, is the most rational and necessary direction for the field to pursue at the time. These high stakes make it imperative to understand the evolving incentives that drive research agenda-setting, particularly in medical fields like immunology, which has wide ranging applications for diseases that affect millions of people daily, including the current COVID-19 outbreak.

Through a series of semi-structured interviews with medical researchers, we gain insights into how research agendas are established, what factors impact decision-making and how an innovative crowdsourcing approach can refocus attention on the patient and their needs. Crowdsourcing offers a potential solution to some of the chaotic and time-consuming problems researchers face. Engaging the public in setting agendas democratizes science, holds researchers accountable to their conduct, and helps them disseminate their results, which are often funded by public contributions. Bringing together and

K. Toeppe et al. (Eds.): iConference 2021, LNCS 12646, pp. 374–383, 2021.
https://doi.org/10.1007/978-3-030-71305-8_31

directly engaging the target patient community echoes user-centered design principles. Thus, crowdsourcing is one potential solution to the problem of end-user involvement in the production of scientific knowledge, while streamlining research agenda-setting, promoting collaboration, and creating open access to knowledge.

2 Literature Review

Research agenda-setting, defined here as the process of determining what projects the researcher will pursue, has been a focus of study in sociology of science for decades [4, 10, 19]. Historically, the emphasis was primarily placed on researchers, but recently an emphasis on other stakeholder groups process has emerged. Medicine serves as an excellent case study for this area of inquiry. Despite the high stakes, medical research often lacks in quality and usefulness, and wastes billions of dollars in investment funds [6, 14, 17, 21]. By engaging patients in the research process, the responsibility of setting research agendas and the risk of potential failure is spread throughout a broader community, and more diverse ideas are considered. We will review the researcher-centric and patient-centric approaches in turn.

2.1 Researcher-Centric

It is commonly understood that research must be based on prior work. Literature reviews guide research agenda-setting by helping to define the theoretical foundation, identify the research problem and justify the value-add [24]. In practice, deciding what to focus on is often shaped by non-scientific constraints and can be a difficult and time-consuming task [13]. Keeping up with relevant research is becoming progressively more difficult as scientific publications have increased 527% annually since 1965 [28, 33]. Tools like Google Scholar make access to literature faster, but there are issues of algorithmic filtering. Though it catalogues between 2 and 100 million records [15], Google Scholar "does not index the majority of the scholarly materials indexed by commercial database vendors and it has never shared with the public their search algorithm" [18]. Difficulties with literature review are exacerbated by what is absent altogether, such as negative results, failed experiments, and disproven hypotheses [27].

It has been long hypothesized, and recently quantitatively illustrated [10], that researcher's agenda-setting choices are shaped by ongoing tensions between two forces: tradition and innovation, which Kuhn refers to as the "essential tension" [19]. Though both approaches are necessary, there are contradictory incentives: researchers are encouraged to focus both on quantity and quality, publishing frequently, while remaining innovative. These incentives are at odds. Scientists who adhere to a research tradition in their domain often achieve more publications, simultaneously limiting their ability to pursue novel ideas that might take the field in new directions. The pressure to publish for achieving tenure may exacerbate researchers' risk-aversity due to the concern that failure will make them seem unqualified or unproductive [9, 10].

Common advice to overcome decision paralysis in research agenda-setting is to follow your interests [3]. But this could lead the researcher to ask unnecessary questions from the perspective of the discipline or common good [2]. Considering these limitations, it seems apt to shift the focus and responsibility away from the researchers exclusively and allow external stakeholders to play a more dominant role.

2.2 Patient-Centric

Kuhn argued that paradigm shifts arising from scientific crises are necessary steps in the evolution of science [20]. Some have suggested that science is due for a paradigm shift from closed, hidden science to open science and data sharing [26]. Traditionally, medical research relies on academic experts, who select methods, gather data, apply for funding and exclusively own the results of their labor. The "open innovation" approach, by contrast, democratizes the scientific process. By utilizing crowdsourcing at different stages in the research process, diverse stakeholders collectively frame questions, prioritize studies, co-create data and raise funds, which creates more open access to intellectual property [31]. The medical research field has begun to explore these strategies, and this shift has started to reshape how the industry functions.

We define crowdsourcing as "an approach to problem solving which involves an organization having a large group attempt to solve a problem or part of a problem, then sharing solutions" [31]. This approach helps solve some of the issues of traditional research approaches and mirrors user-centered design principles, wherein "users have a deep impact on the design by being involved as partners with designers throughout the design process" [1]. Patients have already become contributors to research, resulting in high-quality outcomes and more open science [23, 31, 32, 35]. Crowdsourcing approaches have been used in various medical applications: in clinical trial design [22], data sharing and collaboration [26, 28] development of research questions and data analysis [25, 31, 35], development of antibiotics [7], drugs for lupus [12], antimalarial drugs [30], and the treatment of Castleman disease [8, 35]. There is growing consensus in the industry that without integrating the voices of the patient population, it becomes "impossible to identify the most clinically meaningful questions and research approaches to answering them" [35].

3 Methods

To look at the question of research agenda-setting we drew from a larger project which focuses on the sense-making and tool-use practices of medical immunology researchers. For this analysis, we utilized six 45–60 min interviews which were conducted between 8/10/2020 and 9/17/2020 (Table 1). We focused on three research questions, which were based on the traditional research-centric model:

(1) How do researchers keep up to date on the research in their area of expertise and how do they identify knowledge gaps?
(2) How do researchers make decisions as to their research agendas and how are ideas prioritized?
(3) What role does funding play in research agenda-setting and prioritization?

Interviewees were identified via personal contacts, snowball sampling and Internet searches using keywords related to medical research. The broad scope of experiences, degrees and seniority levels of the interviewees allows for varied approaches to research agenda-setting. The interviews followed a semi-structured protocol and were conducted remotely through Zoom. All interviewees agreed to be audio recorded. The recordings were transcribed with the software Otter.ai and manually edited for accuracy.

Table 1. Interviewee information

Interviewee	Job title	Graduate degrees	Years of experience[a]
1	Associate Professor Biochemistry at a university	PhD Chemistry	30
2	Research Scientist of Ecology and Evolutionary Biology at a university	PhD Biology	12
3	Research Scientist at a clinical laboratory company	PhD Biological and Biomedical Sciences	7
4	Assistant Professor of Medicine at a university	MD, MSc, MBA	8
5	Postdoctoral Researcher at a cancer research center	PhD Microbiology and Immunology	6
6	Lead of Marketing at a grant-giving organization	PhD Microbiology and Immunology	9

[a]Interviewees' subjective estimates of "years working in their area of expertise".

An iterative thematic analysis [5] was conducted by the author, utilizing the qualitative analysis software Atlas.ti. The transcripts were coded into preliminary themes, which were then synthesized into concise categories (Table 2). The quotations in the following section have been minimally cleaned, removing filler language and identifying information.

4 Funding

Traditionally, the question of how researchers set their agendas has utilized a researcher-centric approach. Through the interviews and analysis, we discovered that there is a movement in the medical field towards a patient-led, collaborative approach. The two approaches are reflected in the themes identified in the analysis and Table 2 summarizes the finding elaborated below. The broad topics in Table 2 correspond to the three research questions described prior.

Table 2. Questions and themes

Broad topics	Theme category	Focus	Theme
What do we want to know and what don't we know?	Knowledge Gap	Researcher	Search & Synthesis
			Algorithms
			Academic trust
			Negative Results & Data
		Patient	Crowdsourcing Research Questions
			Crowdsourcing Analysis
How are things normally done around here and what's in it for me?	Incentives & Constraints	Researcher	Academic Freedom
			Lab PI & Mentorship
			External need-driven research
		Patient	Science Communication
Are there resources to do this?	Role of Funding	Researcher	Research Prioritization
			Matching or Avoiding Trends
		Patient	Researcher + Grant Matching

4.1 Researcher-Centric

Knowledge Gap. First, we identified the practices the interviewees utilize to keep up to date on the literature and identify new directions, as a part of research agenda-setting.

Search and Synthesis. All of interviewees expressed a strong preference for receiving literature via automated alerts, from academic databases (e.g., PubMed and SciFinder) and Google Scholar, rather than conducting searches manually. They expressed a feeling of responsibility of staying up to date on their field and demonstrating familiarity with its "core" literature. There were concerns about missing tangentially related papers and lacking the time to pursue them.

Algorithms. A few researchers utilize Google.com directly for literature searches, as "it's way faster, the stuff that's relevant is on the first page, it accounts for my horrible spelling, and it finds stuff that's adjacent to what I'm actually looking for". Another interviewee explained: "If Google missed something, it's probably not important for me to know".

Academic Trust. Due to time constraints and proliferation of literature, interviewees said it is increasingly difficult to rely on heuristics alone to determine what is trustworthy and avoid biases towards big name journals and Western literature.

Negative Results and Data. Another concern is not knowing what a gap in the literature signifies - whether it is a novel idea worth pursuing, or a complicated task that many have tried and failed. This becomes a factor in agenda-setting: "when you find a large gap in the science. And you ask, why does this exist? It seems like a really good question to be asked. It probably means that it's really hard to answer."

Incentives and Constraints. Many of the researchers discussed tensions between the freedom of choosing a research agenda, and the need to contribute to scientific literature, further their own career and potentially benefit society.

Academic Freedom. One interviewee mentioned getting distracted by personal interests: "There are things that I'm interested in, that's absolutely useless. [...] And I got distracted by that, and that's academia, it allowed me to do that", later adding that he regrets not directing his efforts on more significant problems. Another interviewee noted that pressures between the incentives are not equal:

> "If you don't publish, if you don't get grants, you don't maintain your job, and then you don't feed your family. [...] This works exactly how the system is set up −90% personal gain 10% societal benefit."

This includes the pressure to publish work in order to achieve a tenured position: "You can choose easy problems, or you choose the problems that are easy to publish on, just so you can survive."

Lab PI Lead and Mentorship. Researchers' freedom can be constrained by their lab's or direct supervisor's preferences. Many of the researchers discussed the importance of power structures in setting the lab research agenda and the role of a good PI mentor.

External Need-Driven Research Agenda. Problems concerning freedom of choice can also be solved through external factors, such as direct interest from physicians, or a current trend in the world, such as the needs around the COVID-19 pandemic.

Funding.
All interviewees working in academic institutions noted that applying for grants is crucial to their work, and the availability of funds impacts the agenda-setting process. This makes grant-giving organizations responsible for directing the scientific fields to pursue certain projects. One interviewee compared research to building a spider web, with funding prioritizing novel research:

> "Research is spider webbing out and there are more funding agencies interested in adding on to the tips, building new stuff, then filling in the webbing. [...] you can imagine all the spaces in between that you're really missing out on."

Research Prioritization. Another noted that even if they have a prioritized research agenda, if lower-priority studies get funded first, priorities will change. This may not happen if the study was not already planned or does not fit the agenda.

Matching or Avoiding Trends. Most interviewees shared that they have experienced needing to present their work in a way to make it fundable, whether putting it into the context of wider application or making it sound unique and novel.

These themes speak from the researcher-centric lens to the complexity of staying informed on current literature, identifying gaps, making decisions on next steps, prioritizing these steps, and finding the resources to make them a reality. Next, we look at these questions utilizing a patient-centric approach.

4.2 Patient-Centric

Knowledge Gaps. From the patient-centric view, gaps are questions based on community knowledge and personal experience, which are not captured by the researcher-centric approach.

Crowdsourcing Research Questions. An interviewee described their preferred practice of directly asking the physicians, researchers and patients in the community to generate research questions. After gathering the ideas, the community prioritizes them, and a panel of experts in the field selects the most promising to pursue first. Through this collaboration, the broader patient community helps to direct research.

Crowdsourcing Analysis. Another step in the research process that can involve the patient community is data analysis, which distributes the workload, diversifies insights and engages the citizen science community: "Anyone who thinks they might have insights into some data set, can then contribute […] The more minds looking at the data, the better."

Incentives and Constraints. A theme that surfaced throughout the interviews was the responsibility to communicate with the public.

Science Communication. Interviewees argued that many grants are given by government organizations, meaning they are taxpayer-funded, which makes communication with the public "an obligation", whether it's through traditional media - "an invisible importance" one interviewee called it as it indirectly benefits the researcher - or through the community involvement model.

Funding. Another innovative approach is the practice of pitching research projects to the researchers who have the right skillset to complete the project.

Research + Grant Matching. One interviewee described how they integrate this approach into their work: "We say - we're from this foundation, we've got money, we've got samples, and we're willing to coach you through the process, […] you just have to do the research". This approach allows the most qualified minds to engage with research chosen by the patient population and removes the need for the researcher to complete preliminary steps. The interviewee from the grant-giving organization mentioned that they help organizations choose which grants to apply for, minimizing wasted effort.

5 Conclusion

There is no doubt that research should build on prior work, but a sole focus on the researcher during the research agenda-setting process, risks overlooking crucial stakeholders. From the researcher perspective, the gaps that research can fill are found in the literature; for patients, concerns often emerge from community knowledge and personal experience. By incorporating patient voices, researchers can minimize the effects of traditional constraints. By communicating with the public, research institutions can increase transparency and address decreasing levels of trust in science [11, 16, 34]. The patient-centric approach could allow medical research to more quickly improve the quality of people's lives. There are already examples of such approaches impacting their disease fields: one interviewee described the work of the Castleman Disease Collaborative Network as spearheading this novel strategy. Through its integration of patient-centered research agenda setting, they have made incredible progress toward finding a cure for Castleman Disease, and other rare disease fields have started utilizing their approach [35].

5.1 Limitations

The six interviews presented are part of an ongoing project, and the work presented here reflects preliminary findings and theory building. This sample of researchers does not enable us to make generalizable conclusions regarding the breadth of experiences and practices in immunology, medicine, or scientific research more generally. Moreover, the patient-centered approach may not be suited for all forms of scientific work. Potential limitations of crowdsourcing are also relevant to consider [31].

5.2 Future Research

This project will continue to accumulate insights through interviews with various medical researchers, but also expand to analysis of online ethnography of the patient discussion boards that many crowdsourcing-focused research organizations utilize. Finally, we intend to interview representatives of government grant-giving organizations to better understand their research agenda-setting practices and to compare the role of researcher versus patient-raised funds in driving research.

References

1. Abras, C., Maloney-Krichmar, D., Preece, J.: User-centered design. Sage Publications, Berkshire Encyclopedia of Human-computer Interaction (2004)
2. Alahdab, F., Murad, M.H.: Evidence maps: a tool to guide research agenda setting. BMJ Evi. Based Med. **24**(6), 209–211 (2019). https://doi.org/10.1136/bmjebm-2018-111137
3. Booth, W.C., Booth, W.C., Colomb, G.G., Colomb, G.G., Williams, J.M., Williams, J.M.: The Craft of Research. University of Chicago press, Chicago (2003)
4. Bourdieu, P.: The specificity of the scientific field and the social conditions of the progress of reason. Soc. Sci. Inf. **14**(6), 19–47 (1975). https://doi.org/10.1177/053901847501400602

5. Braun, V., Clarke, V.: Using thematic analysis in psychology. Qual. Res. Psychol. **3**(2), 77–101 (2006). https://doi.org/10.1191/1478088706qp063oa
6. Chalmers, I., Glasziou, P.: Avoidable Waste in the Production and Reporting of Research Evidence. **114**(6), 5 (2009). https://doi.org/10.1097/AOG.0b013e3181c3020d
7. Desselle, M.R., et al.: Institutional profile: community for open antimicrobial drug discovery – crowdsourcing new antibiotics and antifungals. Future Sci. OA **3**(2) (2017). https://doi.org/10.4155/fsoa-2016-0093
8. Fajgenbaum, D.C., Ruth, J.R., Kelleher, D., Rubenstein, A.H.: The collaborative network approach: a new framework to accelerate Castleman's disease and other rare disease research. Lancet Haematol. **3**(4), e150–e152 (2016). https://doi.org/10.1016/S2352-3026(16)00007-7
9. Fortunato, S., et al.: Science of science. Science **359**(6379), eaao0185 (2018). https://doi.org/10.1126/science.aao0185
10. Foster, J.G., Rzhetsky, A., Evans, J.A.: Tradition and innovation in scientists' research strategies. Am. Socio. Rev. **80**(5), 875–908 (2015). https://doi.org/10.1177/0003122415601618
11. Gauchat, G.: Politicization of science in the public sphere: a study of public trust in the United States, 1974 to 2010. Am. Socio. Rev. **77**(2), 167–187 (2012). https://doi.org/10.1177/0003122412438225
12. Grammer, A.C., et al.: Drug repositioning in SLE: crowd-sourcing, literature-mining and big data analysis. Lupus **25**(10), 1150–1170 (2016). https://doi.org/10.1177/0961203316657437
13. Grewal, A., Kataria, H., Dhawan, I.: Literature search for research planning and identification of research problem. Ind. J. Anaesthesia **60**(9), 635. (2016). https://doi.org/10.4103/0019-5049.190618
14. Gross, C.P., Anderson, G.F. and Powe, N.R.: The relation between funding by the national institutes of health and the burden of disease. New England J. Med. **340**(24), 1881–1887 (1999). https://doi.org/10.1056/NEJM199906173402406
15. Haddaway, N.R., Collins, A.M., Coughlin, D., Kirk, S.: The role of google scholar in evidence reviews and its applicability to grey literature searching. PLOS ONE **10**(9), e0138237 (2015). https://doi.org/10.1371/journal.pone.0138237
16. Haerlin, B., Parr, D.: How to restore public trust in science. Nature **400**(6744), 499–499 (1999). https://doi.org/10.1038/2286
17. Ioannidis, J.P.A.: Why most clinical research is not useful. PLOS Med. **13**(6), e1002049 (2016). https://doi.org/10.1371/journal.pmed.1002049
18. Jain, V., Raut, D.: Medical literature search dot com. Ind. J. Dermatol. Venereol. Leprol. **77**(2), 135 (2011). https://doi.org/10.4103/0378-6323.77451
19. Kuhn, T.S.: The Essential Tension: Selected Studies in Scientific Tradition and Change. University of Chicago Press, Chicago (1977)
20. Kuhn, T.S.: The Structure of Scientific Revolutions. Original edn., University of Chicago press, Chicago (1962)
21. Macleod, M.R., et al.: Biomedical research: increasing value, reducing waste. Lancet; London **383**(9912), 101–4 (2014). https://dx.doi.org.libezproxy2.syr.edu. https://doi.org/10.1016/S0140-6736(13)62329-6
22. Mullins, C.D., Vandigo, J., Zheng, Z., Wicks, P.: Patient-centeredness in the design of clinical trials. Value Health. **17**(4), 471–475 (2014). https://doi.org/10.1016/j.jval.2014.02.012.
23. Pan, S.W., et al.: Systematic review of innovation design contests for health: spurring innovation and mass engagement. BMJ Innov. **3**(2017), 227–237 (2017). https://doi.org/10.1136/bmjinnov-2017-000203
24. Paré, G., Kitsiou, S.: Methods for literature reviews. In: Handbook of eHealth Evaluation: An Evidence-based Approach. University of Victoria (2017)

25. Ranard, B.L., et al.: Crowdsourcing--harnessing the masses to advance health and medicine, a systematic review. J. General Internal Med. **29**(1), 187–203 (2014). https://doi.org/10.1007/s11606-013-2536-8
26. Rowhani-Farid, A., Allen, M., Barnett, A.G.: What incentives increase data sharing in health and medical research? A systematic review. Res. Integrity Peer Rev. **2**(2017), 4 (2017). https://doi.org/10.1186/s41073-017-0028-9
27. Schooler, J.: Unpublished results hide the decline effect. Nature **470**(7335), 437–437 (2011). https://doi.org/10.1038/470437a
28. Shaw, D.L.: Is open science the future of drug development? Yale J. Biol. Med. **90**(1), 147–151 (2017)
29. Snyder, H.: Literature review as a research methodology: an overview and guidelines. J. Bus. Res. **104**, 333–339 (2019). https://doi.org/10.1016/j.jbusres.2019.07.039
30. Spangenberg, T., Burrows, J.N., Kowalczyk, P., McDonald, S., Wells, T.N.C., Willis, P.: The open access malaria box: a drug discovery catalyst for neglected diseases. PLoS ONE **8**(6), e62906 (2013). https://doi.org/10.1371/journal.pone.0062906
31. Tucker, J.D., Day, S., Tang, W., Bayus, B.: Crowdsourcing in medical research: concepts and applications. PeerJ **7**, e6762 (2019). https://doi.org/10.7717/peerj.6762
32. Wang, C., et al.: Crowdsourcing in health and medical research: a systematic review. Infect. Dis. Poverty **9**(1), 8 (2020). https://doi.org/10.1186/s40249-020-0622-9
33. Wang, Q., Liu, C., Wang, Z.: An advanced and effective literature search algorithm based on analytic hierarchy process. In: 2011 IEEE 10th International Conference on Trust, Security and Privacy in Computing and Communications, November 2011, pp. 1264–1270 (2011)
34. Yarborough, M.: Openness in science is key to keeping public trust. Nat. News. **515**(7527), 313 (2014). https://doi.org/10.1038/515313a
35. Zuccato, M., Shilling, D., Fajgenbaum, D.C.: The collaborative network approach: a model for advancing patient-centric research for Castleman disease and other rare diseases. Emerging Topics Life Sci. **3**(1), 97–105 (2019). https://doi.org/10.1042/ETLS20180178

Conducting Quantitative Research with Hard-To-Reach-Online Populations: Using Prime Panels to Rapidly Survey Older Adults During a Pandemic

Nitin Verma[1](✉), Kristina Shiroma[1], Kate Rich[1,3], Kenneth R. Fleischmann[1], Bo Xie[1,2], and Min Kyung Lee[1]

[1] School of Information, The University of Texas at Austin, Austin, TX 78701, USA
{nitin.verma,kristinashiroma,kfleisch,boxie}@utexas.edu, minkyung.lee@austin.utexas.edu

[2] School of Nursing, The University of Texas at Austin, Austin, TX 78712, USA

[3] Department of Communication, The University of Washington, Seattle, WA 98195, USA
katerich@uw.edu

Abstract. Vulnerable populations (e.g., older adults) can be hard to reach online. During a pandemic like COVID-19 when much research data collection must be conducted online only, these populations risk being further underrepresented. This paper explores methodological strategies for rigorous, efficient survey research with a large number of older adults online, focusing on (1) the design of a survey instrument both comprehensible and usable by older adults, (2) rapid collection (within hours) of data from a large number of older adults, and (3) validation of data using attention checks, independent validation of age, and detection of careless responses to ensure data quality. These methodological strategies have important implications for the inclusion of older adults in online research.

Keywords: COVID-19 · Online data collection · Older adults

1 Introduction

During the COVID-19 pandemic, online data collection has become crucial to safeguard the health of participants, researchers, and society at large. However, online data collection presents challenges for researchers who study hard-to-reach populations such as older adults (65 + years). Traditional sampling methods such as using university subject pools or crowdsourcing platforms (e.g., Amazon's Mechanical Turk (MTurk)) are of limited use in reaching older adults [1, 2]. Another challenge is to design online studies that are comprehensible and usable by older adults.

In this paper, we outline methodological strategies for an online survey study with a large sample of older and younger adults recruited using Prime Panels. This paper is part of a multiphase research project funded by the National Science Foundation (NSF) that aims to explore factors that influence both younger and older adults' trust in public

K. Toeppe et al. (Eds.): iConference 2021, LNCS 12646, pp. 384–393, 2021.
https://doi.org/10.1007/978-3-030-71305-8_32

health information during a pandemic. Empirical data from our NSF project will be reported elsewhere. The strategies outlined here have implications for researchers who study hard-to-reach-online populations such as older adults.

2 Background

As the population and proportion of older adults continue to grow in the U.S., it is increasingly important to collect data that represent this growing population properly [3–5]. In traditional in-person studies, researchers rely on relationships with community partners (e.g., senior centers, healthcare organizations, public libraries) to recruit older adults for data collection [6–9]. Given the high cost of personnel and financial resources to sustain such recruitment, researchers need alternative methods. Additionally, the COVID-19 pandemic has disrupted traditional in-person data collection, making online data collection ever more prominent [10].

In the last decade, crowdsourcing platforms such as MTurk have mitigated some recruitment challenges [11] and driven down the overall cost of conducting online research. While online crowdsourcing has its merits [12], recruiting older adults on crowdsourcing platforms presents many challenges. For example, a research cohort of participants aged 55-and-older is considered a "Premium Qualification" on MTurk and therefore incurs a higher cost than a cohort of younger adults [1, 13]. Also, the availability of older adult crowdworkers may be hindered by factors such as a lack of awareness of the platforms, a lack of access to technology, insufficient technological skills, and a lack of motivation [14–16].

The constant churn of the participant pool in crowdsource markets such as MTurk [17] also makes it hard to generalize research outcomes. Although MTurk and similar platforms help overcome the homogeneity of university participant pools, they tend to yield samples significantly younger and less racially diverse than the American population [2]. This limitation is exacerbated by the fact that Amazon does not publish an age-based breakdown of MTurk workers [18], and that the number of MTurk workers available for a study at any given time is well below the number registered on the platform [19–21]. The relatively small number of active MTurk workers increases the chance that they are familiar with the methods that researchers typically use to ensure data quality [1].

An overarching concern with online research is that recruitment on crowdsourcing platforms may be undermined by bots or malicious actors misrepresenting age, location, or other characteristics, in an effort to obtain tasks and payments [18, 22]. In addition to these concerns, survey studies on online crowdsourcing platforms are vulnerable to response satisficing, with participants paying inadequate attention while they take surveys, thus threatening data validity [23]. These challenges, intensified by the COVID-19 pandemic, motivated us to explore other avenues for online data collection that did not limit us to one monolithic platform such as MTurk. Our search led us to Prime Panels, an aggregation of online research panels maintained by CloudResearch[1] that has been

[1] Formerly known as TurkPrime, and also offering MTurk Toolkit, a platform designed to integrate MTurk into the social science workflow [24].

found to be better at approximating national probability samples than MTurk as reported in previously published comparisons [1].

3 Method

This study was approved by the Institutional Review Board of The University of Texas at Austin. Informed consent was obtained prior to any data collection.

3.1 Survey Design

Our survey study was driven by the following overarching questions: *What differences exist in the information behavior of older and younger adults with respect to COVID-19, and what factors could explain those differences?* We used Qualtrics to design and implement the survey, with two attention checks: one instructional manipulation check [25, 26], and one repeated question to check for consistency of responses. At the end of the survey, we included demographic questions to collect participants' age, gender, race, educational attainment, and political beliefs.

We used a 5-point scale for each Likert-type item, with a visible numeric label for each point from 1 to 5 and provided descriptive labels only at endpoints. This strategy offers the following advantages: (i) the odd number of options avoids forcing neutral participants to choose a side [27, 28]; (ii) the use of 5 points balances the information gained with the cognitive demands on respondents [28]; and (iii) endpoint-only labeling reduces the cognitive load in comparison with labeling all gradations [29]. This strategy permits the use of statistical methods such as correlations and linear regression [28]. To comply with Web Content Accessibility Guidelines (WCAG 2.0)[2] and thus be inclusive of older adults who rely on assistive technologies (e.g., text-to-speech applications), we implemented all Likert-type items using drop-down menus instead of rows of radio buttons.

3.2 Pilot Testing with Older Adults

To ensure the clarity and readability of the survey instrument, we pilot-tested it over the phone with 3 older adults (2 females, 1 male)[3] recruited from participants of our prior studies. We conducted cognitive interviews, a commonly used method to enhance the quality of data collection [30–32], to collect verbal feedback on the design of the instrument. One researcher guided the interview while another took detailed notes. Following each cognitive interview, after participants completed the survey on their computer or tablet, we asked them to re-enter the survey to retrospectively share their thoughts, experiences, and challenges through probing questions and think-aloud prompts. The main

[2] https://www.w3.org/TR/WCAG20/.

[3] We recruited older adult participants from our research team's established relationship with local community partners. We chose to partner with local participants instead of crowdsourced participants due to our established relationship, rapport, level of engagement, and length of the task.

interviewer asked pre-scripted questions and, when necessary, unscripted probing questions to gather requisite feedback [31]. These cognitive interviews enabled us to identify and fix accessibility issues related to clicking fatigue and font legibility. Each interview lasted approximately 45 min. Each participant received a $20 Amazon gift card.

3.3 Collecting Data Using Prime Panels

We deployed the survey using Prime Panels, which provided a price and feasibility estimate based on the approximate duration of the survey, the desired sample size, and demographic parameters. To reach our target of 500 valid responses split into 2 comparable group sizes for older (65 + years) and younger (18–64 years) adults, we conducted 5 batches of data collection between June 26 and July 20, 2020 punctuated by data validation procedures described below. For each batch we used the Prime Panels interface to exclude participants who had taken the survey in a previous batch.

3.4 Validation of Data

To ensure that participants accurately reported their age, we requested Prime Panels to provide us the current age of each participant. The age data provided by Prime Panels is based on the participant's year of birth according to self-report at the time of signing up with one of Prime Panels' providers. Since we did not have access to the years of birth of our participants, we allowed for a difference of 1 year in each participant's age as reported in the 2 data sets. We removed duplicates and used the attention checks to discard invalid responses. To filter out careless responses [33] we used two criteria: (i) long-string responses: all items on a page with 5 or more items were marked with the same option; and (ii) response time: where participants took less than an average of 2 s per item on a page [33–35].

4 Results

Across 5 batches of data collection, we received responses from 669 participants: 272 older adults (40.7%) and 397 younger adults (59.3%; in each batch we got a few more responses than requested; and more responses from younger adults than those from older adults were excluded due to data quality concerns, specified below). Older adults took 10.9 min on average (SD = 12 min) to complete the survey, whereas younger adults took 10.2 min (SD = 21.4 min). The data collection cost $1,173. We paid Prime Panels for all responses even though we discarded some responses during data validation.

Of the 669 responses collected, 185 failed one or more of our validation criteria, leaving 484 responses in our final sample. Table 1 breaks down our data collection by batch, including the time elapsed and responses collected. Table 2 summarizes the number of invalid responses per criterion, broken down by age category.

Table 1. Time elapsed in each batch of data collection

Batch	Prime panels age filter	Time elapsed (Hours)	Responses collected	Valid responses
1	18–64 (Younger adults)	1	55	40
2	18–64 (Younger adults)	7	207	110
3	65–99 (Older adults)	11	258	238
4	General population	2	55	37
5	General population	1	94	59
Total		22	669	484

Table 2. Summary of data clean-up; number of responses discarded for each criterion when considered *independently* of each other

Validation criterion	Number of responses failing a criterion (%)		
	Overall (n = 669)	Older adults (n = 272)	Younger adults (n = 397)
Missing data	11 (1.6)	2 (0.7)	9 (2.3)
Failed at least 1 attention check	114 (17.0)	12 (4.4)	102 (25.7)
Age data validation	52 (7.8)	10 (3.7)	42 (10.6)
Careless responses	81 (12.1)	8 (2.9)	73 (18.4)
Total	258 (38.6)	32 (11.8)	226 (56.9)

5 Discussion

During the COVID-19 pandemic, the older adult population has been especially vulnerable. Compared with younger people, older adults are more likely to develop serious health conditions if infected by the virus, but they are less likely to obtain digital information and services [36, 37]. When data for research must be obtained online, older adults risk being further underrepresented, hindering subsequent decision making based on the data. In this paper, we have outlined methodological strategies for our recently completed online survey study with a large sample of older and younger adults. Our strategies were aimed to ensure the survey's readability and usability, obtain a large, stratified sample of both older adult and younger adults, and validate the data including participants' ages.

Data collection via Prime Panels is far less expensive than the personnel and financial resources required by traditional, in-person studies [6–9]. Thus, this platform offers a viable alternative not only to MTurk, but also to traditional data collection methods. Our initial request for 250 older adult participants cost $660 ($2.64 per participant). Our validation process required us to request additional participants, but the minimal cost per participant was a clear benefit. Indeed, one impressive finding of our study was the stronger performance among older adults with the validation, as 248 of the 272 responses from older adults satisfied all validation criteria (91.2%), while 236 of the 397 responses from younger adults satisfied all validation criteria (59.4%).

It is also important to note that it took very little time to collect data from a large number of older adults. As shown in Table 1, across the 5 batches of data collection, it took less than 14 h to obtain valid responses from 250 older adult participants. As such, Prime Panels provides a rapid means to collect data from older adults compared to other methods. For example, in another research project conducted by our team during the summer of 2020, we conducted a telephone survey to gather data from 200 older adults. Data collection took over 3 months, with a cost of $4000 in participant compensation (this manuscript is available from the authors). Overall, our methodological strategies have significant implications for researchers currently working with online crowdsourcing platforms as well as researchers who have had to shift their data collection from in-person to online due to COVID-19 [10].

6 Limitations

One overarching concern with studying older adults online is that about a third of older adults in the U.S. lack internet access, and only two-fifths of older adults own smartphones [38]. Therefore, despite our rapid and cost-efficient data collection, our sample was likely skewed by technology adoption among older American adults, and thus arguably less representative of this population as a whole than for younger adults.

Moreover, crowdwork remains a largely unregulated part of the internet economy and is vulnerable to exploitative business practices often leading to unfair wages and low quality of work [39, 40]. In our prior studies, we have taken care to ensure that we always pay participants at least the U.S. national minimum wage [41–43]. However, for this study, our ability, as researchers, to control the compensation to our participants to meet our ethical standards was obfuscated by Prime Panels' inability to specify the exact payment received by participants [44]. As a result, it is unclear to us how much compensation our research participants had received, or what form the compensation was in (e.g., monetary compensation or donation to charity, as some crowdsourcing platforms have done [45, 46]). All we were able to do in this circumstance was to ensure that our estimate of the time required was conservative, to ensure that the compensation would be on the higher end of what Prime Panels allows. However, the lack of knowledge about and control of compensation provided by Prime Panels is a limitation of their current implementation.

7 Future Directions

Given the circumstances of the COVID-19 pandemic and the subsequent move to online data collection, more work needs to be done to ensure access to populations traditionally hard to reach online, including, but not limited to, older adults. To facilitate this access, crowdsourcing platforms must be transparent about the demographic make-up of their respective worker populations. Also, to ensure the quality of collected data, crowdsource platforms and online panels should offer researchers a communication channel to allow them to ascertain the validity of demographic data while preserving participants' privacy. An underlying concern in recruiting hard-to-reach-online populations is the availability and reach of the internet in the U.S., as well as the computer literacy of the population of interest. It is, therefore, the responsibility of policymakers, crowdsourcing platforms, and researchers to ensure that hard-to-reach populations are represented in research. Such inclusion efforts will ensure that these populations are represented in research findings and, subsequently, policies developed using the research findings. Finally, these stakeholders need to work together to ensure that participants are compensated fairly for their participation in research.

Acknowledgements. This material is based upon work supported by the National Science Foundation under Grant No. 2027426. We thank John Bellquist, Ph.D., Editor of the Cain Center in the School of Nursing at The University of Texas at Austin, for his professional proofreading of an earlier draft of this manuscript; Le (Betty) Zhou, Ph.D., at Carlson School of Management, the University of Minnesota for helping us with data validation; and the anonymous participants of this study.

References

1. Chandler, J., Rosenzweig, C., Moss, A.J., Robinson, J., Litman, L.: Online panels in social science research: expanding sampling methods beyond Mechanical Turk. Behav. Res. Methods **51**(5), 2022–2038 (2019). https://doi.org/10.3758/s13428-019-01273-7
2. Levay, K.E., Freese, J., Druckman, J.N.: The demographic and political composition of mechanical Turk samples. Sage Open. (2016). https://doi.org/10.1177/2158244016636433
3. He, W., Goodkind, D., Kowal, P.: An aging world: 2015. International Population Reports P95/16–1, United States Census Bureau (2016). https://www.census.gov/content/dam/Census/library/publications/2016/demo/p95-16-1.pdf
4. Weil, J., Mendoza, A.N., McGavin, E.: Recruiting older adults as participants in applied social research: applying and evaluating approaches from clinical studies. Educ. Gerontol. **43**(12), 662–673 (2017)
5. Zickuhr, K., Madden, M.: Older adults and internet use. Pew Research Center (2012). https://www.pewresearch.org/internet/2012/06/06/older-adults-and-internet-use
6. Kwak, J., Xie, B., Champion, J.D., Fleischmann, K.R.: Rural dementia caregivers in Southwest Texas: an exploratory study of advance directives and end-of-life proxy decision making. J. Gerontol. Nurs. **45**(9), 11–17 (2019). https://doi.org/10.3928/00989134-20190530-01
7. Xie, B.: Effects of an e-health literacy intervention for older adults. J. Med. Internet Res. **13**(4), e90 (2011). https://doi.org/10.2196/jmir.1880
8. Xie, B.: Improving older adults' e-health literacy through computer training using NIH online resources. Libr. Inf. Sci. Res. **34**(1), 63–71 (2012). https://doi.org/10.1016/j.lisr.2011.07.006

9. Xie, B., Bugg, J.M.: Public library computer training for older adults to access high-quality Internet health information. Libr. Inf. Sci. Res. **31**(3), 155–162 (2009). https://doi.org/10.1016/j.lisr.2009.03.004
10. Clay, R.A.: Conducting research during the COVID-19 pandemic: advice from psychological researchers on protecting participants, animals and research plans. APA News, American Psychological Association, 19 March 2020. https://www.apa.org/news/apa/2020/03/conducting-research-covid-19
11. Behrend, T.S., Sharek, D.J., Meade, A.W., Wiebe, E.N.: The viability of crowdsourcing for survey research. Behav. Res. Methods **43**(3), 800–813 (2011). https://doi.org/10.3758/s13428-011-0081-0
12. Geldsetzer, P.: Use of rapid online surveys to assess people's perceptions during infectious disease outbreaks: a cross-sectional survey on COVID-19. J. Med. Internet Res. **22**(4), e18790 (2020). https://doi.org/10.2196/18790
13. Lin, S.-Y., Thompson, H.J., Hart, L.A., Fu, M.C., Demiris, G.: Evaluation of pharmaceutical pictograms by older "turkers": a cross-sectional crowdsourced study. Res. Soc. Admin. Pharm. (2020). https://doi.org/10.1016/j.sapharm.2020.08.006
14. Brewer, R., Morris, M.R., Piper, A.M.: Why would anybody do this? Older adults' understanding of and experiences with crowd work. In: CHI 2016: Proceedings of the 2016 CHI Conference on Human Factors in Computing Systems, pp. 2246–2257. ACM Digital Library (2016). https://doi.org/10.1145/2858036.2858198
15. McDuffie, D.: Using Amazon's mechanical Turk: benefits, drawbacks, and suggestions. APS Observer. **32**(2), 34–35 (2019). https://www.psychologicalscience.org/observer/using-amazons-mechanical-turk-benefits-drawbacks-and-suggestions
16. Skorupska, K., Núñez, M., Kopec, W., Nielek, R.: Older adults and crowdsourcing: Android TV app for evaluating TEDx subtitle quality. In: Proceedings of the ACM on Human–Computer Interaction, Article 159. ACM Digital Library (2018). https://doi.org/10.1145/3274428
17. Cheung, J.H., Burns, D.K., Sinclair, R.R., Sliter, M.: Amazon mechanical Turk in organizational psychology: an evaluation and practical recommendations. J. Bus. Psychol. **32**(4), 347–361 (2016). https://doi.org/10.1007/s10869-016-9458-5
18. Ogletree, A.M., Katz, B.: How do older adults recruited using MTurk differ from those in a national probability sample? Int. J. Aging Hum. Devel. (2020). https://doi.org/10.1177/0091415020940197
19. Difallah, D., Filatova, E., Ipeirotis, P.: Demographics and dynamics of mechanical Turk workers. In: WSDM 2018: Proceedings of the Eleventh ACM International Conference on Web Search and Data Mining, pp. 135–143. ACM Digital Library (2018). https://doi.org/10.1145/3159652.3159661
20. Robinson, J., Rosenzweig, C., Moss, A.J., Litman, L.: Tapped out or barely tapped? Recommendations for how to harness the vast and largely unused potential of the mechanical Turk participant pool. PLoS ONE **14**(12), e0226394 (2019). https://doi.org/10.1371/journal.pone.0226394
21. Stewart, N., Ungemach, C., Harris, A.J.L., Bartels, D.M., Newell, B.R., Paolacci, G., Chandler, J.: The average laboratory samples a population of 7,300 Amazon mechanical Turk workers. Judgm. Decis. Making. **10**(5), 479–491 (2015). https://journal.sjdm.org/14/14725/jdm14725.pdf
22. Kennedy, R., Clifford, S., Burleigh, T., Waggoner, P.D., Jewell, R., Winter, N.J.G.: The shape of and solutions to the MTurk quality crisis. Pol. Sci. Res. Methods (2020). https://doi.org/10.1017/psrm.2020.6
23. Kapelner, A., Chandler, D.: Preventing satisficing in online surveys: a "kaptcha" to ensure higher quality data. In: CrowdConf 2010, San Francisco (2010)

24. Litman, L., Robinson, J., Abberbock, T.: TurkPrime.com: a versatile crowdsourcing data acquisition platform for the behavioral sciences. Behav. Res. Methods **49**(2), 433–442 (2016). https://doi.org/10.3758/s13428-016-0727-z
25. Kane, J.V., Barabas, J.: No harm in checking: Using factual manipulation checks to assess attentiveness in experiments. Am. J. Pol. Sci. **63**(1), 234–249 (2019). https://doi.org/10.1111/ajps.12396
26. Oppenheimer, D.M., Meyvis, T., Davidenko, N.: Instructional manipulation checks: detecting satisficing to increase statistical power. J. Exp. Soc. Psychol. **45**(4), 867–872 (2009). https://doi.org/10.1016/j.jesp.2009.03.009
27. Bishop, G.F.: Experiments with the middle response alternative in survey questions. Public Opin. Q. **51**(2), 220–232 (1987). https://doi.org/10.1086/269030
28. Weijters, B., Cabooter, E., Schillewaert, N.: The effect of rating scale format on response styles: the number of response categories and response category labels. Int. J. Res. Mark. **27**(3), 236–247 (2010). https://doi.org/10.1016/j.ijresmar.2010.02.004
29. Aday, L.A., Cornelius, L.J.: Formulating questions about knowledge and attitudes. In: Designing and Conducting Health Surveys: A Comprehensive Guide, 3rd ed., pp. 268–287. Wiley, Jossey-Bass (2006)
30. Beatty, P.C., Willis, G.B.: Research synthesis: The practice of cognitive interviewing. Public Opin. Q. **71**(2), 287–311 (2007). https://doi.org/10.1093/poq/nfm006
31. Drennan, J.: Cognitive interviewing: verbal data in the design and pretesting of questionnaires. J. Adv. Nurs. **42**(1), 57–63 (2003). https://doi.org/10.1046/j.1365-2648.2003.02579.x
32. Jobe, J.B., Mingay, D.J.: Cognitive laboratory approach to designing questionnaires for surveys of the elderly. Public Health Rep. **105**(5), 518–524 (1990). PMCID: PMC1580104
33. Meade, A.W., Craig, S.B.: Identifying careless responses in survey data. Psychol. Methods. **17**, 437 (2012). 10/f399k2
34. Huang, J.L., Curran, P.G., Keeney, J., Poposki, E.M., DeShon, R.P.: Detecting and deterring insufficient effort responding to surveys. J. Bus. Psychol. **27**, 99–114 (2012). 10/dppm46
35. Curran, P.G.: Methods for the detection of carelessly invalid responses in survey data. J. Exp. Soc. Psychol. **66**, 4–19 (2016). 10/f8zmvf
36. Seifert, A., Cotten, S. R., Xie, B.: A double burden of exclusion? Digital and social exclusion of older adults in times of COVID-19. J. Gerontol. B Psychol. Sci. Soc. Sci. (2020). https://doi.org/10.1093/geronb/gbaa098
37. Xie, B., Charness, N., Fingerman, K., Kaye, J., Kim, M.T., Khurshid, A.: When going digital becomes a necessity: ensuring older adults' needs for information, services, and social inclusion during COVID-19. J. Aging Soc. Policy. **32**(4–5), 460–470 (2020). https://doi.org/10.1080/08959420.2020.1771237
38. Anderson, M., Perrin, A.: Tech adoption climbs among older adults. Pew Res. Center (2017). https://www.pewinternet.org/2017/05/17/tech-adoption-climbs-among-older-adults/
39. Semuels, A.: The internet is enabling a new kind of poorly paid hell. The Atlantic, 23 January 2018. https://www.theatlantic.com/business/archive/2018/01/amazon-mechanical-turk/551192/. Accessed 16 Oct 2020
40. Silberman, M.S., Tomlinson, B., LaPlante, R., Ross, J., Irani, L., Zaldivar, A.: Responsible research with crowds: pay crowdworkers at least minimum wage. Commun. ACM **61**(3), 39–41 (2018). https://doi.org/10.1145/3180492
41. Templeton, T.C., Fleischmann, K.R.: The relationship between human values and attitudes toward the Park51 and nuclear power controversies. Proc. Am. Soc. Info. Sci. Tech. **48**, 1–10 (2011). https://doi.org/10.1002/meet.2011.14504801172
42. Verma, N., Fleischmann, K.R., Koltai, K.S.: Human values and trust in scientific journals, the mainstream media and fake news. Proc. Assoc. Inf. Sci. Technol. **54**, 426–435 (2017). 10/ghkc6k

43. Lee, M.K.: Understanding perception of algorithmic decisions: fairness, trust, and emotion in response to algorithmic management. Big Data Soc. **5**, 2053951718756684 (2018). 10/ggsfsp
44. CloudResearch: How are Participants on Prime Panels Compensated? (2019). https://go.cloudresearch.com/knowledge/how-are-participants-on-prime-panels-compensated. Accessed 17 Oct 2020
45. SurveyMonkey: SurveyMonkey Raises $15M for Charitable Causes, One Survey at a Time. SurveyMonkey, 18 February 2020. https://www.surveymonkey.com/newsroom/surveymonkey-raises-15m-for-charitable-causes/
46. Lamb, J.: Clear Outcomes. Americans' COVID-19 preventative practices in April and May 2020. Inter-university Consortium for Political and Social Research (ICPSR). 10/ghfwtc

Collaborative Research Results Dissemination: Applying Postcolonial Theory to Indigenous Community Contexts

Lisa G. Dirks(✉)

University of Washington Information School, Seattle, WA 98195, USA
lgdirks@uw.edu

Abstract. Community engagement in research has become increasingly important; it is essential for research conducted with Indigenous communities. In some cases, community members are receptively engaged in research from beginning to end, but this is inconsistent. Community collaboration during the results dissemination process is an element of engagement that is consistently overlooked or otherwise ineffectively executed. The concept of decolonizing research and the postcolonial theoretical foundations of decolonization are explored in this paper. Decolonizing research involves conducting research with Indigenous communities that places Indigenous voices and epistemologies at the center of the research process. This paper considers a decolonization framework to examine Native American community collaboration in the research results dissemination process including recommendations for applying postcolonial theory in the design of technologies to facilitate collaborative research results dissemination.

Keywords: Indigenous research · Results dissemination · Postcolonial theory · Decolonization · Collaboration

1 Introduction

To date, a large number of research studies conducted in Indigenous communities has excluded significant Indigenous voice [1]. Nevertheless, in the past several decades, there has been an emphasis on decolonizing research conducted in Indigenous communities, which includes active community collaboration. This comprises collaboration that occurs during the entire research process including conceptualizing research topics, developing research questions, collecting, analyzing, and interpreting data, and disseminating study results. However, other than using traditional academic dissemination methods (i.e. journal articles, conference presentations), limited progress has been made to ensure research results are shared with participants or otherwise disseminated to non-academic community audiences [2]. Studies suggest that research participants and community audiences want to receive research results and propose that participants

The original version of this chapter was revised: the main title was corrected. The correction to this chapter is available at https://doi.org/10.1007/978-3-030-71305-8_43

K. Toeppe et al. (Eds.): iConference 2021, LNCS 12646, pp. 394–403, 2021.
https://doi.org/10.1007/978-3-030-71305-8_33

have a right to receive them for contributing to the research [3, 4]. Still, sharing comprehensible research results with non-academic communities is often neglected or done ineffectively [5]. For many researchers, the primary method of disseminating results is to submit manuscripts to peer-reviewed journals or conference presentations which essentially limits their audience to academic scholars [6–8]. Limiting dialogue between community stakeholders and researchers, many of whom may be unacquainted with community perceptions of research or their systems of sharing information, decreases the likelihood that study results will be locally implemented [8].

Internet-based and other technological tools (i.e., social media, podcasts, interactive websites, video, and mobile applications) have potential to increase opportunities for disseminating research results to non-academic audiences more broadly than through conventional means (e.g., journal articles, conference presentations) and possibly with more creativity and collaboration. However, few studies have been conducted that explore using technology tools to share research results with non-academic community audiences let alone with Indigenous communities. Collaborating with Indigenous community members on strategies for sharing research results would be appropriate. However, successful collaboration depends on how strategies are implemented, how accessible the physical technology is (including hardware and software), and how involved community members are in developing strategies for use within their communities. If developed properly, technology tools can potentially be used to facilitate collaborative results dissemination. Proper development and implementation will require regular active community engagement while also acknowledging power relationships. Postcolonial theory can be used to assess these barriers to development and implementation.

This paper examines research results dissemination in an Indigenous context using a postcolonial framework. The first section of the paper provides context for recent disreputable research conduct in Indigenous communities, including a case example of unethical research. This is followed by a grounding in historical foundations of decolonizing research leading into discourse on decolonization's postcolonial theoretical foundations. The paper concludes with a discussion on collaborative dissemination in Indigenous contexts including suggestions for decolonizing the research process by questioning positivist research approaches and promoting dialogue between researchers and Indigenous community members concerning results dissemination and in using postcolonial theory to design technologies to facilitate dissemination collaboration.

1.1 Unethical Research Dissemination Practice

Indigenous communities have experienced a history of unethical research conduct such as data being collected without consent, disclosure of sacred Tribal information, and public reports of research shared without community approval. This disreputable research has imposed stereotypes on Indigenous people (i.e. drunk Indian, spiritual Native motifs) and research that has historically benefitted individual researchers or specific research institutions instead of benefitting the communities where the research was conducted [1]. A prominent example is an early-1990's diabetes research study conducted in the Havasupai community [9]. In this study, Havasupai tribal authorities sought the expertise of a human genetics researcher at the Arizona State University to evaluate the potential genetic causes for high rates of diabetes in their community. The study concluded that there was "too little variation among tribal members' genetics to conclude the incidence of disease among them was genetics-related." [10]. This was a valid result and fulfilled the community's request for research support. However, in addition to using the genetic samples collected for the diabetes study, researchers also used the samples collected as part of the diabetes study to conduct additional research on conditions such as schizophrenia and incest. These additional analyses were performed without consent from individual participants or the Havasupai Tribal leadership. Furthermore, they did not directly share results of these studies with the community or involve community members in the research process. It was only a coincidence that the community discovered further studies utilizing the biospecimen were conducted. These unethical research practices resulted in Havasupai Tribal officials suspending any research requests from Arizona State University in their community. It also prompted systematic changes to informed consent procedures and increased United States Indigenous community involvement in research processes. This example is one of many in which Indigenous community members have been harmed leading to research distrust. As a result of actions like these, Indigenous communities have taken ownership over research by developing systemic practices to ensure research that is conducted in their communities is overseen by Tribal authorities. This oversight often involves direct ongoing collaboration to facilitate transparency in the research process.

2 Decolonizing Research

Unethical research activity such as the Havasupai case happened long before the example above. During worldwide exploration and colonization, Western European colonizers brought ideas, practices and methods for science and technology including how knowledge is recorded (classified and archived). Proponents of decolonization posits that colonizers viewed themselves as central to knowledge and "the arbiter[s] of what counts as knowledge and the source of 'civilized' knowledge" [11]. This view played itself out as colonizers established themselves as "experts" in the places they colonized, including expertise on the Indigenous people and lands they colonized. This was done through knowledge organization (i.e. archives, logs) that put colonizers at the center of knowledge production. Much of this organized knowledge production included appropriated Indigenous knowledge and procedures for assimilating or outlawing Indigenous cultural knowledge and practice deemed uncivilized, non-Christian, or in some way "other" than their own [11]. From a decolonization perspective, these actions were performed to dominate and hold control over colonized societies. What this effectively did was change the knowledge systems of Indigenous communities. Traditional knowledge was lost through forced assimilation, which attempted to extinguish language, culture and belief systems, or the knowledge was reshaped by Western thought, effectually restructuring much of the original intended meaning.

To counter the historical impacts of colonial control, many Indigenous communities have taken a decolonization approach to research. This process involves changing the power dynamics among marginalized communities by utilizing multiple epistemologies, ontologies and axiologies [12]. These practices support Indigenous community wellness by honoring Indigenous epistemologies, philosophies, beliefs, and values that counter the effects of colonialism [13–17]. This is done through challenging ideas that Western scientific methods and epistemologies are the only true scientific methods which essentially depreciates Indigenous epistemologies as simple folklore [1]. Although Indigenous epistemologies are central to decolonization theory and should be incorporated in research practice, Western research methods and theories should not be precluded, but appropriately adapted to specific Indigenous contexts [11, 18]. For example, in some (not all) Indigenous communities, incorporating a narrative approach may be more appropriate than collecting data through using a validated survey instrument or even conducting a content analysis of interview data which parses out different pieces of an interview and by doing so may lose meaning without context found in the interview as a whole. Some Indigenous communities consider the entire story as important for establishing meaning which would be lost if only a part of that story were analyzed through qualitative analysis methods that isolates only selections of a whole interview.

2.1 Self-determination

Self-determination is an integral component for decolonizing research. In the context of decolonization, self-determination involves actions towards shaping a "world according to terms chosen by oneself, or by a people collectively" [19]. For Indigenous communities, this entails constructing their own values and beliefs for what constitutes knowledge [18] and control over expressions of cultural heritage and intellectual property [20]. From a postcolonial lens, through self-determination, Indigenous communities now have more control and ownership over research conducted in their respective communities, how that research is conducted, and who is involved. This is not to say that non-Indigenous researchers are unwelcomed to conduct research in many Indigenous communities, but that they need to acknowledge the community's privilege to determine the research. An insistent reflexivity about who maintains control over decisions is required to effectively respond to self-determination [1, 21].

In the past several decades, Indigenous people have been reclaiming their languages and cultures. Prominent Maori scholar Linda Tuhiwai Smith refers to this as a phase of resistance and survival eventually leading to recovery as Indigenous peoples [11]. She notes that this "sense of optimism… is often criticized by non-indigenous scholars, because it is viewed as being overly idealistic" [11]. Much progress has been achieved in Indigenous communities towards recovering Indigenous identities and showcasing resiliency. This can be seen in Indigenous communities where efforts are being made to revitalize language through educational programming and to learn traditional cultural practices (i.e. tattooing, weaving, traditional food preparation, dance). Related to this, Smith criticizes Western views about what it means to be an authentic Indigenous person and suggests that the "belief that Indigenous cultures cannot change, cannot recreate themselves and still claim to be Indigenous" only encourages marginalization [11]. This leads back to the importance of who determines what Indigenous culture and knowledge are, which, from a self-determination standpoint, should be Indigenous people.

2.2 Postcolonial Theory

Various decolonization processes are rooted in postcolonial theoretical traditions. Postcolonial theory broadly emphasizes systematic approaches to power and dominance over marginalized communities. This includes examining the influences of historical oppression on modern social conditions within marginalized communities [22]. Postcolonial theory situates itself amid other critical theories such as postmodernism, poststructuralism, feminism, and Marxism [23, 24]. These critical theories place emphasis on understanding how power "through advantage and disadvantage, operates based on historical positioning, class, race and gender" and how these power structures intersect to perpetuate control over marginalized communities [25]. Essentially, within an Indigenous community context, postcolonial theory attempts to show how colonization has influenced community deficits, such as racism, historical trauma, and social determinants of health. Some Indigenous scholars believe that postcolonial theory perpetuates Western researchers' control over Indigenous epistemologies [22] and suggest that by focusing on the power relationships using a postcolonial lens, proponents may miss capturing Indigenous values such as those related to family, spirituality, humility and sovereignty [26]. But instead, imply that Western approaches could still be used if done while also honoring Indigenous values and epistemologies [11, 14, 16, 17, 27].

2.3 Indigenous Research Methodologies

Incorporating Indigenous research methodologies is a significant step towards decolonizing research. Many Indigenous scholars assert that Indigenous methodologies focus on collective community needs entrenched in holistic relationships between people, their values, beliefs, and connections to the physical world versus individual needs [11, 17, 18, 27]. Moreover, to explore these holistic relationships, research might instead incorporate traditional empirical approaches, such as observation, but knowledge may also be acquired through generational storytelling, elder-youth apprenticeships, and experiential learning in a practical setting [28]. In this sense, since knowledge is intertwined with daily existence, it can be difficult to categorize and systematically define with empirical approaches [14, 26]. In contrast to positivist social science approaches, where the goal of science is to predict and control behavior, Indigenous research methodologies do not prioritize predicting or controlling behaviors but focus on relationality between constructs such as people and their physical, social, and spiritual environment [1].

Nevertheless, Indigenous communities vary in many ways, such as different linguistic, geographic, legal, cultural, and social differences. No single epistemological approach should be considered as a remedy for all research conducted in all Indigenous communities as each community has its own characteristics, needs, and expectations. For instance, within the United States there are 578 federally recognized tribes [29] many of whom have similar histories of research misconduct, yet there is also considerable variation in social, political, and cultural practices within these communities, making a one-size-fits-all approach to research with Indigenous communities impractical. Moreover, Indigenous knowledge systems have been reshaped through acts of colonization and new knowledge systems were created and continue to be part of each Indigenous community's life. In this regard, decolonizing research also means acknowledging the history and knowledge shifts that resulted from colonization [19, 23]. These acknowledgments highlight the importance of active collaboration with Indigenous communities to determine what works best for their unique circumstances.

3 Reacting to Western Positivist Research – Participatory Approaches

Traditional Western research methodologies, such as those rooted in empiricism and positivism can be used successfully in Indigenous communities depending on the research questions being asked and how they are implemented. However, the underlying epistemologies of these approaches are often counterintuitive to Indigenous epistemologies and are less likely to be successful unless they are adapted [1, 17] which regularly involves engaging Indigenous community members in developing local processes. Participatory research approaches, such as community-engaged research, community-based participatory research, and participatory action research lend themselves to decolonizing these processes. From a postcolonial perspective, participatory approaches can help to address power disputes by challenging dominant systems of knowledge and power [25]. However, not all community-engaged research distributes power equally. On one end of the continuum, research that is exclusively investigator-driven has much less community involvement, while on the other end of the spectrum, research driven by the community, includes much more community control [30]. Between these two points there are varying levels of power-dynamics over who has control over the research.

Conventional positivist research approaches have been criticized for not using Indigenous people's knowledge or empowering communities to create their own new knowledge in order to take action towards goals they determine are important [31]. Engaging community participation in the research process supports the decolonization process by placing the community in the role of knowledge provider, as educators and owners of the knowledge they share [18]. An Indigenous person may take the role of an independent research investigator and may or may not engage the community in the research process. Simply being an Indigenous researcher does not automatically entail the research being conducted engages the community. There may be hidden undercurrents, such as the researcher may not be invested in the community and have priorities that do not align with the community wants or needs. For instance, their actions may be more motivated by research productivity (i.e., manuscripts, funding opportunities) than community expectations. Therefore, transparency measures should be implemented to shine a light on these dynamics.

Although a community-engaged research approach may be appropriate to utilize in Indigenous communities, there are also significant limitations from a postcolonial perspective. Assuming that a community-engaged approach equates giving power to a community is problematic as the concept of empowerment varies across cultures and communities [25]. Community members may be impartial to research or have different priorities to address that may not be addressed through a proposed research study. Moreover, there may be challenges in determining which community members have the authority to speak for the rest of the community. In some cases, the agents chosen to represent the community may establish their own dominance and control reinforcing colonial power relations [25, 32]. In addition to deep-rooted epistemological concerns, there are also technical limitations. Community engagement is a slow process requiring time to develop and maintain relationships which may present time limitations and financial constraints [33–35]. These restrictions are often unrealistic given funding agency expectations of timely progress for study outcomes, which do not often factor in relationship development or community collaboration [25]. Moreover, community priorities may shift and, if the research is no longer a community priority, then continued community engagement may falter [25].

4 Decolonizing Research Results Dissemination

As an outcome of many Indigenous communities' prioritization for decolonizing research, community engagement in research has become increasingly more prominent [33]. Indigenous community members are often engaged throughout the research process. However, effective collaboration during the results dissemination process has been limited [6, 36]. Ongoing multidirectional communication encourages trust and allows communication to flow to encourage information is relevant to the community [36–39]. For Indigenous communities this includes relevance to local context including local values, knowledge, and expertise [35, 40, 41]. Approaching dissemination as a collaborative community-researcher dialogue can strengthen community trust in research while enhancing researchers' understanding of community concerns and perceptions of research [35, 36, 42].

4.1 Potential for Using Postcolonial Theory for Results Dissemination Technology Design

Postcolonial theory has promise to inform active community engagement in the design of research dissemination technologies for Indigenous communities. Design activities could involve an acknowledgement of the social, political, and historical impacts of research on Indigenous communities. This framing can provide transparency and openness to discuss research experiences throughout the design process and provide context for design processes and products. Applying postcolonial theory can also contribute to understanding power dynamics between Indigenous design participants and researchers as well as power relations within their respective community. It may also be useful for understanding how a history of unethical research experience impact Indigenous participants' understanding of the research and dissemination processes. Understanding these relationships may help develop tools that encourage dialogue between researchers and participants. Postcolonial theory can supplement theoretical approaches, such as feminist theory to address how colonial and gender power dynamics impact research results dissemination and access to technology in Indigenous communities. Integrating Indigenous epistemological approaches may also be added to give community-level power to the design process. For example, the design process could include a storytelling component that integrates Indigenous oral traditions giving participants control of what they choose to share while aligning with their interpretations of the world around them. Engaging design participants in these ways may increase parity by supporting their role as educators and owners of the knowledge they choose to share.

Research that honors Indigenous knowledge while acknowledging researchers' learned biases towards what constitutes valid data is imperative for the conduct of ethical research and design with Indigenous communities. Early colonizers did not typically acknowledge Indigenous contribution to scientific foundations [11]; in some ways, this is still the case with modern research. Encouraging transparent communication and collaboration during the research and design activities ideally promotes respect and validity for all knowledge systems (e.g., Western and Indigenous). As each community has its unique characteristics, eliciting feedback from Indigenous communities to provide direct response on how and in what ways they want to receive research results is imperative to make decisions on how technology should be used to disseminate research results to their communities. Community collaboration will undoubtedly uncover other issues that need to be addressed. This will require planning ahead of implementing technology-facilitated collaborative dissemination strategies. These plans should consider evaluating and continually reevaluating community interests in using technology, technological assessment of communities, analysis of potential partnership agencies, and capacity to implement projects.

References

1. Simonds, V.W., Christopher, S.: Adapting western research methods to indigenous ways of knowing. Am. J. Public Health **103**(12), 2185–2192 (2013)
2. Minkler, M., Salvatore, A.L., Chang, C.: Participatory Approaches for Study Design and Analysis in Dissemination and Implementation Research. Oxford University Press, New York (2017)
3. Fernandez, C.V., et al.: The return of research results to participants: pilot questionnaire of adolescents and parents of children with cancer. Pediatr. Blood Cancer **48**(4), 441–446 (2007)
4. Shalowitz, D.I., Miller, F.G.: Communicating the results of clinical research to participants: attitudes, practices, and future directions. PLoS Med. **5**(5), e91 (2008)
5. Ferris, L.E., Sass-Kortsak, A.: Sharing research findings with research participants and communities. Int. J. Occup. Environ. Med. **2**(3), 172–181 (2011)
6. Chen, P.G., et al.: Dissemination of results in community-based participatory research. Am. J. Prev. Med. **39**(4), 372–378 (2010)
7. MacKenzie, C.A., Christensen, J., Turner, S.: Advocating beyond the academy: dilemmas of communicating relevant research results. Qual. Res. **15**(1), 105–121 (2015)
8. Smylie, J., Kaplan-Myrth, N., McShane, K.: Indigenous knowledge translation: baseline findings in a qualitative study of the pathways of health knowledge in three indigenous communities in Canada. Health Promot. Pract. **10**(3), 436–446 (2009)
9. Pacheco, C.M., et al.: Moving forward: breaking the cycle of mistrust between American Indians and researchers. Am. J. Public Health **103**(12), 2152–2159 (2013)
10. Havasupai Tribe of Havasupai Reservation v Arizona Board of Regents (2008)
11. Smith, L.T.: Decolonizing Methodologies: Research and Indigenous Peoples. Zed Books Ltd., London (2013)
12. Sium, A., et al.: Towards the 'tangible unknown': decolonization and the Indigenous future **1**(1) (2012)
13. Saskamoose, J., et al.: Miýo-pimātisiwin developing indigenous cultural responsiveness theory (ICRT): improving indigenous health and well-being. Int. Indigenous Policy J. **8**(4), 1–16 (2017)
14. Battiste, M.: Reclaiming Indigenous Voice and Vision. UBC Press, Vancouver (2011)
15. Lavallée, L.F.: Practical application of an Indigenous research framework and two qualitative Indigenous research methods: sharing circles and Anishnaabe symbol-based reflection. Int. J. Qual. Methods **8**(1), 21–40 (2009)

16. Kovach, M.: Indigenous Methodologies: Characteristics, Conversations, and Contexts. University of Toronto Press, Toronto (2010)
17. Wilson, S.: Research Is Ceremony: Indigenous Research Methods. Fernwood Publishing, Halifax (2008)
18. Datta, R.: Traditional storytelling: an effective Indigenous research methodology and its implications for environmental research. AlterNat. Int. J. Indigenous Peoples **14**(1), 35–44 (2017)
19. Kohn, M., Political Theories of Decolonization: Postcolonialism and the Problem of Foundations. Oxford University Press, New York (2011). McBride, K.D (ed.)
20. Juutilainen, S.A., Jeffrey, M., Stewart, S.: Methodology matters: designing a pilot study guided by indigenous epistemologies. Hum. Biol. **91**(3), 141–151 (2020)
21. Bishop, R.: Addressing issues of self-determination and legitimation in Kaupapa Maori research, pp. 143–160 (1996)
22. Getty, G.A.: The journey between Western and Indigenous research paradigms. J. Transcult. Nurs. **21**(1), 5–14 (2010)
23. Mohammed, S.A.: Moving beyond the "exotic": applying postcolonial theory in health research. ANS Adv. Nurs. Sci. **29**(2), 98–109 (2006)
24. Sharma, M.: 'Can the patient speak?': Postcolonialism and patient involvement in undergraduate and postgraduate medical education. Med. Educ. **52**(5), 471–479 (2018)
25. Darroch, F., Giles, A.: Decolonizing health research: community-based participatory research and postcolonial feminist theory. Can. J. Action Res. **15**(3), 22–36 (2014)
26. Henderson, J.S.Y.: Aboriginal Thought, in Reclaiming Indigenous Voice Vision, p. 248 (2000). Battiste, M. (ed.)
27. Chilisa, B.: Indigenous Research Methodologies. SAGE Publications, Thousand Oaks (2012)
28. Jernigan, V.B.B., D'Amico, E.J., Kaholokula, J.K.: Prevention research with indigenous communities to expedite dissemination and implementation efforts. Prev. Sci. **21**, 74–82 (2018)
29. U.S. Department of the Interior Indian Affairs: Tribal leaders directory (2021). https://www.bia.gov/bia/ois/tribal-leaders-directory/. Accessed 01 Nov 2021
30. Hacker, K.: Community-Based Participatory Research. Sage Publications, Los Angeles (2013)
31. Wallerstein, N., Duran, B.: The conceptual, historical, and practice roots of community based participatory research and related participatory traditions, pp. 27–52 (2003)
32. Cooke, B., Kothari, U.: Participation: The New Tyranny?. Zed Books, London (2001)
33. James, R., et al.: Exploring pathways to trust: a tribal perspective on data sharing. Genet. Med. **16**(11), 820–826 (2014)
34. Rivkin, I., et al.: Disseminating research in rural Yup'ik communities: challenges and ethical considerations in moving from discovery to intervention development. Int. J. Circumpolar Health **72** (2013)
35. Legaspi, A., Orr, E.: Disseminating research on community health and well-being: a collaboration between Alaska Native villages and the academe. Am. Indian Alsk. Native Ment. Health Res. **14**(1), 24–43 (2007)
36. Dillard, D.A., et al.: Challenges in engaging and disseminating health research results among Alaska native and American Indian people in Southcentral Alaska. Am. Indian Alsk. Native Ment. Health Res. **25**(1), 3–18 (2018)
37. Bowen, S., Martens, P.: Demystifying knowledge translation: learning from the community. J. Health Serv. Res. Policy **10**(4), 203–211 (2005)
38. Elsabbagh, M., et al.: Community engagement and knowledge translation: progress and challenge in autism research. Autism **18**(7), 771–781 (2014)
39. Boyd, A.D., Song, X., Furgal, C.M.: A systematic literature review of Cancer communication with indigenous populations in Canada and the United States. J. Cancer Educ., 1–15 (2019)

40. McDonald, M.E., et al.: What do we know about health-related knowledge translation in the Circumpolar North? Results from a scoping review. Int. J. Circumpolar Health **75**, 18 (2016)
41. Timmons, V., et al.: Knowledge translation case study: a rural community collaborates with researchers to investigate health issues. J. Contin. Educ. Health Prof. **27**(3), 183–187 (2007)
42. McDavitt, B., et al.: Dissemination as dialogue: building trust and sharing research findings through community engagement. Prev. Chronic Dis. **13**, E38 (2016)

Studying Subject Ontogeny at Scale in a Polyhierarchical Indexing Language

Chris Holstrom(✉) and Joseph T. Tennis

University of Washington Information School, Seattle, WA, USA
{cholstro,jtennis}@uw.edu

Abstract. Subject ontogeny, the study of how subjects change or do not change during revisions of indexing languages, has added to our understanding of indexing languages through case studies. However, subject ontogeny research to date has been unable to examine key functionality of indexing languages at scale. For example, how do large-scale changes to social and literary warrant affect the utility of indexing languages over time? This paper discusses concrete progress made towards studying subject ontogeny at scale and the challenges presented by studying a large-scale polyhierarchical indexing language, Wikipedia Categories. The paper presents early findings, argues for continued research on subject ontogeny at scale, and suggests possible paths forward for this research.

Keywords: Classification · Subject ontogeny · Large-scale indexing · Polyhierarchy

1 Introduction

Subject ontogeny is the study of how subjects change or do not change during revisions of indexing languages. For example, the subject *Eugenics* changed from *575.1 – Genetics* in the 16th edition of the Dewey Decimal Classification to *363.92 – Social Problems and Population* in DDC 20 [17]. These types of changes in indexing languages reflect social changes, changes in literary and conceptual warrant, changes in preferred vocabulary, and the emergence of new fields of study. Analyzing how these social and linguistic changes are represented over time in indexing languages allows us, on the one hand, to understand the historicity of indexing terms and indexing languages, and to understand the rich stories of different domains [9]. On the other hand, it helps us maintain the utility and value added by indexing languages and controlled vocabularies in information systems [10].

Subject ontogeny has demonstrated its utility with case studies of changes to specific subjects in regularly and discretely updated indexing languages [1,6,8,11]. However, studying subject ontogeny at a larger scale and for indexing languages that are updated continuously has proven more difficult, partly because the types of questions that large-scale subject ontogeny research require are not well supported by the formats in which many indexing languages are

K. Toeppe et al. (Eds.): iConference 2021, LNCS 12646, pp. 404–412, 2021.
https://doi.org/10.1007/978-3-030-71305-8_34

made available. For example, a researcher can page through multiple editions of DDC to understand changes to a specific term like *Eugenics*. However, a researcher can not readily ask broad questions of discovery: Which subjects moved the furthest over time? Which subjects' terms saw the most change in their narrower terms over time? What types of subjects appeared or disappeared from the indexing language, and when? These broad questions could enrich subject ontogeny research by increasing its scope to larger social changes and by enabling researchers to find unexpectedly noteworthy subject changes to study in more detail.

We set out to ask these broad questions by studying subject ontogeny at scale instead of studying individual cases. This paper details our approach to identifying an indexing language to study and to building a system for studying subject ontogeny at scale. This paper does not present a completed system, but describes the properties that such a system would possess; explores the challenges presented by a large, rapidly changing, and polyhierarchical indexing language; and discusses proposed paths forward. The discussion of challenges presented by the Wikipedia Categories indexing language also provides a useful lens for understanding the differences between large, crowd-sourced indexing languages and professionally maintained and controlled indexing languages, such as DDC and Universal Decimal Classification (UDC).

2 Related Work

This research exists at the intersection of two research fields: subject ontogeny and large-scale index analysis, neither of which is complete in addressing our concerns about studying subject ontogeny at scale.

Subject Ontogeny Research. Since its inception in 2002, subject ontogeny research has sparked the interest of knowledge organization scholars and those who study the science of science [2,12,13]. This work, mostly based on case studies, has given rise to a number of constructs useful for understanding the challenges of changing an indexing language over time. Of note are the concepts of collocative integrity [19] and episemantics [20]. The former is how well contemporary versions of indexing languages can pull together, under a single term, all of the resources about a subject, even with changes to the term over time. The latter is a semantic corroboration of the meaning of a term and the impact of its change over time in effecting collocative integrity. There are other constructs that are important to subject ontogeny, but these two drive our desire to study indexing languages at scale because they establish the effects of change across the entirety of the system, not just a single case of a single subject.

Large-Scale Indexing Research. Large-scale analysis of indexing languages surfaced as a popular concern with the rise in popularity of Wikipedia [16]. These studies typically compare Wikipedia to other indexing languages, often focusing on topical coverage [14] or metadata analytics [7]. While we are ultimately interested in quality, atemporal analytics are insufficient for subject ontogeny research. Our desire is to see the effects of change over time, at scale, so that we can first understand the ramifications of change, and then propose design and maintenance ameliorations for large-scale indexing language development.

The outstanding example of subject ontogeny work at a large scale comes from Akdag Salah et al. [2]. This work outlines the quantity of classes added to the UDC, visualizing the entire scheme as a ring with major classes taking up a part of that ring in proportion to the rest of the top-level classes. Building from this, our work-in-progress looks to move beyond case-studies of subject ontogeny and high-level analysis of indexing languages, to document subject ontogeny in its entirety in a large-scale indexing language.

3 Methods

To study subject ontogeny at scale, we identified an appropriate indexing language, developed a basic understanding of its structure, and then downloaded and analyzed multiple revisions of the indexing language. We aimed to build a system that allowed us to ask broad social and linguistic questions about subject ontogeny, collocative integrity, and episemantics.

Identifying an Indexing Language. We began by identifying an indexing language that met the following conditions:

- *Historicity*: The indexing language must be regularly updated and must have a clear record of many past revisions. Those revisions must indicate their date.
- *Breadth*: The indexing language must cover a breadth of subjects representing the complex social universe that makes subject ontogeny at scale useful as a critical and interpretive research approach.
- *Relevance*: The indexing language need not be the most widely used or most authoritative, but it must be well-established and have a critical mass of contributors and users, so that researchers can responsibly connect subject ontogeny in the indexing language with large-scale social changes.
- *Utility*: The indexing language must be available in a digital format that facilitates programmatic manipulation and analysis.

We identified multiple indexing languages with broad coverage, a critical mass of users and contributors, and readily available revision histories. However, identifying an indexing language in a consistent digital format that could be manipulated and analyzed at scale across many revisions proved difficult. For example, DDC revisions are widely available in print and digital format, including scanned PDFs going back to the first edition. However, DDC revisions are not available in a digital format that facilitates programmatic manipulation and analysis. Akdag Salah et al. made available UDC data in an accessible digital format, but not for a sufficient number of revisions to satisfy the historicity required for subject ontogeny research [2]. After ruling out DDC, UDC, and other indexing languages, we found that only Wikipedia Categories met our requirements.

Exploring Wikipedia Categories. After choosing Wikipedia Categories as the best available indexing language to study subject ontogeny at scale, we oriented ourselves by exploring the basic structure of "Category: Main topic classifications" [4] and by running queries on the SPARQL endpoint for DBpedia [15]. We limited the scope of our inquiry to the English language version of Wikipedia.

The main topic classification page displayed top-level subjects in what appeared to be a hierarchical indexing language with 43 broad-ranging top-level subjects, such as *Culture*, *Education*, and *Philosophy*. This top-level page also allowed us to see narrower terms, such as *Philosophy→ Ontology→ Dualism*, by clicking on expandable and collapsible menus. The SPARQL endpoint allowed us to write queries that returned all of the narrower terms under a broader term or to count the narrower terms below a broader term.

Importantly, we observed that the Wikipedia Categories indexing language did not use any numerical notation (such as *539.120.22* for *Conservation laws* in UDC), which meant that the "distance" between terms would be difficult to measure. Furthermore, narrower terms under the same broader term were always displayed in alphabetical order instead of orderings that might indicate semantic or thematic closeness.

Finally, through SPARQL queries, we also observed that many terms had multiple broader terms. For example, *Bonsai* is a narrower term for both *Trees* and *Japanese style of gardening*, indicating a polyhierarchical structure.

```
Belle_%26_Sebastian_albums,prefLabel,Belle & Sebastian albums
Belle_%26_Sebastian_albums,Concept,,
Belle_%26_Sebastian_albums,broader,Albums_by_artist,
Belle_%26_Sebastian_albums,broader,Chamber_pop_albums,
Belle_%26_Sebastian_albums,broader,Indie_pop_albums,
Belle_%26_Sebastian_albums,broader,Singer-songwriter_albums,
Belle_%26_Sebastian_albums,broader,Scottish_albums,
```

Fig. 1. Cleaned RDF data for the term *Belle & Sebastian albums*.

Downloading and Preparing DBpedia Datasets. Having gained a basic understanding of the contents and structure of the indexing language, we downloaded multiple Wikipedia Categories datasets from DBpedia [5]. DBpedia takes regular snapshots of Wikipedia, including the continuously updated Wikipedia Categories. After downloading multiple of these N-Triple RDF datasets, which can be found under "Categories (skos)" headings, we cleaned the data using Python and regular expressions to make each line more readable and to facilitate SQL queries and R analysis. The text serialization in Fig. 1 shows cleaned data for the term *Belle & Sebastian albums*.

The first value in each line is the term itself. The second value indicates the type of label or the relationship that the line defines: a definition of the term (indicated by `Concept`), a preferred label (`prefLabel`), or a broader term relationship (`broader`). The third value indicates the broader term, if applicable. The fourth value indicates the preferred label, if applicable.

3.1 Analyzing the Categories Datasets

With our data cleaned and inserted into a relational database, we began to write SQL queries to ask broad questions of discovery against multiple revisions of the Wikipedia Categories indexing language. For example, we found all of the paths leading from a given term to the top of the indexing language and then programmatically compared those paths to the paths in other revisions. We also counted the narrower terms that were listed directly and indirectly below broader terms and then compared these counts with similar counts for other revisions to find subject areas that experienced significant growth or reductions. However, our early findings suggested that the quantity of terms that were added under particular broader terms had more to do with the interests and propensity for subdivision of the most active editors than with social or literary warrant.

While our approach to building a system for studying subject ontogeny at scale showed some promise, we encountered problems with query performance on these large datasets, and more significantly, issues with the design of the Wikipedia Categories indexing language that made large-scale subject ontogeny analysis infeasible without significant changes to our approach and design.

4 Findings and Discussion

As noted previously, this paper does not present a finished system for studying subject ontogeny at scale. Instead, in this section, we discuss partial progress on such a system through the lens of the challenges presented by studying subject ontogeny at scale in the Wikipedia Categories indexing language.

4.1 Massively Polyhierarchical Structure

Wikipedia's technical and philosophical approaches to categorization explicitly embrace polyhierarchy [21]. Furthermore, the crowd-sourced approach to Wikipedia editing encourages a massive proliferation of terms and relationships in the indexing language. In contrast, traditional indexing languages, such as DDC, require a strict hierarchy, wherein each subject occupies a single location in the hierarchy, and intensive review before terms are added or moved in the language.

These critical differences between hierarchical and polyhierarchical structures proved particularly significant to researching subject ontogeny at scale. For example, we found that the term *Atlantic League mascots* had two broader terms in one revisions of the Wikipedia Categories polyhierarchy: *Atlantic League* and *Baseball team mascots*. However, that seemingly simple bifurcation proved to be the tip of a massive iceberg. By using recursive SQL joins, we found that *Atlantic League Mascots* had 640 unique paths to the top of the polyhierarchy. This structural complexity breaks an assumption of much subject ontogeny research to date—that we can examine the movement of a term from a single point in a hierarchy to another single point over time—and made comparisons between revisions much more complex and difficult to characterize. Also, preliminary analysis found that polyhierarchy led to fundamentally different types of scheme changes, namely that terms were added under more and more broader terms over time instead of being moved from one broader term to another.

4.2 Lack of Numerical Notation

Most traditional indexing languages, such as DDC, UDC, and LCC, use numerical notation to aid retrieval, to indicate the relationship between subjects, and to aid collocation of related information objects on shelves. Previous subject ontogeny research has used numerical notation to measure and visualize the "size" of changes to subjects and indexing languages over time [19].

However, Wikipedia Categories do not use numerical notation and, while some efforts have been made to compare Wikipedia Categories to indexing languages that use numerical notation [3], no robust mapping exists with sufficient hierarchical force to support the type of measurements and visualizations typically used in subject ontogeny research. We superficially overcame this deficiency by analyzing changes to specific subjects, for example finding that *Mixed drinks* was classified under additional broader terms over time, and manually analyzing these changes.

Clearly, however, this case-study approach runs contrary to the goals of studying and visualizing subject ontogeny at scale. We experimented with assigning numerical values to Wikipedia Category terms based on their alphabetical ordering, which provided some convenience, but also misrepresented the semantic distance between terms and made any visual summary of large-scale subject changes misleading. At this point, we are not prepared to recommend a solution for measuring and visualizing subject ontogeny for indexing languages without numeric notation.

4.3 Continuous and Rapid Scheme Change

Much subject ontogeny research to date has analyzed changes to indexing languages that are updated in discrete editions that are released years apart from each other. The relatively slow and well-defined update cycle of indexing languages like DDC is suited to connecting scheme change to large-scale changes in social and literary warrant. For example, attitudes towards eugenics shifted dramatically over decades, and these attitudes are reflected in editions of DDC published decades apart.

In contrast, Wikipedia Categories are updated continuously and rapidly. DBpedia captures periodic snapshots, which function somewhat like editions, but do not offer the clearly discrete updates of other indexing languages and do not necessarily reflect the statement of editorial intent of discretely published editions. Multiple state changes can occur undetected between snapshots [18]. Finally, the rapid, and sometimes back-and-forth, editing of the Wikipedia Categories scheme that we observed in our initial analysis seems less likely to coincide with major changes to social and literary warrant and more likely to reflect less consequential, iterative changes.

Continuously changing schemes force us to update our understanding of historicity and episemantics. This new perspective is unlikely to remain isolated to digital-native indexing languages like Wikipedia Categories as digital publishing becomes more common and technologies like WebDewey become the primary method of delivering indexing languages. Subject ontogeny research must adapt to these updates and systems to study subject ontogeny at scale and must be able to differentiate between significant scheme changes and the small and frequent changes observed in continuously updated schemes.

5 Conclusion and Future Work

To date, most subject ontogeny research has been based on single case studies of particular subjects [1,6,8,11]. The case-study approach leaves us with one particular view on the life of subjects in individual schemes. In contrast, our large-scale approach aims to consider the full ecosystem of subjects over time.

This large-scale approach to studying subject ontogeny is still nascent. However, the challenges discovered in this work-in-progress research motivate and clarify future work. Of particular interest in this study is the analysis of a poly-hierarchical indexing language, something not yet fully described in the subject ontogeny literature. Additionally, the challenges that this study encountered due to the lack of numerical notation in Wikipedia Categories might motivate new means of measuring and visualizing semantic distance. Finally, the rapid and continuous revisions to Wikipedia Categories illustrate the necessity of re-evaluating the assumptions built around edition-based revisions, as iterative updates look set to emerge as the norm.

As we grow our understanding of the particular ecology of subjects found in Wikipedia Categories, we anticipate testing our assumptions about collocative integrity over time [19] and about episemantics, since each category is linked to articles in Wikipedia [20]. These constructs, arising from the subject ontogeny case-study literature, will eventually require large-scale data sources beyond just Wikipedia Categories—possibly from indexing languages such as DDC, UDC, and LCC—to further solidify our understanding of collocative integrity and episemantics as indicators of the value of indexing languages. It is fine to measure particular cases, but to understand at scale will be of the highest importance as we continue to construct large-scale category systems, algorithmically driven classification schemes, and open source information retrieval and browsing systems.

References

1. Adler, M.: Classification along the color line: excavating racism in the stacks. J. Crit. Libr. Inf. Stud. **1**(1), 1–32 (2017). http://journals.litwinbooks.com/index.php/jclis/article/view/17/10
2. Akdag Salah, A.A., Gao, C., Suchecki, K., Scharnhorst, A., Smiraglia, R.P.: The evolution of classification systems: ontogeny of the UDC. In: Neelameghan, A., Raghavan, K.S. (eds.) Proceedings of the Twelfth International ISKO Conference on Categories, Contexts and Relations in Knowledge Organization, Mysore, India, 6–9 August 2012, pp. 51–57. Ergon-Verlag, Würzburg (2012)
3. Akdag Salah, A.A., Gao, C., Suchecki, K., Scharnhorst, A.: Design vs. emergence: visualization of knowledge orders (2011). http://www.scimaps.org/maps/map/design_vs_emergence__127/detail
4. Category: main topic classifications. https://en.wikipedia.org/wiki/Category:Main_topic_classifications. Accessed 14 Oct 2020
5. Datasets | DBpedia. https://wiki.dbpedia.org/develop/datasets. Accessed 19 Oct 2020
6. Fox, M.J.: Subjects in doubt: the ontogeny of intersex in the Dewey decimal classification. Knowl. Organ. **43**(8), 581–593 (2016)
7. Harper, C.A.: Metadata analytics, visualization, and optimization: experiments in statistical analysis of the Digital Public Library of America (DPLA). Code4Lib J. **33** (2016)
8. Higgins, M.: Totally invisible: Asian American representation in the Dewey decimal classification, 1876–1996. Knowl. Organ. **43**(8), 609–621 (2016)
9. Hjørland, B.: Epistemology and the socio-cognitive perspective in information science. J. Am. Soc. Inf. Sci. Technol. **53**(4), 257–270 (2002)
10. Lancaster, F.W.: Indexing and Abstracting in Theory and Practice, 3rd edn. University of Illinois, Champaign-Urbana (2003)
11. Lee, W.-C.: An exploratory study of the subject ontogeny of Eugenics in the new classification scheme for Chinese libraries and the nippon decimal classification. Knowl. Organ. **43**(8), 594–608 (2016)
12. Lee, P.S., West, J.D., Howe, B.: Viziometrics: analyzing visual information in the scientific literature. IEEE Trans. Big Data **4**(1), 117–129 (2017)
13. Scharnhorst, A., Akdag Salah, A.A., Gao, C., Suchecki, K., Smiraglia, R.P.: The evolution of knowledge, and its representation in classification systems. In: Salvic, A. (ed.) Classification & Ontology: Formal Approaches and Access to Knowledge, Proceedings of the International UDC Seminar, The Hague, 19–20 September 2011, pp. 269–282. Ergon-Verlag, Wurzburg (2011)
14. Scharnhorst, A.: Walking through a library remotely. Les cahiers du numérique **11**(1), 103–128 (2015)
15. SPARQL Explorer for http://dbpedia.org/sparql. http://dbpedia.org/snorql/. Accessed 16 Oct 2020
16. Suchecki, K., Akdag Salah, A.A., Gao, C., Scharnhorst, A.: Evolution of Wikipedia's category structure. Adv. Complex Syst. **15**(supp01), 1250068 (2012)
17. Tennis, J.T.: Subject ontogeny: subject access through time and the dimensionality of classification. In: Proceedings of the Seventh International ISKO Conference on Challenges in Knowledge Representation and Organization for the 21st Century: Integration of Knowledge Across Boundaries (2002)
18. Tennis, J.T., Sutton, S.A.: Extending the simple knowledge organization system for concept management in vocabulary development applications. J. Am. Soc. Inf. Sci. Technol. **59**(1), 25–37 (2008)

19. Tennis, J.T.: The strange case of eugenics: a subject's ontogeny in a long-lived classification scheme and the question of collocative integrity. J. Am. Soc. Inf. Sci. Technol. **63**(7), 1350–1359 (2012)
20. Tennis, J.T.: Methodological challenges in scheme versioning and subject ontogeny research. Knowl. Organ. **43**(8), 573–580 (2016)
21. Wikipedia: Categorization. https://en.wikipedia.org/wiki/Wikipedia:Categorization#Category_tree_organization. Accessed 16 Oct 2020

Characterizing Award-Winning Papers in Library and Information Science (LIS): A Case Study of LIS Journals Published by Emerald Publishing

Yi Chen[1], Shengang Wang[2], and Li Yang[2](✉)

[1] School of Information Management, Wuhan University, Wuhan, Hubei Province, China
chenyi@whu.edu.cn

[2] School of Information Studies, University of Wisconsin-Milwaukee, Milwaukee, WI, USA
{shengang,liyang}@uwm.edu

Abstract. This paper explores the characteristics of 106 award-winning papers from the Library and Information Science (LIS) journals published by Emerald Publishing between 2008 and 2019, focusing on collaboration type, paper type, topic, and citation count to illustrate the developmental trends of LIS scholarship. The findings show that the top three topics of the award-winning papers were information service activities, professions and information institutions, and user studies. More than half of the award-winning papers were written by teams, among which inter-institutional collaboration and intradepartmental collaboration accounted for the largest proportion, while interdepartmental collaboration within an institution accounted for the smallest proportion. There were 65 empirical research papers in the sample, among which qualitative studies were dominant, followed by quantitative research and mixed methods research. The award-winning papers had a higher mean and median in citation counts than the average papers concurrently published by the journals. The research results provide implications for researchers and can help them understand the trends in research topics and common analytical types in LIS for their future studies.

Keywords: Award-winning papers · LIS · Authorship · Scientometrics

1 Introduction

Scientific evaluation, especially peer evaluation, of scholarly works is among the cornerstones in academia. Awards for scholars and their works can be seen as one particular form of scientific evaluation. For centuries, academic awards in different forms have been used to publicly recognize and honor individuals and their contributions, thus encouraging further scientific discoveries [1]. Awards for academic publications are worth noting because they signal what kinds of contributions are valued by the scientific community [2]. Understandably, award-winning papers are usually recognized as exemplary works.

K. Toeppe et al. (Eds.): iConference 2021, LNCS 12646, pp. 413–428, 2021.
https://doi.org/10.1007/978-3-030-71305-8_35

Despite the long-established efforts in analyzing LIS literature, a limited number of studies have analyzed award-winning papers in LIS [3, 4]. By analyzing the award-wining LIS papers from Emerald Publishing, this study aims to provide researchers, professionals, and other interested stakeholders with an updated view of the main spheres of LIS.

In this study, we assume that award-winning papers stand out as high-quality ones that represent qualified cases for analyzing the status quo and the intellectual development of LIS. This paper addresses the following research questions:

RQ1. What are the authorship patterns of the award-wining LIS papers from Emerald Publishing?
RQ2. What is the dominant paper type (theoretical article, empirical research, or literature review) within the award-wining LIS papers from Emerald Publishing?
RQ3. What topics have been frequently discussed in award-wining LIS papers from Emerald Publishing?
RQ4. Do the award-winning papers have higher citations than the non-award papers?

2 Literature Review

2.1 Analysis of LIS Literature

There has been long-standing scholarly interest among LIS researchers to utilize bibliometric methods and content analysis to understand the temporal and evolutional landscape of diverse aspects of research and practice in the field. Many studies indicate that scientific collaboration is common in LIS. Blessinger and Frasier [5], for example, pointed out that approximately 54% of 2200 journal articles published between 1994 and 2004 were co-authored by at least two people. Similarly, according to Aharony [6], co-authored works accounted for more than 70% of the publications by 10 LIS journals in 2007 and 2008, with North American and European scholars playing a leading role.

There have been multiple investigations of methodological issues in LIS. For example, Tuomaala et al. [7] analyzed LIS research articles in 2005. They reported that empirical research strategy (76%) was the most frequently used, followed by conceptual research design and other research strategies (e.g., literature review). In terms of types of analysis, quantitative analysis (58.4%) was dominant, while qualitative analysis was used in 14.1% of the sampled LIS articles in 2005. Ullah and Ameen [8] maintained that empirical, descriptive, and quantitative research methodologies have been used in the majority of LIS research. Compared with qualitative and quantitative research, mixed methods research accounts for a small portion of LIS research [9]. Fidel [10] pointed out that 5% of the LIS research applied mixed methods designs during 2005 and 2006.

In terms of topical areas of LIS literature, the classification system originally created by Järvelin and Vakkari [11] has been widely used [7, 12]. There are several main classes in the classification system, including the professions in library and information services, library history, publishing, education in LIS, methodology, analysis of LIS, library and information service activities, information storage and retrieval, information seeking, scientific and professional communication, and other aspects of LIS. Recently, Blessinger and Frasier [5] grouped LIS journal publications from 1994 to 2004 into five major topical categories: library operation (33%), research in library and information science/users (20%), library/information science profession (18%), technology (18%), and publishing/publishing studies (11%).

2.2 Analysis of Award-Winning Papers

There have been a few studies focusing on award-winning articles across disciplines. For example, Tackney et al. [13] conducted a content analysis of 40 best papers from Management Spirituality and Religion Interest Group. Focusing on the history of the Association for Information Systems best publication awards, Ghobadi and Robey [14] developed a framework involving contribution characteristics, demographic patterns, and citation histories of award-winning papers. In addition, some studies have specifically compared the differences between award-winning papers and other papers in citation counts. Sen and Patel [15] examined the citation rates of articles selected for the most prestigious awards of American Society of Civil Engineers. They found that nearly 25% of the award-winning papers were never cited, and over 30% were cited only once. Similarly, Coupé's [16] findings indicated that the papers that won "best paper" prizes in economics and finance journals were rarely the most cited. Wainer et al. [17] compared the citations of award-winning papers and random papers from different computer science conferences, pointing out that award-winning papers had a higher probability to receive more citations. From a different perspective, Mubin and co-authors [18] examined the readability of award-winning papers at the ACM Conference on Human Factors in Computing Systems (also known as the CHI Conference) in comparison to their non-award counterparts. They found that award-winning full papers had a lower readability.

Research analyzing award-winning papers in LIS is, however, relatively scarce. Brooks [3] examined 28 best articles published in the *Journal of the Association for Information Science and Technology (JASIST)* from 1969–1996, reporting that best papers tended to be lengthy and single-authored, and were cited and self-cited more often than average articles. In a similar vein, Zhang et al. [4] expanded the period and compared the citation counts of 45 award-winning articles published between 1969 and 2013 with average papers. They found that most best papers belonged to the top 50% stratum, and there was a wide range of citations among the Best *JASIST* Papers. It should be noted that these two papers have only investigated citation counts of award-winning papers from a single journal (*JASIST*). To have a rather comprehensive view, this study will analyze different aspects of award-winning papers in addition to citation count, including collaboration type, paper type, and topic.

3 Methodology

3.1 Data Collection

There exist various best paper awards in LIS journals (e.g., *JASIST*, *Knowledge Organization*) and conferences (e.g., iConference, the Joint Conference on Digital Libraries Conference). However, information about award-winning papers is often scattered without being organized in one place. Notably, Emerald Publishing, one of the major global academic publishers, has been updating a gallery of award-winning papers on its website since 2008. 16 LIS journals are currently managed by Emerald Publishing, from 15 of which 106 publications were granted outstanding paper awards between 2008 and 2019. Considering the relatively easy data access, we decided to use award-winning papers from Emerald Publisher as a sample. The full list of those 106 publications was first manually created, and the full text of each publication was downloaded to compose the study corpus for further content analysis. Admittedly, the limited dataset of award-winning papers from one publisher may result in bias in terms of reporting or interpreting findings (Table 1).

Table 1. LIS journals published by Emerald Publishing

Journal	No. of papers	Journal	No. of papers
Global Knowledge, Memory and Communication (Previously published as *Library Review)*	12	*The Bottom Line*	8
Reference Services Review	12	*Aslib Journal of Information Management*	5
Library Management	11	*Digital Library Perspectives* (Previously published as *CLC Systems & Services: International Digital Library Perspectives*)	5
Journal of Documentation	10	*Library Hi Tech News*	3
The Electronic Library	10	*Collection and Curation*	2
Library Hi Tech	9	*Data Technologies and Applications*	1
Performance Measurement and Metrics	9	*Information Discovery and Delivery*	1
Online Information Review	8	*Information and Learning Sciences*	0

In addition to full-text files, information of citation counts of both award-winning articles and average paper was retrieved from Scopus. In response to each award-winning paper, citation counts of the remaining average papers published concurrently in the same journal were also kept. In total, citation counts of 826 average papers were collected.

3.2 Data Analysis

The retrieved award-winning papers were examined in four key aspects: collaboration type, paper type, topic, and citation count. The coding schemes for research topic and paper type were developed by the authors based on previous studies [5–7, 19, 20]; they were further refined during the coding procedure.

Specifically, collaboration type was analyzed in accordance with Qin and colleagues' [21] classification of scientific collaboration consisting of no collaboration, collaboration within a department, collaboration between two or more departments within an institution, collaboration between two or more institutions within a country, and international collaboration. Author and institution information of each award-winning paper was manually recorded in a separate spreadsheet for analysis.

This paper analyzed general paper types instead of specific research methods. On the one hand, existing listings and classifications of research methods in LIS follow varying criteria [7, 22, 23]. On the other hand, the authors often found it challenging to assign methods to some research articles due to their implicit method/methodology statements where research methods were not clearly specified. Last but not least, not all the award-winning papers qualify as research "carried out, at least to some degree, by a systematic method with the purpose of eliciting some new facts, concepts, or ideas" [24, p. 251]. Due to the aforementioned concerns, the authors decided to reveal the methodological issues from a macro perspective [7, 19]. Specifically, each article was assigned to one of the following paper types: theoretical paper (i.e., argument and perspective paper, concept analysis), empirical research (i.e., qualitative research, quantitative research, mixed methods research), and literature review.

The coding scheme for analyzing topics comprises the following categories: professions and information institutions; methodology; education in LIS; information systems and technologies; information service activities; information organization and retrieval; information management; information ethics, policy, and security; user studies; scientific communication; analysis of LIS publications; and other aspects. When an article involved many topics, the coders attempted to "identify its main topic" [7, p. 1449]. Meanwhile, the author keywords were recorded to assist in visualizing the topics covered in the study corpus.

Table 2. Inter-coder reliability test

	Simple agreement	Cohen's Kappa
Collaboration type	1	1
Paper type	0.75	0.632
Topic	0.80	0.735

Two co-authors coded each publication independently after a close reading of the full text. A trial coding of 20 articles that were randomly sampled was conducted to assess the intercoder reliability. Both simple agreement (also known as percent agreement) and Cohen's kappa were used. The inter-coder reliability test indicated that the coding process was reliable [25] (Table 2). The two coders discussed and resolved conflicts and continued to code the remaining articles on their own. When all articles were coded, the two coders compared and finalized the coding results.

When analyzing citation counts, we compared the citation count each award-winning paper received with those of average papers published on the same issue. To have a better understanding of the citation counts of both award-winning paper group and average paper group, we also provided the five-number summary for both groups as well as a corresponding box plot with outliers.

4 Results

4.1 Authorship

Excluding the article Research Data Management as a "wicked problem" [26] by 23 authors, the average number of authors per paper was 2. Of the 106 articles, 49 (46%) were single-authored, 27 (25%) were written by two authors, and 17 (16%) were written by three authors. In total, there were 232 authors involved in producing the 106 award-winning articles, among whom five authors were awarded twice for their works: Andrew K. Shento, Kenning Arlitsch, Kerry Wilson, Reijo Savolainen, and Sheila Corrall. The authors across all papers were from 31 different countries; 78 (34%) were from the U.S., 60 (26%) were from the U.K., and the rest were from China, Australia, Germany, Canada, and other countries (Fig. 1). Moreover, it has been found that in addition to faculty members of colleges or universities, practitioners from libraries and other institutions are also key participants in LIS research. More than 55 (24%) authors of award-winning papers were librarians, many of whom were academic librarians. Additionally, more than 35 (15%) authors from other disciplines (e.g., computer science, media studies, sociology, business, and engineering) contributed to the award-winning papers.

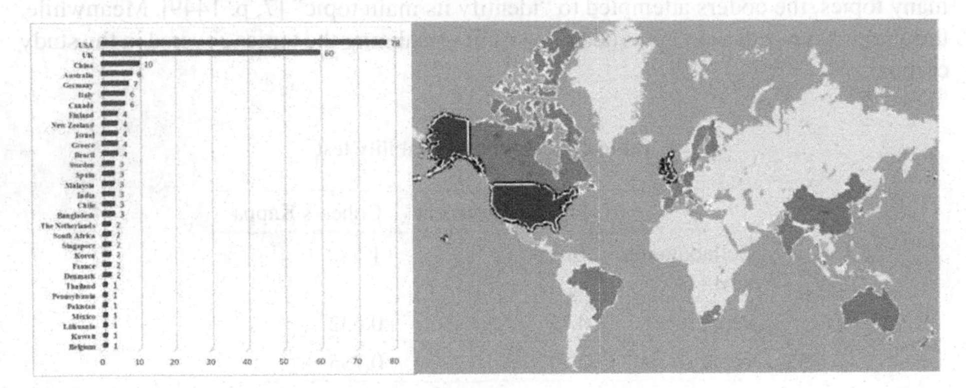

Fig. 1. Countries of affiliation of authors

In terms of collaboration, intradepartmental and inter-institutional collaborations were relatively commonplace, accounting for 45% of all types of collaboration. International collaboration occurred in 8% of all cases. Comparatively, collaboration among two or more departments within an institution only accounted for 1% (Fig. 2).

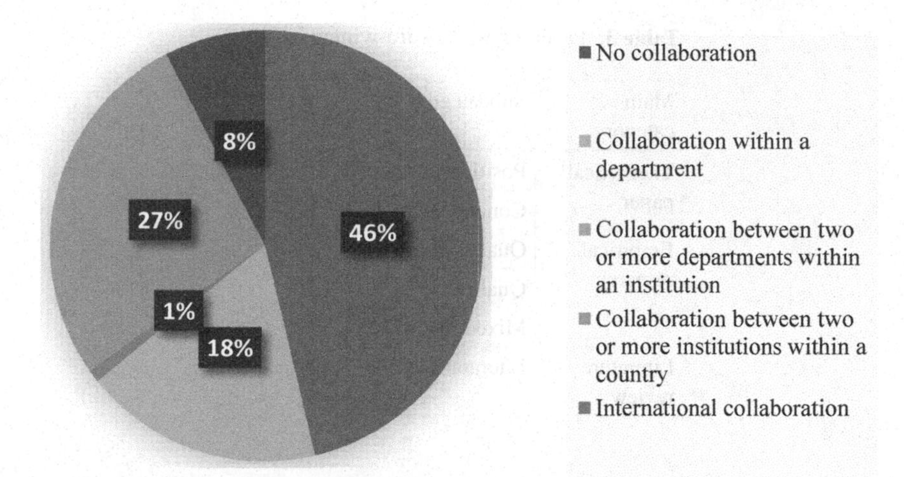

Fig. 2. Distribution of collaboration patterns

4.2 Paper Type

Sixty-five (61.3%) articles were coded as empirical studies, among which qualitative research was dominant, followed by quantitative research and mixed methods research. Notably, no quantitative studies were awarded, while seven qualitative papers were awarded in 2009; no mixed methods research papers were given awards between 2013 and 2016. Theoretical papers accounted for 32.1% of the corpus and mainly consisted of argument/position papers. It is worth mentioning that only four articles, three of which were given awards in 2016, were coded as concept analysis/development articles. Seven literature review papers comprised the smallest proportion of the sample. Figure 3 shows the frequency distribution of the award-wining paper types by award year (Table 3).

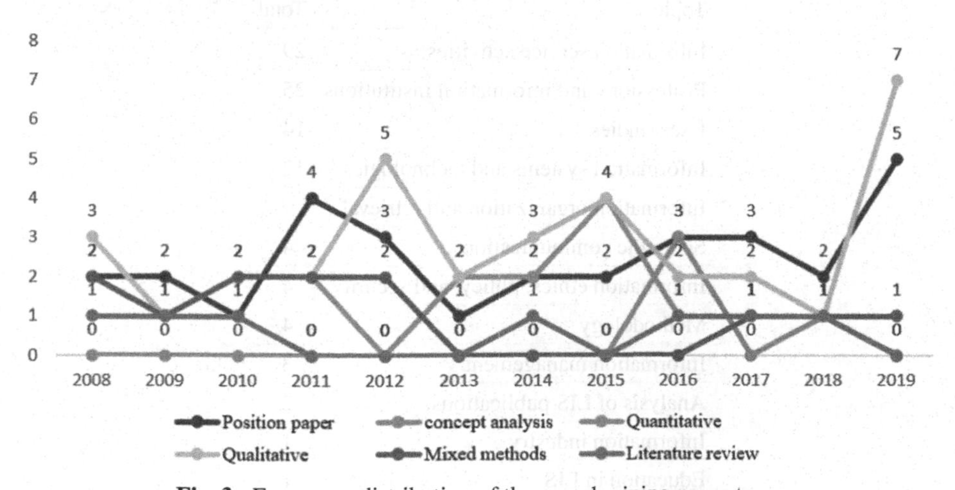

Fig. 3. Frequency distribution of the award-wining paper types

Table 3. Paper types of award-wining papers

Main category	Subcategory	Total
Theoretical paper	Position paper	30
	Concept analysis	4
Empirical research	Quantitative	19
	Qualitative	34
	Mixed methods	12
Literature review	Literature review	7

4.3 Topic

As shown in Table 4, 29 articles concerned information service activities, accounting for the largest proportion (27.4%). The most popular topic within the category of information service activities was user education, and there were ten papers focusing on information literacy. Other popular topics related to information services included digital information resources and reference services. The topic of professions and information institutions accounted for 23.6%, of which more than two in five articles were about library development strategies. For user studies, papers on human information behavior were dominant, and there was notable attention to online communities [27, 28]. Figure 4 shows the frequency distribution of the award-winning paper topics by award year.

Table 4. Topics of award-winning papers

Topic	Total
Information service activities	29
Professions and information institutions	25
User studies	14
Information systems and technologies	12
Information organization and retrieval	7
Scientific communication	4
Information ethics, policy, and security	4
Methodology	4
Information management	3
Analysis of LIS publications	2
Information industry	1
Education in LIS	1

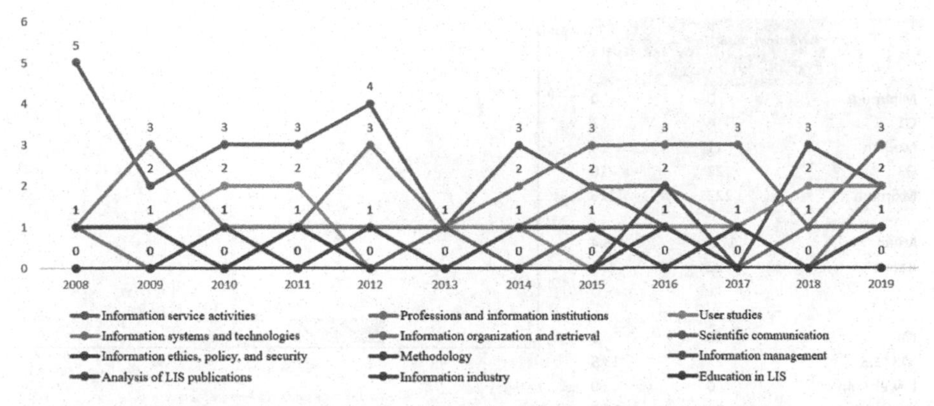

Fig. 4. Frequency distribution of the award-winning papers topics

In terms of keywords, there were originally 538 keywords in total. After data cleaning, 377 distinct keywords were identified, 69 of which appeared twice or more. Figure 5 shows the average number of keywords per paper. Before 2012, the average number of keywords was around 4. It peaked in 2013. In terms of common keywords, academic libraries (19 times), information literacy (13 times), public libraries (12 times), and digital libraries (10 times) were used most frequently. Additionally, the Internet, library instruction, search engines, user studies, and social media appeared more than 5 times (Fig. 6).

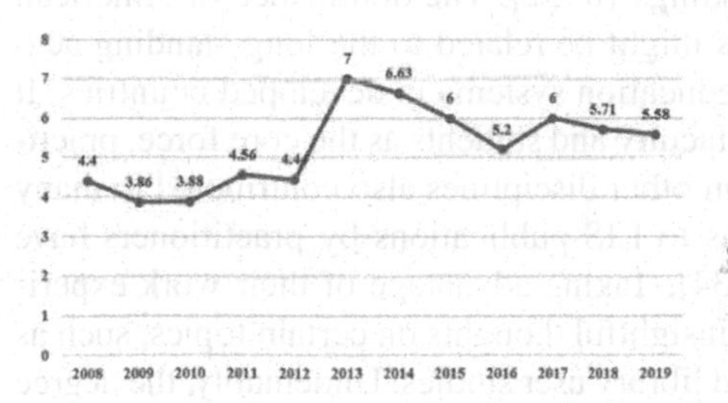

Fig. 5. Average number of keywords per paper

Fig. 6. Word cloud of keywords

4.4 Citation Count

Compared with their average counterparts, 32 award-winning papers were the most cited papers of the journal issues where they got published respectively; the 67 award-winning papers had a higher mean and median in citation counts than the average papers concurrently published by the journals. According to Fig. 7, the citation counts of award-winning paper were generally higher than average papers. However, it is worth noting that some award-winning papers were rarely cited. Specifically, six award-winning papers were never cited, and five were cited only once.

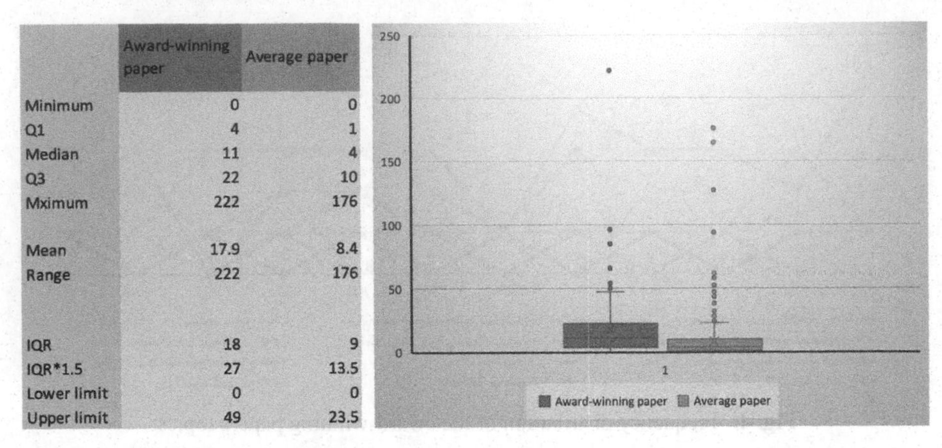

	Award-winning paper	Average paper
Minimum	0	0
Q1	4	1
Median	11	4
Q3	22	10
Mximum	222	176
Mean	17.9	8.4
Range	222	176
IQR	18	9
IQR*1.5	27	13.5
Lower limit	0	0
Upper limit	49	23.5

Fig. 7. Citation counts of award-winning papers and average papers

5 Discussion

5.1 Pervasive Collaboration in LIS Publications

As the findings indicated, more than half of the award-winning papers were coauthored, resonating with the prevalence of collaboration in LIS that has been widely noted in previous studies [5, 29–31]. Not surprisingly, authors from North America (mainly the U.S.) and Europe (mainly the U.K.) contributed to a large portion of the corpus under study, which was in line with previous findings [6, 32]. The dominance of American and European scholars in LIS publications might be related to the long-standing academic tradition and the developed higher education systems in developed countries. It has also been found that in addition to LIS faculty and students as the core force, practitioners (mainly librarians) and authors from other disciplines also contributed to many award-wining papers. In fact, contributions to LIS publications by practitioners have been recognized by previous studies [33, 34]. Taking advantage of their work experience, practitioners are more likely to offer insightful thoughts on certain topics, such as reference services, cataloging practices, and library user studies. Undeniably, the degree of interdisciplinarity in LIS relies on the participation of authors from other disciplines [35]. Authors from other disciplines help extend the scope of LIS research and might provide promising techniques and perspectives to re-examine certain existing topics. In short, scientific collaboration and communication through geopolitical and disciplinary borders are of great importance in supporting the development of one domain, and such practices are currently being nurtured in LIS [31, 36, 37].

5.2 Co-existence of Various Methodological Paradigms in LIS

Reflecting on methodological issues is essential in all disciplines. As previously mentioned, no specific methods were coded for each paper because of the conflicting classification systems of methods in LIS; rather, the present study characterized the sampled papers based on their paper design and associated types of analysis from a macro perspective [7, 19].

In social science research, quantitative research, qualitative research, and mixed methods research are usually seen as three major paradigms [38]. According to the findings, qualitative research accounted for the vast majority of award-winning research papers, echoing the fact that qualitative research is being increasingly recognized and used in LIS studies [12, 39]. It should be noted that qualitative research is not prioritized over quantitative research, and vice versa. Both qualitative and quantitative research have their own advantages in different research scenarios. Generally, qualitative research helps understand complex cultural and social contexts and explore people's lived experiences, while quantitative research is appropriate for examining quantifiable measures of variables mainly by means of descriptive and inferential statistical methods [40–42]. The large number of papers focusing on information service activities and human information behavior in the award-winning papers might explain why qualitative research accounted for a high proportion in the papers to some degree.

Mixed methods research is a relatively young paradigm synthesizing both qualitative and quantitative methods [38]. Although mixed methods research has been discussed and applied in many disciplines [43–45], Fidel [10] found that "mixed methods" was not a familiar term in LIS research, and only 17% of the empirical research articles he surveyed used mixed methods. Similarly, 18% of the empirical award-winning papers in the current study were mixed methods studies. Though Fidel [10] suggested that it is not always easy to identify mixed methods research because of a lack of explicit method statements in many studies, some of the award-winning mixed methods papers in the present study stated clearly that both qualitative and quantitative methods were applied [46–49]. To achieve transparency and replicability, a clear and detailed method statement is expected for all types of research. For instance, Anfara Jr. et al. [50] regard the public disclosure and openness of methods and research processes as one potential way to promote the research quality and rigor in the qualitative research community.

5.3 Diverse yet Consistent Topics in LIS Literature

LIS is an interdisciplinary field that encompasses diverse topics [51]. As shown in the literature review, there have been continuous attempts to categorize topics or areas in LIS [6, 7, 11, 52, 53]. 377 distinct keywords were revealed, with more than 300 of them appearing only once, such as tattooing, HIV, and democracy. The average number of keywords indexed in each article per year could reflect the scope of its topic(s) to some degree [54]. The significant increase in the number of keywords per paper after 2012 might indicate that the award-winning papers are involving more topics. To have a clear understanding of what keywords mean, it is always important to put them in specific contexts. For example, Sundberg and Kjellman [55, p. 18] examined "how tattoos can be considered documents of an individual's identity, experiences, status, and actions" based on the tattoo practices of Russian/Soviet prisoners.

According to Pawley [56], four models dominate LIS research and teaching, namely, science/technology, business/management, mission/service, and society/culture, providing us with a simplified way to examine these diverse topics in LIS. It has been found that the top three topics of the award-winning papers, information service activities, professions and information institutions, and user studies, are all related to the mission/service model. For example, Arlitsch [57] discussed the disruptive implications of the Espresso

Book Machine on library user services, collection development, and special collections; Curry [58] explored the potential of makerspaces to function as new learning spaces within academic libraries in higher education.

The results show that some keywords concerning information technology, such as the Internet, search engines, and social media, were also frequently used. There have been many previous studies demonstrating the importance of information technology. For example, Davarpanah and Asleki [32] claimed that a high number of articles were related to communications and information technology. Liu and Yang [59] pointed out that the most popular research topics in LIS were closely related to social media, data, and information retrieval in the most recent decade. In short, LIS has constantly maintained a focus on the aspects of mission/service and science/technology in the changing social and technological environment. With regard to human and technological dimensions in LIS, Cibangu [60] argued that emphasizing human dimensions does not mean undervaluing the importance of technologies, and LIS scholars are expected to be humanities-prone thinkers, thus making LIS a human science, which provides a potential direction for LIS.

5.4 Complicated Relationship Between Awards and Citation Counts

Awards and citations are often referred to as qualitative and quantitative assessment of quality or impact of publications, respectively. Best paper awards are generally used to acknowledge publications based on their knowledge contribution, paper structure, the rigorousness of the argument/analysis, and paper writing and presentation [61]. However, the processes of selecting award-winning papers are unavoidably subjective; the lack of transparency is also an issue. As for citations, an author may have different reasons when citing other documents, including but not limited to providing background information, giving credit for related works, and criticizing previous works [62, 63]. Previous research shows that compared with positive and neutral citations, negative citations account for a relatively small portion of total citations [63, 64]. Some scholars believe that "even a negative citation makes it clear that the referenced work cannot be simply ignored" [65, p. 2].

Due to the varying nature of awards and citation counts, it is not possible to simply determine that awards lead to high citation counts. Instead, there seems to be a complicated relationship between publication awards and citation counts [66]. We found that the award-winning papers under study did have a higher citation count on average than their average counterparts; however, not all award-winning papers were cited frequently, and 11 of them were either never cited or only cited once. There might be several potential reasons to explain the lower citation counts of some award-winning papers. First, it is likely that one award-winning paper as a "sleep beauty" are so innovative that no one cites it for years to come [67]. Second, people don't cite award-winning papers that merely summarize existing literature and are lacking in interesting or groundbreaking findings. Third, other criteria other than "quality" might be considered during the evaluation processes. There remain other possibilities for sure. The underlying reasons or mechanisms are understudied.

6 Conclusion

This study is descriptive in nature. Instead of arguing what topic and paper type should be given awards, our primary goal is to reveal the characteristics of the award-wining LIS papers under study, and to provide implications for researchers to understand the recent trends in LIS. We found that the authors of more than 60% all of the outstanding papers were from the USA or the UK, 54% of the award-winning papers were co-authored, and intradepartmental collaboration and inter-institutional collaboration within one country accounted for the largest proportion. Furthermore, empirical research accounted for the vast majority of award-winning papers, more than half of which are qualitative. Additionally, there seems to be a wide variety of topics in the LIS literature, with information service activities, professions and information institutions, and user studies topping the list, indicating that the LIS discipline has maintained a focus on users and services. It is also worth noting that LIS has been evolving dynamically, and its foci go beyond the traditional aspects of storage, organization, and use of information, and special attention has been paid to areas such as information literacy, artificial intelligence, and the humanities. We also found that on average, the award-winning papers under study did have a higher citation count than their average counterparts; however, not all award-winning papers were cited frequently.

Admittedly, this study could only partially reflect the disciplinary landscape and development because of the restrained sample of LIS journal publications from Emerald Publishing, which limits the generalizability of the research results. Expanding the scope of research data could help improve the generalizability of the research results. For future research, scholars might be interested in understanding the understudied relationship between publication awards and citation counts.

Acknowledgements. The authors thank the reviewers for their constructive comments.

References

1. Frey, B.S., Neckermann, S.: Awards: a disregarded source of motivation. Perspect. Moral Sci. **0**(11), 177–182 (2009)
2. MacLeod, R.M.: Of medals and men: a reward system in Victorian science, 1826–1914. Notes Rec. R. Soc. Lond. **26**(1), 81–105 (1971)
3. Brooks, T.A.: How good are the best papers of *JASIS*? J. Am. Soc. Inf. Sci. **51**(5), 485–486 (2000)
4. Zhang, P., Wang, P., Wu, Q.: How are the best *JASIST* papers cited? J. Assoc. Inf. Sci. Technol. **69**(6), 857–860 (2018)
5. Blessinger, K., Frasier, M.: Analysis of a decade in library literature: 1994–2004. Coll. Res. Libr. **68**(2), 155–169 (2007)
6. Aharony, N.: Library and Information Science research areas: a content analysis of articles from the top 10 journals 2007–8. J. Librarian. Inf. Sci. **44**(1), 27–35 (2012)
7. Tuomaala, O., Järvelin, K., Vakkari, P.: Evolution of library and information science, 1965–2005: content analysis of journal articles. J. Assoc. Inf. Sci. Technol. **65**(7), 1446–1462 (2014)

8. Ullah, A., Ameen, K.: Account of methodologies and methods applied in LIS research: a systematic review. Libr. Inf. Sci. Res. **40**(1), 53–60 (2018)
9. Hayman, R., Smith, E.E.: Mixed methods research in library and information science: a methodological review. Evid. Based Libr. Inf. Pract. **15**(1), 106–125 (2020)
10. Fidel, R.: Are we there yet?: mixed methods research in library and information science. Libr. Inf. Sci. Res. **30**(4), 265–272 (2008)
11. Järvelin, K., Vakkari, P.: Content analysis of research articles in library and information science. Libr. Inf. Sci. Res. **12**(4), 395–421 (1990)
12. Hider, P., Pymm, B.: Empirical research methods reported in high-profile LIS journal literature. Libr. Inf. Sci. Res. **30**(2), 108–114 (2008)
13. Tackney, C.T., Chappell, S.F., Sato, T.: MSR founders narrative and content analysis of scholarly papers: 2000–2015. J. Manag. Spiritual. Relig. **14**(2), 135–159 (2017)
14. Ghobadi, S., Robey, D.: Strategic signalling and awards: investigation into the first decade of AIS best publications awards. J. Strateg. Inf. Syst. **26**(4), 360–384 (2017)
15. Sen, R., Patel, P.: Citation rates of award-winning ASCE papers. J. Prof. Issues Eng. Educ. Pract. **138**(2), 107–113 (2012)
16. Coupé, T.: Peer review versus citations–an analysis of best paper prize. Res. Policy **42**(1), 295–301 (2013)
17. Wainer, J., Eckmann, M., Rocha, A.: Peer-selected "best papers"—Are they really that "good"? PLoS ONE **10**(3), e0118446 (2015)
18. Mubin, O., Tejlavwala, D., Arsalan, M., Ahmad, M., Simoff, S.: An assessment into the characteristics of award winning papers at CHI. Scientometrics **116**(2), 1181–1201 (2018)
19. Gallardo-Gallardo, E., Nijs, S., Dries, N., Gallo, P.: Towards an understanding of talent management as a phenomenon-driven field using bibliometric and content analysis. Hum. Resour. Manag. Rev. **25**(3), 264–279 (2015)
20. Prebor, G.: Analysis of the interdisciplinary nature of library and information science. J. Librarian. Inf. Sci. **42**(4), 256–267 (2010)
21. Qin, J., Lancaster, F.W., Allen, B.: Types and levels of collaboration in interdisciplinary research in the sciences. J. Am. Soc. Inf. Sci. **48**(10), 893–916 (1997)
22. Chu, H.: Research methods in library and information science: a content analysis. Libr. Inf. Sci. Res. **37**(1), 36–41 (2015)
23. Zimmer, M., Proferes, N.J.: A topology of Twitter research: disciplines, methods, and ethics. Aslib J. Inf. Manag. **66**(3), 250–261 (2014)
24. Peritz, B.C.: The methods of library science research: some results from a bibliometric survey. Libr. Res. **2**(3), 251–268 (1980)
25. Landis, J.R., Koch, G.G.: The measurement of observer agreement for categorical data. Biometrics **33**(1), 159–174 (1977)
26. Awre, C., et al.: Research data management as a 'wicked problem'. Libr. Rev. **64**(4/5), 356–371 (2015)
27. Dai, M.Y., He, W., Tian, X., Giraldi, A., Gu, F.: Working with communities on social media. Online Inf. Rev. **41**(6), 782–796 (2017)
28. Wu, D., Liang, S.B., Yu, W., Baxter, G., Marcella, R., O'Shea, M.: Members of the Scottish Parliament on Twitter: good constituency men (and women)? Aslib J. Inf. Manag. **68**(4), 428–447 (2016)
29. Lipetz, B.A.: Aspects of JASIS authorship through five decades. J. Am. Soc. Inf. Sci. **50**(11), 994–1003 (1999)
30. Terry, J.L.: Authorship in *College & Research Libraries* revisited: gender, institutional affiliation, collaboration. Coll. Res. Libr. **57**(4), 377–383 (1996)
31. Wang, S.: The intellectual landscape of the domain of culture and ethics in knowledge organization: an analysis of influential authors and works. Catalog. Classif. Q. **57**(4), 227–243 (2019)

32. Davarpanah, M., Aslekia, S.: A scientometric analysis of international LIS journals: productivity and characteristics. Scientometrics **77**(1), 21–39 (2008)
33. Finlay, S.C., Ni, C., Tsou, A., Sugimoto, C.R.: Publish or practice? An examination of librarians' contributions to research. Portal: Libr. Acad. **13**(4), 403–421 (2013)
34. Weller, A.C., Hurd, J.M., Wiberley, S.E.: Publication patterns of U.S. academic librarians from 1993 to 1997. Coll. Res. Libr. **60**(4), 352–362 (1999)
35. Chang, Y.W., Huang, M.H.: A study of the evolution of interdisciplinarity in library and information science: using three bibliometric methods. J. Am. Soc. Inf. Sci. Technol. **63**(1), 22–33 (2012)
36. Han, P., Shi, J., Li, X., Wang, D., Shen, S., Su, X.: International collaboration in LIS: global trends and networks at the country and institution level. Scientometrics **98**(1), 53–72 (2014)
37. Sin, S.C.J.: Longitudinal Trends in internationalisation, collaboration types, and citation impact: a bibliometric analysis of seven LIS journals (1980–2008). J. Libr. Inf. Stud. **9**(1), 27–49 (2011)
38. Johnson, R.B., Onwuegbuzie, A.J., Turner, L.A.: Toward a definition of mixed methods research. J. Mixed Methods Res. **1**(2), 112–133 (2007)
39. Togia, A., Malliari, A.: Research methods in library and information science. In: Oflazoglu, S. (ed.) Qualitative versus Quantitative Research, pp. 43–64. IntechOpen, London (2017)
40. Liebscher, P.: Quantity with Quality? Teaching quantitative and qualitative methods in an LIS masters program. Libr. Trends **46**(4), 668–680 (1998)
41. Zhang, J., Wang, Y., Zhao, Y., Cai, X.: Applications of inferential statistical methods in library and information science. Data Inf. Manag. **2**(2), 103–120 (2018)
42. Zhang, J., Zhao, Y., Wang, Y.: A study on statistical methods used in six journals of library and information science. Online Inf. Rev. **40**(3), 416–434 (2016)
43. Guerra, N.G., Williams, K.R., Sadek, S.: Understanding bullying and victimization during childhood and adolescence: a mixed methods study. Child Dev. **82**(1), 295–310 (2011)
44. Ivankova, N.V., Stick, S.L.: Students' persistence in a distributed doctoral program in educational leadership in higher education: a mixed methods study. Res. High. Educ. **48**(1), 93–135 (2007)
45. Mustanski, B., Lyons, T., Garcia, S.C.: Internet use and sexual health of young men who have sex with men: a mixed-methods study. Arch. Sex. Behav. **40**(2), 289–300 (2011)
46. Flavián, C., Gurrea, R.: Perceived substitutability between digital and physical channels: the case of newspapers. Online Inf. Rev. **31**(6), 793–813 (2007)
47. Veinot, T.: A multilevel model of HIV/AIDS information/help network development. J. Document. **66**(6), 875–905 (2010)
48. Voorbij, H.: The use of web statistics in cultural heritage institutions. Perform. Meas. Metr. **11**(3), 266–279 (2010)
49. Zhan, M., Widén, G.: Public libraries: roles in Big Data. Electron. Libr. **36**(1), 133–145 (2018)
50. Anfara Jr., V.A., Brown, K.M., Mangione, T.L.: Qualitative analysis on stage: making the research process more public. Educ. Researcher **31**(7), 28–38 (2002)
51. Gómez-Núñez, A., Vargas-Quesada, B., Chinchilla-Rodríguez, Z., Batagelj, V., Moya-Anegón, F.: Visualization and analysis of SCImago Journal & Country Rank structure via journal clustering. Aslib J. Inf. Manag. **68**(5), 607–627 (2016)
52. Koufogiannakis, D., Slater, L., Crumley, E.: A content analysis of librarianship research. J. Inf. Sci. **30**(3), 227–239 (2004)
53. Zins, C.: Classification schemes of Information Science: twenty-eight scholars map the field. J. Am. Soc. Inf. Sci. Technol. **58**(5), 645–672 (2007)
54. Wang, Z.J., Li, Z.M., Xie, L.N.: Analysis on keywords in the dissertations of Library science abroad. J. Libr. Sci. China **36**(190), 116–123 (2010)
55. Sundberg, K., Kjellman, U.: The tattoo as a document. J. Document. **74**(1), 18–35 (2018)

56. Pawley, C.: Unequal legacies: race and multiculturalism in the LIS curriculum. Libr. Q. **76**(2), 149–168 (2006)
57. Arlitsch, K.: The espresso book machine: a change agent for libraries. Libr. Hi Tech **29**(1), 62–72 (2011)
58. Curry, R.: Makerspaces: a beneficial new service for academic libraries? Libr. Rev. **66**(4/5), 201–212 (2017)
59. Liu, G., Yang, L.: Popular research topics in the recent journal publications of library and information science. J. Acad. Librarian. **45**(3), 278–287 (2019)
60. Cibangu, S.K.: A new direction in information science research: making information science a human science. Inf. Res. **20**(3), paper 686. http://informationr.net/ir/20-3/paper686.html#.X-aHii1h1hE
61. Emerald Publishing. https://www.emeraldgrouppublishing.com/authors/pdf/Choosing%20the%20winners%20of%20the%20Outstanding%20Paper%20Awards.pdf. Accessed 25 Dec 2020
62. Case, D.O., Higgins, G.M.: How can we investigate citation behavior? A study of reasons for citing literature in communication. J. Am. Soc. Inf. Sci. **51**(7), 635–645 (2000)
63. Garfield, E.: Can citation indexing be automated? In: Stevens, M., Guiliano, V., Heilprin, L. (eds.) Statistical Association Methods for Mechanized Documentation: Symposium Proceedings, pp. 189–192. National Bureau of Standards, Washington, DC (1965)
64. Catalini, C., Lacetera, N., Oettl, A.: The incidence and role of negative citations in science. Proc. Natl. Acad. Sci. **112**(45), 13823–13826 (2015)
65. Chen, C.M.: Predictive effects of structural variation on citation counts. http://cluster.ischool.drexel.edu/~cchen/papers/2011/chen2011jasist_preprint.pdf. Accessed 25 Dec 2020
66. Min, C., Bu, Y., Wu, D., Ding, Y., Zhang, Y.: Identifying citation patterns of scientific breakthroughs: a perspective of dynamic citation process. Inf. Process. Manag. **58**(1), 102428 (2021)
67. Braun, T., Glänzel, W., Schubert, A.: On Sleeping Beauties, Princes and other tales of citation distributions…. Res. Eval. **19**(3), 195–202 (2010)

Institutional Management

Becoming Open Knowledge Institutions: Divergence, Dialogue and Diversity

Katie Wilson[1(✉)], Lucy Montgomery[1], Cameron Neylon[1], Rebecca N. Handcock[2], Richard Hosking[2], Chun-Kai (Karl) Huang[1], Alkim Ozaygen[1], and Aniek Roelofs[2]

[1] Centre for Culture and Technology, Curtin University, Perth, Australia
katie.wilson@curtin.edu.au

[2] Curtin Institute for Computation, Curtin University, Perth, Australia

Abstract. The Curtin Open Knowledge Initiative (COKI) is an innovative research project that collects and analyses publicly available research output data to assist and encourage researchers, academics, administrators and executives to understand the actual and potential reach of openness in research, and to assess their progress on the path towards open knowledge institutions. By taking a broad global approach and using multiple data sources, the project diverges from existing approaches, methods and bibliometric measures in the scholarly research environment. It combines analysis of research output, citations, publication sources and publishers, funders, social media events, open and not open access to provide overviews of research output and performance at institutional, funder, consortial and country levels. The project collects and analyses personnel diversity data such as gender, focusing on widening the reach of data analysis to emphasise the importance and value of diversity in research and knowledge production. Interactive visual tools present research output and performance to encourage understanding and dialogue among researchers and management. The path towards becoming open knowledge institutions involves a process of cultural change, moving beyond dominant publishing and evaluation practices. This paper discusses how through divergence, diversity and dialogue the COKI project can contribute to this change, with examples of applications in understanding and embracing openness.

Keywords: Open knowledge institutions · Open access · Research performance · Higher education · Data visualization

1 Introduction

The Curtin Open Knowledge Initiative (COKI) is a strategic research project established at Curtin University, Australia through a critique of global rankings dominating the higher education environment. It challenges the dependence of rankings on impact and citation data from scholarly publications with embedded epistemologies that favour scientific scholarly disciplines and the English language, ignoring a large body of research and knowledge from non-dominant populations and countries. The project embodies a theory of cultural change, focusing on openness and diversity in research production, and new

K. Toeppe et al. (Eds.): iConference 2021, LNCS 12646, pp. 431–440, 2021.
https://doi.org/10.1007/978-3-030-71305-8_36

ways of thinking about institutional research impact. COKI collects, aggregates and analyses data from multiple sources, including information on open access publications, research collaboration, social media events, diversity, policies and infrastructure. The project includes open source software architecture and a cloud platform that gathers, integrates, manages and reports on the data. We undertake research projects that provide critical insights into research performance across countries and institutions, access to knowledge, and the outcomes and interventions of institutional policies [1, 2].

Knowledge is social in terms of its production, its capacity to support public interest, communication, and the different kinds of value that it creates. As discussed by [3], what universities contribute as Open Knowledge Institutions is found in how they support and interact with diverse communities, as well as how they practice communication to encourage dialogue in order to create the diverse values that underlie different conceptions of knowledge. Diversity, inclusion and equity lead to more effective communication, dialogue and knowledge production. Inclusive coordination provides platforms for dialogue among different types of knowledge participants. Broadening traditional measurement of research output to include a larger set of formats such as digital scholarship, short monographs, translations, creative writing, art works, and performance scholarship empowers a wider range of voices and encourages new forms of research and dialogue with extended communities. Diverse models of production and access provide opportunities for broader dialogues in knowledge creation [3]. This involves a process of thinking that diverges from existing publishing and evaluation practices within research institutions.

2 Divergence

Universities are influenced by global market rankings that depend on impact and citation data extracted from dominant scholarly publications primarily in two commercial citation indexes, Scopus from Elsevier and Web of Science from Clarivate Analytics [4]. The ranking systems make assumptions about the usefulness of their proxy data measures that are based on ontologically inconclusive assumptions and reasons [5]. However, the presence of influential world rankings pressure universities and researchers in many countries to aspire to publish in the dominant sources in order for their institutions to compete on the ranking scale. This devalues diverse research from non-dominant countries in the global south, non-English languages, indigenous and minority populations and in disciplines such as the humanities and political activist scholarship where the primary output format is not journal publications [6, 7]. Even universities that explicitly refuse to participate in rankings continue to be included in their listings [8].

The COKI project diverges from dominant scholarly publishing thinking and practices in universities through a critique of the limited bibliometric measures used by global university rankings. It regards information as a community asset and highlights an opportunity to move beyond established, commercial data resources towards the construction of data resources that are governed by the higher education and research community. Universities possess the technical skills and expertise to capture data relating to research and scholarly communication at very large scales, and to build the tools and frameworks needed to interpret and contextualize such data once captured. However, a lack of common approaches and shared datasets has hampered measurement and evaluation, particularly on an international scale. Development of our project has generated intense interest from a wide range of researchers, organisations and institutions, revealing an unfulfilled information need and leading to sharing and dialogue among different communities who come together to exchange knowledge and experiences. These include institutional executives, researchers, librarians, funding bodies, consortia, educational networks and associations from Africa, Arab nations, Europe, Latin America, North America, Oceania and the United Kingdom.

3 Dialogue

COKI collects, stores and analyses publicly available data to enable researchers and universities to track and understand the potential for increasing the reach of their research through open publication formats and options that extend beyond citation chasing. The project has developed a dataset of more than 12 trillion items related to scholarly communication, open access, equity, diversity and inclusion. The dataset draws affiliation, publication, funding and social media events data from Microsoft Academic, Unpaywall, Crossref, Open Citations and ORCID to understand open research output and performance at institutional, funder, publisher and country levels. It includes over 100 million research outputs, 20,000 organisations and 20,000 funders covering a breadth of research output to reach an understanding of who creates research knowledge. This approach aims to encourage dialogue among academics, researchers, library and research staff, administrators and senior executives in universities, research and funding organisations. In order to engage with a wide audience, including those who are not necessarily experts in research measurement, COKI's visual interactive dashboard tools facilitate sharing of data and dialogue among diverse groups and individuals within institutions and communities. The project's data resources and dashboards present open research publication and performance data for research institutions and consortia, funders and publishers.

Existing and new data analysis combine to provide visual, interactive research performance overviews, including open research, funder compliance, collaboration, social media events and diversity. Evidence of the strength and uniqueness of the project emerged from responses to the COKI dashboard developed for the Council of Australian University Librarians (CAUL) and the Council of New Zealand University Librarians (CONZUL). We undertook a research survey of librarians and research staff from the two consortia (CAUL and CONZUL) who used the dashboard in 2020. Respondents commented on the accessibility of the visual format, the value of data such as citation advantages of open access publications (see Fig. 1) and compliance with funder open access mandates, that will enable them to initiate dialogue with researchers, directors and senior executives:

- *Citation data OA/non-OA, averages collated visually in this way very useful for showing OA reluctant researchers the benefit of OA in terms of citations.*
- *Very useful to have all the data together in one place and presented visually.*
- *Reports like these are difficult to get from the IR* [institutional repository], *and cannot get visualisations like these.*

OA Status / Articles / Citations / Average Citations									
	Open Access			Not Open Access			Grand total		
years.publish..	Articles	Citations	Average Citations	Articles	Citations	Average Citations	Articles	Citations	Average Cit...
2007	250	3,188	12.75	557	6,941	12.46	807	10,129	12.61
2008	373	5,040	13.51	672	7,779	11.58	1,045	12,819	12.54
2009	476	6,551	13.76	680	7,137	10.5	1,156	13,688	12.13
2010	574	8,050	14.02	890	9,288	10.44	1,464	17,338	12.23
2011	605	7,392	12.22	1,013	9,682	9.56	1,618	17,074	10.89
2012	686	7,265	10.59	1,074	8,900	8.29	1,760	16,165	9.44
2013	1,098	12,173	11.09	1,156	7,711	6.67	2,254	19,884	8.88
2014	1,263	10,889	8.62	1,213	6,935	5.72	2,476	17,824	7.17
2015	1,419	9,396	6.62	1,431	6,773	4.73	2,850	16,169	5.68
2016	1,635	8,497	5.2	1,749	4,809	2.75	3,384	13,306	3.97
2017	1,692	3,904	2.31	1,929	2,562	1.33	3,621	6,466	1.82
2018	1,473	416	0.28	2,069	411	0.2	3,542	827	0.24
Grand total	12,323	96,964	12.67	16,755	109,078	9.12	29,078	206,042	10.89

Fig. 1. Graph from the CAUL/CONZUL institutional dashboard. Analysis of open access and non-open access article publication citation data for Curtin University 2000–2018 shows overall citation averages for open access article publications are higher than not open. Note that only 2007–2018 data display in the image but the Grand total refers to the extended graph data from 2000–2018. Analysis and graph: COKI.

- *Funder OA/non-OA data useful…The [funder] compliance data is very interesting, especially to some Senior Executives* [9].

COKI offers two publicly available dashboards. The COKI Open Access Dashboard provides country level publication research output by open/non-open access and open/non-open article citation analysis [10]. Figure 2 from this dashboard presents Australia's research output from 2000–2020: analysis of research output by total OA, Gold OA (publisher-mediated), Green OA (repository-mediated), Hybrid OA (published in a journal not listed in the Directory of Open Access Journals (DOAJ), but free to read from publisher with any license), and Bronze OA (free to read online via a publisher but without a license).

The public COKI Research Funding Dashboard provides country level analysis of national and international research funding sources acknowledged in publications from 185 countries: the ratio of domestic to international funding; country of funder; major funders grouped by publication year; and funder name with number of acknowledgements per funder [11]. Figure 3 shows major funders for Australian institutional research output.

The strength of COKI's large dataset is its comprehensiveness, and while the detailed data analysis may be specialized, visual presentations such as those in the figures above provide clear messages about research performance and funding sources. Such messages can initiate further investigation, dialogue and analysis, and assist institutions in charting their progress and publishing options. COKI is working with a range of North American consortia, specifically building dashboards to address consortium level tracking of research publications. Alongside publication and funder data, we analyze public staff institutional demographic data such as gender, indigeneity, ethnicity or nationality and disability, where available, to highlight understanding of diversity within research production and creation. This is discussed in the next section.

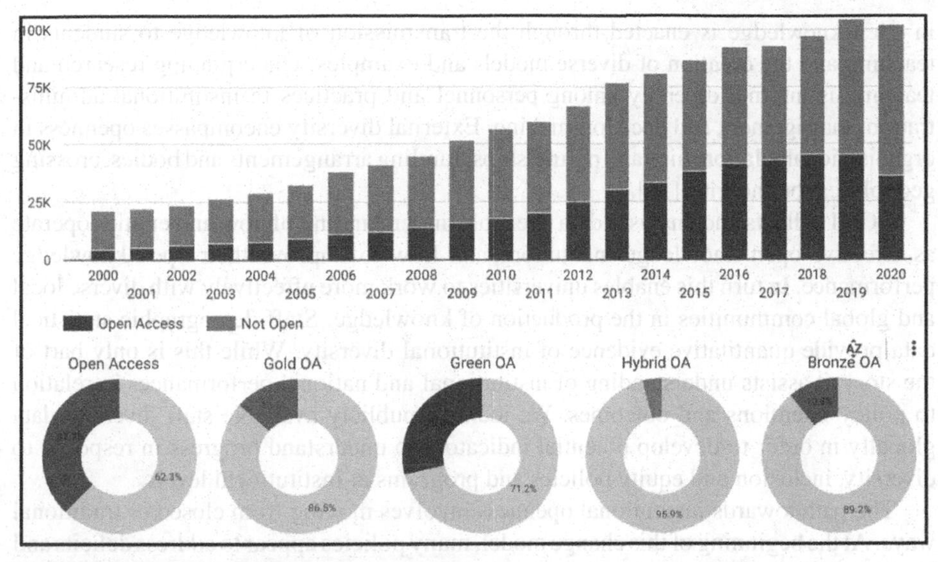

Fig. 2. Australia's research output in terms of open access and not open access, plus percentages of Gold, Green, Hybrid and Bronze OA, 2000–2020, from the public COKI Open Access dashboard. Analysis and image: COKI. (Color figure online)

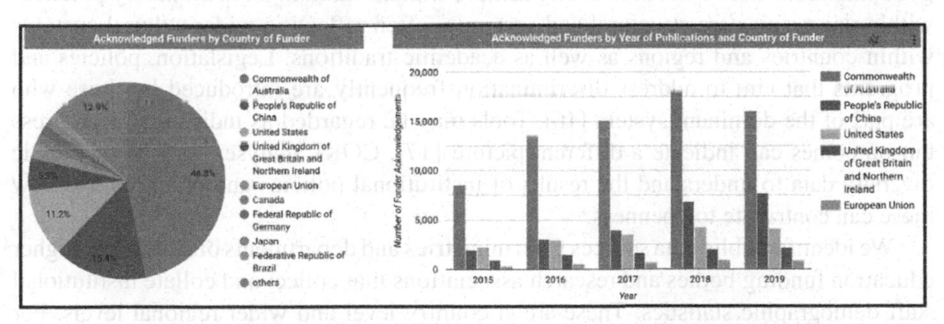

Fig. 3. Australia: acknowledged funders by country and major funders by publication year, 2015–2019, from the public COKI Research Funding dashboard. Analysis and image: COKI.

4 Diversity

Diversity is a key element of open knowledge institutions, and to achieve openness, universities also need to address challenges of diversity in input and output. This means understanding *who* is involved in knowledge creation and research, as well as *how* knowledge is shared within disciplines and scholarly communities, across disciplines, between universities and wider communities. A diversified staff that reflects the diversity of institutional student bodies and communities contributes to openness and the inclusion of diverse ideas, knowledges and languages. Systemic biases within academic research practices and institutions, including colonisation, genderism, sexism and racism perpetuate and legitimize marginalisation of groups of people and thought [12–15]. Diversity in open knowledge is enacted through the transmission of knowledge to students in teaching and the creation of diverse models and examples. Underpinning research and teaching is internal diversity among personnel and practices in institutional administration, management and decision-making. External diversity encompasses openness in organisational relationships and partnerships, funding arrangements and bodies, crossing geopolitical boundaries [16].

COKI collects and analyses data to enable understanding of how universities operate as effective open knowledge institutions and how to improve their open knowledge performance. In turn this enables universities to work more effectively with diverse local and global communities in the production of knowledge. Staff demographic statistical data provide quantitative evidence of institutional diversity. While this is only part of the story, it assists understanding of institutional and national performances in relation to policy intentions and outcomes. We identify publicly available staff diversity data globally in order to develop potential indicators to understand progress in response to diversity, inclusion and equity policies and programs at institutional levels.

The path towards institutional openness involves moving from closed or traditional ways. At the beginning of this change model, many policies appear to address deficits and institutional preservation and reputation. The extent to which an institution translates its intentions into reality is not always visible or transparent in terms of outcomes. Narratives and policy around equity, diversity and inclusion may be associated with good intentions but do not necessarily achieve transformation. Discriminatory practices within universities are structural and systemic, often reflecting wider cultural customs within countries and regions as well as academic traditions. Legislation, policies and programs that aim to address discrimination frequently are introduced by those who are part of the dominant system [16]. Tools may be regarded as indicators of progress but outcomes can indicate a different picture [17]. COKI analyses staff demographic diversity data to understand the results of institutional policies and practices and how these can contribute to openness.

We identify public data sources from ministries and departments of education, higher education funding bodies and research associations that collect and collate institutional staff demographic statistics. These are at country level and wider regional levels. For example, in Europe, the European Tertiary Education Register (ETER) provides detailed data for multiple countries, but statistics are also available from individual country sources. Differences at country levels present challenges in collecting global data. Some countries do not provide detailed statistical collections for public access, for example, those undertaking higher education structural and financial reform. Nations experiencing war and political turmoil may have lower priorities and limited funding directed towards higher education and research [16].

The diversity dimensions collected and reported on most consistently are gender, origin (race, ethnicity or nationality) and age, but detail and availability of the data vary widely by country. Gender or sex of university staff members is the most commonly available dimension across the statistical sources analysed. Limitations exist within this collection however. Data are sometimes only provided for academic staff or faculty. Non-binary sexual preferences are not frequently available, and are only recently recorded in a few countries where numbers are small and are excluded from statistics for privacy and confidentiality.

Where possible, we incorporate staff diversity data into the project's interactive visual data dashboards. For example, the data analysis in Fig. 4 shows the breakdown of women and men in academic and non-academic positions at Curtin University, from 2001 to 2018. The number of women reached 57% in 2018, but only 46% of academic staff were women in the same year.

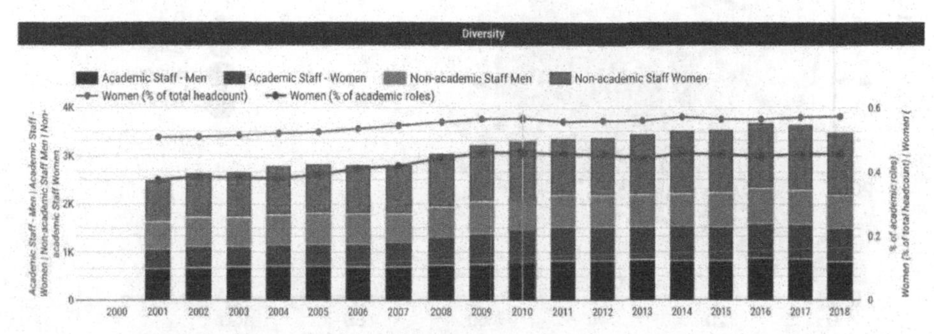

Fig. 4. Gender (men, women) share of academic and non-academic staff, Curtin University, 2001–2018. Source data: Australian Department of Education, Skills and Employment 2019. Analysis and image: COKI.

Diversity also extends to equal inclusion of the knowledges, research and output from researchers and institutions in non-dominant nations. As Fig. 5 illustrates, the highest open access output, both Gold (published in the Directory of Open Access Journals (DOAJ) or free to read via publisher with license) and Green (free to read via an open access repository), emerges from Africa, Asia, Latin America and Europe.

Through a focus on research diversity, the project raises questions about the processes of data collection and data use relating to research output. What are the underlying intentions and epistemologies of the collection process; who determines and legitimizes data collection methods; how are datasets analysed and presented; how do these factors affect the outcomes of analysis, and how the data are used? Concerns about research inequality and data sovereignty include the assumptions and values at many levels, from institutional to governmental, that shape approaches to data collection [18, 19]. For example, research output data from many African-based publications are excluded from the citation sources on which so-called global rankings depend. To extend the project's dataset beyond the global north and English language publications COKI is actively collaborating with partners and organisations on the African continent to incorporate and analyze their research output data.

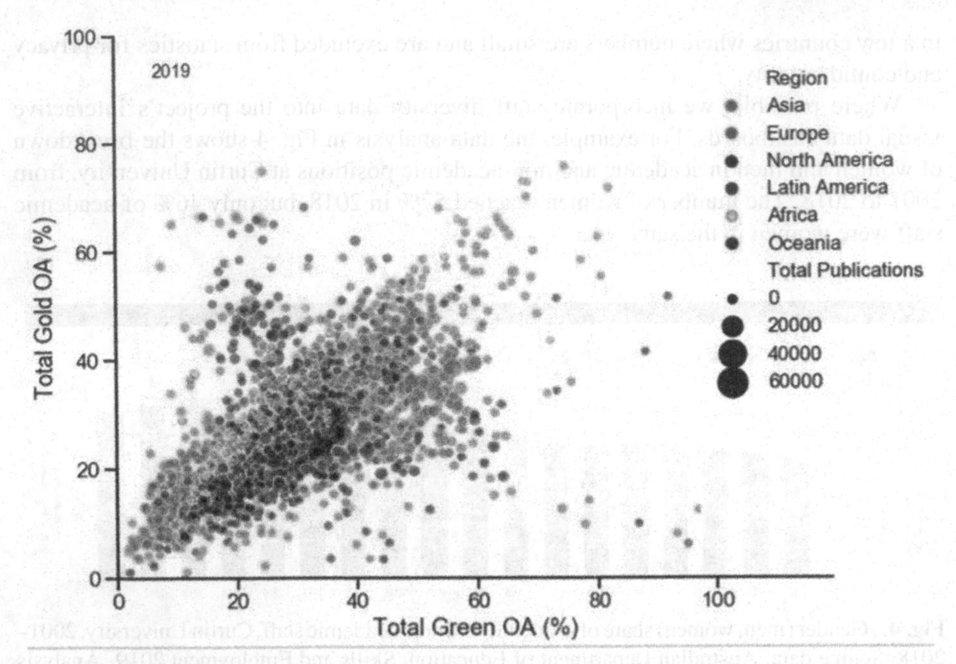

Fig. 5. Regional open access performance shows strong output in Gold (DOAJ and hybrid journals) Latin America and Asia; Green (repository-mediated) OA output for Africa and Europe, 2019. Each dot represents an institution and the size of the dots indicate the number of publications. Data sources: Microsoft Academic, Scopus, Web of Science, CrossRef, UnPaywall. Analysis and image: COKI. (Color figure online)

5 Conclusion

Through data analysis and open practices COKI provides opportunities for researchers, institutions, funding organisations and publishers to understand the reach and potential of open research, and to explore their individual and collective progress in becoming open knowledge institutions. While the project identifies a need for institutional cultural change in this process, it recognizes that this change is complex. The process involves divergence from existing assessment and evaluation practices and associated attitudes within research institutions which reward prestige publishing, follow world university rankings, and ignore open knowledge. It requires multidirectional change from individuals at grassroots levels and management levels within institutions, in the scholarly community and the marketplace. COKI user-centered dashboards are designed to encourage dialogue and assist researchers, managers and students to understand and critique the tools, datasets and methodologies used to measure research performance, as well as the agency that institutions and individuals have to make change within this process.

COKI embraces diversity in bringing together evidence of open research practices from different sources and challenging institutions to review their diversity and inclusion practices and outcomes. We encourage sharing of project data, software and visual dashboard code among researchers, library, research and technical staff and senior executives to enable critical data literacy upskilling [9, 20] and to facilitate dialogue. Options for alternative research publishing practices are well established, but are overshadowed by processes constrained by third party organisations with commercial interests in maintaining dominance in the scholarly communication market. Expanding dialogue among interested members of the scholarly community can include collaboration with commercial sources where relevant and appropriate. With a divergent, dialogic and diverse approach COKI hopes to enhance the possibilities for change within institutions and progress towards openness in research and knowledge sharing.

References

1. Huang, C.-K., et al.: Meta-Research: evaluating the impact of open access policies on research institutions. ELife **9**, e57067 (2020). https://doi.org/10.7554/eLife.57067
2. Wilson, K., et al.: 'Is the library open?': Correlating unaffiliated access to academic libraries with open access support. LIBER Q. **29**(1), 1–33 (2019). https://doi.org/10.18352/lq.10298
3. Montgomery, L., et al.: Open Knowledge Institutions: Reinventing Universities. MIT Press, Cambridge (Forthcoming)
4. Tennant, J.P.: Web of Science and Scopus are not global databases of knowledge. Eur. Sci. Edit. **46**, e51987 (2020). https://doi.org/10.3897/ese.2020.e51987
5. McCormick, A.C.: The intersection of rankings with university quality, public accountability and institutional improvement. In: Hazelkorn, E. (ed.) Global Rankings and the Geopolitics of Higher Education, pp. 205–215. Routledge, London (2017)
6. Shahjahan, R.A., Wagner, A.E.: Unpacking ontological security: a decolonial reading of scholarly impact. Educ. Philos. Theory **51**(8), 779–791 (2019). https://doi.org/10.1080/00131857.2018.1454308
7. Vera Baceta, M.-A., Thelwall, M., Kousha, K.: Web of Science and Scopus language coverage. Scientometrics **121**, 1803–1813 (2019). https://doi.org/10.1007/s11192-019-03264-z
8. McKenna, S.: University rankings don't measure what matters. The Conversation, 15 September 2020 (2020). http://theconversation.com/university-rankings-dont-measure-what-matters-145425
9. Wilson, K., et al.: Extending researcher knowledge of open scholarship performance (Forthcoming)
10. COKI Open Access Dashboard. http://openknowledge.community/dashboards/coki-open-access-dashboard/. Accessed 06 Jan 2021
11. COKI Research Funding Dashboard. http://openknowledge.community/dashboards/funder-countries/. Accessed 06 Jan 2021
12. Ajil, A., Blount-Hill, K.-L.: "Writing the Other as Other": exploring the othered lens in academia using collaborative autoethnography. Decolonization Criminol. Justice **2**(1), 83–108 (2020). https://doi.org/10.24135/dcj.v2i1.19
13. Daza, S.L., Tuck, E.: De/colonizing, (post)(anti)colonial, and indigenous education, studies, and theories. Educ. Stud. **50**(4), 307–312 (2014). https://doi.org/10.1080/00131946.2014.929918
14. Farmer, L.B., Robbins, C.K., Keith, J.L., Mabry, C.J.: Transgender and gender-expansive students' experiences of genderism at women's colleges and universities. J. Divers. High. Educ. **13**(2), 146–157 (2020). https://doi.org/10.1037/dhe0000129
15. Smith, L.T.: *Decolonizing methodologies: Research and indigenous peoples*, 2nd edn. Zed Books, London (2012)
16. Wilson, K., Neylon, C., Montgomery, L., Hosking, R., Huang, C-K., Ozaygen, A.: Global diversity in higher education staffing: towards openness. (Forthcoming)
17. Ahmed, S.: Living a Feminist Life. Duke University Press, Durham (2017)

18. Chan, L.: Open Insights: An interview with Leslie Chan. Open Library of Humanities (2018). https://www.openlibhums.org/news/314/
19. Ríos, C.D., Dion, M.L., Leonard, K.: Institutional logics and indigenous research sovereignty in Canada, the United States, Australia, and New Zealand. Stud. High. Educ. **45**(2), 403–415 (2020). https://doi.org/10.1080/03075079.2018.1534228
20. Carmi, E., Yates, S.J., Lockley, E., Pawluczuk, A.: Data citizenship: rethinking data literacy in the age of disinformation, misinformation, and malinformation. Internet Policy Rev. **9**(2), 1–22 (2020). https://doi.org/10.14763/2020.2.1481

Toward Context-Relevant Library Makerspaces: Understanding the Goals, Approaches, and Resources of Small-Town and Rural Libraries

Soo Hyeon Kim(✉) and Andrea Copeland

Department of Library and Information Science, Indiana University-Purdue University Indianapolis, Indianapolis, IN 46202, USA
skim541@iu.ed

Abstract. While best practices for developing makerspaces in public libraries exist, there is scarce literature that describes how they apply to small-town and rural libraries in alignment with the libraries' existing assets, practices, and constraints. This paper aims to explore the small-town and rural libraries' goals, approaches, and existing resources towards establishing a future makerspace and investigate the extent to which these elements support or hinder the design of the makerspace or maker programming. From the qualitative analysis of cultural probes and interview data with nine librarians, this paper demonstrates two ways that small-town and rural libraries differed from the best practices in the field: a) focusing on attendance and equipping the materials within the makerspace over community building, b) lack of transfer of existing assets and practices to maker programming. Study findings suggest small-town and rural librarians' lack of STEM competencies and knowledge around makerspaces as a critical barrier for applying their existing assets and practices to a new area of maker programming. Our study proposes context-specific recommendations and directions for small-town and rural libraries to design and develop makerspaces.

Keywords: Makerspace · Maker programming · Rural library · Cultural probes · Community engagement

1 Introduction

With the advancement of digital technology and the changing landscape of informal youth learning, public libraries are challenged to transform the library into a hub for "smart and connected communities" [29]. The maker movement—a community of people who make, tinker, and share their processes and products in physical and/or virtual settings—is one way of expanding the role of public libraries towards sites of community engagement [27]. Makerspaces provide opportunities for problem-solving [8, 24] and STEM learning [9, 13, 22, 23]. Making can also empower learners to directly impact the local communities through design [11, 38]. As such, a growing number of libraries

K. Toeppe et al. (Eds.): iConference 2021, LNCS 12646, pp. 441–457, 2021.
https://doi.org/10.1007/978-3-030-71305-8_37

are implementing makerspaces to strengthen community engagement, particularly for youth [26].

Library makerspaces have shown promising results in large urban settings [3, 18]. While the role of public library makerspaces has been discussed [31, 42], it predominantly focused on urban libraries despite that that small-town and rural libraries compose 80.5% of public libraries in the U.S. [36]. Research indicates that rural libraries have different resources, skills, and constraints which impact their librarianship. Rural libraries have fewer full-time employees ranging from 1.3 to 4.2 employees per site [34]. Small-town and rural libraries are often staffed by non-MLIS librarians. Compared to urban libraries that offer STEAM programs that make up 50% of their educational offerings, STEAM programs only comprise 19.7% in rural libraries [34]. Furthermore, the median size of a rural library is only 20% of the median size of an urban library [34]. These figures indicate varied resources within small-town and rural libraries that can impact their ability to shift their practices and reconfigure the library space to accommodate makerspaces or new maker programs.

Best practices provide guidelines for implementing public library makerspaces [1, 20, 40]. However, there is scarce literature that describes how these guidelines apply to small-town and rural libraries. Further work is needed to interrogate the extent to which these guidelines align with the small-town and rural libraries' existing practices. In addition, an explicit effort is needed to consider how the small-town and rural libraries' assets, practices, and constraints can be utilized to strengthen the sustainability and the impact of future makerspaces in rural communities.

Our research, therefore, aims to explore the current status of the small-town and rural libraries that currently do not have makerspaces or consistent maker programs. Given how small-town and rural libraries often do not have the space for permanent makerspaces, we considered supporting the development of consistent maker programs and/or the makerspace. We answer the following research questions:

a) what are the small-town and rural librarians' goals and approaches towards establishing the future makerspace and/or maker programs?
b) what are their assets, practices, and constraints towards designing their regular youth programs?
c) to what extent do these existing assets, practices, and constraints support or hinder the design of the makerspace and/or maker programs?

2 Literature Review

The Maker Movement has quickly spread in public libraries [5], providing numerous reports and practitioner articles. We included both practitioner reports and empirical articles to conduct a literature review of public library makerspace around five themes: a) goals, b) approaches, c) space, tools, and equipment, d) ways to support learning, and e) library professionals' competencies.

In many studies, community building was the overarching goal of the makerspace rather than supporting specific skills (e.g., STEM knowledge) or tinkering with high-tech equipment [6, 18, 26]. Digital media labs (e.g., YOUmedia) that expanded library services through increased technology offerings highlighted that the underlying elements that promoted success were not space or the increased technology, but a purposeful strategy that supported the knowledge creation and community engagement [10]. A review of notable library makerspaces illustrated that the common long-term goal was to create a sense of community in the makerspaces, despite their different short-term goals such as engaging the patrons with activates offered in the space or making the library a fun place to hang out [5]. Similarly, twenty-four learning labs established as creative teen spaces served as community catalysts [2]. For instance, Madison Public Library's maker program, the Bubbler, had the motto "people not stuff." Making Justice, one of the Bubbler programs, occurred at the juvenile detention center to invite participation from court-involved youth to reach out to and include every community member. Consequently, a sense of community grew among youth who participated in the Making Justice program. While goals such as promoting entrepreneurship, innovation, and STEM learning were also noticeable [6, 40], public libraries incorporated makerspaces as community catalysts aligned with libraries' mission to engage with and serve the local community.

Many library makerspaces took varied approaches to develop makerspaces; however, youth and community involvement were recurring themes [2, 3, 18]. Library makerspaces often provided maker-in-residence programs to involve community members with expertise to share and facilitate maker programs [5, 18, 41]. Particularly in rural libraries, library professionals often formed a teen advisory board to help reflect youth's needs and interests, which helped design an environment of belonging for youth [12, 33]. At Pima County Public Library, teens were involved as designers to co-design the community's maker program [2]. As a result of the participatory approach, the design of the maker program was emergent, fluid, and driven by learners who envisioned themselves as active users of the new space and programs.

Notable public library makerspaces generally provide a designated space and a facilitator for the makerspace, along with a range of high-tech and necessary craft equipment [2, 5, 40]. However, when permanent space was not a viable option, libraries often repurposed existing rooms or unused space and utilized rearrangeable mobile shelving to create the makerspace [5, 30]. Emerging examples of rural library makerspaces demonstrate that makerspaces do not solely depend on the space and the tools but on connecting the community needs with the opportunities that arise from the makerspace. For instance, the Simla library started a media lab with a couple of laptops, iPads, and recording equipment [37]. Initially, the community members were intimidated to use the equipment. The library professional actively shared with different organizations and community members to show how they can utilize maker technologies to meet their needs. Slowly, community members used the media lab to design menus for local restaurants, record workshops, and record the local community's oral histories.

Public libraries represent out-of-school contexts ripe for exploration and discovery that are not tied to formal education [35]. As such, library professionals need to consider ways to scaffold learning in makerspaces in addition to providing the tools and equipping the space. Dreessen and Schepers [14] proposed to involve novice learners, who may find high-tech equipment unfamiliar and intimidating, through a combination of an open-door policy, short-term workshops, and long-term community processes. They suggested first familiarizing novice learners via open days that introduce and acquaint participants with the possibilities of tools and equipment, then shift towards combining open days with short-term and long-term community projects that focus on collective prototypes and experimentation. Einarsson and Huertzum [15] demonstrated scaffolding approaches that library makerspace practitioners employ in formal (organized, pre-planned instructions), informal (self-directed, interest-driven), and non-formal (hybrid of formal and informal) activities, which can be taken up by library professionals in other settings. To better support library professionals to support learning in makerspaces, Lee, Recker, and Phillips [28] suggested providing "at-a-glance" program planning materials with visual images that librarians can adapt without extensive preparation. While research has begun to investigate how to support learning in library makerspaces [14, 15], it is less clear how learning is supported in rural library makerspaces, as these studies are in the context of Europe and do not necessarily focus on rural libraries.

To manage and facilitate library makerspaces, scholars emphasized that library professionals reconceptualize their role from an expert to a facilitator or a mentor [12]. Koh, Abbas, and Willett [26] urged future library professionals to be "culturally competent" to connect the culture and the life stories of the community in the makerspace. Phillips, Lee, and Recker [33] described the librarian as "experience engineers" with the characteristics of being user-centered, connected to their communities, and comfortable with risk. Koh and Abbas [27] report the desired competencies for professionals in learning labs and makerspaces: technology, teaching, learning, community partnerships, flexibility, understanding diverse users, management, communication, curiosity, creativity, patience, and subject content knowledge. As such, future library professionals who facilitate maker programming should foster both the sensitivity to understand different cultures and the audacity to take risks to try different approaches towards learning, community engagement, and partnerships.

Our work built upon this literature to explore the extent to which small-town and rural libraries' goals, assets, and practices aligned with the themes that emerged in previously reported cases of public library makerspaces.

3 Methods

This study conducted a case study of nine small-town and rural libraries in the Midwest to identify their motivation for establishing future makerspaces and examine facilitators for and barriers to establishing library makerspaces and/or maker programs. The research includes three phases: cultural probes, semi-structured interviews, and co-design. This paper builds on initial analysis [21] and reports findings from the first two phases of the study. For all methods described in this paper, IRB approval was obtained.

We used purposive sampling to identify libraries in the Midwest that serve a legal service area population that is less than 25,000 and/or is considered either rural fringe, distant, or remote library following the ALA guidelines [36]. Then, we contacted libraries that do not have makerspaces or run consistent maker programs by checking their websites. We also utilized the State Library listserv to send out recruitment emails. The study participant had to be working full-time with at least two years of experience in the youth services department. Saturation of themes was reached after nine participants. The participants were all females with years of experience ranging from 3 to 12. Two had associate's degrees. Two obtained Master of Library and Information Science degrees. Seven had bachelor's degrees in education and had worked as teachers before becoming librarians.

We employed cultural probes [16] (Fig. 1) and semi-structured interviews. Cultural probes have been used widely by researchers in Human-Computer Interaction to elicit information from the users about their daily lives and enable researchers to enter into the participants' local culture [19, 25, 32, 43]. Without the presence of a researcher, participants are encouraged to document their lives by completing activities in the cultural probes, often using a camera. This approach was deemed appropriate given that participants were geographically located far from the research team.

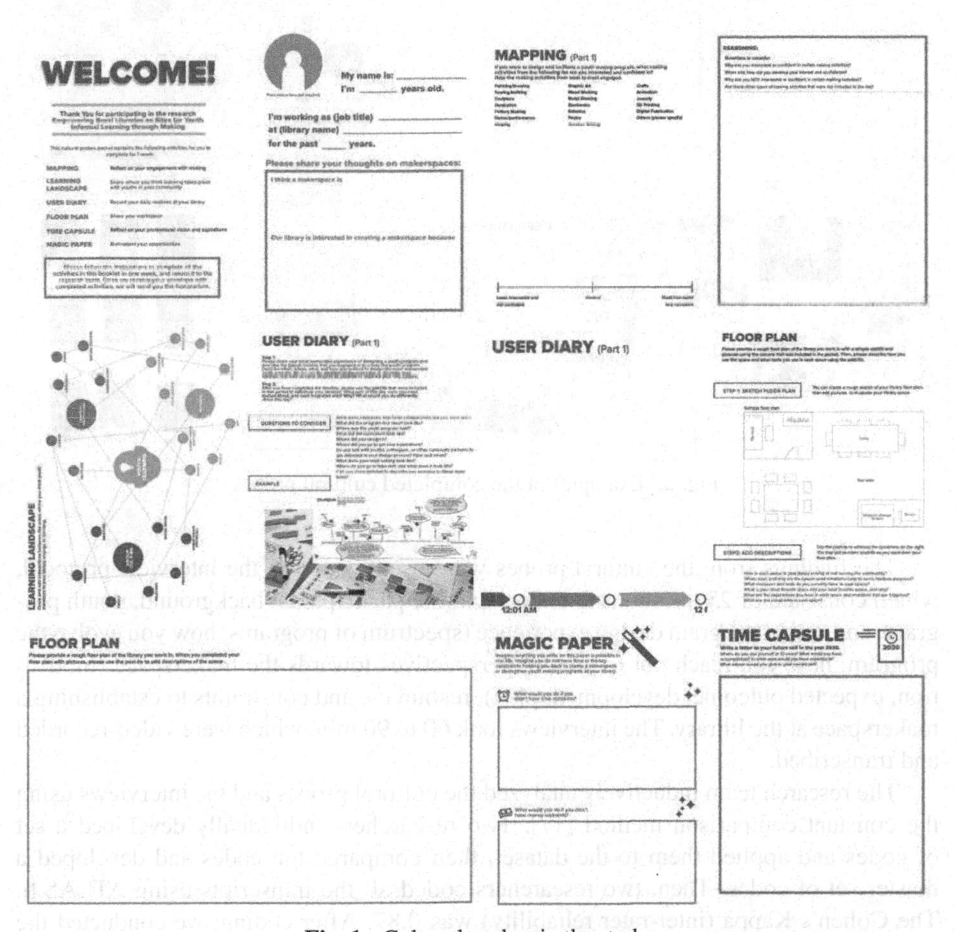

Fig. 1. Cultural probes in the study

The cultural probes included six activities: a) mapping, b) youth learning landscape, c) user diary (part 1: describe routines involved in youth program design, part 2: describe routines involved in maker program design), d) floor plan (describe how library space is serving or not serving the community), e) magic paper (imagine future makerspace), and f) time capsule (professional aspirations). The mapping asked participants to rate the making activities four times from our provided list and provide reasons: a) activities that librarians are interested and confident in, b) activities that youth are interested in, c) activities that community members have expertise in, and d) activities that outside members have expertise. The cultural probes package included the cultural probes, an instant camera, and post-its. The completed cultural probes included descriptions and pictures (Fig. 2). Cultural probes were transcribed for analysis.

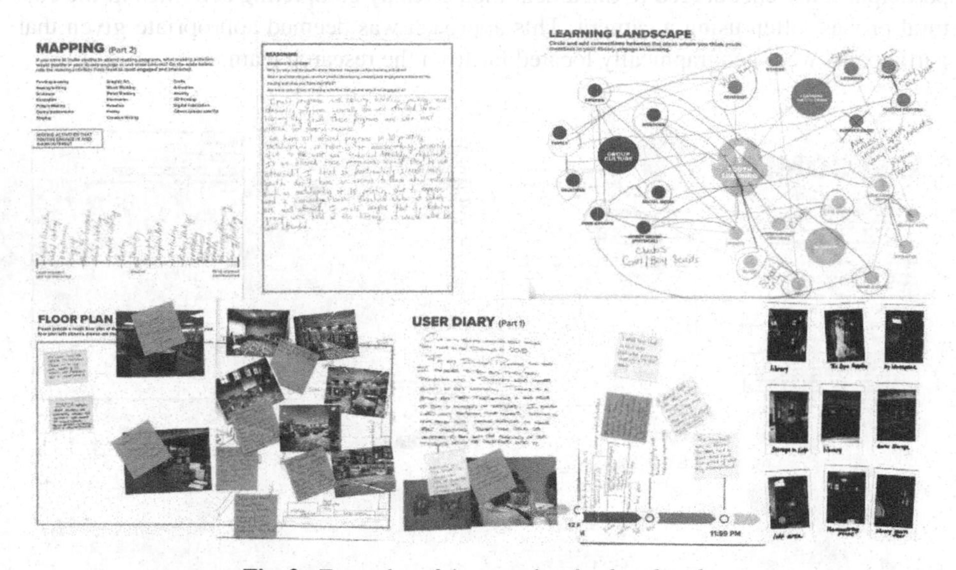

Fig. 2. Examples of the completed cultural probes

The findings from the cultural probes were used to design the interview protocol, which consisted of 25 questions to understand the participants' background, youth program, and maker program design experience (spectrum of programs, how you evolve the program, how you reach out to youth), perspectives towards the makerspace (motivation, expected outcome, development plan), resources, and constraints to establishing a makerspace at the library. The interviews took 60 to 90 min, which were video-recorded and transcribed.

The research team inductively analyzed the cultural probes and the interviews using the constant comparison method [17]. Two researchers individually developed a set of codes and applied them to the dataset, then compared the codes and developed a master set of codes. Then, two researchers coded all the transcripts using ATLAS.ti. The Cohen's Kappa (inter-rater reliability) was 0.87. After coding, we conducted the analysis in two phases. First, we identified participants' goals and approaches towards establishing the makerspace and compared them with previous literature to identify similarities and differences. Second, we identified the participants' assets, practices, and constraints they experienced in youth program design and analyzed how these elements were utilized to design the makerspace or maker programming. Emergent findings were discussed and reviewed iteratively.

We acknowledge that this study, based on nine libraries, is insufficient to provide a complete view of the assets, practices, and constraints of small-town and rural libraries. However, we also note that the goal of qualitative research is not generalizability but transferability. By creating detailed accounts of nine small-town and rural libraries, we argue that the context-specific recommendations and directions that we propose can apply to similar public libraries in the U.S. In future work, we will extend the findings to inform the development of co-design guidelines for small-town and rural libraries to engage in the last phase of the study.

4 Findings

Our findings demonstrate that nine small-town and rural libraries focused on increasing attendance, equipping the materials within the makerspace over community building, and experienced challenges in transferring existing practices related to youth involvement and partnership building to maker programming. Study findings further suggest small-town and rural librarians' lack of STEM competencies and knowledge around makerspaces as a critical barrier for applying their existing assets and practices to a new area of maker programming.

4.1 Goals and Approaches

Participants in this study commonly expressed increasing attendance by providing STEM-related programming in the makerspace as the primary goal. One participant expressed: "*This is almost like... a brand...and any time we have STEM...it will attract them.*" Another participant shared a boost in circulation as a result of increased attendance from the STEM programs: "*That increased right away, the number of people that you have at your program... They start checking books out, or they come for other reasons.*" While providing access to technologies ("*Children and adults in our community have very limited resources and I believe that a makerspace can create a variety of opportunities for everyone*") and promoting personal growth through life-skills, critical thinking, and problem-solving ("*It would help them to solve problems, and it would help them to learn how they solve problems*") were mentioned as important goals, participants anticipated increased attendance as the primary long-term outcome from the makerspace since it would promote the library as a place to engage in STEM learning.

To establish the envisioned makerspace, participants focused on providing the space and the tools within the makerspace. Many considered securing a designated area for the makerspace as a necessary step, followed by furnishing it with appropriate equipment through funding opportunities. One participant mentioned: "*And so that is one of our biggest issues, just finding the space in our library and maybe redeveloping what we have so that we could have something where we can set it up. So, the money for what we would need and just space trying to figure out.*" When libraries already had a designated space for a makerspace, participants considered the equipment they planned to purchase.

However, their approach to establishing the makerspace lacked articulation of how they plan to facilitate the kind of engagement and learning they expressed as the learning goals for the makerspace (e.g., problem-solving, critical thinking). Many participants expected that an organized space with high-tech materials would naturally trigger community members to engage in making independently. They expected youth to independently engage with making, without facilitation, as many participants struggled with a busy schedule.

There's electrical kits and engineering kits…just have things there and organized and it would be open and clean and ready for someone to walk in… pull it out and work on it.

I typically don't have a lot of free time to just go over there one-on-one and try and do something with them. So, they would get a lot more exposure to being able to make, not on my schedule, but on theirs.

The interview excerpts highlight the participants' expectation that learning would occur once youths were given the space and the tools. Although some participants emphasized the importance of first understanding the community members' interests towards the makerspace, overall, there was evidence that participants focused more on the material aspect rather than considering the role of facilitation in the makerspace to support community building.

4.2 Resources for Youth Program Design

Assets. Notably, many (seven out of nine) participants had degrees in education with professional teaching experience at the K-12 level. Having an education background was considered an asset: "*Having a degree in education and having that curriculum background helps a lot when I'm planning storytime because I run a curriculum for that. And I kind of use those lesson plans in the same manner to do that.*" Also, many shared the passion for supporting youth to learn and grow, as evident in the remark below:

I originally, from the jump, wanted to work with teens. That's always been kind of my mission, why I started with the secondary English education degree…And being in the library setting has been fun…but exposing our rural kids to maybe socially conscious materials, more challenging materials, things that they may not have been exposed to…so that they can have better well-rounded thoughts. Not just so ag [agriculture] focused and small-minded. I mean, not small-minded, small-town focused.

Interestingly, several participants described that becoming a librarian was "unplanned." Many started as a teacher but found that teaching was not aligned with their goals. Finding a library position was unintentional but fit well with their life situations. Several remarks expressed the unintentionality of finding their current positions: "*I was offered this job when my kids went away to college, so the timing was perfect and I've always loved it*"; "*Yeah, so the lady that actually did it, she went on maternity leave…So, I went ahead and I applied for the position and got it.*" Our analysis illustrated that many participants shared a similar trajectory of realizing that the teaching profession did not live up to their expectations and found library positions in unanticipated ways, which aligned closely with how they want to support learning. The following remark highlights the participants' motivation to promote youth learning in authentic learning settings, rather than focusing on the assessment.

> *But the whole, the formalness of the, not that the standards are bad, but just that there was a lot more to it than just what I felt was important to the kids as far as getting them to learn what they wanted to learn…So, when the opportunity came to work in the library, I was like, this would be great…it felt a lot, it was less formal and it was more authentic, I guess more authentic learning.*

Practices. Analysis findings further illustrated that many participants actively engaged youth to find out their interests and connected the youth programs to their interests. Several mentioned getting an idea about a new program by talking to the youth informally when visiting the library. Participants also utilized their regular school visits to gauge interests from the youth on what has been trending, as reflected in several participants:

> *I think for some reason they were big into Harry Potter and Doctor Who. So, we were kind of dealing every week a different theme and we were talking about different books so that we would do activities that were related to those books.*

> *So, it's trying to find out what their needs are, what their interests are, and how to bring kids in, youth in general… We're willing to try just about anything, we've tried lots.*

It was noticeable that all participants actively sought public school partnerships to support youths' literacy from early years. They provided regular visits to daycares and elementary schools to promote storytimes or summer reading programs. Some established specific partnerships with the school to go in during designated times to start a book club with interested students. A few participants reached out even further to invite community members with various professions during a guest reader's week: "*I invited, you know, bankers people worked in banks. I invited the jailer from our, our jail, the EMS people, all of them. And I had the mayor as well. They came and I let them pick out their own books so they feel a little more comfortable…They would read to the students.*" As such, participants engaged in regular practices of reaching out and involving youth to make their library programming relevant to youth's interests and needs.

Constraints. Lastly, our analysis demonstrated that participants frequently experienced constraints related to lack of time, staff, budget, which hindered them from investing time in programming design. One mentioned the challenge of not having enough time because of the number of responsibilities she would take on, given the low number of staff: "*It's not that we don't have the money to purchase the books or information or look it up. Time is a big issue…my biggest issue is time. You know, I would love to do a whole lot of things, but by keeping track of everything else…the program development right* *now for me is not high on my priority list.*" When participants were asked about their experience of completing the cultural probes, several shared the lack of time in their daily routine to think and reflect. The following remark reflects this: "*This is probably some of the longest time that I ever just sat at my desk... I don't, I don't have a day where I just sit and plan. I'd love a day where I just sit and plan.*" Also, most participants mentioned the lack of budget as a barrier towards providing new programs to patrons, as highlighted in one participant's remark: "*A very small budget. I think this year our budget for programs is $1,400 total for the whole year.*"

4.3 Supporting Maker Programs with Existing Resources

Misalignment Between Assets and Youth Interests. The last stage of our analysis investigated how the participants' assets, practices, and constraints experienced during youth programming were utilized to design the makerspace and/or maker programs. Findings from the cultural probes mapping activities illustrated that librarians' areas of interest and confidence in making (e.g., crafts, sewing) were the opposite of what youth in their communities may be interested in (e.g., robotics). Interestingly, the librarians' area of expertise mapped closely with the community members' perceived expertise while making activities in which youth might be interested in mapped closely with what the outside members might offer. This finding highlights the misalignment between librarians' assets and youth' interests.

Lack of Transfer of Youth Involvement Practices. Our findings previously showed that participants routinely engaged in practices that were supportive towards designing youth programs (i.e., youth involvement and active partnership with schools). However, within the domain of maker programming, there was no evidence of youth involvement. As many experienced the lack of time, participants frequently used available lesson plans or crafts activities from Pinterest, State library training materials, and summer reading program manual: "*I try to just not recreate the wheel and steal someone else's programming.*" As a result, the maker programs that participants shared were generic, lacking connection to community interests. For instance, the most memorable maker programs that participants shared in the cultural probes related to arts and crafts (i.e., marshmallow toothpick challenge, knitting, tie-dye, canvas painting, gingerbread making program).

In contrast, the most memorable youth programs described in the cultural probes included creative ideas relevant to the community. One participant described the Karen Land program in which she invited the veteran racer of the Iditarod, Karen Land, around the time of the Iditarod itself. Iditarod is a sled dog race in Alaska that many elementary-aged youths learn. The program had high attendance, bringing out of county attendees as well. Other examples included the Candyland program, teen game day, summer-long maker event, and summer reading programs with activities that tied with community interest. This finding demonstrates that participants' regularly-engaged practice of youth involvement did not transfer to maker programming.

Lack of Transfer of Partnership Practices. Participants' prior knowledge and experience in establishing partnerships to develop regular youth programs also did not transfer to maker programming because it required creating new connections in STEM areas.

We posit that because STEM was not their "wheelhouse," participants found it challenging to recognize the social capital and resources that may already be available in their connections, as highlighted by one participant:

> *But, again, I struggle with the technological side of making. Where, my younger brother... he has a 3D printer and he is awesome on it...And I'm like, gosh, you should come into my library and show these kids...I need to do that. It's like, again, one of those things where it was like, I thought, I didn't think about, I should just really pull in more people.*

Given the lack of STEM professionals in the community, most STEM partnerships were one-off programs by borrowing STEM resources from the State Library or the museum. A few participants sought out STEM expertise from the university, but it was challenging to get connected without knowing anyone personally ("*I've tried to make connection there a few times, and I just get sent to the next person, then the next person, then the next person and it fizzles out...So, unless you know somebody, it's hard to get into that*"). Also, being geographically far, these small-town and rural libraries experienced challenges to invite outside STEM experts to come in: "*We keep trying to send out letters... it's just frustrating because it's such a small area and everything was so far away from us.*"

Limited STEM Competencies. Importantly, our findings suggest that participants' limited STEM competencies and experience with makerspaces potentially influence the lack of transfer of the participants' existing practices to maker programming. Overall, participants expressed their lack of knowledge on makerspaces. When asked to share any makerspaces that they considered to be exemplars and describe the characteristics of the exemplar makerspaces that they wanted to adopt, the majority expressed that either they do not know or have not seen a makerspace: "*I see things through email and stuff of what people have in their makerspaces, but I personally haven't seen any real phenomenal ones*"; "*There aren't very many in our area, I don't know a whole lot.*" One participant expressed the difficulty in keeping herself up-to-date given her location and personal situation: "*We're kind of sequestered out here. And don't have little kids anymore, so my, I'd have to go out and seek those opportunities.*"

In addition to their limited experience with makerspaces, all participants expressed that their limited STEM competencies would hinder them from offering more STEM programs.

> *Our biggest response was that they wanted STEM or STEAM-related activities. And for us, it's uncomfortable doing it.*

> *But I will be honest that this is where I feel a lack of confidence in that area, so I fear the technology so I don't lead it. It's terrible, I really need to step out of my comfort zone, learn something new for them, but the way I generally choose the programming is what I feel comfortable leading, what I feel like I can provide some expertise.*

Participants mentioned seeking outside expertise rather than leading STEM programs due to their limited STEM competencies: "*There are certain activities that I or other staff members are not comfortable with, such as digital fabrication, robotics, or even animation. Therefore, if we were to do something with one of those, it would make more sense for us to bring someone from the outside to present a program in order to better serve our patrons.*" However, seeking out STEM expertise outside the community was challenging, as mentioned by many participants earlier.

We found that the lack of STEM competencies also influenced some participants to label themselves as "non-STEM" and establish that as part of their identity.

I'm okay with Legos, but I am not a building kind of person.

Those types of things [robotics, engineering] I'm not necessarily, my husband is very much into. And my husband and my oldest son are very, they just have that ability. I am not. I'm a books person, I'm not a hands-on person by nature…if we have a makerspace on robotics, I've got to be trained in robotics. You know, I have to know more than the kids to be able to do it, basically.

As expressed through participants' description of themselves as "a books person" and "not a building kind of person," the lack of STEM confidence influenced them to avoid taking risks and try out a new technology-oriented program. One participant described her encounter with library staff at a conference who described using a robotic kit to build a walking robot using electrical tape. The participant shared the fear she had towards the technology, which hindered her from running a similar program.

If it's electrical, I'm going to cut the wire. You don't just cut it… I can't do this… I'm going to have kids that are frying themselves. But she's [librarian at the conference] like, "it's really simple and then you just stick it on here and plug the two ends in. And then, the robot just walks on the table." I'm like, "I can't, I just have visions of people being electrocuted."

Findings show that participants' limited STEM competencies and their perception of technology as too complicated hindered them from designing and introducing STEM programs with technological components (e.g., robotics, animation). We suggest that participants' limited STEM competencies and perception of technology are critical barriers to applying their existing assets and practices to maker programming.

5 Discussion

Our study investigated the small-town and rural librarians' goals, approaches, and resources (i.e., assets, practices, constraints) towards establishing their future makerspaces or maker programming through a case study of nine libraries. From the qualitative analysis of cultural probes and interviews, this paper illuminates two ways that small-town and rural libraries differed from the best practices in the field. First, they focused on attendance and equipping the materials within the makerspace over community building. Second, they experienced lack of transfer of existing assets and practices to maker programming.

Compared to notable makerspaces that put a strong emphasis on community building, the nine small-town and rural libraries in our dataset focused more on increasing attendance through providing STEM programs and technologies in the makerspace. In addition to providing equitable access to technology and a social space for collaboration, which were reported as goals of rural library makerspace [6], the study finding demonstrates that small-town and rural librarians emphasized increasing the attendance for the library—contrary to the community-oriented, collective goal of notable makerspaces [2, 40]. The emphasis on STEM programming also illustrates that nine librarians in this study did not attend to diverse learning practices that emerge from the makerspace, such as collaboration [1] or empowering youth to impact the local community [2]. We posit that the emphasis put on increasing attendance through STEM programming may relate to the decrease in visitation [36] and the lack of makerspace examples from rural public libraries. This finding points towards the need to support rural library professionals to reimagine the potential of the makerspace (or maker programs) towards community building and broaden their conceptualization of making beyond STEM. For instance, the Bubbler makerspace was an arts-based makerspace that moved fluidly into the neighborhood and community spaces to serve various needs of the community [18]. Small-town and rural library makerspaces must consider different ways to connect with and meet community members.

Our study also suggests that small-town and rural librarians' lack of STEM competencies potentially hinders them from transferring their assets and practices related to youth involvement and partnership to a new area of maker programming. Since the adoption of the makerspace requires a shared vision from the library professionals who will design and develop the makerspace, it is crucial to examine how the social norms of the makerspace align with current practices of rural librarians. Makerspaces generally require a different style of facilitation and tools than literacy programs. As such, librarians' compatibility with these less familiar STEM tools and maker practices may be low, particularly when the librarian finds high-tech equipment and STEM knowledge to be too complicated, as illustrated by multiple interview excerpts in this study. Public librarians' lack of STEM competencies and confidence in technology have been reported [4, 7]. However, our findings add to the literature that their lack of confidence in STEM and technology can be a critical barrier for applying their assets and practices that support the design of makerspace. Given the extent to which prior experience and knowledge influence STEM programming in libraries and an apparent lack of competencies related to facilitating learning and community-based collaborations, iSchools should consider including in their curriculum elements of design thinking, facilitation in formal and informal learning settings, technical and tangible making, as well as methods for collaboration and partnering. The authors advocate a facilitated learning curriculum, similar to the one designed by the first author [39], to address the future needs of rural librarians.

This finding suggests several implications for rural library practices and public library makerspace research. Although a makerspace facilitator's competencies depend on individual librarian's prior experience and knowledge [27], strategies such as augmenting existing programs with maker elements [30], while applying their existing practices of youth involvement and school partnership, could potentially support small-town and rural library professionals to incrementally develop technological fluency and increase confidence towards maker programming. Library professionals could begin with smaller-scale maker programs and observe how the patrons react to reimagine different ways of incorporating maker elements into their regular programs. As they engage in trial-and-error, they can also learn to shift their conceptualization of mentorship from an instructor to a facilitator [12].

Similar to how youth were positioned as designers for the learning lab [2] or how the teen advisory board could actively advocate youth's voices [12, 33], small-town and rural librarians could involve youth and patrons, in ways that they have done in the past, to brainstorm the context-relevant makerspace for the community. If the library does not have the budget to create a designated space for the makerspace, the librarian and the community members could imagine different forms of hybrid spaces (i.e., physically meeting at the library teen room two days and virtually collaborating for three days). Such efforts should also continue within the community of practice—consisted of similar-sized rural and small-town libraries—that share similar norms and practices to inform one another. The community of practice can help share similar challenges and possible solutions based on tested cases in the field. This would also support small-town and rural libraries to continuously motivate themselves with examples that they can model and adopt instead of cases that may seem irrelevant to their context. Importantly, large-scale efforts to systematically support small-town and rural libraries with STEM resources and expertise, utilizing networks between informal learning institutions, research institutions, and government agencies, may be needed to strengthen STEM partnerships. Finally, public library makerspace research needs to continue the empirical research to design, develop, and investigate different configurations of rural library makerspaces to suggest models, frameworks, and practices that rural library professionals can adopt to build their capacity to design and facilitate learning in makerspaces.

6 Conclusion

Traditionally, the small-town and rural public library has been recognized as a trusted institution in the community to access information services and support [34]. As such, small-town rural libraries are well-positioned to support their community members by leveraging available library resources and services. To support community members in the changing landscape of digital technology and informal learning, small-town and rural librarians must continue to enact their unique assets and practices while acknowledging the barriers to developing context-relevant rural library makerspaces.

Acknowledgment. The project is supported by the IUPUI Arts and Humanities Institute. We thank the participants and Hana Jun for supporting the initial data analysis.

References

1. ALA: makerspaces. http://www.ala.org/tools/atoz/makerspaces. Accessed 25 Feb 2020
2. Association of science-technology centers and urban libraries council: learning labs in libraries and museums: transformative spaces for teens., Washington, D.C. (2014)

3. Austin, K., et al.: Reimagining Learning, Literacies, and Libraries. YOUmedia, Chicago (2011)
4. Baek, J.Y.: The accidental STEM librarian: an exploratory interview study with eight librarians. A National Center for Interactive Learning Education/Research Report, Space Science Institute, Boulder, CO (2013)
5. Bagley, C.A.: Library makerspace profiles. In: Makerspaces: Top Traiblazing Projects, A LITA Guide, pp. 19–102. Library and Information Technology Association (2014)
6. Barniskis, S.C.: STEAM: science and art meet in rural library makerspaces. In: Conference 2014 Proceedings, pp. 834–837 (2014). https://doi.org/10.9776/14158
7. Beck, D.M., Callison, R.: Becoming a science librarian. Sci. Technol. Libr. **27**(1–2), 121–134 (2006). https://doi.org/10.1300/J122v27n01
8. Bevan, B., et al.: Learning through STEM-rich tinkering: findings from a jointly negotiated research project taken up in practice. Sci. Educ. **99**(1), 98–120 (2015)
9. Blikstein, P., Worsley, M.: Children are not hackers: building a culture of powerful ideas, deep learning, and equity in the maker movement. In: Peppler, K., et al. (eds.) Makeology: Makerspace as Learning Environments (Volume 1), pp. 64–80. Routledge, New York (2016)
10. Blowers, H.: Supporting the knowledge continuum through technology: from consumption to fabrication. Comput. Libr. **32**(9), 30–32 (2012)
11. Calabrese Barton, A. et al.: Mobilities of criticality: space-making, identity and agency in a youth-centered makerspace. In: Transforming Learning, Empowering Learners: The International Conference of the Learning Sciences (ICLS), pp. 290–297. International Society of the Learning Sciences, Singapore (2016)
12. Clegg, T., Subramaniam, M.: Redefining mentorship in facilitating interest-driven learning in libraries. In: Lee, V.R., Philips, A.L. (eds.) Reconceptualizing Libraries: Perspectives from the Information and Learning Sciences, pp. 140–157, Routledge (2018)
13. Dougherty, D.: The maker mindset. In: Honey, M., Kanter, D.E. (eds.) Design, Make, Play: Growing the Next Generation of STEM Innovators, pp. 7–16, Routledge (2013)
14. Dreessen, K., Schepers, S.: Three strategies for engaging non-experts in a fablab. In: NordiCHI 2018, pp. 482–493 (2018). https://doi.org/10.1145/3240167.3240195
15. Einarsson, Á.M., Hertzum, M.: Scaffolding of learning in library makerspaces. In: Proceedings of FabLearn Europe 2019. ACM, New York (2019). https://doi.org/10.1145/3335055.3335062
16. Gaver, B., et al.: Design: cultural probes. Interactions **6**(1), 21–29 (1999)
17. Glaser, B., Strauss, A.: The constant comparative method of qualitative analysis. In: The Discovery of Grounded Theory: Strategies for Qualitative Research, pp. 101–115. Aldine Transaction, Piscataway, NJ (1967)
18. Halverson, E.R., et al.: The Bubbler as systemwide makerspace: a design case of how making became a core service of the public libraries. Int. J. Des. Learn. **8**(1), 57–68 (2017)
19. Hutchinson, H. et al.: Technology probes: inspiring design for and with families. In: Proceedings of the SIGCHI Conference on Human Factors in Computing Systems, pp. 17–24. ACM (2003)
20. Ito, M., et al.: Connected learning: an agenda for research and design. Digital Media and Learning Research Hub (2013)
21. Kim, S.H., Copeland, A.: Rural librarians' perspectives on makerspaces and community engagement. Proc. Assoc. Inf. Sci. Technol. **57**(1), e351 (2020). https://doi.org/10.1002/pra2.351
22. Kim, S.H., Zimmerman, H.T.: Collaborative argumentation during a making and tinkering afterschool program with squishy circuits. In: Proceedings of the 12th International Conference on Computer Supported Collaborative Learning (CSCL), pp. 676–679, Philadelphia, PA (2017)

23. Kim, S.H., Zimmerman, H.T.: Towards a stronger conceptualization of the maker mindset: a case study of an afterschool program with squishy circuits. In: ACM International Conference Proceeding Series (2017). https://doi.org/10.1145/3141798.3141815
24. Kim, S.H., Zimmerman, H.T.: Understanding the practices and the products of creativity: making and tinkering family program at informal learning environments. In: Proceedings of the 18th ACM International Conference on Interaction Design and Children, pp. 246–252. ACM (2019). https://doi.org/10.1145/3311927.3323117
25. Kjeldskov, J., et al.: Using cultural probes to explore mediated intimacy. Aust. J. Inf. Syst. **22**(3), 102–115 (2004)
26. Koh, K., et al.: Makerspaces in libraries. In: Lee, V.R., Philips, A.L. (eds.) Reconceptualizing Libraries: Perspectives from the Information and Learning Sciences, pp. 17–36. Routledge, New York (2018)
27. Koh, K., Abbas, J.: Competencies needed to provide teen library services of the future: a survey of professionals in learning labs and makerspaces. J. Res. Libr. Young Adults **7**(2), 1–22 (2016)
28. Lee, V. et al.: Conjecture mapping the library: Iterative refinements toward supporting maker learning activities in small community spaces. In: International Conference of the Learning Sciences (ICLS), pp. 320–327. ISLS, London, UK (2018)
29. Lee, V.: Libraries will be essential to the smart and connected communities of the future. In: Lee, V.R., Philips, A.L. (eds.) Reconceptualizing Libraries: Perspectives from the Information and Learning Sciences, pp. 9–16. Routledge, New York (2018)
30. Lee, V.R. et al.: Supporting interactive youth maker programs in public and school libraries: design hypotheses and first implementations. In: Proceedings of the 2017 Conference on Interaction Design and Children, pp. 310–315 (2017). https://doi.org/10.1145/3078072.3079741
31. Moorefield-Lang, H.M.: Makers in the library: case studies of 3d printers and maker spaces in library settings. Libr. Hi Tech. **32**(4), 583–593 (2014)
32. Obrist, M., et al.: Interactive TV for the home: an ethnographic study on users' requirements and experiences. Int. J. Hum. Comput. Interact. **24**(2), 174–196 (2008). https://doi.org/10.1080/10447310701821541
33. Philips, A.L. et al.: Small-town libraries as experience engineers. In: Reconceptualizing Libraries: Perspectives from the Information and Learning Sciences. Routledge, New York, NY (2018)
34. Real, B., Norman, R.R.: Rural libraries in the United States: recent strides, future possibilities, and meeting community needs (2017)
35. Subramaniam, M., et al.: Using technology to support equity and inclusion in youth library programming: current practices and future opportunities. Libr. Q. **88**(4), 315–331 (2018). https://doi.org/10.1086/699267
36. Swan, D.W., et al.: The state of small and rural libraries in the United States, Washington, DC (2013)
37. Techsoup: digital media labs and makerspaces in small and rural libraries. https://www.techsoup.org/community/events-webinars/digital-media-labs-makerspaces-small-rural-libraries-2014-02-26. Accessed 20 Feb 2020
38. Thanapornsangsuth, S.: Using human-centered design and social inventions to find the purposes in making. In: 6th Annual Conference on Creativity and Fabrication in Education, pp. 17–25. ACM (2016)
39. University T. of I: Facilitated Learning specialization in Master of Library and Information Science – School of Informatics and Computing : IUPUI. https://soic.iupui.edu/lis/master-library-science/facilitated-learning/. Accessed 10 Jan 2021
40. Wang, F. et al.: The state of library makerspaces. Int. J. Librariansh. **1**, 1–2 (2016). https://doi.org/10.23974/ijol.2016.vol1.1.12

41. Willett, R., et al.: Democratizing the maker movement: a case study of one public library system's makerspace program. Ref. User Serv. Q. **58**(4), 235–245 (2019)
42. Williams, R.D., Willett, R.: Makerspaces and boundary work: The role of librarians as educators in public library makerspaces. J. Libr. Inf. Sci. **51**(3), 801–813 (2019). https://doi.org/10.1177/0961000617742467
43. Wyeth, P., Diercke, C.: Designing cultural probes for children. In: Proceedings of the 18th Australia conference on Computer-Human Interaction: Design: Activities, Artefacts and Environments, pp. 385–388. ACM (2006)

The Historical Development of Library Policy in the State of Oregon: Discussions on Library Management by Special Districts

Issei Suzuki[1](✉) and Masanori Koizumi[2]

[1] Graduate School of Library, Information and Media Studies, University of Tsukuba, 1-2 Kasuga, Tsukuba, Ibaraki 305-8550, Japan
s1830503@s.tsukuba.ac.jp

[2] Faculty of Library, Information and Media Science, University of Tsukuba, 1-2 Kasuga, Tsukuba, Ibaraki 305-8550, Japan
koizumi@slis.tsukuba.ac.jp

Abstract. Current literature has discussed various arguments regarding library districts. However, these have not detailed the basis of approval for each state's special public library management districts. Therefore, this study clarifies the historical discussions on the formation of library districts that were permitted in 1981 by Oregon state law. Regarding the research method, we conducted an extensive literature search centered on primary materials. From the late 1950s to the early 1960s, there were concerns that if the state allowed the formation of library districts, it would hinder the formation of larger and more efficient library systems and fragment public services. During this period, citizens' groups demanded that special districts manage libraries under state law, and the state government was concerned that intergovernmental relationships would become more complicated. As a solution to this problem, the state government enabled county service districts to provide library services in 1973. However, this management model was not widespread in Oregon; in spite of the county service district's tax authority, it did not allow residents to elect board members. The formation of library districts was later approved in 1981 to address financial difficulties. However, the 1981 law was problematic, and the advantages and disadvantages of library districts in the local administration had not been carefully discussed due to financial deterioration in governmental sectors.

Keywords: Public libraries · Financial difficulties · Special districts · Policy analysis · Oregon

1 Introduction

1.1 Financial Deterioration of Government Sectors and Diversification of Public Library Management

Various public sectors, such as public libraries, face global financial issues. Further, public libraries are experiencing challenges in responding to the demands for services

K. Toeppe et al. (Eds.): iConference 2021, LNCS 12646, pp. 458–465, 2021.
https://doi.org/10.1007/978-3-030-71305-8_38

required by society and communities. In such a situation, library districts in special districts, which tend to increase as a legal basis for public libraries, are attracting attention in the United States. A special district is a local government that provides services when a general-purpose government, such as a county or city, is unable to provide some public services and is approved by a referendum. These governments have substantial power and independence from other governments over fiscal policy. They are also managed by independent boards of residents. In 1942, there were approximately 8,300 special districts; this number increased by 2017 to 38,500. Additionally, the number of library districts—for managing libraries within special districts—has increased since the latter half of the 20th century.

Public libraries were divided into three main legal bases in 2017 [1]. The first is a general-purpose government, which encompasses 64.0% of public libraries; the second is a library district, which accounts for 15.4%; and the third is a non-profit organization, which accounts for 13.9%. General-purpose governments are local, such as those found in counties and cities, and provide library services as a part of their administrative services. Library districts are a special district for the single purpose of library management, with tax levies and bond authority, and are formed through a referendum. Non-profit organizations that operate public libraries are exempt from federal corporate income tax.

The proportion of library districts formed as a legal basis for managing public libraries in the United States has increased since 1990, and the proportion of library districts increased by 8.9%, from 6.5% in 1992 to 15.4% in 2017. Consequently, library districts surpassed non-profit organizations in 2010 to become the second-largest in position after general-purpose governments.

1.2 Current Arguments on Public Library Management by Special Districts

How have policymakers evaluated library districts? Generally, such evaluations can be divided by authority, or those in charge of library management and general administration. For example, representatives of the chief officers of state library agencies in New York and Oregon praised the formation of library districts for their ability to obtain stable funding based on taxes levied [2]. The Colorado State Library also noted that library districts' tax levies could provide more stable financial resources [3]. These perspectives reveal the advantages of library districts' management as a way to obtain stable funds, while public library finances have tightened due to government sectors' financial deterioration.

Next, we illustrate evaluations from the general administration's perspective. Mullin [4] discussed the fragmentation of local governments in the United States through an analysis of water districts, a form of special district, as well as library districts. She observed special districts regarding the increase: "Governance of American communities is becoming more specialized."

Regarding this "fragmentation of local administration," Governor Andrew Cuomo of New York stated the following in his 2014 policy address [5]:

> "So why are our property taxes so high? Because we have too many local governments and we have had them for too long. 10,500 local governments, these are

towns, villages, fire district, water district, library, sewage district, one district just to count the other districts in case you missed a district. We have a proliferation of government that is exceedingly expensive and costly."

In this way, Governor Cuomo pointed out that residents' property taxes have increased because of the large number of local governments. In other words, it is problematic that the taxable amount has collectively increased because each independent local government conducts its own taxation processes. As Governor Cuomo's remarks indicate, the "fragmentation of local administration" has a negative effect of increasing the burden on residents.

Additionally, and based on this theory, it is possible to manage with a high degree of autonomy for such special districts as library districts, as these purportedly and simultaneously increase independence and autonomy.

While various arguments regarding library districts have been discussed, literature has failed to provide a detailed analysis of the basis for approval of special districts for public library management in each state. Therefore, this study clarifies the historical discussions on the formation of library districts that were permitted in 1981 by Oregon state law. By analyzing this policy-making process, this study can present more arguments that are overlooked in current discussions. This will provide an essential perspective for future research examining the actual conditions of library district management through case analyses.

Suzuki and Koizumi's [6] research establishes a theoretical foundation for this work, as it corresponds to a part of the state government's legislation as analyzed in our research project (Fig. 1). Based on these research results, we plan to clarify the actual conditions in library districts.

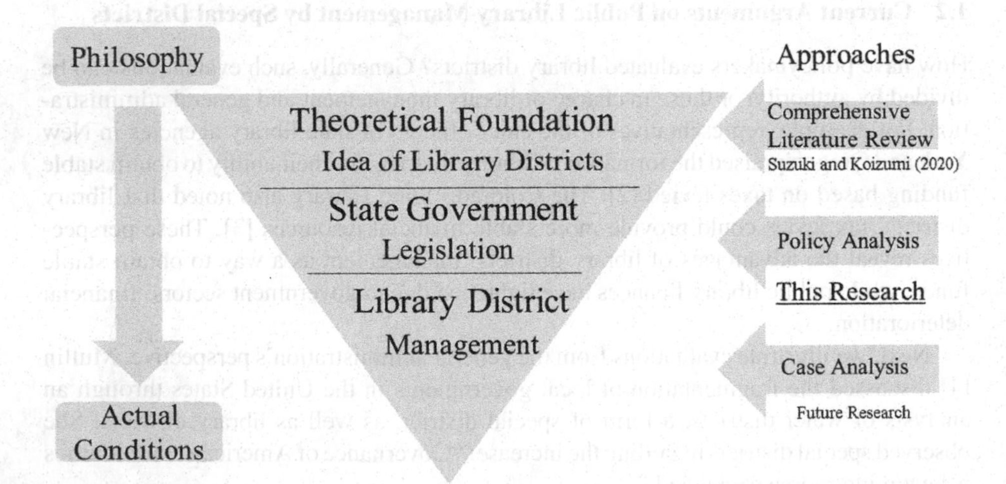

Fig. 1. The research project's structure

2 Method

We analyze the historical development of library policy in the state of Oregon. This study targeted Oregon given the following. The 2005 *Handbook for Trustees of Oregon Public Libraries* published by the Oregon State Library describes the library trustee's roles and responsibilities; library districts are evaluated according to the legal basis of public libraries in the state [7]:

> "In the early years of the 21st century, most Oregon public libraries are either city, county, special district, or county service districts. And, while most of those are city or county libraries today, funding considerations suggest that many are likely to find conversion to special district or county service district status an attractive option in the coming years."

This demonstrates that the approval of library management through special districts in 1981 influenced the current state library policy. In the 24 years from 1992 to 2017, the legal basis changed. Further, the proportion of general-purpose governments decreased by 20% and that of library districts increased by 20.3% [1]. The library state agency of Oregon promoted the formation of library districts [2], even at the state policy level, and the proportion of library districts has increased as a legal basis for public libraries.

Regarding our research method, we conducted an extensive literature review centered on primary materials as collected from the State Library of Oregon Digital Collections. These materials include internal materials from the State Library, a guidebook on library districts published by the State Library, and a newsletter from the Oregon State Library Association.

However, a review of the materials' references, citations, and contents revealed that only some of the materials were digitized at the State Library. Therefore, the first author visited the State Library of Oregon and collected materials while referring to the card catalog. Additionally, he collected newspaper articles from the Multnomah County Service Library District in Oregon containing discussions on the formation of library districts using a full-text database by *The Oregonian*, a regional newspaper in Oregon. The author also surveyed the *PNLA Quarterly*—a journal published by the Pacific Northwest Library Association that regularly reports on the latest status of library policy in Oregon—in the Keio University Libraries (慶應義塾大学メディアセンター).

3 Results and Discussion

In the late 1950s and early 1960s, in addition to state and local governments, the federal government became involved in library policy in the United States under the Library Services Act of 1956 and the Library Services and Construction Act of 1964. Regarding public libraries at that time, President John F. Kennedy referred to the term "library" more than six times in his inaugural speech [8], and stated the following in a special message on education in January 1963 [9]:

> "The public library is also an important resource for continuing education. But 18 million people in this nation still have no access to any local public library service and over 110 million more have only inadequate service."

As indicated by this quote, while public libraries' significance in the United States was recognized in the 1960s, people could not receive their services. Under such circumstances, bills were submitted by citizen groups in 1959 and 1963 aiming to enact state laws to enable library management by special districts in Oregon [10]. These bills would allow voters to form library districts at their discretion.

The 1959 bill was submitted by a citizen group in the Rogue River area, which hoped to become independent of the Jackson County library system. The 1963 bill was submitted by the St. Helens Community Achievement Council, as the county lacked a public library. This council hoped to expand the public library's service and taxation areas without being tied to the existing city's geographical territory. It was estimated that forming a library district would double the service population and quadruple the taxable amount.

In this way, discussions on the formation of library districts were held with the intention of not only gaining independence from a general-purpose government, but also expanding the area in which library services could be provided.

The Oregon State Library and Oregon Library Association have recommended that county governments manage public libraries to provide residents with high-quality library services [10]. However, county governments in the 1950s formed a library system in only one such county, while 18 counties still lacked library systems. Therefore, citizens sought a method to promote the development of public libraries, and library management by special districts became an option. The State Library of Oregon's *Special Taxing Districts for Public Libraries: An Evaluation* as published in 1964 notes five features of special districts, as Table 1 illustrates.

Table 1. Five features of special districts [10]

	Description
1.	It is located in an unincorporated area (i.e., outside of any existing city boundaries)
2.	It is smaller than a county and conforms to local population and geographic patterns
3.	It levies its own taxes and may incur indebtedness
4.	It is independent of city or county government
5.	It is in business solely to operate a specific public service

However, these bills were not passed by the state legislature. During this time, an interim legislative committee expressed concerns about the increase in special districts [10]:

"The increased utilization of the special service district has led to fragmentation of governmental jurisdiction and authority of local government in Oregon resulting in a (1) lack of coordination of governmental activity, (2) duplication of activity, (3) overlapping of jurisdictions, and (4) inefficient operations resulting in wasteful expenditures of tax dollars."

The state government was concerned that intergovernmental relationships would be complicated by the increase in special districts rather than library management by special districts. Additionally, the interim legislative committee expressed the following views [10]:

> "Local government structure in Oregon presents a bewildering pattern as a result of the fragmentation, the overlapping, and the duplication. The result has often been that local government in our urban areas operates without democratic control."

The increase in special districts demonstrated that local governments providing public services operated outside the control of democracy. Further, librarians from California, Idaho, and Illinois—where library districts have already been formed—stated their views on library districts, as evident from the following comment from an Illinois librarian [10]:

"Library districts in Illinois have been formed on a smaller scale than a county but this is not the intent of the act. Such districts would seem to inhibit the formation of larger systems because once formed by referendum there is a natural reluctance to change. . . Some districts have been formed that are too small to be adequately financed. . ."

Simultaneously, the Advisory Committee on Intergovernmental Relations [11] also concerns the succession of special districts being formed across the United States. A major national-level issue has involved determining whether increasing numbers of special districts are properly managed under citizens' control.

Regarding the increase of special districts and the complexity of intergovernmental relations, California's 1951 Community Service District Act allowed the formation of a limited multi-purpose government, including county service districts. The Oregon State Library indicated that such county service districts included library management as a means of providing library services to residents [10].

Therefore, the Oregon state government considered amending its 1963 County Service District Law, which was promulgated to form a multipurpose government in Oregon, to enable library management. This is because County Service Districts was expected to be efficiently managed in a unified manner under the county government's control, and to respond well to citizens' needs.

However, although County Service Districts were allowed to manage public libraries in 1973, it did not become popular in Oregon [12]. County service districts allowed for governance by the county government's board of directors, and these districts were also entitled to the county government's administrative services. Additionally, library boards differ from special districts, in that county governments appoint the former.

Such county service districts were described as "less than a model of clarity" [12], and critics were concerned with the divided authority between financing and the responsibility for governance. From the late 1970s to the early 1980s, a national-level economic crisis and tightening of public spending in the United States created financial difficulties for the entire public library sector, while discussions on library districts began in earnest. Specifically, an economic recession occurred in the early 1980s, in that a declining demand for housing and timber products devastated its core timber industry, with unemployment rates reaching 12.5% in Oregon [13]. Meanwhile, Governor Victor Atiyeh of Oregon, who took office in January 1979, demanded budget cuts of 30% from all state agencies, in response to a 200 million dollar plus revenue deficit [14]. In addition to the state government, the county government's finances were also deteriorating.

The Oregonian reported on the financial challenges surrounding public libraries on September 8, 1981, as follows [15]:

> "County and city governments which dole out money to libraries have undergone budget squeezes in the past several years. Forced to vie with essential services such as fire and police, library budgets in some cases have taken more than their share of cutbacks."

For example, the Jackson County Library in southwestern Oregon had significantly decreased its personnel and book budgets [16]. Additionally, the Washington and Multnomah County libraries in the northwestern part of the state held referendums to introduce a library purpose tax amid financial deterioration.

The idea has gained support that public libraries as special districts are financially independent to solve city and county governments' financial problems. Public library management by library districts gained support from advocates of the Washington, Jackson, and Wasco County libraries, in addition to the Library Friends Association at the Multnomah County public library in Portland [15]. Moreover, the Deschutes County librarian described the severe financial cuts of the 1980s as "dreadful," and observed the following regarding such fiscal deterioration's impacts on the library [15]:

> "Our full-time employees all went on part-time status, part-time employees were dismissed, hours of the Bend library were cut more than 50% and we eventually even lost our book-buying budget."

Under such circumstances, the state bill H.B. 2823, which permitted the formation of library districts in 1981, passed the House [17]. Additionally, the Oregon State Library Association sponsored the development of legislation-governing library districts [18], which permitted the formation of library districts [19].

4 Conclusion

From the late 1950s to the early 1960s, there were concerns that allowing the formation of library districts would hinder the formation of larger, more efficient library systems, and cause fragmentation among public services. During this period, citizens' groups demanded that special districts manage libraries under state law, and the state government was concerned that intergovernmental relationships would become more complicated. As a solution to this problem, the state government enabled county service districts to provide library services in 1973. However, the management model was not widespread in Oregon because the county service district, despite its tax authority, did not allow residents to elect board members. Later, the formation of library districts was approved in 1981 to solve financial problems, although this law was problematic, in that government-sector finances' deterioration prevented a careful discussion of the advantages and disadvantages of library districts in local administration.

5 Future Research

This study provided the following three arguments from an analysis of the policy-making process regarding library districts in Oregon. The first issue is whether the library districts' geographical boundaries were properly defined. The second is whether the formation of library districts complicates intergovernmental relations. The third is whether library districts can conduct efficient management. We plan to clarify these arguments through a detailed case analysis in the future.

References

1. Institute of museum and library services: public libraries survey fiscal year 2017: supplementary tables. Institute of Museum and Library Services, Washington, DC (2019)
2. Owens, P.L., Sieminski, M.L.: Local and State Sources of Funding for Public Libraries: The National Picture. RPA, Williamsport (2007)
3. Lietzau, Z.: Colorado library districts thrive while other library types face big cuts. Fast Facts. ED3/110.10/No. 193 (2003)
4. Mullin, M.: Governing the Tap: Special District Governance and The New Local Politics of Water. MIT Press, Cambridge (2009)
5. New York State Government: Transcript: Governor Cuomo's 2014 State of the State address. https://www.governor.ny.gov/news/transcript-governor-cuomos-2014-state-state-address. Accessed 15 Oct 2020
6. Suzuki, I., Koizumi, M.: Theoretical bases of public library management by special-purpose governments in the United States (in Japanese). In: Proceeding of Japan Society of Library and Information Science 68th Conference, pp. 1–4 (2020)
7. Lidman, R.: Handbook for Trustees of Oregon Public Libraries. Oregon State Library, Salem (2005)
8. Ladenson, A.: American Library Laws. American Library Association, Chicago (1982)
9. Fry, J.W.: LSA and LSCA, 1956–1973: a legislative history. Libr. Trends **24**(1), 7–26 (1975)
10. Loeber, T.S.: Special Taxing Districts for Public Libraries: An Evaluation. Oregon State Library, Salem (1964)
11. Advisory Committee on Intergovernmental Relations: The Problem of Special Districts in American Government. ACIR, Washington, DC (1964)
12. Ginnane, M.: Library Districts in Oregon: A Planning Sourcebook. Oregon State Library, Salem (1991)
13. Guggemos, E.: Atiyeh! The Governor Victor Atiyeh Collection. Pacific University Libraries, Forest Grove (2013)
14. Pacific Northwest Library Association: Oregon. P.N.L.A. Quart. **45**(1), 19 (1980)
15. Brennan, T.: Libraries also may find financial salvation in assessment district idea. The Oregonian, p. 23, 8 September 1981
16. Pacific Northwest Library Association: Oregon. P.N.L.A. Quart. **45**(4), 40–41 (1981)
17. Oregon State Library: Watermark 8. Oregon State Library, Salem (1981)
18. Pacific Northwest Library Association: Oregon-Library Development. P.N.L.A. Quart. **47**(3), 9 (1981)
19. Oregon State Library: Watermark 5. Oregon State Library, Salem (1981)

Importance of Digital Library Design Guidelines to Support Blind and Visually Impaired Users: Perceptions of Key Stakeholders

Iris Xie[1(✉)], Rakesh Babu[2], Shengang Wang[1], Tae Hee Lee[1], and Hyun Seung Lee[1]

[1] School of Information Studies, University of Wisconsin-Milwaukee, Milwaukee, WI, USA
{hiris,shengang,taehee,lee649}@uwm.edu
[2] Envision, Wichita, KS, USA
Rakesh.Babu@envisionus.com

Abstract. This study was conducted to understand how key stakeholders perceive the importance of design guidelines that specifically target the needs of blind and visually impaired (BVI) digital library (DL) users. An in-depth survey questionnaire was distributed among 150 participants representing three stakeholder groups: BVI users, DL developers, and scholars/experts. Participants were informed about different help-seeking situations that BVI users encountered when interacting with a DL non-visually using screen readers. They were then presented with a set of design guidelines to address each situation. Finally, they were asked to rate the importance of each set of guidelines in remediating each corresponding situation. Both quantitative analysis and qualitative analysis were applied to analyze the data. The results show that all key stakeholders agree it is critical to develop DL design guidelines to support BVI users. On the one hand, the three groups share some similarities in rating the importance of guidelines for these help-seeking situations; on the other hand, the disparities mainly lie in the fact that DL developers and the scholars/experts focused more on the guidelines addressing the accessibility-related situations, while BVI users emphasized that DL design guidelines need to take into consideration both accessibility and usability-related situations.

Keywords: Digital library design guidelines · Assessment · Blind and visually impaired users

1 Introduction and Literature Review

Blind and visually impaired (BVI) users are more likely to encounter accessibility and usability problems than their sighted counterparts mainly because of the sight-centered design of information retrieval (IR) systems [1, 2]. In this paper, BVI users are defined as those relying on screen readers to interact with IR systems. Accessibility and usability problems contribute to help-seeking situations for BVI users in their interactions with IR systems. A help-seeking situation is characterized by a problem that arises during users' interaction with IR systems, leading them to seek assistance in order to fulfill their tasks.

K. Toeppe et al. (Eds.): iConference 2021, LNCS 12646, pp. 466–474, 2021.
https://doi.org/10.1007/978-3-030-71305-8_39

Based on Power, Freire, Petrie, and Swallow's [3] user study, the help-seeking situations related to accessibility problems can be summarized as follows: difficulty accessing content in expected locations, difficulty loading pages fast, difficulty accessing alternative document formats, difficulty accessing complex information architectures, broken links, functionality not working as expected, inconsistent organization of content, etc. Similarly, Rømen and Svanæs [4] investigated the accessibility of 47 websites and found that frequent accessibility-related situations for BVI users were caused by unidentifiable and redundant links. As for associated factors, Borodin, Bigham, Dausch, and Ramakrishnan [5] pointed out that complex web pages and dynamic and automatically-refreshing content contributed to BVI users' help-seeking situations. Help-seeking situations also link to usability problems. According to Lazar, Al-len, Kleinman, and Malarkey [6], screen reader users may encounter situations affected by the following usability problems: confusing screen reader feedback, conflict between the screen reader and the application, poorly designed forms, and disorientation due to misleading links.

In recent years, there has been growing attention to BVI users' situations in DL environments. Xie and her colleagues [2, 7] identified 17 types of problematic situations that BVI users faced in nine categories: difficulty accessing information; difficulty evaluating information; difficulty with help; difficulty locating information or features; difficulty refining collections or results; difficulty identifying current status or paths; confusion about multiple programs or structures; avoidance of format, approach, or input fields; and difficulty constructing search statements.

Continuous efforts have been made to ensure universal accessibility and usability. The Web Accessibility Initiative (WAI) of the World Wide Web Consortium (W3C) has developed accessibility guidelines to support equal access. The revised version of WCAG 2.0 named WCAG 2.1 was released in 2018 [8]. However, these guidelines were not created directly based on the help-seeking situations that users have when interacting with systems.

Research has been conducted to investigate the validity of the design guidelines for BVI users. Some studies have assessed design guidelines by analyzing their relevance in relation to real-life situations [9, 10]. This line of research examines how closely design guidelines correspond to the situations relate to accessibility and usability issues that users face in using websites and applications. Based on user testing, Clegg-Vinell, Bailey, and Gkatzidou [10] evaluated accessibility guidelines WCAG 2.0 and the W3 Mobile Web Best Practices (MWBP) 1.0., and they discovered that some of the accessibility and usability-related situations considered important to participants are not addressed in the guidelines. Their research showcased the weaknesses of these guidelines, such as lacking details in providing information about content scalability on mobile platforms. In their preliminary result, they noted that the guidelines provided: (1) inadequate detail regarding the purpose of icons or text; (2) small sized icons or text; and (3) undetectable elements when using VoiceOver. Along the same line, Calvo, Seyeda-rabi, and Savva [9] highlighted several problems WCAG has not addressed: "Hide information incorrectly," "Do not use common design patterns," "Wide gaps between related information," "Use of custom components," "Buttons and text size are small," "Colour contrast ratio between icons and background is not enough," and "Important information is not shown at the top." Usefulness of guidelines is another research area. Rømen and Svanæs [4] designed a study to identify whether WCAG 1.0 and WCAG 2.0 consider users' situations caused by accessibility problems. The participants were asked to think aloud while performing assigned tasks, and they were interviewed after completing the tasks. The results showed that both WCAG 1.0 and WCAG 2.0 had poor performance in identifying accessibility problems in websites.

Studies have also reviewed design guidelines and provided suggestions for the future development of guidelines [8, 11–13]. Termens, Ribera, Porras, Boldú, Sulé, and Paris [12] identified the difference between WCAG 1.0 and WCAG 2.0 and illustrated how to migrate from WCAG 1.0 to WCAG 2.0 to adjust to changes. Cooper [14] stressed the importance of user need documentation in relation to the content, authoring tools, and user agents, all of which are interconnected with each other. He stated that comprehensive user need documentation is critical for revising design guidelines to adjust to technological changes. A recent publication by Spina [8] addressed the additional aspect of WCAG 2.1 and its potential to apply to library web accessibility.

While previous research has focused on the assessment of guidelines based on feedback from either users or experts, none of the existing studies has evaluated guidelines from multiple stakeholders' perspectives in DL environments. The authors of this project first created DL design guidelines for accessibility and usability based on the types of help-seeking situations BVI users encounter in their interactions with DLs. This paper reports the initial results of the assessment of the importance of the created DL guidelines[1] from perspectives of BVI users, scholars/experts, and DL developers. The research question and associated hypothesis are:

What are the similarities and differences in perceptions of the importance of guidelines for different types of help-seeking situations among BVI users, scholars/experts, and DL developers?

H0 (1-37): There is no significant difference in rating the levels of perceived importance of guidelines for different types of help-seeking situations by BVI users, scholars/experts, and DL developers.

2 Methodology

An in-depth survey was administered to 150 participants representing three key groups of stakeholders to provide feedback for the draft of guidelines created by the research team based on user studies. As there are 37 situations with associated guidelines, it is a challenge for each participant to assess all the guidelines. Therefore, 25 participants in each group assessed one half of the guidelines. The BVI users were mainly recruited through the National Federation of the Blind; the scholars/experts were primarily solicited based on their publications on the topic and accessibility and usability experts through related listservs; the DL developers were recruited from academic libraries across the United States. Table 1 presents demographic data of the 150 participants. In the user group, 15 participants self-identified as having intermediate information searching skills; 27 and 8 thought they had advanced and expert skills, respectively. The purpose of the survey was to solicit qualitative and quantitative feedback on the guidelines created by the research

[1] https://sites.uwm.edu/guidelines/.

team based on user studies. Each participant was instructed to review the current DL guidelines and offer feedback. Quantitatively, participants were instructed to fill out a survey consisting of four questions for each guideline associated with their related situation. The questions used a 1-to-7 Likert scale (1 = not at all, 7 = extremely) and were designed to assess the importance, clarity, relevance, and usefulness of the guidelines that address BVI users' help-seeking situations. Qualitatively, participants were asked to specify the reasons for their ratings. In addition, suggestions to enhance the guidelines were collected. Since importance serves as a key variable, this paper focuses only on the analysis of the importance ratings and associated reasons. Descriptive analysis was performed to identify the guidelines rated the most important and the least important related to each type of help-seeking situation. In addition, a stock chart graphic was created to show the highest ranges and lowest ranges of ratings among the three groups. Most importantly, ANOVA tests were performed to examine the similarities and differences among the ratings of the three groups of stakeholders using the 7-point scale. In general, ANOVA is used to determine if any significant mean difference exists among the multiple groups. By applying ANOVA tests, the research team was able to identify the significant results related to the study hypothesis.

Table 1. Demographic data of participants

Demographic characteristics		User (N = 50)	Scholar/Expert (N = 50)	Developer (N = 50)	Total (N = 150)
Age	18–29	16%	14%	2%	11%
	30–39	26%	38%	44%	36%
	40–49	30%	22%	42%	31%
	50–59	14%	16%	10%	13%
	60–69	10%	10%	2%	7%
	70+	4%	0%	0%	1%
Gender	Female	66%	44%	60%	57%
	Male	32%	56%	38%	42%
	Other	2%	0%	2%	1%
Highest degree earned	Some college Associate's	32%	14%	2%	16%
	Bachelor's	30%	34%	20%	28%
	Master's	28%	38%	70%	45%
	Doctorate	10%	14%	8%	11%

3 Results and Discussion

The results show that BVI users, scholars/experts, and DL developers shared similarities and differences in their perceived ratings of the importance of DL guidelines. In this section, each situation is numbered, such as S1, S2, etc. Figure 1 presents the three highest and lowest importance ratings of the guidelines for specific situations as perceived by the three stockholder groups. Both the scholar/expert group and the DL developer group considered the guidelines for *S1 Difficulty accessing alternative text for an image* (scholars/experts: M = 6.72, DL developers: M = 6.76) as the most important while the BVI user group ranked these guidelines at the sixth place (M = 6.72). For BVI users, they cared the most about the guidelines for *S28 Difficulty locating a play or stop button* (M = 6.88), and the scholar/expert group ranked the associated guidelines at the third place (M = 6.68) and DL developers at the fourth place (M = 6.56). The guidelines for *S6 Difficulty recognizing page loading status* were considered the least important by all the three groups (BVI users: M = 5.44, scholars/experts: M = 5.32, DL developers: M = 5.32).

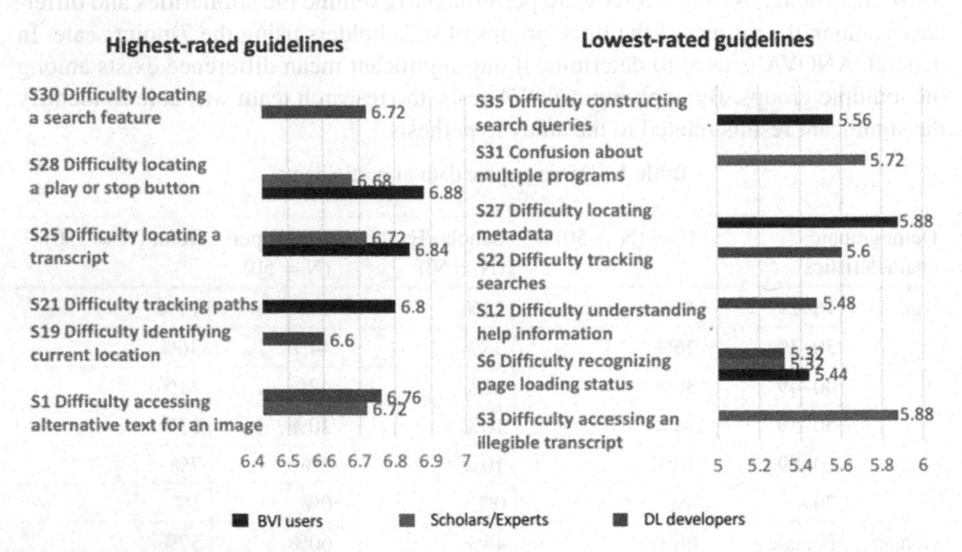

Fig. 1. Highest-rated and lowest-rated guidelines in terms of importance perceived by each group of the stakeholders

Following the same color key of Fig. 1, Fig. 2 highlights various ranges of perceived importance of guidelines for each situation by the three groups of participants. In the figure, "D" represents "difficulty" to avoid repetition. The **similarities of ratings** among the three groups are highlighted by low ranges of perceived importance (<0.10) for guidelines of four situations, including *S1 Difficulty accessing alternative text for an image* (Range: 0.04, BVI users: M = 6.72, scholars/experts: M = 6.72, DL developers: M = 6.76), *S16 Difficulty assessing relevance of search results, subjects, or collections* (Range = 0.08, BVI users: M = 6.20, scholars/experts: M = 6.16, DL developers: M = 6.12), *S18 Difficulty assessing format of an item* (Range = 0.08, BVI users: M = 6.12, scholars/experts: M = 6.16, DL developers: M = 6.20), and *S30 Difficulty locating a search feature* (Range: 0.08, BVI users: M = 6.76, scholars/experts: M = 6.68, DL developers: M = 6.72). In contrast, the **differences of ratings** among the three groups are shown by high ranges of perceived importance (>0.75) for guidelines of three situations, including *S32 Confusion about digital library structure* (Range: 0.8, BVI users: M = 6.72, scholars/experts: M = 6.16, DL developers: M = 5.92), *S34 Difficulty understanding browsing structure* (Range: 0.8, BVI users: M = 6.84, scholars/experts: M = 6.16, DL developers: M = 6.04), and *S21 Difficulty tracking paths* (Range: 0.76, BVI users: M = 6.80, scholars/experts: M = 6.04, DL developers: M = 6.04). While the above high ranges of perceived importance show the differences between BVI users and the other two groups, it is also worth noting that there are high ranges (0.6–0.75) of perceived importance for two situations between scholars/experts and DL developers, including *S22 Difficulty tracking searches* (Range: 0.72, scholars/experts: M = 6.20, DL developers: M = 5.48) and *S25 Difficulty locating a transcript* (Range: 0.60, scholars/experts: M = 6.72, DL developers: M = 6.12). The difference between the two groups lies in their own background and work responsibilities. For example, regarding the guidelines for *S22 Difficulty tracking searches*, one scholar/expert confirmed the importance of relevant guidelines, commenting that "If a user is unable to track a search it will be difficult for them to find what they are looking for" (S7); however, DL developers tend to comparatively underestimate the importance, and one participant stated, "Search is not necessarily the primary means of access in every digital collection. Not every digital collection is expansive enough where complex searching will be regularly used, or a search history feature would be worth the expense" (D9).

Moreover, the ANOVA test results show that there are significant differences in the perceived importance of the DL design guidelines for the following two situations among the three groups: *S21 Difficulty tracking paths* and *S34 Difficulty understanding browse structure*. The differences are mainly between the BVI user group and the other two groups. The guidelines for *S21 Difficulty tracking paths* were considered more important by the BVI user group (M = 6.80) than the other two groups (scholars/experts: M = 6.04, DL developers: M = 6.04, $F(2, 72) = 3.613$, $p < .05$). Below is one typical explanation from a BVI user participant: "Tracking paths is probably one of the most difficult things to do as a BVI user. Being able to tell where you came from and where you have been is really important" (U11). Also, the BVI user group perceived the guidelines for *S34 Difficulty understanding browsing structure* as more important (M = 6.84) than the other groups (scholars/experts: M = 6.16, DL developers: M = 6.04, $F(2, 72) = 6.017$, $p < .05$). Another BVI user participant shared her thoughts on the importance of creating guidelines related to the situation, "I consider this situation where a user is confused about how to browse a DL, or the best approach for browsing digital collections, a substantial challenge to BVI users" (U19). The results indicate that, while DL developers and scholars/experts cared more about creating guidelines to address accessibility-related situations (e.g., *S1*), BVI users were concerned not only about developing guidelines on accessibility but also usability issues (e.g., *S21*, *S32*, and *S34*).

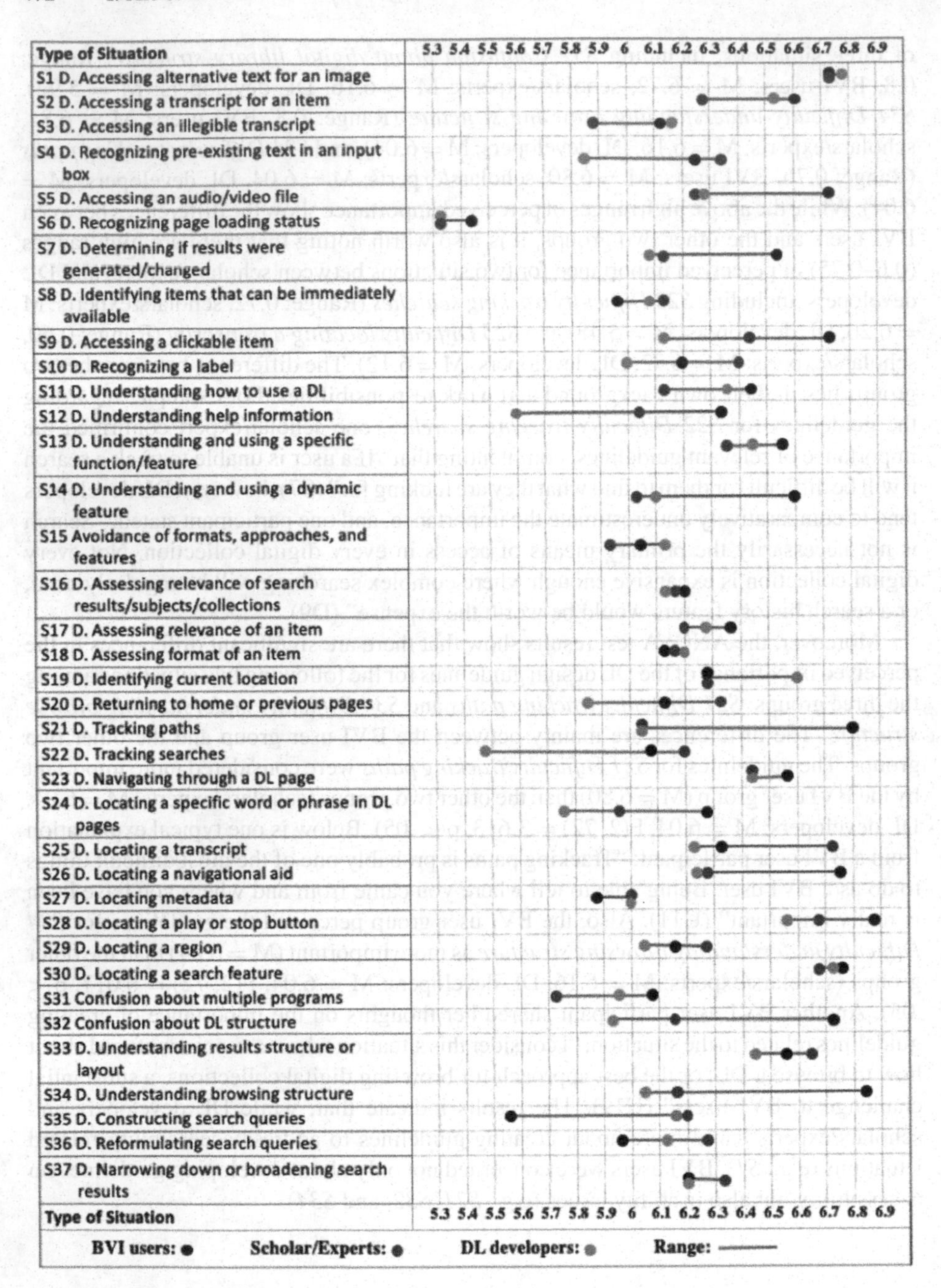

Fig. 2. Importance rating of guidelines for each situation by three groups

4 Conclusion

Results of this study indicate that all key stakeholders recognize that creating DL design guidelines to support BVI users is critical. Comparative analyses across groups show varying degrees of perceived importance of these guidelines for different types of stakeholders. At one end of the spectrum are four situations—*difficulty accessing alternative text for an image, difficulty assessing relevance of search results/subjects/collections, difficulty assessing format of an item, and difficulty locating search features*—whose guidelines received nearly unanimous ratings by all stakeholder groups. At the other end of the spectrum are three situations—*confusion about DL structure, difficulty understanding browsing structure, and difficulty tracking path*—whose guidelines received quite diverse ratings from different stakeholders. The variations in ratings for these guidelines become significant between BVI users and others (DL developers and scholars/experts). The difference in ratings for guidelines relevant to all other situations fell in the middle of the spectrum. This variance in ratings points to a difference in perceptions based on first-hand knowledge of the non-visual DL experience versus perceptions based on second-hand knowledge. It also highlights the fact that, while DL developers and scholars/experts believe designing for accessibility is paramount, BVI users would desire both accessibility and usability in DL design. Thus, there is a need to re-evaluate existing accessibility guidelines such as the WCAG for usability issues using screen readers. Accessibility and usability are the two interrelated concepts, and we need to take both into consideration when developing design guidelines. An important takeaway from this research is the need for more user-centered design guidelines where the emphasis is placed on non-visual enablement of an IR task using an assistive technology such as a screen reader. By keeping the DL design guidelines focused on the help-seeking situations faced by BVI users, it was possible to obtain a comparative view of the value of these design guidelines for creating an accessible and usable DL. Having established the need for creating design guidelines targeting the BVI users of DLs, the research team is in the process of developing and validating guidelines on accessibility and usability of DLs to support this group of users.

Acknowledgements. The authors thank IMLS Leadership Grants for Libraries for funding for this project.

References

1. Leuthold, S., Bargas-Avila, J.A., Opwis, K.: Beyond Web content accessibility guidelines: design of enhanced text user interfaces for blind internet users. Int. J. Hum. Comput. Stud. **66**(4), 257–270 (2008)
2. Xie, I., Babu, R., Castillo, M.D., Han, H.: Identification of factors associated with blind users' help-seeking situations in interacting with digital libraries. J. Assoc. Inf. Sci. Technol. **69**(4), 514–527 (2018)
3. Power, C., Freire, A., Petrie, H., Swallow, D.: Guidelines are only half of the story: accessibility problems encountered by blind users on the web. In: CHI'12 Proceedings of the SIGCHI Conference on Human Factors in Computing Systems, pp. 433–442. ACM, New York (2012)

4. Rømen, D., Svanæs, D.: Validating WCAG versions 1.0 and 2.0 through usability testing with disabled users. Univ. Access Inf. Soc. **11**(4), 375–385(2012)
5. Borodin, Y., Bigham, J.P., Dausch, G., Ramakrishnan, I.V.: More than meets the eye: a survey of screen-reader browsing strategies. In: Proceedings of the 2010 International Cross Disciplinary Conference on Web Accessibility, pp. 1–10. ACM, New York (2010)
6. Lazar, J., Allen, A., Kleinman, J., Malarkey, C.: What frustrates screen reader users on the web: a study of 100 blind users. Int. J. Hum.-Comput. Interact. **22**(3), 247–269 (2007)
7. Xie, I., Babu, R., Joo, S., Fuller, P.: Using digital libraries non-visually: understanding the help-seeking situations of blind users. Inf. Res. **20**(2), n2 (2015)
8. Spina, C.: WCAG 2.1 and the current state of web accessibility in libraries. Weave J. Libr. User Exp. **2**(2) (2019). https://doi.org/10.3998/weave.12535642.0002.202
9. Calvo, R., Seyedarabi, F., Savva, A.: Beyond web content accessibility guidelines: expert accessibility reviews. In: Proceedings of the 7th International Conference on Software Development and Technologies for Enhancing Accessibility and Fighting Info-exclusion, pp. 77–84. ACM, New York (2016)
10. Clegg-Vinell, R., Bailey, C., Gkatzidou, V.: Investigating the appropriateness and relevance of mobile web accessibility guidelines. In: Proceedings of the 11th Web for All Conference, pp. 1–4. ACM, New York (2014)
11. Reid, L., Snow-Weaver, A.: WCAG 2.0: A web accessibility standard for the evolving web. In: Proceedings of the 2008 International Cross-disciplinary Conference on Web Accessibility, pp. 109–115. ACM, New York (2008)
12. Termens, M., Ribera, M., Porras, M., Boldú, M., Sulé, A., Paris, P.: Web content accessibility guidelines: From 1.0 to 2.0. In: Proceedings of the 18th International Conference on World Wide Web, pp. 1171–1172. ACM, New York (2009)
13. Schmutz, S., Sonderegger, A., Sauer, J.: Implementing recommendations from web accessibility guidelines: a comparative study of nondisabled users and users with visual impairments. Hum. Factors J. Hum. Factors Ergon. Soc. **59**(6), 956–972 (2017)
14. Cooper, M.: Web accessibility guidelines for the 2020s. In: Proceedings of the 13th Web for all Conference, pp. 1–4. ACM, New York (2016)

Usage of E-books During the COVID-19 Pandemic: A Case Study of Kyushu University Library, Japan

Mei Kodama(✉), Emi Ishita, Yukiko Watanabe, and Yoichi Tomiura

Kyushu University, Fukuoka 819-0395, Japan
kodama.mei.415@s.kyushu-u.ac.jp

Abstract. The COVID-19 pandemic has had an impact on education and research in universities throughout the world. Many academic libraries have been closed, and users have had to use e-books instead of printed books in libraries. As e-books are not yet commonly used in Japan, this paper examines the impact of the pandemic on the use of e-books in academic libraries in Japan. As a case study, the usage data for each e-book platform in Kyushu University in Japan were analyzed, and the overall trends in each period before and during the pandemic were revealed. The access counts of e-books from January to June 2019 and those for 2020 were examined. The number of total access counts in 2020 was higher than that in 2019 on all nine platforms. The platform that saw the highest access count growth rate was JSTOR, with an 846% increase. The usage of e-books in the Maruzen eBook Library, which holds many Japanese textbooks, was examined in detail. E-books on mathematics, medical science, and programming languages were in constant use both before and during the pandemic. E-books in law and economics started to be used during the pandemic. These results indicate that the pandemic has evoked new needs for e-books in certain fields.

Keywords: Usage of e-books · COVID-19 · Academic libraries

1 Introduction

According to a survey conducted in 2015, 94% of public libraries in the United States provide e-books to their users [1]. However, a survey conducted in 2020 showed that only 7.2% of Japanese public libraries provide electronic library services, including e-book lending [2]. It can thus be said that Japanese libraries are lagging behind with regard to the provision of e-books.

Various measures have been taken around the world due to the COVID-19 pandemic. Many universities in Japan closed their campuses and introduced online education. Following a nationwide state of emergency issued on April 16, the start of the semester, which normally begins in April, was postponed.

We focus on the impact of the pandemic on the use of e-books in university libraries. As a case study, we analyzed the use of e-books in Kyushu University, which is a national research university and has 18,566 students and 2,088 faculty members as of May 1,

K. Toeppe et al. (Eds.): iConference 2021, LNCS 12646, pp. 475–483, 2021.
https://doi.org/10.1007/978-3-030-71305-8_40

2020 [3]. Kyushu University postponed the start of the spring quarter from April 8 to May 7 and decided to introduce online education.

Many university libraries were also closed or provided limited services following the lockdown of university campuses. According to a survey conducted by saveMLAK, 74 of the 86 national university libraries were closed as of May 1 [4]. Various types of information and services were provided via library websites. Kyushu University Library was closed from April 11 to the end of May, with limited service resuming after June. While the library was closed, they provided a service to send faculty members, graduate students, and senior undergraduate students books and copies of articles on request. In addition, the library launched a new website, "Library Response to Novel Coronavirus (COVID-19)," to provide information.

Many providers and publishers, including KinoDen, EBSCO, and Oxford University Press (OUP), provided special trials and free access for a limited period. It has been difficult to use materials in libraries in the usual way due to the pandemic. As many journal articles were already provided online and Kyushu University provides remote access services, it would be expected that the impact of library closures on electronic journals would be small. However, we can assume that there has been an impact on the use of books, as when the library is closed, only e-books are available. This could be a turning point in the transition from the use of printed books to e-books. Forecasts say that online education and remote learning will continue after the end of the pandemic. By analyzing the usage of e-books under these circumstances, this paper hopes to contribute to creating a comfortable and useful environment for student learning.

In this paper, we analyze the usage data of each e-book platform and clarify the overall trends in each period before and during the pandemic using usage data from Kyushu University Library. We also examine implicit and explicit needs by analyzing the usage data for e-books on specific platforms.

2 Related Works

Digital content, including e-books, has already been used in higher education institutions in the US. In 2019, *Library Journal* conducted a survey of 199 faculty members in higher education institutions in the US who were engaged in a wide range of research fields [5]. The results showed that 84% of faculty members used some digital resources (e-books, 54%) for education. The result indicated that many faculty members in the US use digital content such as e-books for their courses.

However, the introduction of e-books in Japanese educational institutions has been delayed. According to a survey conducted by the Ministry of Education, Culture, Sports, Science and Technology in 2019, the total cost of library materials in Japanese universities was 70.8 billion yen, of which only 1.5 billion (2.2%) was spent on e-books [6].

The use of e-books for education is not progressing in Japan. For example, we examined how many courses included links to e-books in their 2020 syllabi on the "Syllabus at Kyushu University" website [7]. We obtained 612 results by searching for the keyword "http" and examined each link in the search results to confirm whether e-books or other resources. There were only 35 courses that had links to e-books. The total number of courses on this site in 2020 was 6,507. Few textbooks or reference books were explicitly indicated as being available as e-books in the syllabi.

Three Japanese national university libraries conducted a survey of patron-driven acquisitions using the Maruzen eBook Library (MeL) e-book platform in 2015 [8]. The survey aimed to provide user-oriented Japanese book content, as there has been a delay in the digitization of books in Japan. At these universities, the ratio of Japanese e-books was very low. The results for Ochanomizu University showed that e-books on science, especially mathematics and information science, were mainly used.

3 Analysis of the E-book Usage Data of Each Publisher's Platform

We submitted an application to Kyushu University Library to obtain usage data for two terms (before and during the pandemic). On approval of our application, we obtained usage data for nine e-book platforms in 2019 and 2020. In addition, data on library circulation from the same period and a list of textbooks referred to in the 2019/2020 syllabi were also obtained.

3.1 Overview

Table 1 shows the number of e-books accessed and the number of total accesses in 2019/2020 for each platform and other kinds of information. The OUP access statistics are for the period from January to May 2019/2020, those of MeL are from January to July 2019/2020, and others are from January to June in 2019/2020. The column "Language" indicates the language(s) of the e-books held by each platform. "Trials" means whether each publisher provided special trials to Kyushu University during the pandemic. "Data Format" is the format that has been used in each platform to count the number of e-books and accesses. The way of counting the number of e-books differs depending on the data format, but the number of e-books in this table shows the number of e-books that have been used, even once, during these periods [9]. "Number of total accesses" indicates the total number of times e-books were accessed. Japan Knowledge Lib and MeL count access to each e-book. However, on platforms using COUNTER5 and ProQuest Central, access to sections (e.g., chapters and encyclopedia entries) in each e-book was counted [10, 11]. The way of counting the number of accesses is different, but we apply the same method for each platform in 2019 and 2020. "YoY growth rate" is calculated by dividing "Number of total accesses in 2020" by "Number of total accesses in 2019".

The number of total accesses in 2020 was higher than that in 2019 on all platforms. The growth rate of JSTOR (846%) was the highest among all platforms. The number of accesses to foreign e-books increased considerably. In platforms holding Japanese e-books, the growth rate of MeL (268%) was the highest. In general, Japanese books are used for textbooks in many courses. We are also interested in the impact of online education on the usage of e-books. We analyze access data for MeL in detail in the next section.

3.2 Analysis of MeL Access Data

The Top 10 Most Accessed E-books in 2019. Table 2 shows the top 10 most accessed e-books in 2019. We show only the "Simplified title (English)," which is a simplified

Table 1. Number of e-books accessed and total number of accesses on each platform

Platform	Language	Trials	Number of e-books accessed		Number of total accesses		YoY growth rate	Data format
			2019	2020	2019	2020		
Elsevier Science Direct	Foreign	✓	300	408	1,524	4,558	299%	COUNTER5
JSTOR		✓	112	433	364	3,080	846%	
Springer Link		✓	4,676	4,578	9,763	10,392	106%	
OUP		✓	108	422	692	3,552	513%	
Wiley Online Library		–	189	223	3,254	3,263	100%	
Maruzen eBook Library	Foreign/Japanese	✓	2,816	5,751	11,588	31,010	268%	COUNTER4
JapanKnowledge Lib		–	81	86	58,715	69,256	118%	
ProQuest Central	Foreign	✓	68	140	1,841	5,813	316%	
EBSCO e-book Collection	Foreign/Japanese	✓	501	655	1,918	2,363	123%	Own

translation of the Japanese title provided by the author. "Total access in 2019/2020" is the total number of accesses of each e-book in 2019. "Total accesses in 2019" is shown as a comparison. "Circulation" is the number of times printed books were borrowed from the library in 2019/2020. "Subject" is extracted from the classes of the Nippon Decimal Classification (NDC), which is widely used in Japanese libraries. "Syllabus" is whether this printed book or e-book was referred to in any syllabus. We examined the list of books referred to in the 2019 syllabi obtained from Kyushu University Library and the Kyushu University syllabus site. As the syllabus for some schools is not included in the site, we also searched for syllabi using the Kyushu University website full-text search form, using book titles as keywords. In this search, we checked only the top ten ranked search results.

As shown in Table 2, e-books in the "Mathematics," "Medical science," and "General works" fields were often used. Both of the e-books in the "General works" field (top 6 and 10) were about programming (information science). This result supports the results of Ochanomizu University's survey [8]. In addition, e-books in the "Medical science" field use colored graphical explanations or videos; as medical and pharmacy students prefer multimedia resources, this is an advantage of e-books. As a result, there was a trend toward e-book usage before the pandemic (normal situation). We identified a demand for books on mathematics, medical science, and programming languages in normal situations.

The printed book circulation counts for 2019 and 2020 are shown in the table. There is no correlation between the number of accesses of e-books and printed book circulation counts. This indicates that analyzing e-book usage data is an effective way of determining the need for e-books.

Table 2. Top 10 Most Accessed E-books in 2019

	Simplified title	Publisher	Total access		Circulation		Subject	Syllabus in 2019
			2019	2020	2019	2020		
1	Conquering the complex function for engineering students	Morikita	165	50	11	5	Mathematics	✓ printed
2	Electromagnetic field and vector calculus	Iwanami	107	14	21	3	Mathematics	✓ printed
3	Neuroanatomy lecture notes in colored graphic explanations	Kunpodo	104	145	5	5	Medical science	
4	Case study exercises written by Tokyo University students with 50 carefully selected frameworks	Toyo Keizai	81	11	6	2	Psychology	
5	Introduction to Economics for 99% of people (2nd ed.)	Otsuki	81	24	1	0	Economics	✓ e-book
6	An introduction to programming, starting with Python	Corona	76	11	4	1	General works	
7	Postoperative nursing 【Video】	Institute of A-V Medical Education	74	23	0	0	Medical science	
8	An introduction to statistics for life sciences realized with R and graphs	Yodosha	70	11	0	0	Biology	
9	The world of labor law (12th ed.)	Yuhikaku	69	0	9	0	Sociology	
10	New deep learning textbooks learning with Python	Shoeisha	67	171	7	2	General works	

The Top 10 Most Accessed E-books in 2020. Table 3 shows the top 10 most accessed e-books in 2020. The number of accesses in 2020 increased significantly compared with that in 2019. Books in the “Construction. Civil engineering” and “General History of North America” fields did not appear in Table 2. Three e-books in “Economics” ranked in the top 10. None of the e-books apart from that ranked 9th were used in 2019. The access count in 2019 for eight of the ten books was 0. These books except for the second-ranked book were not referred to in syllabi, but we assume that they were mentioned by faculty in classrooms or seminars or were used for student assignments. These results indicate different trends of e-book usage during the pandemic.

Table 3. Top 10 most accessed e-books in 2020

	Simplified title	Publisher	Total access		Circulation		Subject	Syllabus in 2020
			2019	2020	2019	2020		
1	International tax law (3rd ed.)	Univ. of Tokyo Press	0	350	1	0	Public finance	
2	Soil mechanics from basic to advanced level	Kyoritsu	0	315	7	1	Construction. Civil engineering	✓ printed
3	National Library of Medicine classification 2016, in Japanese	Japan Medical Library Association	0	302	0	0	Libraries, Library & information sciences	
4	American history: Study from society, culture and history (2nd ed.)	Keio Univ. Press	0	264	0	0	General history of North America	
5	Japanese economic history	Yuhikaku	N/A	259	4	3	Economics	
6	An introduction to international management	Yuhikaku	0	224	3	1	Economics	
7	A book to learn the basics of Java SE Bronze in 2 weeks	Impress R&D	0	215	0	0	General works	
8	Japanese economic history (2nd ed.)	Univ. of Tokyo Press	0	178	2	0	Economics	
9	New deep learning textbooks by Python	Shoeisha	67	171	7	2	General works	

(continued)

Table 3. (*continued*)

	Simplified title	Publisher	Total access		Circulation		Subject	Syllabus in 2020
			2019	2020	2019	2020		
10	Natural Language Processing	Kyoritsu	0	166	0	0	General works	

Books in "Mathematics" and "Medical science," which were used frequently in 2019, do not appear in the table. The e-book on "Medical science," which was ranked in the top 3 in 2019, had 145 accesses in 2020. Although the e-book was used more during the pandemic, other books were accessed more than it in 2020, so it did not appear in the top 10 rankings. It can be said that the pandemic has evoked new needs.

3.3 Subjects of E-books with the Top 100 Usage Growth Rates

Table 4 shows the subjects of e-books that were among the top 100 e-books with the highest increases in usage and the number of e-books referred to in syllabi in 2020. The number of e-books is 106 because there were 8 e-books jointly ranked 99th. Note that the usage count for e-books that were not used in 2019 was considered as 1 to calculate the rate of increase.

"Law" dominates the rankings, followed by "Mathematics." In "Medical sciences," e-books were used, in spite of not having been referred to in the syllabi. E-books on "Mathematics" were used previously, as shown in Table 2, but their usage increased further under the pandemic. There is a stable demand for the mathematics field which is not affected by either normal or emergency situations.

Table 4. Subjects of e-books with the top 100 growth rate of usage

Subject	Number of e-books	Syllabus in 2020
Law	23	6
Mathematics	12	3
Medical sciences	10	0
Economics	10	3
General works	7	1
Social sciences	6	1
Physics	4	0
Electrical engineering	4	0
Others	30	4
	106	18

According to the subject classifications of MeL, in the subject distribution of all 55,346 titles in their collection as of September 2020, 2% of e-books are in "General," 55% are in "Humanities and Social Sciences," and 43% are in "Science, Technology, and Medicine (STM)" [12]. Almost half of the e-books are in the "Humanities and Social Sciences." Compared to e-books on STM topics, they had been little used before. However, e-books on law, economics, and social sciences started to be used during the pandemic.

4 Conclusion

In this survey, e-books in STEM fields, especially mathematics, medical sciences, and programming languages that were normally used before the pandemic were further used during it. This shows that there is a constant and stable demand for e-books in these fields. The use of e-books on law and economics increased. Printed books in these fields have been used frequently. Library closures and online education have influenced this increase in demand for e-books. At this point, it can be said that the potential need for e-books in the humanities and social sciences has also been indicated. Many e-books not listed on syllabi were also used. Faculty members might think of introducing them during classes. However, we were not able to identify the reasons users used those particular books from the usage data. In the future, we plan to conduct interviews with faculty members and students to inquire regarding the reasons.

We compared usage data from before and during the pandemic. In a future study, we plan to analyze the usage of e-books after the pandemic. Moreover, examining not only the number of accesses but also the characteristics of e-books can inform the selection and purchase of e-books adapted to the needs of courses and self-study. As a goal of this future study, we plan to design online research and learning environments, including a reasonable e-book collection.

Acknowledgments. We would like to thank Kyushu University Library for providing usage data. This work was supported by JST AIP Grant Number JPMJCR19U1, Japan.

References

1. Library Journal, School Library Journal: 2015 Survey of Ebook Usage in U.S. Public Libraries. http://www.thedigitalshift.com/research/ (2015). Accessed 14 Oct 2020
2. Denshi shuppan seisaku ryutsu kyogikai: Denryukyo, denshi toshokan o donyu shiteiru kokyo toshokan joho o koshin. https://aebs.or.jp/pdf/E-library_introduction_press_release20200701.pdf (2020). Accessed 15 Oct 2020
3. Kyushu daigaku koho shitsu: 2020 nendo Kyushu daigaku gaiyo shiryo hen. https://www.kyushu-u.ac.jp/f/41532/2020kyudaigaiyou_all_A4_compressed.pdf (2020). Accessed 19 Dec 2020
4. saveMLAK. Covid-19-survey. https://savemlak.jp/wiki/covid-19-survey (2020). Accessed 15 Oct 2020
5. Library Journal: Academic Faculty: Textbook & Course Materials Affordability Survey Report. https://www.libraryjournal.com/?page=academic-faculty-textbook-course-materials-affordability-download-confirmation (2019). Accessed 14 Oct 2020

6. Monbu kagakusho: Reiwa gannendo gakujutsu joho kiban jittai chosa no kekka o kohyo shimasu. https://www.mext.go.jp/content/20200721-mxt_jyohoka01-000005810.pdf (2020). Accessed 15 Oct 2020
7. Kyushu daigaku: Kyushu daigaku shirabasu. https://syllabus.kyushu-u.ac.jp/ (2020). Accessed 15 Oct 2020
8. Tateishi, A., Etori, N., Shoji M.: PDA de kawaru sensho no mirai: Chiba daigaku, Ochanomizu daigaku, Yokohama kokuritsu daigaku, sandaigaku renkei purojekuto no torikumi. Joho no Kagaku to Gijutsu. **65**(9), 379–385 (2015). https://doi.org/10.18919/jkg.65.9_379
9. HIto2014COUNTER4 ni tsuiteYakugaku Toshokan.594265269
10. Hendry, J.: Release 5 Manual for Librarians. Books: Understanding Metrics and Standard Views. https://www.projectcounter.org/wp-content/uploads/2020/04/Release_5_Librarians_PDF_20200428.pdf (2020). Accessed 14 Oct 2020
11. Mellins-Cohen, T.: The Friendly Guide to Release 5 for Librarians. https://www.projectcounter.org/wp-content/uploads/2019/05/Release_5_Librarians_20190509-Revised-Edition.pdf (2018). Accessed 14 Oct 2020
12. Maruzen eBook Library. https://elib.maruzen.co.jp/ Accessed 14 Oct 2020

Image-Building of Public Library from Readers' Perspective: A Case Study on the Northern Haidian Library

Tianji Jiang[1](✉) and Linqi Li[2]

[1] Department of Information Studies, University of California, Los Angeles, CA 90024, USA
tianji008@ucla.edu
[2] School of Information Management, Nanjing University, Nanjing 210023, China
mg1914004@smail.nju.edu.cn

Abstract. Organization image management is regarded as an effective way for libraries to rebuild their public images through their own endeavors and interactions with the public. However, the previous studies usually explore the issue from the perspective of library staff, yet seldom from the perspective of readers. This article examined reader's needs for library service, and set up a model of readers' hierarchy of needs. We also explored a method to build a positive image for libraries from the perspective of readers' demands, and set up the model of Image-building for public libraries. We hoped to provide a new approach to learn readers' needs for library, as well as a new sight on library image construction.

Keywords: Readers' needs · Library image-building · Grounded theory · The Northern Haidian Library · Guest book

1 Introduction

Organization image is often viewed as a cognitive construct signifying the perception of an organization and defined as an impression created in the mind of an audience [1]. The concept was first put forward in management studies [2], but soon introduced to studies in other fields, such as communication [3], public relations [4] and marketing [5]. In the current well-informed society, we witness an increasing importance of organization image, which is considered as an indispensable part of competitiveness of an organization. Moreover, not only commercial organizations but also many non-profit organizations (NPOs) and public service agencies are paying much attention to manage their organization images [6]. A good organization image can bring them more awareness, higher reputation and sometimes more donations.

As Internet technology upgrading rapidly today, the way people learn and educate themselves, the way they search for information, and the way they read are also changing. These changes seriously weaken the traditional roles of libraries, such as collectors, organizers, keepers and disseminators of information resources [7]. Moreover, libraries are seldom recognized for the unique and irreplaceable service they provide for their

K. Toeppe et al. (Eds.): iConference 2021, LNCS 12646, pp. 484–492, 2021.
https://doi.org/10.1007/978-3-030-71305-8_41

communities [8]. And previous studies have indicated that the use of the Internet would replace the use of public libraries for studying, work, or leisure activities. [9, 10]. There are reasons to believe that libraries are at the risk of being marginalized if they keep standing still and do nothing to rebuild people's impression on them.

Since libraries in China are usually playing supplementary roles and used to working behind the scenes, the public know little about libraries and their work [11]. Furthermore, libraries used to pay little attention to their public image, thus stereotype images and misperceptions about libraries do exist among people. These ideas are often falsely accusing libraries of their value and status. Therefore, it is vital for libraries to better utilize its human resources, collections and environment to find new ways for development in the new times.

Organization image management is regarded as an effective way for libraries to rebuild their public images through their own endeavors and interactions with the public [12]. Nowadays, image of library has already become a hot topic in library science studies in China. However, most current researches on library images just focus on analyzing its contents or interpreting its concepts. These works usually explore the issue in the sight of library operators, yet lack of considerations from the readers' perspective. Naturally, it is difficult to form a full and in-depth understanding of the process of forming the image of the library, as well as to have specific guidance on how the library constructs a public image. Therefore, we try to explore readers' understanding of the library's image, and clarify how various components of the library's service influence the overall image of a public library in the eyes of readers. This can provide some new suggestions and references for management of public library images.

2 Methodology and Research Design

2.1 Data Collection

A guestbook is an approach for visitors to acknowledge their visits to a site and comment on it. In Chinese public libraries, it is an important way for library managers to evaluate their performance, know the needs of the readers, and advance their work. We collected the comments from the guest books of Northern Haidian Library, which was a recently opened public library in the suburban area of Beijing, China. It was chosen as our sample because of the diverse population it served, which came from various economic and educational backgrounds.

Each comment consisted of two parts. The first part tagged the customers' attitude, including positive, negative or neutral. The second part was texts about what the reader wanted library managers to know, which was regarded as the main part of the message. We come to work on these words left by readers, trying to dig out further information from them.

We used the guest books which contain all the readers' comments. Due to privacy concerns, we could only get access to data which were made three years ago, therefore we only analyzed contents of the guest books created from July 2016 to October 2017 in this study. In addition, we got a copy of the working log of the same period from the library with the help of the deputy director. A total of 209 readers' comments were collected and 206 were checked to be valid.

2.2 Data Analysis and Research Process

With the data we collected, we tried to build a model for public library image management. Qualitative research methods, including grounded theory and process data analysis, were conducted in the analysis. We read all the user comments and coded those about users' needs. Inductive coding was adopted to code the data. It referred to a data analysis process whereby the researcher read and interpreted raw textual data to develop concepts, themes or a process model through interpretations based on data [13]. We first did open coding on readers' needs, then did the axial coding based on the open coding results, and finally did selective coding based on hierarchical relationships in the axial coding results. The coding result is showed in Table 1.

Table 1. Coding result of the readers' messages

Safety	Environment	Service			Knowledge	Participation	Care	Dignity
		Staff	Equipment	The ohters				

In addition, we found that issues about children's behavior regulation had always been an important topic that frequently appeared in the readers' comments, and the contents of the comments showed significant changes over time. Moreover, we learnt from the library staff that they had always put emphasis on children's behavior regulation. The library had done much about it, including setting up specified reading areas for children, strengthening education on behavior in library, and restricting children's access to some places in the library. Therefore, we attempted to combine the changes in readers' comments with the corresponding measures taken by the library to look further into user's needs, and thinking about how library can improve their organizational image by satisfying the needs.

3 Findings

3.1 Analysis of Demand Characteristics of Readers

Readers often showed unambiguous emotional tendencies in their comments such as praise, suggestions and criticisms (Table 2). In order to have a deeper analysis, reader comments are divided into positive ones and negative ones according to their emotional tendency. For the rigor of division, sentimental dictionary was created and words with emotional characteristics were put into positive dictionary and negative dictionary separately. Next, programs were written to determine the emotional tendency through word segmentation and counting the frequency of positive words and negative words. At last, a manual inspection of the analysis results was conducted to correct unreasonable divisions.

Through analysis, it turns out that a gap between how readers feel about the library service provided and what they comment do exists. Bad services are likely to bring critical opinions, but good services are less likely to get praises on it. For example, when

Table 2. Statistics table indicating readers' emotional tendencies on different issues

	Safety		Environment		Librarian Service		Equipment Service		General Service		Knowledge		Participation		Care		Dignity	
	Positive	Negative	Positive	Negative	Positive	Negative	Positive	Negative	Positive	Negative	Positive	Negative	Positive	Negative	Positive	Negative	Positive	Negative
Mostly positive	0	0	22	3	21	1	2	0	2	0	9	1	2	0	4	0	25	0
Mostly neutral	0	2	5	59	5	17	1	16	0	7	1	58	0	4	0	1	2	0
Mostly negative	0	2	2	40	1	11	1	8	0	4	0	8	0	0	0	0	0	7
Total	0	4	29	102	27	29	4	24	2	11	10	67	2	4	4	1	27	7

readers hold the opinion that the library does not satisfy their environment needs, they tend to criticize the library; however, when those needs are satisfied, their tendency to praise in guest book is relatively weaker. Thus, the study draws on dual-factor theory, putting forward hygiene needs and motivation needs to make detailed measurement on readers' needs.

Hygiene needs refer to basic needs that readers think the library should meet. Readers don't make positive comments on the library when this type of needs are satisfied, however, if the library fail to satisfy those needs, readers are likely to make negative comments. Motivation needs refer to optional needs that readers think the library may or may not meet. If this type of needs is not satisfied, readers will not make negative comments, but once those needs are satisfied, readers tend to make positive comments.

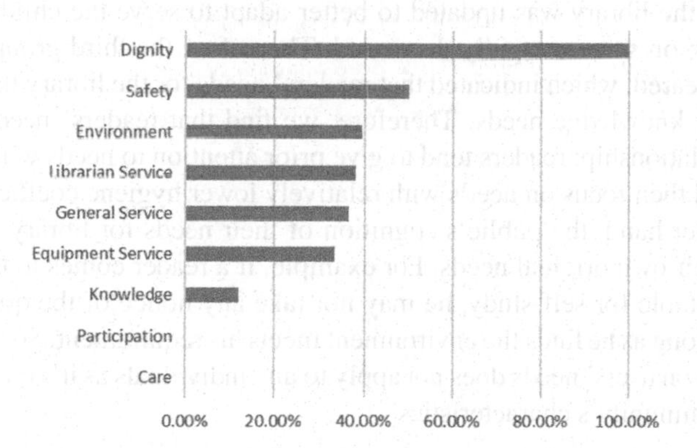

Fig. 1. Hygiene coefficient for different reader needs

Through interviews with some librarians and readers, we find that readers are more inclined to make criticisms or suggestions regarding needs that are not satisfied rather than praise the satisfied needs in the guest book. Thus, it is effortless to realize what the unsatisfied needs are while the already satisfied needs cannot be identified. As a result, in the following analysis, we can only draw upon the concept of hygiene factor from dual-factor theory and the motivation factor cannot be used. According to this, a hygiene coefficient is defined, which means that of all readers' comments, the probability that a comment's overall attitude is negative when certain needs are not satisfied. Based on

Fig. 1, we find that needs for respect, safety and environment are the main contents of hygiene factors.

3.2 Analysis of Needs Hierarchy

We then selected 72 comments concerning children's behavior regulation from the guest books as sample, and combined the corresponding measures taken by the library to further clarify the connections among readers' different needs of the library. The comments were divided into 3 groups by their topics. The first group were about needs of environment. The readers were writing the comments to complain about children's disturbance to the quiet reading environment and ask library staff to stop it. The second group focused on improving children-oriented services, such as increasing child-friendly self-service equipment and holding more specific events for children. The last group were talking about needs of knowledge, suggesting that the library increase their children-oriented collection.

Notably, the three groups of comments didn't appear at the same time in the guest books, but in different time period separately. The first group were mainly written down in the first 6 months, when the library was in its trail period. It was before the library opened its children's library, and children were accessible to all reading areas without any supervisions. Since the library opened its children's library and increased restrictions on children's movements, the problem of reading environment changed a lot and the comments on environment issues almost disappeared. After that, the topics of the comments shifted to service needs gradually, and the second group of comments arose. In May 2017, the library was updated to better adapt to serve the children, thus comments focusing on service rapidly decreased. Thereafter, the third group of comments frequently appeared, which indicated that readers' needs for the library had stepped into a higher level: knowledge needs. Therefore, we find that readers' needs have a clear hierarchical relationship: readers tend to give prior attention to needs with high hygiene coefficient and then focus on needs with relatively lower hygiene coefficient.

On the other hand, the public's cognition of their needs for library service is constrained by their own original needs. For example, if a reader comes to the library only for a place suitable for self-study, he may not take any notice of the quality of library collections as long as he finds the environment meets his requirement. So the hierarchical analysis above on users' needs does not apply to any individuals as it's the generalization of a reader community's characteristics.

3.3 Reader's Hierarchy of Needs Model

Based on the above analysis on readers' needs characteristics and hierarchy, we concluded the 9 categories into 6 core decisive categories and reveals their hierarchical relationships to construct reader's hierarchy model of needs (Fig. 2). The model indicates that the readers' needs consist of respect, environment, service, knowledge, care and participation hierarchically.

- Respect needs: needs for respectful and equal treatment.
- Environment needs: needs for clean, quiet and comfortable environment.

- Service needs: needs for various basic services, such as services from librarians, equipment, etc.
- Knowledge needs: needs for knowledge from books and activities.
- Care needs: needs for kind treatment regarding personal special requirements.
- Participation needs: needs for being engaged in library's activities.

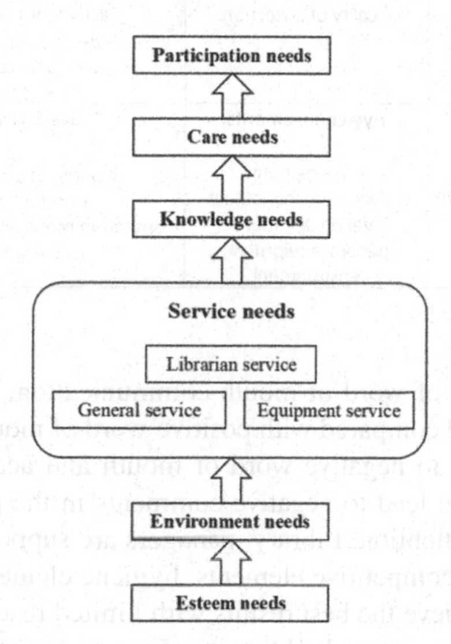

Fig. 2. Reader's hierarchy of needs model

Generally, readers tend to pay attention first to low-level needs, and then focus on high-level needs when the former has been satisfied. Readers are not likely to make positive comments if the low-level needs are satisfied, but on the contrary, they tend to make negative comments if the low-level needs are not satisfied, thus harming the library image. When the high-level needs are satisfied, readers are likely to make positive comments and it brings benefits to the library image. Readers rarely make negative comments if the high-level needs are not satisfied. Combined with the results of analysis on readers' needs characteristics above, we can come to a conclusion that the higher the level of needs is, the lower the hygiene coefficient will be.

Notably, as what we have mentioned above, the analysis and summary of readers' needs hierarchy are based on masked data, only reflecting characteristics of a group. As a result, analysis and generalization in this section are aimed at reader group's characteristics, which reflects overall regularities of a group and does not apply to any individuals. This point must be noted in practice.

3.4 Model of Public Library Image-Building

We associated readers' needs with elements of library public image and refers to readers' feedback under different circumstances. The matrix of library public image elements was constructed (Table 3).

Table 3. The matrix of library public image elements

Good Performance \ Bad Performance	Criticize	Not Criticize
	Competetive Elements	**Motivation Elements**
Praise	librarian litarcy, librarian service, quality of collection	care for special groups, reader's participations, activies for readers, high-tech equipment, voluntary service, other service
	Hygiene Elements	**Edge Elements**
Not Praise	management, collection management, librarian sttendence general equipment, environment	quantity of collections, collection types, operation management system, niche system

According to the law of word of mouth communication, negative word of mouth always has a wider spread compared with positive word of mouth (Wei & Wang, 2010). Users are more sensitive to negative word of mouth and accept it easier. Therefore, elements that are likely to lead to negative comments in the process of library public image-building deserve attention. Library managers are supposed to build library public image in the order of competitive elements, hygiene elements, motivation elements and edge elements to achieve the best results with limited resources. Based on this, we integrate elements of library image-building into 5 core categories, reveals their relationships, and then forms the model of library public image-building. The model indicates that library public image-building should be adapted to readers' needs, conducted in the order of librarian image, equipment and environment, service, collection and activities, care and participation, and with the outermost element prioritized (Fig. 4).

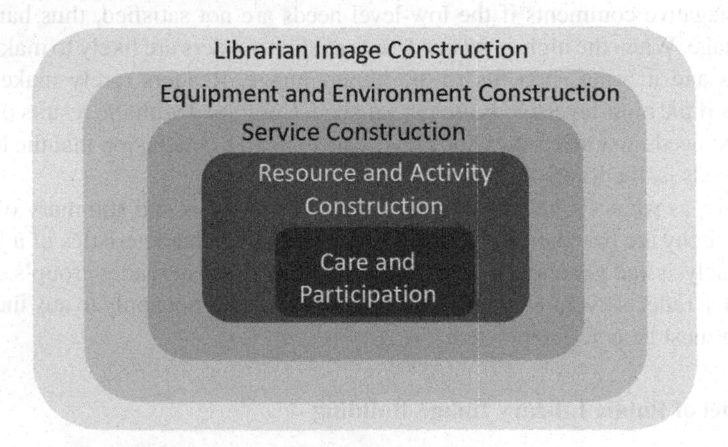

Fig. 3. The model of library public image-building

4 Conclusion and Implications for Future Research

4.1 Results

What makes up the public image of a library, and how can library build a positive public image? This article took the Northern Haidian Library as an example to answer these questions. Through analysis of all comments in readers' book since their opening and interviews with librarians, we generalized readers' needs and demand characteristics, put forward readers' hierarchy model of needs and the model of public library image-building.

From the perspective of readers' needs, we find that readers' needs for the library are hierarchical. Whether the needs are satisfied will affect readers' comments on the library and needs in different levels have different influence mechanisms on the readers' perception of the library.

From the perspective of the building of library public image, managers should take readers' needs into account and realize that only with a clear goal can library achieve optimal results with limited resources. In the operation of public libraries, it is necessary to strengthen the cultivation of librarians and the maintenance of the environment, especially the construction of service attitude. Good service attitude and favorable environment are the basis for libraries to compete.

In addition, we find that readers' perception to the library image is a dynamic process. Every time readers come to the library, they are perceiving it and the library is improving and changing simultaneously. As the library continues to meet readers' needs, on one hand, readers have a better perception of the library image; on the other hand, they put forward demand with higher level. If higher-level needs are not satisfied over a period, readers' perception of the library may get worse. As a result, clarifying requirements and making response to them are of vital importance for the library to keep a favorable public image, and with the level of readers' needs upgrading, library service will step into a higher level to make further developments. In the continual interaction between the library and readers, the quality of library service keeps progressing.

4.2 Significance

From the point of academics, this research applies grounded theory and case study to find out reader's needs with the messages they left on the library service, and then explore a method to build up a positive image for libraries from the perspective of reader's demands. The method is very different from those current ones which are designed from the standpoint of libraries themselves. In addition, it brings us a new approach to understand the organizational image of public libraries. On the other hand, from the point of practice, this research uses the Northern Haidian Library as an example to summarize readers' demands for library services, as well as help librarians to recognize their misunderstandings on identifying what readers really want. This will enable the librarians to have better recognition and understanding on readers' demands. Besides, the article also puts forward some solutions to promote the library's images, and these solutions will also provide some inspirations or have some reference value for other libraries.

4.3 Implication for Further Research

Restricted by the case we selected in the research, the model proposed in this paper has a limited range of application and can't interpret some special cases, such as Tianjin Binhai Library, which was one of the most famous libraries in China last year. Besides, although readers' messages are rich and reliable in expressing their views on library, we found in our field investigation that most readers, especially the young, tend to simply regard the guest book in a library as a complaint book. Most people leave messages on the guest books only for complaining bad services, while few are for praising anything good in the library. So it lacks typicality only using readers as analysis objects. What's more, the article used readers' messages to scoop out the readers' needs without any further interviews or investigations. This leads to a weak point in matching readers' needs with the elements of library image in the article. We hope that we can do some further researches on this topic to refine our research work in the future.

References

1. Frandsen, S.: Organizational image. In: The International Encyclopedia of Organizational Communication, pp. 1–10. American Cancer Society (2017). https://doi.org/10.1002/9781118955567.wbieoc103
2. Dowling, G.R.: Managing your corporate images. Indus. Mark. Manage. **15**, 109–115 (1986). https://doi.org/10.1016/0019-8501(86)90051-9
3. Yanan, J.: Public Relation. Fudan University Press, Shanghai (2001)
4. Nekmat, E., Gower, K.K., Ye, L.: Status of image management research in public relations: a cross-discipline content analysis of studies published between 1991 and 2011. Int. J. Strat. Commun. **8**, 276–293 (2014). https://doi.org/10.1080/1553118X.2014.907575
5. Thøger Christensen, L., Askegaard, S.: Corporate identity and corporate image revisited: A semiotic perspective. Eur. J. Mark. **35**, 292–315 (2001). https://doi.org/10.1108/03090560110381814
6. Rho, E., Yun, T., Lee, K.: Does organizational image matter? Image, identification, and employee behaviors in public and nonprofit organizations. Public Adm. Rev. **75**, 421–431 (2015). https://doi.org/10.1111/puar.12338
7. Bertot, J.C., Jaeger, P.T., McClure, C.R., Wright, C.B., Jensen, E.: Public libraries and the internet 2008-2009: issues, implications, and challenges. First Monday **14**, 11 (2009). https://doi.org/10.5210/fm.v14i11.2700
8. Smith, M.: Top ten challenges facing public libraries. Public Libr. Q. **38**, 241–247 (2019). https://doi.org/10.1080/01616846.2019.1608617
9. Foundation, Benton: Buildings, Books and Bytes. Benton Foundation, Washington, DC (1996)
10. Robinson, J.P., Martin, S.: IT and activity displacement: behavioral evidence from the U.S. general social survey (GSS). Soc. Indic. Res. **91**, 115 (2009). https://doi.org/10.1007/s11205-008-9285-9
11. Fang, W.: The construction of the public library image from the perspective of public relations. Fujian Libr. Theory Pract. **3**, 44–46 (2009)
12. Huang, X.: The basic goal of library public relations: build a good library image. Res. Libr. Sci. **6**, 27–29 (1992)
13. Chandra, Y., Liang, S.: Qualitative Research Using R: A Systematic Approach. Inductive Coding, pp. 91–106. Springer, Singapore (2019)

Development and Evaluation of a Digital Museum of a National Intangible Cultural Heritage from China

Xiao Hu(✉), Jeremy Tzi-Dong Ng, and Ruilun Liu

The University of Hong Kong Shenzhen Institute of Research and Innovation, Shenzhen, China
xiaoxhu@hku.hk, {jntd,laualan}@connect.hku.hk

Abstract. Intangible cultural heritage (ICH) such as traditional craftsmanship lacks a physical form and often originates from minority groups with little documentation. Digital technologies can be leveraged for documenting and archiving these assets of humanity. In particular, digital museums are established for promoting public understanding and appreciation of cultural heritage. Despite the richness of ICH in China, the development of digital museums of ICH is still in an early stage and mostly from government endeavours. As part of an inter-disciplinary collaborative project involving academic researchers, information professionals, and a private not-for-profit museum, this paper described the development of Gifts from Lanmama, a digital museum of Miao embroidery as a unique ICH from Guizhou ethnic minorities in China. This paper also reported a preliminary evaluation of the digital museum with 78 users, in terms of its usability and affordance for learning about cultural heritage. Results revealed the strengths of the digital museum in terms of the rigor of metadata and its impact on improving users' understanding and appreciation of Miao embroidery. Some issues and challenges were also identified, such as the lack of channels for user-system communication. These evaluation results offer insights for further improving the digital museum and other end-user oriented digital presentations of similar ICH.

Keywords: Digital museum · Intangible cultural heritage · User evaluation

1 Introduction

Cultural heritage is a group of invaluable assets that inherit and embody human activities passed down from previous generations. It is a responsibility in our citizenship to safeguard cultural heritage for ourselves and future generations [1]. However, intangible cultural heritage (ICH) such as traditional craftsmanship lacks a physical form and often originates from minority groups whose languages may not have written scripts. These make it very challenging to preserve and promote these ICHs [2]. Digital technologies have been leveraged for documenting and archiving ICH [3]. China is well-recognized for its rich history and culture yet digitizing ICH in China is at an initial stage and is mostly through government endeavours [4]. Guizhou province in southwest China houses multiple ethnic minorities, among which the traditional craft of artisan embroidery of Miao

K. Toeppe et al. (Eds.): iConference 2021, LNCS 12646, pp. 493–501, 2021.
https://doi.org/10.1007/978-3-030-71305-8_42

people was listed in the inaugural register of the National Intangible Cultural Heritage of China in 2006 [5]. With a long history of migration, the Miao are skillful at documenting their socio-cultural history and folk customs through their clothing, creating a vibrant cultural heritage [6]. For instance, the needlework of Miao women in history was emphasized as they produced various festival and daily live items with highly detailed patterns, including costumes, purses, baby-carrying belts, hats, bibs, and so on. Many of these exquisite traditional crafts that have ever emerged in Chinese history are intactly preserved by Miao embroidery techniques, including weaving, embroidering, and dyeing. They are of exceptionally high value in historical, anthropological, artistic, aesthetic, iconic and inspirational terms.

ICH is mostly rooted in people's traditions and daily lives. Therefore, effective conservation and preservation of ICH cannot only rely on government efforts. Grass-root initiatives and collaboration across sectors are desirable for improving the effectiveness, efficiency, and sustainability of ICH preservation [2]. As part of an inter-disciplinary collaborative project involving academic researchers, library professionals, and a private not-for-profit museum, a digital museum was built for digitizing, exhibiting, and preserving these unique ICH of Guizhou ethnic minorities. The ultimate goal is to facilitate access of a global audience, enhancing their understanding and appreciation of Chinese cultural heritage. As one of the first collaborative efforts between academia, information professionals and the private not-for-profit sector in China, it is essential to evaluate the usability of the digital museum and its affordance for facilitating the audience to learn about the Miao embroidery as a cultural heritage. The evaluation can also help us understand users' interactions with the digital museum and areas needing improvement from users' perspectives. This paper reports the development and evaluation of the 'Gifts from Lanmama' digital museum[1] (Fig. 1). Preliminary evaluation results will offer insights for further improving the digital museum, as well as other end-user oriented digital presentations of similar ICH, such as interactive multimedia exhibitions.

2 Related Work

2.1 Digital Museums of Cultural Heritage

To embrace the evolving potential of digital environments and multimedia in creating and distributing cultural content, more and more digital museums have been established worldwide to exhibit collections of digitized cultural heritage [7]. Over the recent two decades, intangible cultural heritage (ICH) has been receiving increasing attention after the UNESCO acknowledged its significance via the Convention for the Safeguarding of the Intangible Cultural Heritage in 2003 [8]. While people living in modern cities may have limited opportunities of experiencing ICH due to the often small numbers of its successors, digital museums help not only make institutionalized documentation of ICH but also promote public understanding and appreciation [9]. Despite the richness of cultural heritage in China, the development of digital museums in China is still in an early stage [10]. Xiong (2020) [11] conducted a review of 466 digital museum related articles published and indexed in the China National Knowledge Infrastructure, concluding that

[1] https://lanmama.lib.hku.hk/.

there were insufficient collaborations between universities and museums as well as a lack of evaluation of digital museums in China.

2.2 Evaluation of Cultural Heritage Digital Museums

Prior studies on evaluating cultural heritage digital museums often focus on usability. Usability of a digital resource refers to the extent to which the resource meets the users' needs [12], also reflecting the quality of user experience [13]. Usability evaluation is one of the most frequently used tools for assessing how a digital museum supports its users' needs [7], offering designers and developers insights into the usefulness of the digital museum and revealing design problems and erroneous elements [14]. As one of the purposes of memory institutions is educating visitors, it is important to evaluate to what extent a cultural heritage digital museum helps visitors understand this cultural heritage and arouse their interests. Online questionnaire surveys are one of the most common methods of digital museum evaluation [15], while other methods include direct observation and log analysis [16]. There have been numerous studies on evaluating the usability of cultural heritage digital museums. For instance, Pallas et al. [15] evaluated the websites of 210 art museums worldwide in terms of usability and other aspects such as content and presentation. Their findings suggest that easy and simple navigation, return-to-Home and Help buttons, and absence of page errors are essential features for visitors' perceived ease of use. In a more recent study, Hu et al. [17] involved fourteen professional and layman users for evaluating the digital museum of cultural heritage in Dunhuang, China, yielding that introductions to the cultural heritage materials can be added for users with little background knowledge, and more communication channels such as a forum can be established for enhancing user-to-user interactions.

3 The Digital Museum: *Gifts from Lanmama*

Our collaborator, the Lanmama Ethnic Minority Apparel Museum, is a private, not-for-profit museum located in Guizhou, China. It houses about 2,000 pieces of apparels from ethnic minorities in Guizhou province, including Miao, Dong, Buyi, and those with small populations yet rich histories. The embroidery of Miao people contributes to a significant portion of the collections in the museum, with a diversity of items from festival dresses to daily household items such as aprons and bags. The dates of these items span from late Qing dynasty to the 1960s, reflecting the lives and traditions of ethnic minority people in a systematic and comprehensive manner. To ensure the quality of the digital museum, the items to be digitized was selected by this museum given their expertise, such that selected items cover all representative techniques of weaving, dyeing and embroidery, from various branches of major Guizhou Miao tribes. A total of 109 items were photographed by a professional photographer, resulting in 668 high-resolution digital images. Together with the photos were detailed descriptions of the items. To serve an international audience, the descriptions, originally in Chinese, was carefully translated into English. The translation process referred to a detailed monograph on Miao embroidery from Guizhou, whose content is aligned in Chinese and English [18].

Dublin Core (DC) was adopted as the metadata schema [19], with the following elements chosen for describing each item: *Title*, *Subject*, *Description*, *Creator*, *Date*, *Rights*, *Identifier* and *Coverage*. After reviewing the available software for implementing the digital museum, Omeka.org, the open-source online exhibit tool for memory institutions (i.e., galleries, libraries, archives, museums), was utilized for its core functionality of publishing collections and exhibits, ease of use and flexibility, and extensibility and scalability enabled by a suite of plug-ins [9]. Users of the 'Gifts from Lanmama' digital museum can freely browse the 109 items individually, three collections organized based on the procedure of making the crafts ("Weaving", "Dyeing", "Embroidery"), 28 exhibits of different embroidery techniques (e.g., "Knot stitch", "Counted thread stitch"), or browse items on a map, i.e., according to the geographic locations of their origins (e.g., Qianxi county in Guizhou). The digital museum also supports searching by metadata and full-text, where users can apply different query types such as Boolean search and phrase search. Notably, the interface has a responsive web design that is compatible with both computers and mobile devices, entailing a browsing experience with minimal resizing and optimal scrolling.

Fig. 1. Homepage of the Gifts from Lanmama digital museum

4 Research Method

A questionnaire survey in both English and Chinese versions was carried out to collect users' responses. It includes evaluation criteria adapted from the usability evaluation model devised by Jeng [16], the scale of a website's interactivity with users proposed by Liu [20], and those related to the ultimate purpose of the digital museum: facilitating 'learning about cultural heritage' [21]. Hence, the framework of evaluation criteria in this study (Table 1) comprises dimensions including effectiveness for learning about cultural heritage, user satisfaction (ease of use, organization, labeling, visual appearance, content), and user-system interactivity (active control, two-way communication). It is noteworthy that effectiveness of the digital museum on 'learning about cultural heritage' was operationalized as its impact on users' understanding of, interest, and engagement in the exhibited ICH [21]. In the questionnaire, participants were asked to rate the digital museum against these criteria on a 6-point Likert scale from 1 (Strongly Disagree) to 6 (Strongly Agree). They were also asked, though not obligatory, to provide relevant comments and suggestions in an open-ended question. Participation in this study was voluntary without any remuneration.

Participants were recruited by a mix of convenience and snowball sampling of the researchers' networks and official dissemination channels of the hosting institution (e.g., websites, mailing lists). Besides descriptive statistics, Spearman's rank-order correlation analyses were conducted to reveal possible inter-relationships between these evaluation criteria [24]. Participants' free-text comments were analyzed thematically and quoted to supplement quantitative results.

5 Evaluation Results

78 valid responses were collected, with 77% (n = 59) from female participants. The ages of participants ranged from 18 to 64 (mean = 30.6). All of them were native Chinese speakers. More than half (61%) were students in higher education institutions. A considerable portion (80%) of the participants reported their studies or work were not relevant with cultural heritage, whereas the remaining (20%) deemed the opposite. The majority of them (87%) were interested in intangible cultural heritage. 71 (91%) respondents provided free-text comments and suggestions.

Rating scores on the various criteria of effectiveness, satisfaction and interactivity were aggregated across participants and are presented in Table 1. Among all criteria and variables, the consistency of information presented in the digital museum received the highest rating (mean = 5.19, on a six-point Likert scale), echoed by the comment that the "style of [information] presentation was uniform and made sense" (Participant#46). This indicates the rigor of metadata displayed to the audience, attributable to the adoption of the Dublin Core metadata standard which is well-known for its compatibility with cultural heritage information [22]. The lowest rated criterion was two-way communication, likely due to the lack of any explicit functions in the digital museum for users to air their feedback. The decision of not having such interactive features stemmed from our past experience in developing digital exhibitions which attracted spam comments resulting in disrupted operations of our server. Similar issues were also reported in the literature [23]. Focusing on the effectiveness of the digital museum on 'learning about cultural heritage', the perceived impacts of the digital museum on improving participants' understanding of (mean = 4.97) and interest in Miao embroidery (mean = 4.83) were rated higher than their engagement with this ICH (mean = 4.69). This might be due to the fact that only static images were included as visual surrogates of the embroidery items. Being described as "delicate" (Participant#15), "having spirituality" (Participant#68) and "lively" (Participant#34), these high-resolution images work well in demonstrating the items and the craftsmanship. However, they may not be sufficient in providing an engaging *experience*. Virtual reality (VR), being an immersive technology, can deliver the virtual experience for users to engage and interact with the ICH [8]. In fact, a multimodal exhibition for this project involving the display of Miao embroidery items in a physical venue, the launch of this digital museum, and a station for an immersive VR experience of the exhibits were planned to be conducted in Spring 2020. These were unfortunately cancelled due to the COVID-19 pandemic. In addition, the making of the embroidery items (i.e., craftsmanship) is a process which would be better captured by media with temporal information such as videos [6]. In this regard, we have noted participants' suggestions on presenting "videos that show how the [embroidery] work is done" (Participant #2), catering to users' interests in the "making process" of the embroidery techniques (Participant #76).

Spearman's correlation analysis revealed that the better participants' understanding of Miao embroidery was, the higher rating they gave on the visual appearance of the digital museum (correlation coefficient r = 0.500; $p = 0.000$). This seems to confirm the relationship between understanding and appreciation of ICH [25]. Similarly, as expected, the more participants were interested in the cultural heritage, the more positive the rating they gave on their browsing experience (r = 0.593; $p = 0.000$). Besides, there was also a significant correlation between participants' engagement in the Miao embroidery and their rating on the searching functionality of the digital museum (r = 0.609; $p = 0.000$). This seems to suggest that users who were more engaged with the ICH would prefer using the search function to locate specific items in their mind [26]. Last but not least, participants' perceived two-way communication with the digital museum also quite strongly correlated with their rating on searching experience (r = 0.713; $p = 0.000$). This is not surprising as searching is an interactive process [27].

Table 1. Evaluation criteria and participants' ratings (N = 78)

Criterion	Variable	Rating mean (SD)
Learning about cultural heritage	Understanding of the cultural heritage	4.97 (1.04)
	Interest in the cultural heritage	4.83 (1.05)
	Engagement in the cultural heritage	4.69 (1.14)
Ease of use	Navigation	4.96 (0.99)
	Browsing of items	4.81 (1.13)
	Searching of items	4.85 (0.99)
Organization	Organization of information	4.91 (0.96)
	Organization of items through collections	5.10 (0.82)
	Organization of items through exhibitions	5.00 (0.95)

(*continued*)

Table 1. (*continued*)

Criterion	Variable	Rating mean (SD)
Labeling	Comprehensibility of information	5.08 (0.85)
	Consistency of information	5.19 (0.74)
Visual appearance		4.74 (1.16)
Content	Accuracy of information	5.09 (0.79)
	Comprehensiveness of information	4.99 (0.90)
Interactivity	Active control	4.86 (1.07)
	Two-way communication	4.47 (1.21)

There were other noteworthy suggestions extracted from the open-ended comments of the participants. For instance, one specifically expressed his/her expectation on a "warmer lighting" of the photos that could "simulate a virtual visit" to a physical museum (Participant #19), which denotes the importance of considering not only the quality of photos [28] but also visitors' hedonic or affective experience during the digitization process [29].

6 Conclusion and Future Work

In this study, we presented the development and user evaluation of a digital museum of Miao embroidery as a unique intangible cultural heritage of ethnic minorities in Guizhou, China. Evaluation results have helped us identify the strengths of this digital museum (e.g., rigor of metadata), elements of weakness (e.g., two-way communication), and suggestions for further improvement. Further additions of the digital museum may include videos of the craftsmanship and VR scenes of the physical museum. This study is exploratory in nature and the participant sample was limited to the Chinese population. Towards the goal of promoting the Miao embroidery to an international audience, our ongoing work involves recruiting non-Chinese participants for the evaluation. Future research will also include more comprehensive evaluation methods such as usability testing and system log analysis.

Acknowledgement. This study is supported by a Knowledge Exchange grant sponsored by the University of Hong Kong and a grant (No. 61703357) by National Natural Science Foundation of China. We thank Ms. Szeto Chui for her assistance in building the digital museum and thank the University of Hong Kong Libraries for hosting the digital museum as part of its permanent collections.

References

1. Reed, B.S., Said, F., Davies, I.: Heritage schools: a lens through which we may better understand citizenship and citizenship education. Citizensh. Teach. Learn. **12**(1), 67–89 (2017)

2. Logan, W.S.: Closing Pandora's box: human rights conundrums in cultural heritage protection. In: Silverman, H., Ruggles, D.F. (eds.) Cultural Heritage and Human Rights, pp. 33–52. Springer, New York (2007). https://doi.org/10.1007/978-0-387-71313-7_2
3. Idris, M.Z., Mustaffa, N.B., Yusoff, S.O.S.: Preservation of intangible cultural heritage using advance digital technology: issues and challenges. Harmonia: J. Arts Res. Educ. **16**(1), 1–13 (2016)
4. Zhou, Y., Sun, J., Huang, Y.: The digital preservation of intangible cultural heritage in china: a survey. Preserv. Digit. Technol. Cult. **48**(2), 95–103 (2019)
5. Chen, P.: Research on the technological innovation of Miao pile embroidery against the modern aesthetic demand. In 5th International Conference on Arts, Design and Contemporary Education. Atlantis Press (2019)
6. Torimaru, T.: Similarities of Miao embroidery and ancient Chinese embroidery and their cultural implications. Res. J. Text. Apparel **15**(1), 52–57 (2011)
7. Kiourexidou, M., Antonopoulos, N., Kiourexidou, E., Piagkou, M., Kotsakis, R., Natsis, K.: Websites with multimedia content: a heuristic evaluation of the medical/anatomical museums. Multimod. Technol. Inter. **3**(2), 42 (2019)
8. Kim, S., Im, D.U., Lee, J., Choi, H.: Utility of digital technologies for the sustainability of intangible cultural heritage (ICH) in Korea. Sustainability **11**(21), 6117 (2019)
9. Hardesty, J.L.: Exhibiting library collections online Omeka in context. New Library World. **115**, 75–86 (2014)
10. Pei, S., Zhu, J.: Art education in museum adolescent education activities taking Shaanxi history museum as an example. In: 3rd International Conference on Art Studies: Science, Experience, Education, Atlantis Press (2019)
11. Xiong, J.: Research evolution of digital museums in China. Sci. Insights **34**(2), 183–190 (2020)
12. Folmer, E., Bosch, J.: Architecting for usability: a survey. J. Syst. Softw. **70**(1–2), 61–78 (2004)
13. Petrie, H., Bevan, N.: The evaluation of accessibility, usability, and user experience. In: The Universal Access Handbook, Vol. 1, pp. 1–16 (2009
14. Pallas, J., Economides, A.A.: Evaluation of art museums' web sites worldwide. Inform. Serv. Use **28**(1), 45–57 (2008)
15. Cunliffe, D., Kritou, E., Tudhope, D.: Usability evaluation for museum web sites. Mus. Manag. Curatorship **19**(3), 229–252 (2001)
16. Jeng, J.: Usability evaluation of digital library. In: Theng, Y. L., Foo, S., Goh, D. H. L., Na, J. C. (eds.) Handbook Of Research On Digital Libraries: Design, Development, And Impact, pp. 278–286. IGI Global (2009)
17. Hu, X., Ho, E.M., Qiao, C.: Digitizing Dunhuang cultural heritage: a user evaluation of Mogao cave panorama digital library. J. Data Inform. Sci. **2**(3), 49–67 (2017)
18. Torimaru, T.: One Needle, One Thread: Guizhou Miao (Hmong) Embroidery and Fabric Piece Work From Guizhou. China. China Textile & Apparel Press, Beijing, China (2011)
19. Koutsomitropoulos, D.A., Solomou, G.D., Papatheodorou, T.S.: Metadata and semantics in digital object collections: a case-study on CIDOC-CRM and Dublin Core and a prototype implementation. J. Digit. Inform. **10**(6), 1 (2009)
20. Liu, Y.: Developing a scale to measure the interactivity of websites. J. Advert. Res. **43**(2), 207–216 (2003)
21. Pérez, R.J., López, J.M.C., Listán, D.M.F.: Heritage education: exploring the conceptions of teachers and administrators from the perspective of experimental and social science teaching. Teach. Teach. Educ. **26**(6), 1319–1331 (2010)
22. Bonfigli, M.E., Cabri, G., Leonardi, L., Zambonelli, F.: Virtual visits to cultural heritage supported by web-agents. Inf. Softw. Technol. **46**(3), 173–184 (2004)

23. Kidd, J., Cardiff, R.: 'A space of negotiation': visitor generated content and ethics at tate. Mus. Soc. **15**(1), 43–55 (2017)
24. Joo, S.: How are usability elements-efficiency, effectiveness, and satisfaction-correlated with each other in the context of digital libraries? Proc. Am. Soc. Inform. Sci. Technol. **47**(1), 1–2 (2010)
25. Srinivasan, R., Boast, R., Furner, J., Becvar, K.M.: Digital museums and diverse cultural knowledges: moving past the traditional catalog. Inform. Soc. **25**(4), 265–278 (2009)
26. Walsh, D., Hall, M.M., Clough, P., Foster, J.: Characterising online museum users: a study of the National Museums Liverpool museum website. Int. J. Digit. Libr. **21**(1), 75–87 (2020)
27. Ruthven, I.: Interactive information retrieval. Ann. Rev. Inf. Sci. Technol. **42**, 43–92 (2008)
28. Artese, M.T., Ciocca, G., Gagliardi, I.: Evaluating perceptual visual attributes in social and cultural heritage web sites. J. Cult. Herit. **26**, 91–100 (2017)
29. Hylland, O.M.: Even better than the real thing? Digital copies and digital museums in a digital cultural policy. Cult. Unbound **9**(1), 62–84 (2017)

Correction to: Collaborative Research Results Dissemination: Applying Postcolonial Theory to Indigenous Community Contexts

Lisa G. Dirks

Correction to:
Chapter "Collaborative Research Results Dissemination: Applying Postcolonial Theory to Indigenous Community Contexts" in: K. Toeppe et al. (Eds.):
***Diversity, Divergence, Dialogue*, LNCS 12646,**
https://doi.org/10.1007/978-3-030-71305-8_33

The chapter was inadvertently published with an incorrect wording of the main title: "Collaborative Research Results Dissemination: Applying Postcolonial Theory to Indigenous Community Collaboration in Research Results Dissemination" whereas it should read: "Collaborative Research Results Dissemination: Applying Postcolonial Theory to Indigenous Community Contexts".

The updated version of this chapter can be found at
https://doi.org/10.1007/978-3-030-71305-8_33

K. Toeppe et al. (Eds.): iConference 2021, LNCS 12646, p. C1, 2021.
https://doi.org/10.1007/978-3-030-71305-8_43

Author Index

Printed in the United States
by Baker & Taylor Publisher Services

Printed in the United States
by Baker & Taylor Publisher Services